D1526746

Error-Control Coding for Data Networks

THE KLUWER INTERNATIONAL SERIES
IN ENGINEERING AND COMPUTER SCIENCE

Error-Control Coding for Data Networks

by

Irving S. Reed

University of Southern California
Los Angeles, California

Xuemin Chen

General Instrument Cooperation
San Diego, California

KLUWER ACADEMIC PUBLISHERS
Boston / Dordrecht / London

Distributors for North, Central and South America:
Kluwer Academic Publishers
101 Philip Drive
Assinippi Park
Norwell, Massachusetts 02061 USA
Tel: 781-871-6600
Fax: 781-871-6528
E-mail: kluwer@wkap.com

Distributors for all other countries:
Kluwer Academic Publishers Group
Distribution Centre
Post Office Box 322
3300 AH Dordrecht, THE NETHERLANDS
Tel: 31 78 6392 392
Fax: 31 78 6546 474
E-mail: orderdept@wkap.nl

 Electronic Services: http://www.wkap.nl

Reed, Irving S.
    Error-control coding for data networks / by Irving S. Reed, Xuemin Chen.
        p. cm. -- (Kluwer international series in engineering and computer science ; SECS 508)
    Includes bibliographical references.
    ISBN: 0-7923-8528-4
    1. Error-correcting codes (Information theory)   2. Data transmission systems. 3. Coding theory. I. Chen, Xuemin, 1962- . II. Title.   III. Series.
TK5102.96.R44 1999
005.7'2--dc21   99-24310
                                                                         CIP

Copyright © 1999 by Kluwer Academic Publishers.

All rights reserved. No part of this publication may be reproduced, stored in a retrieval system or transmitted in any form or by any means, mechanical, photo-copying, recording, or otherwise, without the prior written permission of the publisher, Kluwer Academic Publishers, 101 Philip Drive, Assinippi Park, Norwell, Massachusetts 02061

The senior author had the most wonderful privilege of meeting with Claude Shannon, the discoverer of information theory, as well as on other occasions with the inventors of the first error-correction codes, Richard Hamming and Marcel Golay. Thus the authors wish first to dedicate this book to Claude Shannon and then to Richard Hamming and Marcel Golay in memorial for the inspiration they provided to all students, and to the practitioners of information theory and coding. Also the senior author dedicates this work in memory of his early associates, David Muller and Gustave Solomon, both of whom contributed along with the senior author to the development of error-correction codes.

The authors also dedicate this book to their families and friends.

Irving S. Reed, Xuemin Chen

NOTE TO INSTRUCTORS

Exercises suitable for classroom instruction or self testing are available free of charge by contacting the authors at milly@rcf.usc.edu or xchen@gi.com.

Contents

Preface	xiii
1 Error-Control Mechanisms	1
1.1 Introduction	1
1.2 Error-Handling Processes in Communication Systems	4
1.3 Examples of Error-Control Strategies	7
1.4 Basic Principles of Error-Control Codes	10
1.5 Noisy Channel Coding Theorem	13
1.6 Decoding Complexity and Coding Performance	17
1.7 Channel Bandwidth and Signal Power	21
1.8 Approaching Shannon Capacity	24
Problems	28
Bibliography	29
2 Elements of Algebra	31
2.1 Groups	31
2.2 Rings	37
2.3 Basic Structures of Fields	39
2.4 Vector Spaces	40
2.5 Finite Fields	43
2.6 Euclid's Algorithm	48
2.7 Binary Field Arithmetic	51
2.8 Arithmetic Operations in $GF(q)$	59
Problems	70
Bibliography	71
3 Linear Block Codes	73
3.1 Error-Control Block Codes	73
3.2 Definition of (n,k) Linear Codes over GF(q)	76
3.3 Decoding of Linear Block Codes	90

	3.4 Performance of Algebraic Decoding	100
	3.5 Hamming Codes	104
	3.6 Reed-Muller Codes	106
	3.7 Linear Block Codes for Burst-Error Correction	119
	3.8 Product Code	121
	Problems	130
	Bibliography	136
4	**Linear Cyclic Codes**	**139**
	4.1 Description of Linear Cyclic Codes	140
	4.2 Shift-Register Encoders and Decoders of Cyclic Codes	151
	4.3 Binary Quadratic Residue Codes and Golay Code	162
	4.4 Error Detection with Cyclic and Shortened Cyclic Codes	167
	4.4.1 Error Detection with Cyclic Codes	168
	4.4.2 Applications of CRC in Industry Standards	177
	Problems	183
	Bibliography	186
5	**BCH Codes**	**189**
	5.1 Definition of the BCH Codes	190
	5.2 The BCH Bound on the Minimum Distance d_{min}	196
	5.3 Decoding Procedures for BCH Codes	199
	5.4 Algebraic Decoding of Quadratic Residue Codes	213
	5.5 BCH Codes as Industry Standards	215
	Problems	227
	Bibliography	230
6	**Reed-Solomon Codes**	**233**
	6.1 The State of RS Coding	234
	6.2 Construction of RS Codes	243
	6.3 Encoding of RS codes	247
	6.4. Decoding of (n,k) RS Codes	256
	6.4.1 Errors-only Decoding of RS Codes	257
	6.4.2 Error-and-Erasure Decoding of RS codes	269
	Problems	281
	Bibliography	283
7	**Implementation Architectures and Applications of RS Codes**	**285**
	7.1 Implementation of RS Codes	285
	7.2 RS codes in industry standards	299
	7.2.1 The CCSDS Standard	299

Contents

	7.2.2 The Cross-Interleaved Reed-Solomon Code(CIRC)	300
	7.2.3 RS Product Code in the Digital Video Disk(DVD Specification	307
	Bibliography	312

8 Fundamentals of Convolutional Codes — 315

  8.1 Convolutional Encoder — 316
  8.2 State and Trellis-Diagram Description of Convolutional Codes — 320
  8.3 Nonsystematic Encoder and Its Systematic Feedback Encoder — 323
  8.4 Distance Properties of Convolutional Codes — 328
  8.5 Decoding of Convolutional Codes — 333
  8.6 Performance Bounds — 352
  8.7 Punctured Convolutional Codes of Rate $(n-1)/n$ and Simplified Maximum-Likelihood Decoding — 357
  Problems — 364
  Bibliography — 366

9 ARQ and Interleaving Techniques — 369

  9.1 Automatic Repeat Request — 369
  9.1.1 A Stop-and-Wait Protocol — 374
  9.1.2 A Sliding-Window Protocol — 378
  9.1.3 A Protocol Using Go-Back-n — 382
  9.1.4 A Protocol Using Selective Repeat — 387
  9.2 Interleavers — 392
  9.2.1 Definitions of Interleaver Parameters — 394
  9.2.2 Periodic interleavers — 395
  9.2.3 Comparison of the Two Interleavers — 399
  9.2.4 Integration of an Interleaver in a Communication System — 400
  9.2.5 A C-Language Implementation — 401
  Problems — 403
  Bibliography — 405

10 Applications of Convolutional Codes in Mobile Communications — 407

  10.1 Convolutional Codes used in the GSM systems — 407
  10.1.1 Channel Coding of the Speech Data in a GSM System — 408
  10.1.2 Channel Coding of the Data Channel — 413
  10.1.3 Channel-Coding of the Signaling Channels — 415
  10.2 Convolutional Codes Specified in CDMA Cellular Systems — 416
  10.2.1 Error Protection for Reverse CDMA Channel — 418
  10.2.2 Error Protection for Forward CDMA Channel — 424
  Bibliography — 427

11   Trellis-Coded Modulation — 429

    11.1 M-ary Modulation, Spectral Efficiency and Power Efficiency — 431
    11.2 TCM Schemes — 436
    11.2.1 Encoding of Trellis Codes — 436
    11.2.2 Decoding of Trellis Codes — 441
    11.2.3 Coding Gain — 444
    11.3 Set Partitioning and Construction of Codes — 448
    11.3.1 Set Partitioning — 449
    11.3.2 The Error-Event Probability — 453
    11.4 Rotational Invariance — 456
    11.5 Unequal Error Protection(UEP) Codes and the Pragmatic
        Approach to TCM Systems — 464
    Problems — 467
    Bibliography — 468

12   Concatenated Coding Systems and Turbo Codes — 469

    12.1 Concept of Concatenated Coding System — 469
    12.2 Concatenated Coding Systems with Convolutional
        (or Trellis) Codes and RS Codes — 471
    12.3 Turbo Codes — 497
    12.3.1 Parallel Concatenated Coding — 499
    12.3.2 The Constituent Encoder — 500
    12.3.3 Interleaver Design — 502
    12.3.4 Puncturing the Output — 505
    12.3.5 Decoding of Turbo Codes — 507
    12.3.6 Some Applications of Turbo Coding in Mobile
        Communications — 511
    Bibliography — 513

Appendix A   Some Basics of Communication Theory — 517

    A.1 Vector Communication Channels — 518
    A.2 Optimal Receivers — 521
    A.3 Message Sequences — 524

Appendix B   C-programs of Some Coding Algorithms — 527

    B.1 Encoding and Decoding for Hamming Codes — 527
    B.2 Compute Metric Tables for A Soft-Decision Viterbi Decoder — 533

Index — 537

About the Authors — 547

Preface

The purpose of *Error-Control Coding for Data Networks* is to provide an accessible and comprehensive overview of the fundamental techniques and practical applications of the error-control coding needed by students and engineers. An additional purpose of the book is to acquaint the reader with the analytical techniques used to design an error-control coding system for many new applications in data networks.

Error-control coding is a field in which elegant theory was motivated by practical problems so that it often leads to important useful advances. Claude Shannon in 1948 proved the existence of error-control codes that, under suitable conditions and at rates less than channel capacity, would transmit error-free information for all practical applications. The first practical binary codes were introduced by Richard Hamming and Marcel Golay from which the drama and excitement have infused researchers and engineers in digital communication and error-control coding for more than fifty years. Nowadays, error-control codes are being used in almost all modern digital electronic systems and data networks. Not only is coding equipment being implemented to increase the energy and bandwidth efficiency of communication systems, but coding also provides innovative solutions to many related data-networking problems.

The subject of error-control coding bridges several disciplines, in particular mathematics, electrical engineering and computer science. The theory of error-control codes often is described abstractly in mathematical terms only for the benefit of other coding specialists. Such a theoretical approach to coding makes it difficult for engineers to understand the underlying concepts of error correction, the design of digital error-control systems, and the quantitative behavior of such systems. In this book only a minimal amount of mathematics is introduced in order to describe the many, sometimes mathematical, aspects of error-control coding. The concepts of error correction and detection are in many cases sufficiently straightforward to avoid highly theoretical algebraic constructions. The reader will find that the primary emphasis of the book is on practical matters, not on theoretical problems. In fact, much of the material covered is summarized by examples of real developments and almost all of the error-correction and detection codes introduced are related to practical applications.

This book takes a structured approach to channel-coding, starting with the basic coding concepts and working gradually towards the most sophisticated coding systems. The most popular applications are described throughout the book. These applications include the channel-coding techniques used in the mobile communication systems such as the global system for mobile communications (GSM) and the code-division multiple-access (CDMA) system; coding schemes for High-Definition TeleVision (HDTV) system, the Compact Disk(CD), and Digital Video Disk(DVD); and also the error-control protocols for the data-link layers of networks, and much more. The book is compiled carefully to bring engineers, coding specialists, and students up to date in many important modern coding technologies. We hope that both college students and communication engineers can benefit from the information in this, for the most part, self-contained text on error-control system engineering.

The chapters are organized as follows :

Every course has its first lecture a sneak preview and overview of the techniques to be presented. Chapter 1 plays such a role. Chapter 1 begins with an overview that is intended to introduce the broad array of issues relating to error-control coding for digital communications and computer storage systems.

Chapter 2 provides an introduction to elementary algebra such as the finite fields and polynomials with coefficients over a finite field. This algebraic framework provides most of the tools needed by the reader to understand the theory and techniques of error-control codes. Although the mathematical treatment in this chapter is somewhat rigorous, it is limited in scope to materials that could be used in succeeding chapters. Also discussed in this chapter is the implementation of finite-field arithmetic.

Chapter 3 gives a complete survey of linear block codes and the analysis and elementary design encoders and decoders for block codes. Some important codes such as the Hamming and Reed-Muller codes are introduced in this chapter. Also discussed are the burst-error correction and product codes and the means for analyzing the performance of block codes.

Chapter 4 is a practical treatment of the cyclic codes including the implementation of the encoding and decoding algorithms. Also studied are the Golay code and quadratic-residue codes. Error-detection with cyclic codes, and the applications of cyclic-redundant check codes to data networks are also discussed in detail.

In Chapter 5, BCH codes and algebraic decoding algorithms, such as Peterson's algorithm, are introduced and studied. Some BCH codes as industry standards for digital-video and high-speed data transmission are also discussed in this chapter.

Chapters 6 and 7 are devoted to Reed-Solomon(RS) codes and their realization and applications. A comprehensive overview of RS codes is given in Chapter 6. This chapter provides a detailed coverage of the principles, structures, and encoding-decoding methods of RS codes. The commonly-used encoding and decoding algorithms, e.g. Berlekamp-Massey and Euclidean algorithms, are discussed in detail for error- and erasure-correction. To further demonstrate the power of the RS codes, certain implementation architectures, and applications of RS codes to the CD and DVD are illustrated in Chapter 7.

Chapters 8 and 10 are devoted to convolutional codes and their applications. Chapter 8 introduces the fundamental properties of convolutional codes, coupled with the encoder state and trellis diagrams that serve as the basis for studying the trellis code structure and distance properties. Maximum-likelihood decoding using the Viterbi algorithm for hard and soft decisions are intensively covered in this chapter. Two important applications of convolutional codes to wireless communications are given in Chapter 10. They are the Global System for Mobile communication (GSM) system and the Code-Division Multiple Access (CDMA) system.

Automatic re-transmission request(ARQ) is investigated in Chapter 9. Several protocols of ARQ and their applications to modern networks are presented in this chapter. Also an introduction is given of the interleaving techniques.

Chapter 11 is mainly concerned with trellis-coded modulation(TCM) techniques for bandlimited channels. A simple way of describing trellis-coding technique is covered in this chapter. Key techniques for the design of TCM systems are also included.

Concatenated and Turbo codes are introduced in Chapter 12. Applications of concatenated codes to deep-space communication and digital television are discussed in this chapter. The near-capacity performance of turbo code is demonstrated and studied also in this chapter. This is, perhaps, the most active area in research and development of error-control coding during recent years. This research could bring us finally to the promised land of Shannon's noisy channel coding theorem.

This book has arisen from various lectures and presentations in graduate-level courses on error-control codes. It is intended to be an applications-oriented text in order to provide the background necessary for the design and implementation of error-control systems of data networks.

Although this text is intended to cover most of the important and applicable coding techniques, it is still far from complete. In fact, we are still far from a fundamental understanding of many new coding techniques, e.g. Turbo codes, nor has coding power been fully exploited in the modern communication networks.

We wish to acknowledge everyone who helped in the preparation of this book. In particular, the reviewers have made detailed comments on parts of the book which guided us in our final choice of content. We would also like to thank Professors Robert J. McEliece, Stephen B. Wicker, G. L. Feng, Robert M. Gray, Lloyd Welch, Solomon W. Golomb, Robert A. Scholtz, Dr. Ajay Luthra and Dr. Paul Moroney for their continuing support and encouragement. We also gratefully acknowledge Professors Tor Helleseth and T.K. Truong for their contributions in the many joint papers which are reflected in this book. We would also like to thank Mr. Mark Schmidt, Mr. Bob Eifrig, and Dr. Ganesh Rajan for their reading of parts of the manuscript and for their thoughtful comments. It was important to be able to use many published results in the text. We would like to thank the people who made possible of these important contributions.

Support for the completion of the manuscript has been provided by the University of Southern California and the General Instrument Corporation, and to all we are truly grateful. In particular at USC we truly appreciate the attentiveness that Milly Montenegro and Mayumi Thrasher have given to the preparation of the manuscript.

Finally, we would like to show our great appreciation to our families for their constant help, support and encouragement.

1 Error-Control Mechanisms

1.1 Introduction

We are now in the age of digital Information and processing. Digital information has become an asset not only to the business community, but also in all of society. The entire economy and the daily lives of people are more and more dependent on data provided by computers or microprocessors, the digital audio played by compact-disk players and over the internet, the digital television signals broadcast from satellites, etc.. Just as there are techniques on how best to handle physical commodities, there are also efficient methods to manage, transmit, store and retrieve digital information. Among these techniques, no one need be reminded of the importance, not only of the speed of transmission, but also of the accuracy of the information- transfer process. Error-control mechanisms are now an integral part of almost all information-transfer processes. This text addresses the issues of error-control coding in modern digital electronic systems and data networks.

The general term, error-control, involves the imposition of certain restrictions on the characteristics of the source (information) signals so that the received or retrieved signals can be processed in such a manner that the effects of "noise", i.e. the errors of transmission, are reduced or even eliminated. The term "error-control coding" implies a technique by which redundant symbols are attached "intelligently" to the data by the error-control encoder of the system. These redundancy symbols are used to detect and/or correct and/or interpolate the erroneous data at the error-control decoder. In other words error-control coding is achieved by restrictions on the characteristics of the encoder of the system. These restrictions make it possible for the decoder to correctly extract the original source signal with high reliability and fidelity from the possibly corrupted received or retrieved signals.

The purposes of error-control coding for a digital-electronic system are expressed generally as follows: (1) To increase the reliability of noisy-data communication channels or data-storage systems, (2) To control errors in such a manner that a faithful reproduction of the data is obtained, (3) To increase the overall signal-to-noise energy ratio (SNR) of a system, (4) To reduce the noise effects within a system and/or (5) To meet the commercial demands of efficiency, reliability and the high performance of economically practical digital transmission and storage systems. These objectives must be tailored to the particular application.

The subject of error-control coding often is described abstractly in mathematical terms only for the benefit of other coding specialists. Such a theoretical approach to coding often makes it difficult for engineers to understand the underlying concepts of error correction, the design of digital error-control systems, and the quantitative behavior of such systems. In this text only a minimal amount of mathematics is introduced to describe the many, sometimes mathematical, aspects of error-control coding. The concepts of error correction and detection are in many cases sufficiently straightforward to avoid highly theoretical algebraic constructions.

To illustrate the concepts of error-control coding, some examples from everyday experience are given. Suppose first that a supervisor is planning to meet you within your department. He tells you the time and location of the meeting. In order to assure himself that you received this message, he repeats the same exact message over again. This *repetition* of the message is to achieve error avoidance. He repeats his message with the intent to add redundancy and reduce the chance that you heard the message incorrectly.

Again suppose that you get a phone call from your supervisor about this meeting, and you didn't clearly catch the time of the meeting because of background phone noise. You might ask, "What is the correct time?". He repeats the information about the time. Such *information on* the *"re-transmission"* also is a form of error correction.

Next suppose at eight o'clock in the morning that you are in a large auditorium with many other people trying to listen to an announcement of another meeting that morning. You hear your supervisor say, " The meeting

Error-Control Mechanisms

will be held in the Faculty Center at seven o'clock this morning". This is not a typo; you heard "...seven..." Common sense tells you that he must have said "eleven". Why ? You didn't hear the actual spoken word "eleven" since there was noise in the auditorium. However, you heard "seven" and your common-sense "decoding algorithm" finds that "seven" is not a valid time for this meeting since it is already 8:00 in the morning. The English language already has considerable redundancy so that some words are not used in certain sentences or in the same context. Your common sense "decoding algorithm" needs to come up with a list of suspected, possible, valid words which sound close to "seven" and have the correct meaning to create a valid sentence. The proper selection would be the correct word, "eleven". This example illustrates the rather complex process of *error-correction* that is performed by a human being.

Now imagine that the above communication occurred between two robots instead of between you and your supervisor. If these two robots did not communicate by sound, but only by passing coded data through an electronic media or channel, such as a twisted pair of wires, the system should be able to perform a similar process of *error-correction* and *information interpolation*. But by what means ? The answer is to develop and design a digital electronic error-control coding system to correct the transmissions between these robots.

Actually, the above daily-life examples show intuitively where error-control coding is needed, and how it might work in practice. The error-handling strategies, demonstrated in the above examples of human communication, are developed for robots by using digital-electronic systems. In order to develop error tolerance between electronic systems, error-control codes are used. The purpose of error-control coding is to reduce the chance of receiving messages which differ from the original message. To obtain such an error-control capability, redundancy is added in a systematic, controlled, manner to the message so that the decoding of the received data yields the correct original message. This text discusses how to design and specify the error-control strategies and codes needed for digital-electronic communication and storage systems.

1.2 Error-Handling Processes in Communication Systems

It is more than fifty years since the first error-correction and detection coding engines were introduced to digital computers and transmission systems. Since then, a variety of error-control coding sub-systems have evolved for almost all communication and processing systems. One is the magnetic-disk storage system shown in Figure 1.1, which requires physically reliable access to the data stored on a magnetizable surface. Other subsystems include telecommunication networks as shown in Figure 1.2.

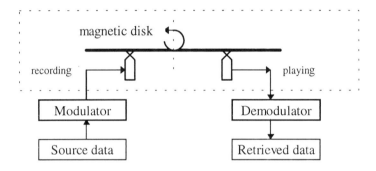

Figure 1.1 A magnetic disk storage system.

Error-control processes have evolved from the design of simple error-correction and (or) detection codes for the implementation of complex error-control protocols. Here an error-control protocol is defined to be the rules that are used to combine multifunctional error-correction and detection coding into a total system. One may ask ,"What is needed to design or implement such a system of error-control codes and/or protocols ?". The answer is to be found in the study of information communications.

Before attempting to design or specify any particular error-control procedure, it is necessary to understand the problems faced by the digital-electronic systems associated with information communications and to know how effectively these problems can be overcome. A simplified coding model of a digital-communication system is shown in Figure 1.3. The same model can be used to describe an information storage system if the storage medium is considered to be a channel.

Error-Control Mechanisms

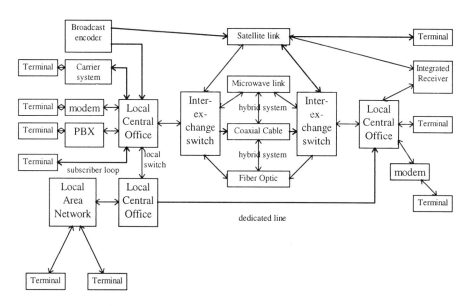

Figure 1.2 A modern communication network

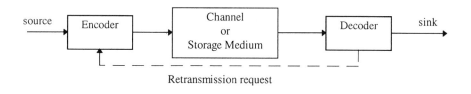

Figure 1.3 A simplified model of a communication or storage system

The source information usually is composed of binary or alpha-numeric symbols. The encoder converts these information messages into electrical signals that are acceptable to the channel. Then these modified signals are sent to the channel (or storage medium) where they may be disturbed by "noise". Next the output of the channel is sent to the decoder, which makes a decision to determine which message was sent. Finally this message is delivered to the recipient (data sink).

Typical transmission channels are twisted-pair telephone lines, coaxial-cable wires, optical fibers, radio links, microwave links, satellite links, etc.. Typical storage media are semiconductor memories, magnetic tapes or disks, compact disks(CDs), optical memory units, digital video disks(DVDs), etc.. Each of these channels or media is subject to various types of noise disturbances. For example, the disturbance on a telephone line may come from impulsive circuit-switching noise, thermal noise, crosstalk between lines or a loss of synchronization. The disturbances on a CD often are caused by surface defects, dust, or a mechanical failure. Therefore, the problems a digital communication system faces are the possible message errors that might be caused by these different disturbances. To overcome such channel "noise", "good" encoders and decoders need to be designed to improve the performances of these channels. Note here that the encoder and decoder are assumed to include both error-control coding and modulation.

The purpose of error-correcting and detecting codes is to combat the noise in digital information transmission and storage systems. Figure 1.4 shows the block diagram of a typical error-handling system.

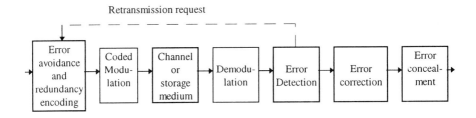

Figure 1.4 The major processes in an error-handling system

In an ideal system the symbols that are obtained from the channel (or storage medium) should match the symbols which originally entered the channel (or storage medium). In any practical system often there are occasional errors, and the purpose of error-control coding is to detect and, perhaps, correct or conceal such errors.

The first stage in Figure 1.4 is concerned with the encoding for error avoidance by the use of redundancy. This includes such processes as the pre-coding of the data for modulation, the placing of digital data at an appropriate position on the tape in certain digital formats, the rewriting of a

Error-Control Mechanisms

read-after-write error on a computer tape, error-correction and detection encoding, etc.. Following such preprocessing, the encoded data are delivered to the modulator in the form of a signal vector or code. Then, the modulator transforms the signal vector into a waveform that can be accommodated by the channel. After being transmitted through the channel this waveform possibly is disturbed by noise. Then the demodulator can produce corrupted signal vectors which in turn cause possible errors in the message data. The occurrence of errors is detected after the receipt of this data. The detection of such error results in some course of action. For example, in a bi-directional link, a retransmission might be requested. In other cases correctable error patterns can be eliminated effectively by an error-correction engine. Also some data that are corrupted by uncorrectable errors can be estimated by, what are called, concealment algorithms. Finally in some error-control protocols, used for packetized data communications, the detection of an error can result in the dropping of a data packet.

1.3 Examples of Error-Control Strategies

In a real communication link the error-control techniques, that are needed for an error-handling system, are shown in Figure 1.4. The type of error-control system depends primarily on the application. This error-control strategy depends on the channel properties of the particular communication link and the type of error-control codes to be used. The determination of an error-control strategy is illustrated next by some examples.

Without the dashed feedback line, shown in Figure 1.4, communication channels are one-way channels. The codes for this case are designed primarily for error-correction and/or error-concealment. Error control for a one-way system usually is accomplished by the use of *forward error-correction* (FEC). That is, by employing error-correcting codes that can be used to automatically correct errors which are detected at the receiver.

Frequently communication systems employ two-way channels, an approach that should be considered in the design of an error-control system. With a two-way channel both error-correction and detecting codes can be used. When an error-detecting code is used and an error is detected at one terminal, a request for a repeat transmission can be given to the transmitting terminal. Most errors are corrected quite effectively by this procedure.

There are examples of real one-way channels, in which the error probabilities are reduced by the use of an error-correcting code, but not by an error detection and retransmission system. For example, with a magnetic-tape storage system it is too late to ask for a retransmission after the tape has been stored for any significant period of times, say for a day or even a week. In such a case, errors are detected when the record is read. Encoding with, what are called, forward error-correction codes is often no more complex than with error-detecting codes. It is the decoding that requires complex digital equipment. Another example is the transmission from a spacecraft. For this case the amount of hardware and power at the transmitter is limited and the encoding equipment for an error-correcting system is likely to be less extensive than it is for a long-distance remotely controlled retransmission system. The complex error-correcting (decoding) procedures are performed on earth, where the equipment limitations are not as severe. This latter example illustrates the fact that there do exist system problems which call for error-correcting codes that have no feedback.

However, there are good reasons for using error detection and retransmission for some applications when it is possible. Error detection is by its nature a much simpler computational task than error correction and requires usually much simpler decoding equipment. Also, error detection with retransmission tends to be adaptive. In a retransmission system, redundant information is utilized only in the retransmitted data when errors occur. This makes it possible under certain circumstances to obtain better performance with a system of this kind than it is theoretically possible over a one-way channel. Error control by the use of error detection and retransmission is called *automatic repeat request* (ARQ) . In an ARQ system, when an error is detected at the receiver, a request is sent to the transmitter to repeat the message. This process continues until it is verified that the message was received correctly. Typical applications of ARQ include the protocols for telephone and fax modems.

There are definite limits to the efficiency of a system which uses simple error detection and retransmission by itself. First, short error-detecting codes are not efficient detectors of errors. If very long codes are used, retransmission would have to be performed too frequently. It can be shown that a combination of the correction of the most frequent error patterns in combination with a detection and a retransmission of the less frequent error patterns is not subject to such a limitation. Such a mixed error-control process is usually called a *hybrid error-control* (HEC) strategy. In fact, HEC is often more effective than either a forward error-correction system or a detection and retransmission system. Many present-day digital systems use a combination of forward error correction and detection with or without feedback.

Other than error correction and detection, what is called, error concealment can be very important in some systems. To illustrate the process of error concealment, consider the following : Errors in audio and video recorders are often at a disadvantage compared with recording in a computer in that the former are not able to preformat and verify the medium. For such systems, there is no time for a return and a retry. However, these systems often have the advantage that there is usually a certain amount of natural redundancy in the information conveyed. If an error cannot be corrected, then it often can be concealed. Again, in an audio system if a sample is lost, it is possible to obtain an approximation to it by interpolating between the adjacent values of the data. Momentary replacement interpolations of errors in digital music usually are not serious, but sustained use of interpolation can restrict bandwidth and cause serious distortion if the music has a number of dominant high-frequency components. In the advanced system a short-term spectral analysis of the sound is made, and if the sample values are not available, then the samples that have the same or similar spectral characteristics are inserted. This concealment method is often quite successful since in music the spectral shape changes relatively slowly. If there are many corrupted examples for concealment, the only other recourse is to mute the music.

Another example is the video recorder. In such a recorder, interpolation is used in a manner similar to that used in audio, but there are additional possibilities, such as the taking of data from adjacent lines or from a previous field (or frame). Video concealment is a complex subject which is discussed in details in [25]. In general if use is to be made of concealment on replay, the data must generally be reordered or shuffled prior to recording. To take a simple example, it may happens that the odd-numbered samples are subject to a delay whereas the even-numbered samples are undelayed. On playback this process needs to be reversed. If a gross error occurs on the tape, following the reverse shuffle, this must be broken down into two separate error zones, one that has corrupted the odd samples and the other in which the even samples are corrupted. Interpolation is then necessary only if the power of the correction system is exceeded.

In the communication systems discussed so far all bits of the input information sequence are assumed usually to have the same significance or importance. However, to specify the error-control strategies of many systems one must also consider what types of data need to be protected most strongly in the transmission. In other words, the sensitivity of information to errors must be assessed.

In a computer system, for example, there is a much greater sensitivity to errors in the machine instructions than in the user's data. For audio and video data the sensitivity to errors is usually subjective. In both cases, the effect of an error in a single bit depends on the significance of the particular bit. If the least significant bit of a sample is wrong, the chances are that the effect is lost in the noise. On the other hand, if the most significant bit is in error, a significant transient is added to the sound or picture that can cause serious distortion. For instance, the effect of the uncorrected significant-bit errors of the pixels in a picture is to produce dots in the picture which are visible as white dots in the dark areas and vice versa. For packetized-data communications, the header data of a packet are usually much more sensitive to errors than the payload data of the packet in a transmission.

If the error rate demanded by the application cannot be met by the unaided channel or storage medium, some form of error handling becomes necessary.

1.4 Basic Principles of Error-Control Codes

It is seen above that the performance of an error-handling system relies heavily on the error-correction and/or detection codes which are designed for the given error-control strategy. There are two different types of codes that it is common to use in these designs. These are the block and convolutional codes. It is assumed for both of these types of codes that the information sequence is encoded with an alphabet set Q of q distinct symbols, called a q-ary set, where q is a positive integer.

In general a code is called a block code if the coded information sequence can be divided into blocks of n symbols that are decoded independently. These blocks are called the codewords. If there exist a total of M distinct codewords, then it is customary to say that the code has block length n and size M, or it is an (n, M) block code. The *rate* of a q-ary (n, M) code is $R = \dfrac{\log_q M}{n}$ in symbols per symbol times. For $q = 2$ the block code is called a binary block code, and its rate is measured in bits per bit-time.

Error-Control Mechanisms

Another simpler description of the most useful block codes is given as follows: The encoder of a block code divides the information sequence into message blocks, each of k information bits. The message block, called the *message word*, is represented by a binary k-tuple or vector of form, $\bar{m} = (m_0, m_1, \ldots, m_{k-1})$. Evidently, there are a total of $M = 2^k$ different possible message words(vectors). The encoder transforms each message word $\bar{m}$ independently into a codeword(code vector) $\bar{c} = (c_0, c_1, \ldots, c_{n-1})$. Therefore, corresponding to the 2^k different possible messages, there are 2^k different possible code words at the encoder output. This set of 2^k code words of length n is called an (n, k) block code. The code rate of an (n, k) block code is $R = \dfrac{k}{n}$ in bits per symbol(or bit)time. Since the n-symbol output codeword depends only on the k-bits of the message (or in general the $\log_q M$-bits), the encoder is, what is called, memoryless and can be implemented with a combinational logic circuit(or a sequential circuit that stops after a finite number of steps).

The encoder of a convolutional code also accepts k-bit blocks of an information sequence and produces an encoded sequence of n-symbol blocks. In convolutional coding, the symbols are used to denote a sequence of blocks rather than a single block. However, in the latter case, each encoded block depends not only on the corresponding k-bit message block at the same time unit, but also on m previous message blocks. Hence, the convolutional encoder is said to have a *memory of order* m. The set of encoded sequences produced by a k-input and n-output encoder of *memory of order* m is called an (n, k, m) convolutional code. Again the ratio $R = \dfrac{k}{n}$ is the code rate of the convolutional code. Since the convolutional encoder contains memory, it is implemented as a sequential logic circuit.

As mentioned in Section 1.1, a basic principle of error-control coding is to add redundancy to the message in such a way that the message and its redundancy are interrelated by some set of algebraic equations. When the message is disturbed, the message with its redundancy is decoded by the use of these relations. In other words the error-control capability of a code comes from the redundancy which is added to the message during the encoding process. One may ask, "what type of algebraic equations between the message and its redundancy are used ?". Answers to this question are contained in several chapters of this book.

To illustrate the principle of error-control coding, we begin first by discussing an example of a binary block code of length 3. In this case, there are a total of 8 different possible binary 3-tuples : (000), (001), (010), (011), (100), (101), (110), (111). First, if all of these 3-tuples are used to transmit messages, one has the example of a (3,3) binary block code of rate 1. In this case if a one-bit error occurs in any codeword, the received word becomes another codeword. Since any particular codeword may be a transmitted message and there are no redundancy bits in the codeword, errors can neither be detected nor corrected.

Next, let only four of the 3-tuples, namely (000), (011), (101), (110) be chosen as codewords for transmission. These are equivalent to the four 2-bit messages, (00), (01), (10), (11) with the 3rd bit in each 3-tuple equal to the "exclusive or " (XOR) of its 1st and 2nd bits. This is an example of a (3,2) binary block code of rate 2/3. If a received word is not a codeword, i.e. the 3rd bit does not equal the XOR of the 1st and 2nd bits, then an error is detected. However, this code is not able to correct an error. To illustrate what can happen when there are errors, suppose the received word is 010. Such errors can not be corrected even if there is only one bit in error since, in this case, the transmitted codeword could have been any of the three possibilities : (000), (011), (110).

Finally, suppose only the two 3-tuples, namely, (000), (111) are chosen as codewords. This is a (3,1) binary block code of rate 1/3. The codewords (000) , (111) are encoded by duplicating the source bits 0, 1 two additional times. If this codeword is sent through the channel, and one or two bit errors occur, the received word is not a codeword. Errors are detected in this scenario. If the decision (rule of decoder) is to decide the original source bit as the bit which appears as the majority of the three bits of the received word, a one-bit error is corrected. For instance, if the received word is (010), this decoder would say that 0 was sent. Chapters 3 through 8 are devoted to the analysis and design of block codes for controlling the errors introduced in a noisy environment.

Consider next the example of a (2,1,2) binary convolutional code. Let the information sequence be $\bar{m} = (m_0, m_1, \ldots, m_6) = (1011100)$ and the encoded sequence be $\bar{c} = (c_0^{(1)} c_0^{(2)}, c_1^{(1)} c_1^{(2)}, \ldots, c_6^{(1)} c_6^{(2)}) = (11,10,00,01,10,01,11)$. Also assume the relations,

$$c_i^{(1)} = m_{i-2} + m_{i-1} + m_i,$$
$$c_i^{(2)} = m_{i-2} + m_i,$$

between $\bar{m}$ and $\bar{c}$ are given, where $m_{-2} = m_{-1} = 0$ and "+" means sum modulo 2(see Sec.2.7). Suppose the third digit of the received sequence is in error. That is, let the received sequence begin with $\bar{c} = (11,00,00,...)$. The following are the eight possible beginning code sequences (00,00,00,...), (00,00,11,...), (00,11,10,...), (00,11,01,...), (11,10,11,...), (11,10,00,...), (11,01,01,...), (11,01,10,...). Clearly, the 6-th path, which differs from the received sequence in but a single position, is intuitively the best choice. Thus, a single error is corrected by this observation. Next, suppose digits 1,2 and 3 were erroneously received. For this case the closest code sequence would be (00,00,00,...) and the decoder would make an undetectable error. Convolutional codes are discussed in Chapters 8 and 10.

If the codes introduced above are applied to the data communication or storage system shown in Figure 1.1, what quantities can be used to measure the error-performance improvement due to coding ? Obviously, certain statistical measures, such as the probability of error, are good candidates since most errors are caused by random noise and interference in the system. Other commonly used quantities include the error probabilities from the decoder, e.g. the bit-error rate of the system, the probability of incorrectly decoding a codeword, the probability of an undetected error, etc.. In the physical layers of a communication system these error probabilities usually depend on the particular code, the decoder, and, more importantly, on the underlying channel/medium error probabilities.

1.5 Noisy Channel Coding Theorem

In the study of communication systems one of the most interesting problems is to determine for a channel how small the probability of error can be made by a decoder of a code of rate R. A complete answer to this problem is provided to a large extent by a specialization of an important theorem, due to Claude Shannon in 1948, called the *noisy channel coding theorem* or the *channel capacity theorem*. This fundamental theorem is described heuristically in the ensuing discussion.

First, a general decoding rule is derived for a coded system. A block diagram of a coded system for an additive noise channel is shown in Figure 1.5. In such a system the source output $\bar{m}$ is encoded into a code sequence (codeword) $\bar{c}$. Then $\bar{c}$ is modulated and sent to the channel. After demodulation, the decoder receives a sequence $\bar{r}$ which satisfies $\bar{r} = \bar{c} + \bar{e}$, where $\bar{e}$ is the error sequence. The final decoder output $\hat{m}$ represents the encoded message.

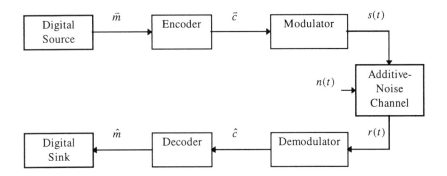

Figure 1.5 Coded system on an additive noise channel.

The primary purpose of a decoder is to produce an estimate $\hat{m}$ of the transmitted information sequence $\bar{m}$ that is a function of the received sequence $\bar{r}$. Equivalently, since there is a one-to-one correspondence between the information word(or sequence) $\bar{m}$ and the codeword(or sequence) $\bar{c}$, the decoder can produce an estimate $\hat{c}$ of the code word $\bar{c}$. Clearly, $\hat{m} = \bar{m}$ if and only if $\hat{c} = \bar{c}$. A decoding rule is a strategy for choosing an estimated code word $\hat{c}$ of each possible received sequence $\bar{r}$. If the code word $\bar{c}$ is transmitted, a decoding error occurs if and only if $\hat{c} \neq \bar{c}$. Given that $\bar{r}$ is received, the conditional probability of an error of the decoder, given the received codeword $\bar{r}$, is defined by

$$P(E|\bar{r}) = P(\hat{c} \neq \bar{c}|\bar{r}). \tag{1.1}$$

Then the error probability of the decoder is given by

$$P(E) = \sum_{\bar{r}} P(E|\bar{r})P(\bar{r}), \tag{1.2}$$

where $P(\bar{r})$ denotes the probability of receiving the codeword $\bar{r}$. Evidently, $P(\bar{r})$ is independent of the decoding rule used since $\bar{r}$ is produced prior to the decoding process. Hence, the optimum decoding rule must minimize $P(E|\bar{r}) = P(\hat{c} \neq \bar{c}|\bar{r})$ for all $\bar{r}$. Since minimizing $P(\hat{c} \neq \bar{c}|\bar{r})$ is equivalent to the maximization of $P(\hat{c} = \bar{c}|\bar{r})$, $P(E|\bar{r})$ is minimized for a given $\bar{r}$ by choosing $\hat{c}$ to be a code word $\bar{c}$ that maximizes

$$P(\bar{c}|\bar{r}) = \frac{P(\bar{r}|\bar{c})P(\bar{c})}{P(\bar{r})}. \tag{1.3}$$

That is, $\hat{c}$ is chosen to be the most likely code word, given that $\bar{r}$ is received. If all information sequences, and hence all code words, are equally likely, then the probabilities $P(\bar{c})$ and $P(\bar{r})$ are the same for all $\bar{c}$ and $\bar{r}$. This is the case of a uniform codeword probability so that maximizing Eq. (1.3) is equivalent to the maximization of the conditional probability $P(\bar{r}|\bar{c})$.

If each received symbol in $\bar{r}$ depends only on the corresponding transmitted symbol, and not on any previously transmitted symbol, the channel is called a discrete memoryless channel (DMC). Thus for a DMC

$$P(\bar{r}|\bar{c}) = \prod_i P(r_i|c_i) \tag{1.4}$$

since each received symbol depends only on the corresponding transmitted symbol. A decoder that chooses its estimate to maximize Eq. (1.4) is called a maximum likelihood decoder (MLD). If the code words are not equally likely, an MLD is not necessarily optimum, since the conditional probabilities $P(\bar{r}|\bar{c})$ must be weighted by the apriori codeword probabilities $P(\bar{c})$ and $P(\bar{r})$ to determine which code word maximizes $P(\bar{c}|\bar{r})$. However, in most systems the codeword probabilities are not known exactly at the receiver, thereby making optimum decoding nearly impossible. Thus MLD becomes the best, feasible, decoding rule.

The important physical and statistical properties of a channel for a communication system are studied next. Usually, the description of a channel includes the received power, the available bandwidth, the noise statistics, and other possible channel impairments such as fading, etc.. Two somewhat typical communication systems are examined here in order to provide some typical representatives. The two primary resources of a communication system are the received power and the available transmission bandwidth. In many communication systems one of these resources may be more precious than the other, and most systems can be classified as either bandwidth limited or power limited. In a bandwidth-limited system, spectrally-efficient modulation techniques can be used to save bandwidth at the expense of power. In a power-limited system, power-efficient modulation techniques can be used to save power at the expense of bandwidth. In both the bandwidth limited and power-limited systems error-correction coding is used to save power or to improve the error performance at the expense of bandwidth or sometimes even without any increase in bandwidth.

The relationship of channel capacity with noise and the other physical parameters of a channel is of considerable interest. It is important to know the ultimate limitations of a channel to efficiently and reliably transmit information without error. Such a limit is characterized by the channel capacity which is defined as the maximum number of bits that can be sent per second reliably over the channel.

In Figure 1.6 the ratio of bit-energy to the noise-power spectral density, E_b/N_0 in decibels, is shown as a function of the ordinate as the ratio of encoder input rate in the information or bits, that can be transmitted per Hertz(Hz) in a given bandwidth W, measured in Hertz(cycles/sec). Let R denote the code rate. Then the ratio R/W is called the bandwidth efficiency since it reflects how efficiently the bandwidth resource is utilized. In the particular case of a transmission over an additive white-Gaussian-noise (AWGN) channel, Figure 1.6 plots the Shannon-Hartley formula for channel capacity given by

$$C = W \log_2\left(1+\frac{S}{N}\right), \qquad (1.5)$$

where $S/N = (E_b R)/(N_0 W)$ is the ratio of the received average-signal power to the noise power. When the logarithm is taken to the base 2, the capacity, C, is given in bits/second. The capacity of a channel defines the maximum number of bits that can be sent reliably per second over the channel. When the information rate, R, equals the channel capacity C, the curve separates the region of reliable communications from that region where reliable communications is not possible.

The central question of noisy channels was answered by Shannon in his landmark paper in 1948 [1]. Roughly speaking, Shannon's noisy-channel coding theorem states : For every memoryless channel of capacity C there exists an error-correcting code of rate $R<C$, such that the error probability $P(E)$ of the maximum-likelihood decoder for a power-constrained system can be made arbitrarily small. If the system operates at a rate $R>C$, the system has a high probability of error, regardless of the choice of the code or decoder.

The noisy-channel coding theorem is extremely powerful and general and is not restricted to Gaussian channels(see [1][2]).

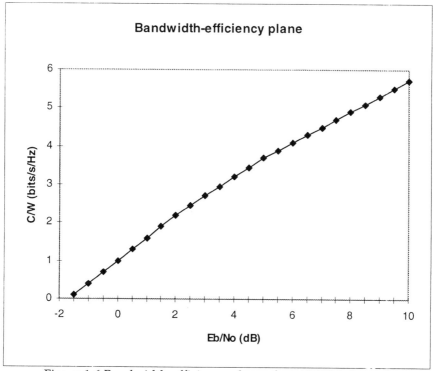

Figure 1.6 Bandwidth-efficiency plane of an AWGN channel

1.6 Decoding Complexity and Coding Performance

In the last section, we briefly discussed the noisy channel coding theorem. The good news about this theorem is that there exist "good" error-correction codes for any rate $R < C$ such that the probability of error in a ML decoder can have an arbitrary small value. However, the proof of this theorem is nonconstructive. It leaves open the problem of how to search for specific "good" codes. Also Shannon assumed exhaustive ML decoding, where the complexity is proportional to the number of words in the code. It is clear that long codes are needed to approach capacity and, therefore, that more practical decoding methods would be needed. These problems, left by Shannon, have kept researchers occupied for more than 50 years.

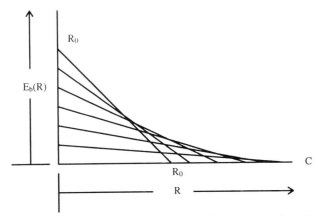

Figure 1.7 The error exponent $E_b(R)$ for a very noisy channel

Gallager[4] showed that the probability of error of the "good" block code of length n and rate $R<C$ is bounded exponentially with block length as follows:

$$P(E) \le e^{-nE_b(R)}, \tag{1.6}$$

where the error exponent $E_b(R)$ is greater than zero for all rates $R<C$. A typical plot of $E_b(R)$ vs. R for a very noisy channel, such as a power-limited Gaussian channel, is illustrated in Figure 1.7, where R_0 is a parameter called the exponential bound parameter[24]. Like Shannon, Gallager continued to assume a randomly chosen code and an exhaustive ML decoding. The decoding complexity $\hat{K}$ is then on the order of the number of codewords, i.e. $\hat{K} \cong e^{nR}$. Therefore, the probability of error $P(E)$ is bounded only algebraically with decoding complexity $\hat{K}$ as follows:

$$P(E) \le \hat{K}^{-E_b(R)/R}. \tag{1.7}$$

Gallager's bounds for block codes extend[7] also to an exponential error bound for convolutional codes of memory order m as follows:

$$P(E) \le e^{-(m+1)nE_c(R)}, \tag{1.8}$$

where $(m+1)n$ is called the constraint length of the convolutional code, and the convolutional error exponent $E_c(R)$ is greater than zero for all rates

$R < C$. A typical plot of $E_c(R)$ vs. R for a very noisy channel is illustrated in Figure 1.8. Both $E_b(R)$ and $E_c(R)$ are positive functions of R for $R < C$ and are completely determined by the channel characteristics.

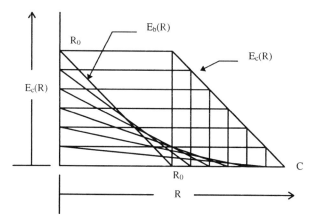

Figure 1.8 The error exponent $E_c(R)$ for a very noisy channel

It is shown in [6] that the complexity of the ML decoding algorithm for convolutional codes, now called the Viterbi algorithm, is exponential in the constraint length, i.e. $\hat{K} \cong e^{(m+1)nR}$. Thus, the probability of error again is bounded only by an algebraic function of complexity of the form,

$$P(E) \leq \hat{K}^{-E_c(R)/R}.\tag{1.9}$$

Both of the bounds (1.6) and (1.8) imply that an arbitrarily small error probability is achievable for $R < C$ by increasing the code length n for block codes or by increasing the memory order m for convolution codes. For codes to be effective they must be long in order to average the effects of noise over a large number of symbols. Such a code might have as many as 2^{200} possible codewords and many times the number of possible received words. While exhaustive ML decoding conceptually exists, such a decoder with increasing block-code length becomes impossible to realize. It is clear that the key obstacles to a practical approach to channel capacity are not only the construction of specific "good" long codes, but also the problem of their decoding complexity.

It turns out that some simple mathematical algebraic structures are needed to make it possible to find classes of "good" codes. Even more importantly such algebraic structures make it feasible to obtain the encoding and decoding operations that need to be implemented in electronic equipment. Thus, there are three primary aspects of coding design : (1) To find codes that have the required error-correcting ability. This usually demands that the codes be long. (2) To find practical methods for encoding. (3) To find practical methods of making decisions at the receiver, that is, the procedures for error correction. A typical attack on the problem is to find codes that can be proved mathematically to satisfy the required error-correction capability. Such codes must have the proper mathematical framework to make this possible. Mathematical algorithms are exploited in order to satisfy and realize the later two requirements, the ability to encode and decode.

As it is seen in Section 1.4, error-control codes need to insert redundancy into the message sequence in such a manner that the receiver can detect and possibly correct errors that occur in a transmission. The amount of added redundancy is expressed usually in terms of the code rate R. The code rate is the ratio of k, the number of data symbols transmitted per code word, to n, the total number of symbols transmitted per code word. In some systems such symbols can be the sample values of the modulation signals. In telecommunication environments where noise is present, the demodulated received-signal values invariably contains errors.

Errors in sampled received data are usually characterized in terms of a bit-error rate(BER). Assuming that the data-symbol transmission rate R_s (in symbols/second) remains constant, the added redundancy forces one to increase the overall symbol-transmission rate by the rate R_s/R. Also if the transmitter power level is constant, then the received energy-per-symbol is reduced from E_s to $R \cdot E_s$. The demodulated BER value is thus increased in comparison to its original BER. However, when the degraded, demodulated, data symbols and redundant symbols are sent to the error-control decoder, redundancy can be used to correct some fraction of the errors, thereby improving the reliability of the demodulated data.

If the code for modulation redundancy is selected properly, the BER at the output of the decoder can be an improvement over that at the output of the demodulator of the original, uncoded, system. The amount of improvement in BER is usually discussed in terms of the additional transmitted power that is needed to obtain the same performance without coding. This difference in power is called the coding gain. Since the uncoded and coded systems usually have the same noise power, the coding gain is often measured by the difference of signal-to-noise ratio in decibels(dB) as follows:

$$\text{Coding Gain} = 10\log_{10}\left(\frac{S}{N}\right)_{coded} - 10\log_{10}\left(\frac{S}{N}\right)_{uncoded}.$$

Often coding gains of 6 to 9 dB are achievable in standard error-correcting systems, thereby allowing for a reduction in transmitted power by a rate of 4 to 8. The use of error-control coding thus introduces an important positive factor into the power budget used in the design of a communication system.

1.7 Channel Bandwidth and Signal Power

It was shown in 1928 [8] by Nyquist that a channel of bandwidth W (in hertz or Hz or cycles per second) is capable of supporting pulse-amplitude-modulated (PAM) signals at a rate of 2W samples without causing intersymbol interference. In other words, there are approximately 2W independent signal dimensions per second. If the two carriers $\sin(2\pi f_0)$ and $\cos(2\pi f_0)$ are used in quadrature, as in double-sideband, suppressed-carrier, amplitude modulation (DSB-SC), there are W pairs of dimensions (or complex dimensions) per second. The ubiquitous quadrature-amplitude-modulated (QAM) formats are introduced in Appendix A.

The parameter that characterizes how efficiently a system uses its allotted bandwidth is defined by the bandwidth efficiency η as follows

$$\eta = \frac{\text{Bit rate}}{\text{Channel bandwidth } W} \quad \text{(bits/s/Hz)} \tag{1.10}$$

In order to compare different communication systems, a second parameter, which expresses the signal-power efficiency, also needs to be considered. This parameter is the information bit-error probability P_b. Many practical systems typically require $10^{-3} \geq P_b \geq 10^{-5}$, though P_b often needs to be lower as it is in digital television, i.e. High-Definition Tele-Vision(HDTV).

The maximum bandwidth efficiency η_{max} of an additive white-Gaussian noise(AWGN) channel is found by dividing the channel capacity C in Eq.(1.5) by the bandwidth W to obtain:

$$\eta_{max} = \log_2\left(1 + \frac{S}{N}\right) \quad \text{(bits/s/Hz)}. \tag{1.11}$$

To calculate η, the bandwidth W must be properly defined. Some commonly used definitions are given as follows:

1. The Gabor bandwidth: $W = \sqrt{\int_{-\infty}^{\infty} fS(f)df}$, where $S(f) = \dfrac{|X(f)|^2}{\int_{-\infty}^{\infty}|X(f')|^2 df'}$ is a normalized energy spectrum of signal $x(t)$, and $X(f)$ is the Fourier Transform of $x(t)$.

2. The z% bandwidth: W is defined so that some percentage z of the transmitted signal power falls within a frequency band of width W. This z% bandwidth corresponds to an out-of-band power of $10\log_{10}(1-\dfrac{z}{100})$ dB. For example, the 3dB bandwidth is the value of W for $W>0$ such that 50% of the transmitted signal power falls within the band of width W.

Note next that the average signal power S in watts can be expressed by

$$S = \frac{kE_b}{T} = RE_b, \quad (1.12)$$

where E_b is the energy per bit, k is the number of bits transmitted-per-symbol, and T is the duration of that symbol. The parameter $R = k/T$ is the transmission rate of the system in bits/s. Also R is called the spectral bit rate. The signal-to-noise power ratio S/N can be re-written as $S/N=(RE_b)/(N_0W)$, where N_0 is the single-sided noise-power spectral density. Then, the Shannon limit in terms of the bit energy and noise-power spectral density is obtained as follows:

$$\eta_{max} = \log_2\left(1 + \frac{RE_b}{N_0W}\right). \quad (1.13)$$

Since $\dfrac{R}{W} = \eta_{max}$ is the limiting spectral efficiency, from Eq. (1.13) a bound is obtained on the minimum bit energy required for reliable transmission at a given spectral efficiency, namely the bound,

$$\frac{E_b}{N_0} \geq \frac{2^{\eta_{max}} - 1}{\eta_{max}}. \quad (1.14)$$

This bound is called the Shannon bound. This is the bound plotted in Figure 1.6 for a spectral efficiency of $\eta_{max} = 2$ bits/s/Hz and is also listed in Table 1.1 as a function of E_b/N_0.

Error-Control Mechanisms

If spectral efficiency is not at a premium, and a large amount of bandwidth is available for transmission, one can use bandwidth, rather than power, to increase the channel capacity in (1.5). In the limit as the signal is allowed to occupy an infinite amount of bandwidth, that is, as $\eta_{max} \to 0$, one obtains

$$\frac{E_b}{N_0} \geq \lim_{\eta_{max} \to 0} \frac{2^{\eta_{max}} - 1}{\eta_{max}} = \ln(2) = -1.59 dB.$$

which is the minimum bit-energy to noise-power spectral density required for reliable transmission. This minimum is given by $E_b / N_0 = -1.59$dB.

Table 1.1 shows the power and bandwidth efficiencies of some popular uncoded quadrature constellations and of a number of coded transmission schemes. This table clearly demonstrates the benefits of coding for given bandwidth efficiencies.

For all of the coding systems shown in Table 1.1, it can be seen that, for a given bandwidth efficiency η, the closer one gets to the Shannon limit, the more complexity needs to be expended to achieve the desired operating point. Even the most complex schemes are still about 2 dB away from the Shannon limit. Note also for the smaller error probabilities that this discrepancy increases, thereby requiring an even larger decoder complexity.

From basic communication theory it is known [2] that a signal of duration T and of bandwidth occupancy W spans approximately $2.4WT$ orthogonal dimensions. Since a finite-duration signal in time theoretically has an infinite bandwidth occupancy, the number of dimensions is somewhat difficult to describe precisely. For this reason, one often expresses Shannon's capacity formula (1.5) in terms of the maximum-rate per dimension. That is, if R_d is the rate in bits/dimension, then the capacity of an AWGN channel per dimension is the maximum rate at which reliable transmission is possible. It is given by (see [2])

$$C_d = \frac{1}{2} \log_2 \left(1 + 2 \frac{R_d E_b}{N_0}\right) \quad \text{(bits/dimension)} \quad (1.15)$$

transmission methods	E_b/N_0 (dB)	Number of states	η (bits/s/Hz)	Shannon Bound (C/W)
uncoded BPSK	8.5	----	1.0	5.2
uncoded QPSK	8.5	----	1.7	5.2
uncoded 8-PSK	12.2	----	2.6	6.7
uncoded 16-QAM	12.5	----	3.5	7
(2,1,7) QPSK	3.2	128	1.0	2.7
(4,1,14) QPSK	1.8	16,384	0.4	2.0
8-PSK TCM	6.7	4	1.7	4.4
8-PSK TCM	6.2	16	1.7	4.1
8-PSK TCM	5.5	64	1.7	3.7
8-PSK TCM	5.0	256	1.7	3.5
16-QAM TCM	7.3	4	2.6	4.6
16-QAM TCM	6.6	16	2.6	4.3
16-QAM TCM	6.0	64	2.6	4.0
16-QAM TCM	5.4	256	2.6	3.7

Table 1.1 (Approximate) spectral efficiencies vs E_b/N_0 achieved by various coded and uncoded transmission methods.

By an application of a similar substitution to that used above, the Shannon bound, normalized per dimension, is obtained as

$$\frac{E_b}{N_0} \geq \frac{2^{2C_d} - 1}{2C_d}. \tag{1.16}$$

Equation (1.16) is useful when the question of waveforms and pulse shaping is not a central issue, since it allows one to eliminate these considerations by treating signal dimensions rather than the signal itself (see also Appendix A).

1.8 Approaching Shannon Capacity

A combination of Reed-Solomon codes with convolutional codes achieved the ultimate success in the application of error-correction codes to the deep-space probes launched from the 1960's to the present. For a long time, error-control coding was considered to be a curiosity of deep-space communications, its only viable application. This was the first important application of the power-limited case, and it was a picture-book success story for error-correction codes.

Error-Control Mechanisms

To understand the possible gain of coding for deep-space communications, consider an uncoded transmission with a binary phase-shift keying (BPSK) modulation (see Appendix A) and coherent detection. A bit-error rate of $P_b = 10^{-5}$ can be achieved at a bit energy to a noise-power spectral-density ratio of $E_b/N_0 = 9.6$ dB and a spectral efficiency of 1 bit/dimension. It is known from the inequality in (1.16) that 1 bit/dimension is theoretically achievable with $E_S/N_0 = 1.76$ dB. This indicates that a power savings of nearly 8 dB is theoretically possible by the use of coding.

Coding techniques	E_S/N_0 (dB)	Spectral Efficiency (bits/symbol)
uncoded BPSK	9.6	1.0
(32,6) Reed-Muller code with ML decoding	6.4	0.187
(255,123) BCH code with algebraic decoding	5.7	0.5
(2,1,31) convolutional code with sequential decoding	2.5	0.5
(2,1,6) convolutional code with ML decoding	4.5	0.5
(4,1,14) convolutional code with ML decoding	1.75	0.25
(4,1,14) convolutional code with ML decoding and (255,223) Reed-Solomon code	0.8	0.22
Turbo code	0.7	0.5

Table 1.2 Milestones in the drive toward channel capacity achieved by the space systems.

One of the earliest attempts to close this signal-energy gap was the use of the (32, 6) first-order Reed-Muller(RM) code [9]. This code was used on the *Mariner* Mars and Viking missions in conjunction with BPSK and soft-decision maximum-likelihood decoding. This system had a spectral efficiency of 6/32=0.1875 bits/symbol and achieved $P_b = 10^{-5}$ with $E_S/N_0 = 6.4$ dB. Thus, the (32,6) RM code required 3.2dB less power than BPSK at the cost of a fivefold increase in the bandwidth. The performance of this RM code is given in Table 1.2, as are all of the other systems, discussed here.

In 1967, a new algebraic decoding technique was discovered for the Bose-Chaudhuri-Hocquenghem (BCH) and Reed-Solomon(RS)[12, 13] codes, which enabled an efficient hard-decision decoding of both classes of block codes. For example, the (255,123) BCH code had the rate $R_d \approx 0.5$ bits/symbol and achieved $P_b = 10^{-5}$ with $E_b / N_0 = 5.7$ dB using algebraic decoding.

The next step was taken with the introduction of a probabilistic soft-decision decoding, called (soft) sequential decoding. Sequential decoding was used first on the Pioneer 9 mission [14] and achieved more than a 3dB gain in E_b / N_0, a first remarkable step. The Pioneer 10 and 11 missions in 1972 and 1973 both used a half-rate (1/2), long-constraint-length (31), nonsystematic convolutional code [15]. A sequential decoder was used to achieve $P_b = 10^{-5}$ with $E_b / N_0 = 2.5$ dB and $R_d = 0.5$. This was only 2.5 dB away from the capacity of the channel.

Sequential decoding had the disadvantage that the computational load was variable, and that it grew exponentially the closer the operation point moved towards capacity. For this and other reasons, the next generation of space systems employed the newly-developed Viterbi maximum-likelihood decoding algorithm. With such a decoder, a concatenated scheme, namely an inner convolutional code combined with an outer RS code, showed a great performance and became the standard for deep space missions both of NASA and ESA. The *Voyager* spacecraft launched in 1977 used a half-rate (1/2-rate) with short-constraint-length (6) convolutional code in conjunction with a soft-decision Viterbi decoder, achieving $P_b = 10^{-5}$ with $E_b / N_0 = 4.5$ dB and a spectral efficiency of $R_d = 0.5$ bits/symbol. The outer (255,223) RS code [9][10][11], which is a code of length 255 bytes each of 8 bits, was used to reduce the required signal-to-noise ratio by 2.0 dB for the *Voyager* system. This concatenated code was also only 2.5 dB away from the capacity of the channel.

The biggest Viterbi decoder built to date [18] was for a rate-1/4 convolutional code of constraint-length 15. Such a decoder was the inner-code used on the *Galileo* mission. This yielded a spectral efficiency of $R_d = 0.25$ bits/symbol and achieved $P_b = 10^{-5}$ with $E_b / N_0 = 1.75$ dB. The performance of this system was also 2.5 dB away from the capacity limit. The system for *Galileo* was further enhanced by the concatenation of an outer RS code with the convolutional inner code. This same outer (255,223) RS code reduced the required signal-to-noise ratio by 0.8 dB for the *Galileo* system. In this case, Shannon's limit was only 1.6 dB away !

Motivated by these successes, a similar coding scheme also became the standard for digital TV, which transmits MPEG-2 compressed video via broadcast channels. Only the RS code was modified to have a shorter block-length.

More recently, Turbo codes [17] using iterative decoding have virtually closed the gap towards Shannon capacity by achieving $P_b = 10^{-5}$ at the spectacularly low $E_b/N_0 = 0.7$ dB with $R_d = 0.5$ bits/symbol. It appears that the 40-year effort to reach capacity has been achieved with this latest discovery. Turbo codes are treated briefly in Chapter 12.

For deep-space communication binary PSK was the best choice for the modulation format because bandwidth was not limited. For satellites, the bandwidth requirements are more stringent. Therefore, 4-QPSK was used, allowing uncoded 2 bits/symbol, and less for coded systems, because the redundant bits could be spread out in time. Then came the invention of the bandwidth-efficient coded modulation by Ungerboeck[18].

Space applications of error-control coding met with spectacular successes, and for a long time the belief was prevalent that coding was useful only to improve the power efficiency of a digital transmission. This attitude was overturned in 1980's by the successes of trellis coding on voice-band data-transmission systems. Here it was not the power efficiency : the spectral efficiency was the issue. That is, given a standard telephone channel with an essentially fixed bandwidth and SNR, what was the maximum practical rate for a reliable transmission?

The first commercially available voice-band modem in 1962 achieved a transmission rate of 2400 bits/s. Over the next 10 to 15 years these rates improved to 9600 bits/s, which was then considered to be the maximum achievable rate, and efforts to push this rate higher were frustrated. Ungerbock's invention of TCM in the late 1970s, however, opened the door to further, unexpected, improvements. The modem rates jumped to 14.400 bits/s and then to 19.200 bits/s, using sophisticated TCM schemes [18].

One recent development in voice-band data modems was the establishment of the CCITT V.34 modem standard [19]. These modems were specified to achieve a maximum transmission rate of 28,800 bits/s, and its extensions to V.34 to cover two new rates 31,200 bits/s and 33,600 bits/s. However, at these high rates, modems operate successfully only over a small percentage of the connections. It seems that the limits of the voice-band telephone channel have been reached (see [20]). These results need to be compared to estimates of the channel capacity for a voice-band telephone channel, which are about 30,000 bits/s. The application of TCM represents one of the fastest migrations of an experimental laboratory system to an international standard (V.32,V.34) [19, 20]. Hence, what once was believed to be impossible, achieving capacity on band-limited channels, has become a commonplace reality within about one and a half decades after the invention of TCM [22].

In many ways the telephone-voice-band channel was an ideal application for error-control coding. Its limited bandwidth of about 3kHz (400-3400 Hz) implies relatively low data rates by modern standards. It therefore provided an ideal experimental field for high-complexity error-control methods, which can be implemented without much difficulty, using the current digital signal-processing (DSP) technology. It is thus not surprising that coding for voice-band channels was the first successful application of bandwidth-efficient error control.

Nowadays, trellis coding in the form of bandwidth-efficient TCM as well as more conventional convolutional coding is being considered for satellite communications, both for geostationary and low-earth-orbiting satellites, for land-mobile and satellite-mobile services, cellular communications networks, personal communications services (PCS), and high frequency (HF) troposphere long-range communications, among many others. It seems that the age of a widespread application of error-control coding has only just begun.

Problems

1.1 Compute Shannon's limit on E_b / N_0 for $R_d = 0.5$ and $R_d = 0.25$.

1.2 Conventional modems, such as V.34, treat the telephone network as a pure analog channel. Assume that the bandwidth of this channel is $W = 4,000 Hz$ and that the peak signal-to-noise ratio which the V.34 modem can be operated is $S/N=26dB$. Estimate the bit-rate limit of the V.34 modem.

Error-Control Mechanisms 29

1.3 Assume that the spectral efficiency of the V.34 modem is $R/W=8.96$ $bits/s/Hz$. Determine the minimum bit energy to noise-power spectral density needed for reliable transmission and for estimating the number of bits, transmitted per symbol, if $S/N=26dB$.

Bibliography

For books and special issues of journals devoted to error-control coding :

[1] C. E. Shannon, " A mathematical theory of communication", Bell Sys. Tech. J., vol. 27, pp.379-423 and 623-656, 1948.
[2] J. M. Wozencraft and I. M. Jacobs, Principles of Communication Engineering, John Wiley & Sons, New York, 1965.
[3] G. D. Forney, Jr., Concatenated Codes, MIT Press, Cambridge, MA, 1966.
[4] R. G. Gallager, "A simple derivation of the coding theorem and some applications", IEEE trans. Information Theory, vol. IT-11, pp.3-18, Jan. 1965.
[5] G. D. Forney, Jr., "The Viterbi algorithm", Proc. IEEE, vol. 61, pp.268-278, March 1973.
[6] G. D. Forney, Jr., "Convolutional codes II. Maximum-likelihood decoding", Information and Control, vol. 25, pp.222-266, 1974.
[7] A. J. Viterbi, "Error bounds for convolutional codes and an asymptotically optimum decoding algorithm", IEEE Trans. Information Theory, vol. IT-13, pp.260-269, April, 1967.
[8] H. Nyquist, "Certain topics in telegraph transmission theory," *AIEE Trans.*, pg.617, 1946.
[9] F. J. MacWilliams and N. J. A. Sloane, The Theory of Error Correcting Codes, North-Holland, Amsterdam, 1988.
[10] I.S.Reed and G. Solomon, "Polynomial codes over certain finite fields", SIAM Journal of Applied Mathematics, Vol. 8, pp.300-304, 1960.
[11] I. S. Reed, L.J. Deutsch, I.S. Hsu, T.K. Truong, K. Wang, and C.-S. Yeh, "The VLSI implementation of a Reed-Solomon encoder using Berlekamp's bit-serial multiplier algorithm", IEEE Trans. on Computers, Vol. C-33, No.10, Oct. 1984.
[12] E. R. Berlekarnp, Algebraic Coding-Theory, McOraw-llill, New York, 1968.
[13] J. L. Massey, "Shift register synthesis and BCH decoding" *IEEE Trans. Inform. Theory,* Vol. IT-15, pp. 122-127, 1969.

[14] G. P. Forney, "Final report on a study of a sample sequential decoder" Appendix A, Codex Corp., Watertown, MA, U.S. Annlay Satellite Communication Agency Contract DAA B 07-68-C-0093, April 1968.

[15] J. L. Massey and D.J. Costello, Jr., "Nonsystematic convolutional codes for sequential decoding in space applications", *IEEE Trans. Commun.*, Vol. COM-19, pp. 806-813, 1971.

[16] D. M. Collins, "The subtleties and intricacies of building a constraint-length 15 convolutional decoder", *IEEE Trans. Commun.*, Vol. COM-40, pp. 1810-1819, 1992.

[17] C. Berroti, A. Glavieux, and P. Thitimajshima, "Near Shannon limit error-correcting coding and decoding: Turbo-codes," *Proc. 1993 IEEE int. Conf. on Comm.*, Geneva, Switzerland, pp. 106-107, 1993.

[18] G. D. Forney, "Coded modulation for bandlimited channels:' *IEEE Information Theory Society Newsletter,* December 1990.

[19] CCITT Recommendations V34, 1995.

[20] R. Wilson, "Outer limits," *Electronic News*, May 1996.

[21] U. Black, The V Series Recommendations, Protocols for Data Communications Over the Telephone Network, McGraw-Hill, New York, 1991.

[22] G. Ungerbock, "Channel coding with multilevel/phase signals", IEEE Trans. Inform. Theory, Vol. 28, No.1, pp.55-67, 1982.

[23] I. S. Reed and X. Chen, article Channel Coding, Networking issue of Encyclopedia of Electrical and Electronic Engineering, John Wiley & Sons, Inc. New York, Feb., 1999.

[24] G. C. Clark, Jr. and J. B. Cain, Error-Correction Coding for Digital Communications, Plenum Press, New York, 1981.

[25] J. Watkinson, The Art of Digital Video, Focal Press, London & Boston, 1990.

2 Elements of Algebra

Error-control coding -- the magic technology that enables reliable digital communications -- is described usually in terms of some special mathematics. The purpose of this chapter is to present briefly these mathematical tools. Some basic structures of algebra, such as groups, rings, fields, and vector spaces, are introduced. These mathematical entities will play an important role in error-control coding theory as well as in encoder-decoder implementations. Only those properties of algebra which are needed for the study of error-correction codes are discussed in any detail. It is assumed that the reader already has some familiarity with these topics. Hence, the discussion given here is not intended to be either systematic or complete.

2.1 Groups

A group is an elementary structure which underlies many other algebraic structures, such as rings, fields, etc. It is also a fundamental concept used in the study of error-correction codes. In this section some basic knowledge about groups is introduced. Then, the properties of subgroups and cyclic groups, etc., are discussed.

2.1.1 Basic Concepts

Definition 2.1. Let **G** be a non-empty set with an algebraic operation $\circ$ defined for each pair of its elements. Then, **G** is called a group if and only if for all $a, b, c \in \mathbf{G}$, the operation $\circ$ satisfies the four following axioms :
(1) $a \circ b \in \mathbf{G}$ (algebraic closure).

(2) There exists an element $e \in G$ such that $e \circ a = a \circ e = a$ (e is the identity element).
(3) There exists an element $a^{-1} \in G$ such that $a \circ a^{-1} = a^{-1} \circ a = e$ (existence of an inverse element a^{-1}).
(4) $a \circ (b \circ c) = (a \circ b) \circ c$ (associativity).
If the group G satisfies : $a \circ b = b \circ a$, then G is called a commutative or Abelian group. A group G is denoted by $(G, \circ)$.

Example 2.1. Let Z and R be, respectively, the set of all integers and the set of all real numbers. Then, (Z,+) and (R-{0}, •) are two Abelian groups with the identity elements 0 and 1.

Each of the above two groups contains an infinite number of elements. Such groups are called infinite groups. There also exist groups which have a finite number of elements. The latter group is called a finite group. Some of these groups are useful algebraic structures that can be used to define error-control codes. When the group G is finite, the number of elements in G, denoted by |G| (the cardinality of G), is called the order of G.

Example 2.2. The set $I_2 = \{0,1\}$ with the modulo-2 addition $\oplus$ is a finite Abelian group ($I_2, \oplus$) of order 2.

2.1.2 Subgroup

Definition 2.2. Let $(G, \circ)$ be a group. Also let H be a non-empty subset of G. If $(H, \circ)$ is a group, then it is called a subgroup of G. $(H, \circ)$ is called a normal subgroup if $a \circ H \circ a^{-1} = H$ for all $a \in G$.

Proposition 2.1. Let H be a subset of G, then H is a subgroup of $(G, \circ)$ if and only if one of the following statements is satisfied :
(1) $(H, \circ)$ is a group.
(2) If $a, b \in H$, then $a \circ b \in H$, and $a^{-1} \in H$.
(3) if $a, b \in H$, then $a \circ b^{-1} \in H$.
Proof. The truth of statements (1) and (2) is verified directly by Definition 2.1. If (2) is true and if $a, b \in H$, then $a, b^{-1} \in H$ so that $a \circ b^{-1} \in H$. Therefore, (3) is true. If (3) is true, then $e \in H$ so that $a^{-1} \in H$ for all $a \in H$, and (2) is true. Q.E.D.

Example 2.3. Consider the set of even integers $H = \{\cdots, -4, -2, 0, 2, 4, \cdots\}$. Then (H,+) is a subgroup of Z, where Z is the set of all integers defined in *Example*

2.1. That **H** is a subgroup is verified by any of the statements in Proposition 2.1.

If **G** is a finite group, the following proposition is given without proof (see any book on modern algebra) :

Proposition 2.2. Let **H** be a subgroup of the finite group **G** and $|H| = m$ and $|G| = n$. Then, $m|n$ (m divides n), i.e. the order of a subgroup must be a factor of the order of the original group.

2.1.3 Cyclic Groups

The cyclic group is an important algebraic structure in the theory of error-correction codes. It is defined as follows :

Definition 2.3. A group **G** is called a cyclic group if each of its elements is equal to a power of some element $a \in G$. Such a cyclic group **G** is denoted by $G(a, \circ)$ or more simply by $G(a)$.

From Definition 2.3 the cyclic group $G(a) = \{a^0 = e = 1, a^1, a^2, a^3, \cdots\}$ has $a^0 = 1$ as its unit element e. Element a is called a generating element of $G(a)$. If $a^i \neq a^j$ for all $i \neq j$ and $a^i, a^j \in G(a)$, then such a cyclic group is called an infinite cyclic group. Otherwise, it is called a finite cyclic group. In a finite cyclic group **G**, there exists a positive integer $n \neq 0$ for any element $a \in G$ such that $a^n = 1$. The smallest positive integer which satisfies this condition is the order of the element a.

Proposition 2.3. Let n be the order of the element $a \in G$. Then, it has the following properties :
(1) If $a^m = 1$, then $n|m$.
(2) The order of a^i is $\dfrac{i}{(i,n)}$, where (i,n) denotes the great common divisor of integers i and m.
(3) If the element $b \in G$ has order m and $(n,m) = 1$, then the order of the element ab is nm.
Proof : If (1) is not true, then by the division algorithm for positive integers $m = kn + j$ and $0 < j < n$. Thus, $a^m = a^{kn+j} = a^j = 1$. This contradicts the fact that n is the order of a. Q.E.D.

Example 2.4. Consider $G = \{1,2,3,4\}$. Then, $(G, \bullet \mod 5)$ is a cyclic group. The generating element is $a = 2$, i.e. $2^0 = 1, 2^1 = 2, 2^2 = 4$, $2^3 = 8 \equiv 3 \pmod 5, 2^4 = 16 \equiv 1 \pmod 5$. The order of 2 is 4. Such a cyclic group can also be generated by another element 3 of order 4, i.e. $3^0 = 1, 3^1 = 3, 3^2 = 9 \equiv 4 \pmod 5, 3^3 = 27 \equiv 2 \pmod 5, 3^4 = 81 \equiv 1 \pmod 5$. Therefore, the generator of G is not unique. Any element of order n can be a generator of a cyclic group of order n.

It is seen from Proposition 2.2 that if element a generates a finite cyclic group $G(a)$ such that $a^n = 1$ and $n = ik$, then a^i is a generator element of a subgroup $H(a^i)$. In *Example 2.4.* $2^2 = 4$ is a generator element of the subgroup $H = \{1, 4\}$.

2.1.4. Remainder Groups

It is known from Euclid's division algorithm that any two integers a and m can be expressed uniquely as follows:
$$a = qm + r, \qquad 0 \leq |r| \leq |m| - 1,$$
where q and m are the quotient and the remainder, respectively, and $|\bullet|$ denotes absolute value.

Let m be a positive integer. Consider the set of non-negative integers $G = \{0, 1, 2, \cdots, m-1\}$. Let $+$ denote real addition. Define modulo-m addition on G as follows: For any integers i and j in G, $i + j = r \pmod m$, i.e. r is the remainder that results from a division of $i + j$ by m.

Proposition 2.4. The set $G = \{0, 1, 2, \cdots, m-1\}$ is a Abelian group of order m under modulo-m addition. Such a group is called the additive remainder group.
Proof: First, we see that 0 is the identity element. For $0 < i < m$, both i and $m - i$ are in G. Since
$$i + (m - i) = (m - i) + i = m,$$
it follows from the definition of modulo-m addition that
$$i + (m - i) \equiv 0 \pmod m.$$
Therefore, i and $m - i$ are additive inverses of each other with respect to modulo-m addition. It is also clear that the inverse of 0 is itself. Since real addition is commutative, it follows from the definition of modulo-m addition

for any i and j in **G** that $i+j \pmod{m} = j+i \pmod{m}$. Therefore, modulo-m addition is commutative.

Next, it is shown that modulo-m addition is associative. Let $i, j,$ and k be three integers in **G**. Since real addition is associative, one has
$$i+j+k = (i+j)+k = i+(j+k).$$
Dividing $i+j+k$ by m, one obtains
$$i+j+k = qm+r,$$
where q and r are the quotient and the remainder, respectively, and $0 \leq r < m$. Now, a division of $i+j$ by m yields
$$i+j = q_1 m + r_1,$$
with $0 \leq r_1 < m$. Therefore, $i+j \equiv r_1 \pmod{m}$. Next a division of $r_1 + k$ by m produces
$$r_1 + k = q_2 m + r_2$$
with $0 \leq r_2 < m$. Hence, $r_1 + k \equiv r_2$ and
$$(i+j)+k \equiv r_2 \pmod{m}.$$
Combining the above equations, yields
$$i+j+k = (q_1 + q_2)m + r_2.$$
This implies that r_2 is also the remainder when $i+j+k$ is divided by m. Since the remainder resulting from the division of an integer by another integer is unique, one must have $r_2 = r$. As a result, it is shown that
$$(i+j)+k \equiv r \pmod{m}.$$
Similarly, one can show that
$$i+(j+k) \equiv r \pmod{m}.$$
Therefore, $(i+j)+k \pmod{m} = i+(j+k) \pmod{m}$ and modulo-m addition is associative. Thus, the set $\mathbf{G} = \{0,1,2,\cdots,m-1\}$ is a group under modulo-m addition. Q.E.D.

the modulo-5 addition	0 1 2 3 4
0	0 1 2 3 4
1	1 2 3 4 0
2	2 3 4 0 1
3	3 4 0 1 2
4	4 0 1 2 3

Table 2.1 Modulo-5 Addition

Example 2.5. The addition table for the remainder group under modulo-5 addition is shown in Table 2.1.

Next let m be a prime integer, then m is an integer from the set $\{2,3,5,7,11,...\}$. Consider the set of nonzero integers, $\mathbf{G} = \{1, 2, \cdots, m-1\}$ and let $\bullet$ denote real-number multiplication. Define modulo-m multiplication on $\mathbf{G}$ as follows : For any integers i and j in $\mathbf{G}$, let $i \bullet j = r \bmod m$, i.e. r is the remainder that results from a division of $i \bullet j$ by m.

Proposition 2.5. If m is a prime integer, then the set $\mathbf{G} = \{1, 2, \cdots, m-1\}$ of nonzero integers is an Abelian group of order m under modulo-m multiplication. Such a group is called the multiplicative remainder group.

Proof : First, assume that $i \bullet j$ is not divisible by m. Then, $i \bullet j = k \bullet m + r$, where $0 < r < m$ and r is an element in $\mathbf{G}$. The set $\mathbf{G}$ is closed under the modulo-m multiplication. It is easy to check that modulo-m multiplication is commutative and associative. The identity element is 1. The next thing needed is to prove that every element in $\mathbf{G}$ has an inverse. Let i be an element in $\mathbf{G}$. Since m is a prime integer and $i < m$, integers i and m must be relatively prime (i.e., i and m do not have any common factor greater than 1). Thus by Euclid's theorem there exist two integers a and b such that
$$a \bullet i + b \bullet m = 1.$$
Rearranging this equation, one has
$$a \bullet i = -b \bullet m + 1. \tag{2.1}$$
This means, when $a \bullet i$ is divided by m, that the remainder is 1. If $0 < a < m$, a is in $\mathbf{G}$, and it follows from (2.1) and the definition of modulo-m multiplication that $a \bullet i \,(\bmod m) = i \bullet a \,(\bmod m)$ and $a \bullet i \equiv 1 \,(\bmod m)$. Therefore, a is the inverse of i. However, if a is not in $\mathbf{G}$, one can divide a by m to obtain,
$$a = q \bullet m + r. \tag{2.2}$$
Since m is a prime integer, the remainder r cannot be 0, and it must be between 1 and $m-1$. Therefore, r is an element of $\mathbf{G}$. By combining (2.1) and (2.2), one obtains
$$r \bullet i = -(b + q \bullet i) \bullet m + 1.$$
Therefore, $r \bullet i \,(\bmod m) = i \bullet r \,(\bmod m)$ and $r \bullet i \equiv 1 \,(\bmod m)$ so that r is the inverse of i. Hence, any element i in $\mathbf{G}$ has an inverse with respect to modulo-m multiplication. The set $\mathbf{G} = \{1, 2, \cdots, m-1\}$ under modulo-m multiplication is a group. Q.E.D.

Example 2.6. For $m = 2$ one has the group $\mathbf{G} = \{1\}$ with only one element under modulo-2 multiplication. *Example 2.4* is another example of the multiplicative remainder group.

2.2 Rings

2.2.1 Basic Structure

Definition 2.4. A non-empty set **R** with two algebraic operations, written $*$ (called "multiplication") and $+$ (called "addition"), is called a ring if and only if these two operations satisfy the following axioms for all $a,b,c \in \mathbf{R}$:
(1) $(\mathbf{R},+)$ is an Abelian group with identity element 0.
(2) $a*b \in \mathbf{R}$ (closed under multiplication).
(3) $a*(b*c) = (a*b)*c$ (associativity of multiplication).
(4) $a*(b+c) = a*b+a*c$, and $(b+c)*a = b*a+c*a$ (distributive laws).
Usually, the ring **R** is denoted by $(\mathbf{R},+,*)$. A ring **R** is said to be commutative if and only if $a*b = b*a$ for all $a,b \in \mathbf{R}$.

Example 2.7. The set of all integers **Z** and the set of all real numbers are the commutative rings $(\mathbf{Z},+,*)$ and $(\mathbf{R},+,*)$, respectively, under real addition and multiplication. It is easy to verify from Section 2.14 that the set $\mathbf{R} = \{0,1,2,\cdots,m-1\}$, where m is a positive integer greater than 1, is also a commutative ring under modulo-m addition and multiplication. Such a ring is called a remainder or residue-class ring.

2.2.2 Subrings and Ideals

There are two basic substructures in a ring : subrings and ideals. The former play the same role as subgroups; the latter, however, correspond to the normal subgroups, defined in Definition 2.2. A subring is defined as follows :

Definition 2.5. Let $(\mathbf{R},+,*)$ be a ring and **S** be a non-empty subset of **R**. If $(\mathbf{S},+,*)$ is a ring, then **S** is called a subring of **R**.

Definitions 2.4 and 2.5 imply the following proposition :

Proposition 2.6. The set $\mathbf{S} \subset \mathbf{R}$ is a subring if and only if
(1). $(\mathbf{S},+)$ is a subgroup of $(\mathbf{R},+)$,
(2). $a*b \in \mathbf{S}$ for all $a,b \in \mathbf{S}$.

Example 2.8. Consider subsets of the integer ring **Z**. Then, the subset $\mathbf{Z}(m) = \{0,\pm m,\pm 2m,\pm 3m,\cdots\}$ for some integer m is a subring of **Z** , e.g. if m=3, $\mathbf{Z}(m) = \{0,\pm 3,\pm 6,\pm 9,\cdots\}$ is a subring of **Z**.

Definition 2.6. Let **R** be a commutative ring and **I** be a non-empty subset of R. Then **I** is called an ideal of **R** if
(1) $a - b \in \mathbf{I}$, for all $a, b \in \mathbf{I}$.
(2) $a \bullet r = r \bullet a \in \mathbf{I}$ for $a \in \mathbf{I}, r \in \mathbf{R}$.

From Definition 2.6 it is evident that an ideal **I** is a subring of **R**. Another equivalent definition of an ideal is the following:

Proposition 2.7. **I** is an ideal of **R** if
(1) (**I**,+) is a subring of (**R**,+),
(2) For all $a \in \mathbf{R}$, and for all $b \in \mathbf{I}$, one has $a \bullet b \in \mathbf{I}$ and $b \bullet a \in \mathbf{I}$.

Example 2.9. Let **R** be a commutative ring and $a \in \mathbf{R}$. Then, the set of multiples of the element a (called the generator) is an ideal, denoted by (a). Such an ideal is called a **principal ideal**.

Example 2.10. Every ideal **I** of the integer ring **Z** is a principal ideal. This fact can be shown as follows : Let **I** be a non-zero ideal of **Z**. Among the non-zero elements of **I**, consider an element a with the minimum absolute value. It follows from Euclid's equation that every $b \in \mathbf{I}$ satisfies $b = q \bullet a + r$, where r is the remainder and $r \in \mathbf{I}$. If $r \neq 0$, then $|r| < |a|$. This contradicts the fact that $a \in \mathbf{I}$ has a minimal absolute value. Thus, $r = 0$. Therefore, **I** is the set of multiples of a and as a consequence the principal ideal (a).

2.2.3 Polynomial Rings, Polynomial Ideals, and Quotient Rings.

Let *R(x)* be the set of all polynomials with real coefficients. It follows from Definition 2.4 that *R(x)* is a commutative ring. The ideals constructed in *R(x)* are called polynomial ideals. One has the following proposition :

Proposition 2.8. Every polynomial ideal $I(x)$ of the ring *R(x)* is principal.
Proof : The proof of this proposition is similar to that given in *Example 2.10* if one replaces "the absolute value of an integer" by "the degree of the polynomial" and uses Euclid's division algorithm for polynomials given in the next paragraph.

Next assume that $I(x) = (g(x))$, where $g(x) \in R(x)$. Then, it is well known by Euclid's division algorithm for polynomials that any polynomial $f(x) \in R(x)$ can be expressed as follows:

Elements of Algebra

$$f(x) = q(x)g(x) + r(x) \equiv r(x) \mod g(x)$$

for $0 \leq \deg(r(x)) < \deg(g(x))$, where $\deg(r(x))$ denotes the degree of $r(x)$ and $r(x)$ is the remainder polynomial. Hence, for a given $g(x)$ such that $\deg(g(x)) = n$, one has that $\deg(r(x)) < n$ for any $f(x) \in R(x)$. Thus, $R(x)$ can be expressed as $R(x) = \{\overline{r(x)} : \deg(r(x)) < n\}$, where $\overline{r(x)} = \{f(x) : f(x) \equiv r(x) \mod g(x)\}$ is called a remainder (or residue) class (set) of $g(x)$. Hence, any polynomial $f(x) \in R(x)$ must be in one of these remainder classes of the polynomials, modulo $g(x)$.

2.3 Basic Structures of Fields

Other than the structures of groups and rings discussed above, the concept of a field, especially a finite field, is very important in error-correction theory. Roughly speaking, a field is a set of elements in which one can perform addition, subtraction, multiplication, and division without leaving the set. Also in a field, additions and multiplications satisfy the commutative, associative, and distributive laws. A formal definition of a field is given next.

Definition 2.7. Let **F** be a non-empty set with the two algebraic operations + and * defined for each pair of its elements. Then, **F** is a field if and only if the following conditions are satisfied :
(1) (**F**,+) is an Abelian group. The identity element with respect to addition is called the zero element or the additive identity of **F** and is denoted by 0.
(2) (**F**-{0}, *) is an Abelian group. The identity element with respect to multiplication is called the unit element or the multiplicative identity of **F** and is denoted by 1.
(3) For all $a, b, c \in \mathbf{F}$, $a*(b+c) = a*b + a*c$, and $(b+c)*a = b*a + c*a$. i.e. multiplication is distributive over addition.

Example 2.11. The set of all rational numbers, the set of all real numbers, and the set of all complex numbers constitute respectively the rational field, the real field, and the complex field with respect to addition and multiplication. These are infinite fields since they have an infinite number of elements.

A number of basic properties of fields can be derived from the definition of a field.

Proposition 2.9. Let **F** be a field. Then, for all $a, b, c \in \mathbf{F}$, one has the following relations :
(1) $a*0 = 0*a = 0$,
(2) $a*b \neq 0$ if $a \neq 0$ and $b \neq 0$,
(3) $a*b = 0$ and $a \neq 0$ imply that $b = 0$,
(4) $-(a*b) = (-a)*b = a*(-b)$,
(5) For $a \neq 0$, $a*b = a*c$ implies that $b = c$.

Proof : For (1) first, note that $a = a*1 = a*(1+0) = a + a*0$. Adding $-a$ to both sides of the above equality, one has $-a + a = -a + a + a*0$, $0 = 0 + a*0$, so that $0 = a*0$. Similarly, one can show that $0*a = 0$. Therefore, one obtains $a*0 = 0*a = 0$. Relation (2) is a direct consequence of condition (2) in Definition 2.7. Also (3) is implied by Condition (2). For (4) note that $0 = 0*b = (a+(-a))*b = a*b + (-a)*b$. i.e. $(-a)*b$ is the additive inverse of $a*b$ so that $-(a*b) = (-a)*b$. Similarly, one can prove that $-(a*b) = a*(-b)$. Finally, since a is a nonzero element in the field, it has a multiplicative inverse a^{-1}. Multiplying both sides of $a*b = a*c$ by a^{-1}, one obtains $a^{-1}*(a*b) = a^{-1}*(a*c)$, $(a^{-1}*a)*b = (a^{-1}*a)*c$, $1*b = 1*c$. Thus, $b = c$. and relation (5) is true. Q.E.D.

Definition 2.8. A non-empty subset **F'** of a field **F** is a subfield if and only if **F'** is a field with the same algebraic operations of **F**. The field **F** is called an extension field of **F'**.

Example 2.12. The real field is a subfield of the complex field. The real field is an extension field of the field of rational numbers.

2.4 Vector Spaces

To study linear codes one needs the concept of a vector space. A vector space is constructed over fields. To introduce this concept we first illustrate with some important spaces. Let R denote the field of real numbers. The space R^2 consists of all row vectors with two real-number components and is represented by the usual (x, y) plane. That is, the two components of the position vector (x, y) are the x and y coordinates of the corresponding point. R^3 is equally familiar with the three components designating a point in three-dimensional space. Within these spaces, two operations are possible: one can add any two vectors, and one can multiply vectors by real-number scalars. The valuable thing about these spaces is that the extension to n

Elements of Algebra

dimensions over an arbitrary field is quite straightforward. These examples motivate the following definition of a vector space.

Definition 2.9. Let $(V,+)$ be an Abelian group. Let $\mathbf{F}$ be a commutative field with the identity elements, 0 and 1 for the operations $+$ and $*$, respectively. A multiplication operation, denoted by $\bullet$, between the elements in $\mathbf{F}$ and the elements in $\mathbf{V}$, is also defined. The set $\mathbf{V}$ is called a vector space over the field $\mathbf{F}$ if it satisfies the following conditions:
(1) For any element $a \in \mathbf{F}$ and any element $v \in \mathbf{V}$ one has $a \bullet v \in \mathbf{V}$.
(2) (Distributive laws) For any elements $u, v \in \mathbf{V}$ and any elements $a, b \in \mathbf{F}$ one has
$$a \bullet (u+v) = a \bullet u + a \bullet v,$$
$$(a+b) \bullet v = a \bullet v + b \bullet v.$$
(3) (Associative law) For any $v \in \mathbf{V}$ and any $a, b \in \mathbf{F}$ one has
$$(a * b) \bullet v = a \bullet (b \bullet v).$$
(4) For any $v \in \mathbf{V}$ one has $1 \bullet v = v$.

The elements of $\mathbf{V}$ are called vectors, and the elements of the field $\mathbf{F}$ are called scalars. The addition on $\mathbf{V}$ is called vector addition. The multiplication, which maps a scalar in $\mathbf{F}$ and a vector in $\mathbf{V}$ into a vector in $\mathbf{V}$, is called scalar multiplication. The additive identity(zero) of $\mathbf{V}$ is denoted by $\mathbf{0}$.

Example 2.13. R^n is the n-dimensional vector space with scalars in R that is equipped with vector addition and multiplication given, respectively, by
$$(u_1, u_2, \ldots, u_n) + (v_1, v_2, \ldots, v_n) = (u_1 + v_1, u_2 + v_2, \ldots, u_n + v_n)$$
and
$$a \bullet (u_1, u_2, \ldots, u_n) = (au_1, au_2, \ldots, au_n).$$

Some basic properties of a vector space $\mathbf{V}$ over the field $\mathbf{F}$ are derived next from Definition 2.9.

Proposition 2.10. For any scalar $c \in \mathbf{F}$ and any vector $u \in \mathbf{V}$,
(1) $c \bullet 0 = 0$,
(2) $(-c) \bullet u = c \bullet (-u)$.
Proof .(1) Since $1+0=1$ in $\mathbf{F}$, one has $1 \bullet u = (1+0) \bullet u = 1 \bullet u + 0 \bullet u$. It follows from Definition 2.9 that $u = u + 0 \bullet u$. Let $-u$ be the additive inverse of u. Then, one has that $0 = 0 + 0 \bullet u = 0 \bullet u$ by adding $-u$ to both sides of $u = u + 0 \bullet u$. Therefore, $c \bullet 0 = c \bullet (0 \bullet u) = (c*0) \bullet u = 0 \bullet u = 0$.

(2) From the proof of (1) and Definition 2.9 one has $0 = 0 \bullet u = (c-c) \bullet u = c \bullet u + (-c) \bullet u$. Combining this with the relations, $c \bullet u + c \bullet (-u) = c \bullet (u + (-u)) = c \bullet 0 = 0$, yields finally the desired result, $(-c) \bullet u = c \bullet (-u)$. Q.E.D.

A vector space V over a field F may contain a subset S of V which is also a vector space over the field F. Such a subset is called a (vector) subspace of V.

Proposition 2.11. S is a subspace of V if and only if one of the following conditions is satisfied :
(1) For any two vectors $u, v \in S$ and any two scalars $a, b \in F$, $a \bullet u + b \bullet v \in S$.
(2) For any two vectors $u, v \in S$ and any scalar $a \in F$, one has $u - v \in S$ and $a \bullet u \in S$.

Proof. First, prove that condition (1) is equivalent to condition (2). If (1) is true, then it follows from $a \bullet u + b \bullet v \in S$ that $u - v \in S$ by simply choosing $a = 1$ and $b = -1$. Also $a \bullet u \in S$ by choosing $a = 1$ and $b = 0$. Therefore, condition (2) is true. If (2) is true, then for any scalars $a, b \in F$, one has $a \bullet u \in S$ and $b \bullet u \in S$. This implies that $(-b) \bullet u \in S$. Therefore, $a \bullet u - (-b) \bullet v = a \bullet u + b \bullet v \in S$, i.e. condition (1) is true. Next, prove that condition (2) is true if and only if S is a subspace of V over F. If S is a subspace of V over F, it follows directly from Definition 2.9 that condition (2) is true. Condition (2) says simply that S is closed under vector addition and scalar multiplication in V. Now for any scalar $a \in F$, one has $a \bullet u \in S$. This fact ensures that, for any vector $u \in S$, that its additive inverse satisfies $(-1) \bullet u \in S$. Thus, one has $u + (-1) \bullet u = 0 \in S$. Therefore, S is a subgroup of V. Since a vector in S is also a vector of V, the associative and distributive laws also must hold for S. Hence, S is a vector space over F and is a subspace of V. Q.E.D.

Let $v_1, v_2, \ldots, v_k$ be k vectors in a vector space V over a field F. Also let $a_1, a_2, \ldots, a_k$ be k scalars in F. Then the sum $a_1 \bullet v_1 + a_2 \bullet v_2 + \cdots + a_k \bullet v_k$ is called a linear combination of $v_1, v_2, \ldots, v_k$. Clearly, the sum of two linear combinations of the vectors $v_1, v_2, \ldots, v_k$, satisfies

$$(a_1 \bullet v_1 + a_2 \bullet v_2 + \ldots + a_k \bullet v_k) + (b_1 \bullet v_1 + b_2 \bullet v_2 + \ldots + b_k \bullet v_k)$$
$$= (a_1 + b_1) \bullet v_1 + (a_2 + b_2) \bullet v_2 + \ldots + (a_k + b_k) \bullet v_k,$$

which is also a linear combination of the vectors $v_1, v_2, \ldots, v_k$. Finally the product of a scalar c in F with a linear combination of the $v_1, v_2, \ldots, v_k$ satisfies

$$c \bullet (a_1 \bullet v_1 + a_2 \bullet v_2 + \ldots + a_k \bullet v_k) = (c \cdot a_1) \bullet v_1 + (c \cdot a_2) \bullet v_2 + \ldots + (c \cdot a_k) \bullet v_k,$$

Elements of Algebra

which is also a linear combination of the vectors $v_1, v_2, \ldots, v_k$. Thus the following proposition follows by a use of Proposition 2.11 :

Proposition 2.12. Let $v_1, v_2, \ldots, v_k$ be k vectors in a vector space **V** over a field **F**. The set of all linear combinations of $v_1, v_2, \ldots, v_k$ forms a subspace **S** of **V**.

If $a_1 \bullet v_1 + a_2 \bullet v_2 + \cdots + a_k \bullet v_k = 0$ only when all of the α_i are zero, $v_1, v_2, \ldots, v_k$ are called linearly independent. Vectors $v_1, v_2, \ldots, v_k$ are linearly dependent if there exist $a_1, a_2, \ldots, a_k$ which are not all zero such that $a_1 \bullet v_1 + a_2 \bullet v_2 + \cdots + a_k \bullet v_k = 0$. If $v_1, v_2, \ldots, v_k$ are linearly independent and all linear combinations of $v_1, v_2, \ldots, v_k$ forms the vector space **S**, then $v_1, v_2, \ldots, v_k$ is, what is called, a basis (basis set) for **S**, and the number of vectors k in a basis is called the dimension of **S** over the field **F** (Finally note that the number of vectors in any basis set of **S** is the same).

2.5 Finite Fields

There exist fields which have only a finite number of elements. Such a field is called a **finite** or **Galois field**. The finite field is perhaps the most useful algebraic structure for error-control coding. When the field **F** is finite, the number of elements in **F** is called the order of **F**. A finite field of order q is denoted by $GF(q)$.

Example 2.14. The set $F_2 = \{0,1\}$ under modulo-2 addition and multiplication forms a finite field $GF(2)$. The modulo-2 addition and multiplication tables are given in Tables 2.2 and 2.3.

+	0	1
0	0	1
1	1	0

Table 2.2 Modulo-2 Addition

*	0	1
0	0	0
1	0	1

Table 2.3 Modulo-2 Multiplication

Finite fields are used in most of the known constructions of error-correcting codes. This subsection provides a study of finite fields and some of their elementary properties. We begin with the following simple example of a field.

Example 2.15. Let p be a prime number. Then the integers, modulo p, form a field of order p, denoted by $GF(p)$. The elements of $GF(p)$ are $\{0,1,2,\ldots,p-1\}$, and the additive and multiplicative operations are carried out mod p, e.g. $GF(3)$ is the ternary field $\{0,1,2\}$, with $1+2=3 \equiv 0 \pmod{3}$, $2 \cdot 2 = 4 \equiv 1 \pmod{3}$, $1-2=-1 \equiv 2 \pmod{3}$, etc.

Next define the characteristic of a field. Let **K** be an arbitrary finite field of order q. Since **K** is a field, it contains the unit element 1. Since **K** is finite, the elements $1, 1+1=2, 1+1+1=3, \ldots$ cannot all be distinct. Therefore, there is a smallest integer p such that $p = 1+1+\cdots+1(p \text{ times}) = 0$. Such a number p is called the *characteristic* of the field.

Proposition 2.13. The characteristic p of a finite field is a prime integer.
Proof : Suppose that p is not a prime and is equal to the product of two smaller integers k and m, i.e. $p = km$. Since the field is closed under multiplication, the product,

$$\left(\sum_{i=1}^{k} 1\right) \bullet \left(\sum_{i=1}^{m} 1\right),$$

is also a field element. It follows from the distributive law that

$$\left(\sum_{i=1}^{k} 1\right) \bullet \left(\sum_{i=1}^{m} 1\right) = \sum_{i=1}^{km} 1.$$

Since $\sum_{i=1}^{km} 1 = 0$, then either $\sum_{i=1}^{km} 1 = 0$ or $\sum_{i=1}^{m} 1 = 0$. However, this contradicts the definition that p is the smallest positive integer for which $\sum_{i=1}^{p} 1 = 0$. Therefore, one concludes that p is a prime integer. Q.E.D.

It follows from the definition of the characteristic of a finite field $GF(q)$ that for any two distinct positive integers k and m less than p that

$$(\sum_{i=1}^{k} 1) \neq (\sum_{i=1}^{m} 1).$$

Suppose that $\sum_{i=1}^{k} 1 = \sum_{i=1}^{m} 1$. Then one has $\sum_{i=1}^{m-k} 1 = 0$ on the assumption that $m > k$. However, this is impossible since $m - k < p$. Therefore, the sums,

$$1 = \sum_{i=1}^{1} 1, \sum_{i=1}^{2} 1, \sum_{i=1}^{3} 1, \ldots, \sum_{i=1}^{p-1} 1, \text{ and } \sum_{i=1}^{p} 1 = 0,$$

constitute p distinct elements in $GF(q)$. In fact, this set of sums itself is a field of p elements, namely $GF(p)$, under the addition and multiplication operations of $GF(q)$. Since $GF(p)$ is a subset of $GF(q)$, $GF(p)$ is a subfield of $GF(q)$. Therefore, any finite field $GF(q)$ of characteristic p contains a subfield of p elements. Also, it can be proved that if $p \neq q$, then q is a power of p, i.e. $q = p^n$ for some positive integer n.

Now let a be a nonzero element in $GF(q)$. Since the set of nonzero elements of $GF(q)$ is closed under multiplication, the following powers of a, namely,

$$a^1 = a, a^2 = a \cdot a, a^3 = a \cdot a \cdot a, \ldots$$

must also be nonzero elements in $GF(q)$. Since $GF(q)$ has only a finite number of elements, the powers of a, given above, cannot all be distinct. Therefore, at some point in the sequence of powers of a, there must be a repetition; that is, there must exist two positive integers k and m such that $m > k$ and $a^k = a^m$, Let a^{-1} be the multiplicative inverse of a. Then $(a^{-1})^k = a^{-k}$ is the multiplicative inverse of a^k. Multiplying both sides of $a^k = a^m$ by a^{-k}, we obtain

$$1 = a^{m-k}.$$

This implies that there must exist a smallest positive integer n such that $a^n = 1$. This integer n is called the order of the field element a. Therefore, the sequence $a^1, a^2, a^3, \ldots$ repeats itself after $a^n = 1$. Also, the powers $a^1, a^2, a^3, \ldots, a^{n-1}, a^n = 1$ are all distinct. In fact, they form a group under the multiplication operations of $GF(q)$. Note first that this set contains the unit element 1. Next consider $a^i \cdot a^j$. If $i + j \leq n$, then

$$a^i \cdot a^j = a^{i+j}.$$

On the other hand If $i + j > n$, one has $i + j = n + r$, where $0 < r \leq n$. Hence,

$$a^i \cdot a^j = a^{i+j} = a^n \cdot a^r = a^r.$$

Therefore, the powers $a^1, a^2, a^3, \ldots, a^{n-1}, a^n = 1$ are closed under the multiplications of $GF(q)$. For $1 \le i < n$ the power a^{n-i} is the multiplicative inverse of a^i. Since the powers of a are nonzero elements in $GF(q)$, they satisfy the associative and commutative laws. Therefore, one concludes that the elements $a^n = 1, a^1, a^2, \ldots a^{n-1}$ form a group under the multiplication operation of $GF(q)$. Recall that a group is a cyclic group if there exists an element in the group whose powers constitute the entire group.

Proposition 2.14. Let a be a nonzero element of a finite field $GF(q)$. Then $a^{q-1} = 1$.

Proof. Let $b_1, b_2, \ldots, b_{q-1}$ be the $q-1$ nonzero elements of $GF(q)$. Clearly, the $q-1$ elements, $a \cdot b_1, a \cdot b_2, \ldots a \cdot b_{q-1}$, are nonzero and distinct. Thus,

$$(a \cdot b_1) \cdot (a \cdot b_2) \cdots (a \cdot b_{q-1}) = b_1 \cdot b_2 \cdots b_{q-1}$$
$$a^{q-1} \cdot (b_1 \cdot b_2 \cdots b_{q-1}) = b_1 \cdot b_2 \cdots b_{q-1}$$

Since $a \ne 0$ and $(b_1 \cdot b_2 \cdots b_{q-1}) \ne 0$, one must have $a^{q-1} = 1$. Q.E.D.

Proposition 2.15. Let a be a nonzero element in a finite field $GF(q)$. Also, let n be the order of a. Then n divides $q-1$.

Proof. Suppose that n does not divide $q-1$. Then a division of $q-1$ by n yields

$$q - 1 = kn + r,$$

where $0 < r < n$. Then

$$a^{q-1} = a^{kn+r} = a^{kn} \cdot a^r = (a^n)^k \cdot a^r.$$

Since $a^{q-1} = 1$ and $a^n = 1$, one must have $a^r = 1$. This is impossible since $0 < r < n$. Hence $r=0$, and n must divide $q-1$. Q.E.D.

In a finite field $GF(q)$ a nonzero element a is said to be primitive if the order of a is $q-1$. Therefore, the powers of a primitive element generate all of the nonzero elements of $GF(q)$. The following important proposition about primitive elements is given next without proof.

Proposition 2.16. Any finite field contains a primitive element.

Example 2.16. Consider the prime field $GF(7)$ illustrated by Tables 2.5 and 2.6. The characteristic of this field is 7. If one takes the powers of the integer 3 in $GF(7)$ using the multiplication table, one has

$3^1 = 3$, $3^2 = 3 \cdot 3 = 2$, $3^3 = 3 \cdot 3^2 = 6$, $3^4 = 3 \cdot 3^3 = 4$, $3^5 = 3 \cdot 3^4 = 5$, $3^6 = 3 \cdot 3^5 = 1$.
Therefore, the order of integer 3 is 6, and integer 3 is a primitive element of $GF(7)$. The powers of the integer 4 in $GF(7)$ are
$$4^1 = 4, \ 4^2 = 4 \cdot 4 = 2, \ 4^3 = 4 \cdot 4^2 = 1.$$
Clearly, the order of the integer 4 is 3, which is a factor of 6.

Next, we construct a very useful vector space over $GF(2)$ which plays a central role in coding theory. Consider an ordered sequence of the m components, $(u_0, u_1, \ldots, u_{m-1})$, where each component u_i is an element from the binary field $GF(2)$. Since there are two choices for each u_i, there are exactly 2^m distinct m-tuples. Let $V_m = [GF(2)]^m$ denote this set of 2^m distinct m-tuples. Now, define an addition "+" on V_m as follows: For any $\bar{u} = (u_0, u_1, \ldots, u_{m-1}) \in V_n$ and $\bar{v} = (v_0, v_1, \ldots, v_{m-1}) \in V_n$,
$$\bar{u} + \bar{v} = (u_0 + v_0, u_1 + v_1, \ldots, u_{m-1} + v_{m-1}), \tag{2.1}$$
where $u_i + v_i$ is carried out in modulo-2 addition. Clearly, $\bar{u} + \bar{v}$ is also an m-tuple over $GF(2)$. Hence, V_m is closed under the addition defined by Eq. (2.1). It is verified that V_m is a commutative group under the addition defined by Eq. (2.1). First it is seen that the all-zero m-tuple $\bar{0} = (0,0,\ldots,0)$ is the additive identity. For any $\bar{u} \in V_m$, $\bar{u} + \bar{u} = (u_0 + u_0, u_1 + u_1, \ldots, u_{m-1} + u_{m-1}) = (0,0,\ldots,0) = \bar{0}$. Hence, the additive inverse of each m-tuple in V_m is itself. Since modulo-2 addition is commutative and associative, one can check easily that the addition defined in Eq. (2.1) is also commutative. Therefore, V_m is a commutative group under the addition in Eq. (2.1).

The scalar multiplication of an m-tuple $\bar{u} \in V_m$ by an element $a \in GF(2)$ is given as follows:
$$a \cdot (u_0, u_1, \ldots, u_{m-1}) = (a \cdot u_0, a \cdot u_1, \ldots, a \cdot u_{m-1}), \tag{2.2}$$
where $a \cdot u_i$ is carried out in modulo-2 multiplication. Clearly, $a \cdot (u_0, u_1, \ldots, u_{m-1})$ is also an m-tuple in V_m. If $a = 1$, $1 \cdot (u_0, u_1, \ldots, u_{m-1}) = (1 \cdot u_0, 1 \cdot u_1, \ldots, 1 \cdot u_{m-1}) = (u_0, u_1, \ldots, u_{m-1})$. It is shown easily that the vector addition and scalar multiplication defined by Eqs. (2.1) and (2.2), respectively, satisfy the distributive and associative laws. Therefore, one obtains:

Proposition 2.17. The set V_m of all m-tuples over $GF(2)$ forms a vector space over $GF(2)$.

2.6 Euclid's Algorithm

Given two integers or polynomials a and b Euclid's algorithm is a recursive technique for finding the greatest common divisor (a,b) of a and b. The genesis of the Euclidean algorithm is Euclid's division algorithm $a = q \cdot b + r$ defined in Sec. 2.1.4 for $a \geq b, b > r$ if they are integers or $\deg(a) \geq \deg(b)$, $\deg(b) > \deg(r)$ if they are polynomials. The origin of the recursive steps of Euclid's algorithm is based on the proposition given next.

Proposition 2.18. Let a and b be two positive integers (or polynomials). Then if $a = q \cdot b + r$ for $0 \leq r < b$ $(0 \leq \deg(r) < \deg(b))$, one has $(a,b) = (b,r)$, where (a,b) denotes greatest common divisor.

Proof : Let $d_1 = (a,b)$ and $d_2 = (b,r)$. Thus since d_1 divides both a and b, d_1 must divide $(a - q \cdot b) = r$. Hence, $d_1 \leq d_2$. Conversely, since d_2 divides both b and r, d_2 must divide $(q \cdot b + r) = a$. Thus d_2 divides both a and b as well as r. Hence $d_2 \leq d_1$. Therefore $d_1 = d_2$ and the proposition is proved.

By Proposition 2.18 if r_1 has the remainder of a divided by b, r_2 is the remainder of b divided by r_1, ..., r_n is remainder of r_{n-2} divided by r_{n-1}. Finally if zero is the remainder of r_{n-1} divided by r_n, one has the following chain of equalities : $(a,b) = (b,r) = \cdots = (r_{n-1}, r_n) = (r_n, 0) = r_n$. This leads one to conclude immediately that the recursive system of equations,

$$a = q_1 \cdot b + r_1, 0 < r_1 < b$$
$$b = q_2 \cdot r_1 + r_2, 0 < r_2 < r_1$$
$$r_1 = q_3 \cdot r_2 + r_3, 0 < r_3 < r_2$$
$$\vdots \qquad \vdots$$
$$r_{n-2} = q_n \cdot r_{n-1} + r_n, 0 < r_n < r_{n-1}$$
$$r_{n-1} = q_{n+1} \cdot r_n,$$

(2.3)

generates the greatest common divisor,

$$(a,b) = r_n \qquad (2.4)$$

This algorithm is illustrated next in a simple example.

Elements of Algebra

Example 2.17. Suppose that the two integers are $a=186$ and $b=66$. Then
$$186=(66)*2+54$$
$$66=(54)*1+12$$
$$54=(12)*4+6$$
$$12=6*2+0.$$
By Eq.(2.4) the greatest common divisor is 6.

The above process of finding the greatest common divisor $d=(a,b)$, the algorithm calculates two numbers, or polynomials f and g such that
$$fa+gb=d. \qquad (2.5)$$
One way to accomplished this is to work backward through the system of equations in Eq.(2.3) successively eliminating the remainders $r_{n-1}, r_{n-2}, \ldots, r_2, r_1$, until one finds $r_n = (a,b)$ as the linear combination of a and b in the form given in Eq.(2.5).

Another more efficient method to find (a,b) as a linear combination of a and b of the form in Eq. (2.5) is to find in a recursive fashion two coefficients f_i and g_i such that
$$f_i a + g_i b = r_i \qquad (2.6)$$
for $i=1,2,\ldots,n$, where by Eq.(2.4) $r_i = d$. To find the required recursion that f_i and g_i satisfy write those steps of Eq.(2.3) as follows:
$$f_{i-2}a + g_{i-2}b = r_{i-2},$$
$$f_{i-1}a + g_{i-1}b = r_{i-1},$$
$$f_i a + g_i b = r_i,$$
and use the relations in Eq.(2.3) to obtain
$$r_i = r_{i-2} - q_i r_{i-1}$$
$$= f_{i-2}a + g_{i-2}b - q_i(f_{i-1}a + g_{i-1}b)$$
$$= (f_{i-2} - q_i f_{i-1})a + (g_{i-2} - q_i g_{i-1})b$$
which yields the identity,
$$f_i a + g_i b = (f_{i-2} - q_i f_{i-1})a + (g_{i-2} - q_i g_{i-1})b.$$
Since a and b are arbitrary, the equating of coefficients yields the following recursions for the coefficients f_i and g_i as follows:
$$f_i = f_{i-2} - q_i f_{i-1}, \text{ and } g_i = g_{i-2} - q_i g_{i-1} \qquad (2.7)$$
the same recursive equation that the r_i satisfy. For Example 2.18 one has
$$1*(186)-2*(66)=54.$$
$$-(186)+3*(66)=12.$$
$$5*(186)-14*(66)=6,$$
and

$-11*(186)+31*(66)=0.$

These recursive relationships are now stated formally for polynomials since all of the subsequent discussions of the Euclidean algorithm are restricted to polynomials over some finite field.

Define $q_i(x)$ to be the quotient polynomial that is produced by dividing $r_{i-2}(x)$ by $r_{i-1}(x)$. That is

$$q_i(x) = \left[\frac{r_{i-2}(x)}{r_{i-1}(x)}\right]_0^\infty$$

where $[\bullet]_0^\infty$ denotes the polynomial of the nonnegative powers of x. Then, the recursive Euclidean algorithm is given by

$$r_i(x) = r_{i-2}(x) - q_i(x)r_{i-1}(x),$$
$$f_i(x) = f_{i-2}(x) - q_i(x)f_{i-1}(x),$$

and

$$g_i(x) = g_{i-2}(x) - q_i(x)g_{i-1}(x).$$

Initial conditions for the algorithm are

$$f_{-1}(x) = g_0(x) = 1,$$
$$f_0(x) = g_{-1}(x) = 0,$$
$$r_{-1}(x) = a(x),$$

and

$$r_0(x) = b(x).$$

The nature of this algorithm is such that

$$\deg[r_i(x)] < \deg[r_{i-1}(x)]$$

for all i. Thus, the algorithm always converges to a zero remainder in a finite number of steps.

Example 2.18

Consider two polynomials over the integer ring : $a(x) = x^3 + 1$ and $b(x) = x^2 + 1$. The procedure of Euclid's algorithm for $a(x)/b(x)$ is shown in Table 2.4. The last nonzero remainder $r_i(x) = x+1$ is the greatest common divisor(GCD).

Elements of Algebra

i	$f_i(x)$	$g_i(x)$	$r_i(x)$	$q_i(x)$
-1	1	0	x^3+1	---
0	0	1	x^2+1	---
1	1	x	$x+1$	x
2	$x+1$	x^2+x+1	0	$x+1$

Table 2.4 Example of Euclid's algorithm with $a(x)$ and $b(x)$.

2.7 Binary Field Arithmetic

Error-control codes can be constructed in general with symbols from any finite field $GF(q)$, where q is either a prime integer p or a power of p. However, for practical reasons the codes with symbols from the binary field $GF(2)$ or its extension field $GF(2^m)$ are the most widely used in digital-data transmission and storage systems. This follows quite simply from the fact that the information in these systems is coded almost universally in binary form. Binary codes and codes with symbols from the field $GF(2^m)$ are the symbols of primary concern in this book. Most of the results presented here can be generalized to codes with symbols from any finite field $GF(q)$ with $q \neq 2^m$. This subsection discusses the arithmetic in the binary field $GF(2^m)$, used in the remaining chapters.

It is a well known fact that modulo-2 additions and multiplications are basic to binary arithmetic. These operations are defined in Tables 2.2 and 2.3, respectively. This arithmetic is actually equivalent to ordinary arithmetic, except that any even number is equivalent to zero (i.e., 1+1=2=0), and any odd number is equivalent to one. The implementation of these elementary operators is given in Figure 2.1 in the form of standard logic gates.

XNOR Gate for "+"   AND Gate for " • "

Figure 2.1 Logic gates for binary arithmetic, "+" and " • "

Next, consider computations with polynomials whose coefficients are from the binary field $GF(2)$. A polynomial $f(x)$ of the single variable x with coefficients from $GF(2)$ is of the following form:

$$f(x) = f_0 + f_1 x + f_2 x^2 + \cdots + f_n x^n,$$

where $f_i = 0$ or 1 for $0 \leq i \leq n$. The degree of a polynomial is the largest power of x with a nonzero coefficient. For the above polynomial if $f_n = 1$, $f(x)$ is a monic polynomial of degree n; if $f_n = 0$, $f(x)$ is a polynomial of degree less than n. The degree of $f(x) = f_0$ is zero.

In the following the phrase, "a polynomial over $GF(2)$", means "a polynomial with coefficients from $GF(2)$." There are two polynomials over $GF(2)$ of degree 1: x and $1+x$. There are four polynomials over $GF(2)$ of degree 2: $x^2, 1+x^2, x+x^2$ and $1+x+x^2$. In general, there are 2^n polynomials over $GF(2)$ of degree n. Polynomials over $GF(2)$ can be added (or subtracted), multiplied, or divided in the usual manner. Let

$$g(x) = g_0 + g_1 x + g_2 x^2 + \cdots + g_m x^m$$

be another polynomial over $GF(2)$. To add polynomials $f(x) = f_0 + f_1 x + \cdots + f_m x^m$ and $g(x) = g_0 + g_1 x + \cdots + g_n x^n$, one just simply adds the coefficients of the same power of x in $f(x)$ and $g(x)$ as follows (assuming that $m \leq n$):

$$f(x) + g(x) = (f_0 + g_0) + (f_1 + g_1)x + \cdots + (f_m + g_m)x^m + f_{m+1}x^{m+1} + \cdots + f_n x^n,$$

where $f_i + g_i$ is carried out in a modulo-2 addition which is implemented by the logic given in Figure 2.1.

A multiplication of the above polynomials $f(x)$ and $g(x)$ is given by the following product:

$$f(x) \cdot g(x) = c_0 + c_1 x + c_2 x^2 + \cdots + c_{n+m} x^{n+m},$$

where

$$c_0 = f_0 g_0,$$
$$c_1 = f_0 g_1 + f_1 g_0,$$
$$c_2 = f_0 g_2 + f_1 g_1 + f_2 g_0,$$
$$\vdots$$
$$c_i = f_0 g_i + f_1 g_{i-1} + f_2 g_{i-2} + \cdots + f_i g_0,$$
$$\vdots$$
$$c_{n+m} = f_n g_m.$$

Elements of Algebra

(Multiplication and addition of coefficients are modulo-2.) It is clear from the above equations that if $g(x) = 0$, then $f(x) \cdot 0 = 0$.

Example 2.19

The multiplication of $f(x) = f_0 + f_1 x$ and $g(x) = g_0 + g_1 x + g_2 x^2$ yields the following result:

$$f(x)g(x) = f_0 g_0 + (f_0 g_1 + f_1 g_0)x + (f_0 g_2 + f_1 g_1)x^2 + f_1 g_2 x^3.$$

The implementation of such a multiplication is performed by binary arithmetic operations "+" and " • " and also by delay units. The delay unit is realized usually by, what is called, the flip-flop(FF). A FF operation stores a binary symbol for one clock step(or time unit). If at clock time t a symbol s is presented to the input of a FF, then at time $t+1$ the symbol will appear at the output of this FF. The implementation logic is shown in Figure 2.2.

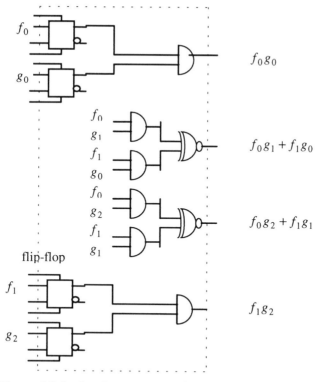

Figure 2.2 An implementation of polynomial multiplication

It is verified readily that the polynomials over $GF(2)$ satisfy the following conditions:

(I) Commutative law :
$$a(x)+b(x) = b(x)+a(x)$$
$$a(x) \cdot b(x) = b(x) \cdot a(x)$$

(ii) Associative law :
$$a(x)+[b(x)+c(x)] = [a(x)+b(x)]+c(x)$$
$$a(x) \cdot [b(x) \cdot c(x)] = [a(x) \cdot b(x)] \cdot c(x).$$

(iii) Distributive law:
$$a(x) \cdot [b(x)+c(x)] = [a(x) \cdot b(x)]+[a(x) \cdot c(x)].$$

Suppose that the degree of $g(x)$ is not zero. When $f(x)$ is divided by $g(x)$, it can be shown in the same manner used for real polynomials that a unique pair $q(x)$ and $r(x)$ of polynomials over $GF(2)$ can be obtained such that
$$f(x) = q(x)g(x)+r(x),$$
where $q(x)$ is called the quotient, $r(x)$ is the remainder such that the degree of $r(x)$ is less than that of $g(x)$. This is Euclid's division algorithm for polynomials over $GF(2)$.

If a polynomial $f(x)$ over $GF(2)$ has an even number of nonzero terms, it is divisible by $x+1$. A polynomial $p(x)$ over $GF(2)$ of degree m is said to be irreducible over $GF(2)$ if $p(x)$ is not divisible by any polynomial over $GF(2)$ of degree less than m but greater than zero.

An important proposition about irreducible polynomials over $GF(2)$ is given next without proof.

Proposition 2.19. Any irreducible polynomial over $GF(2)$ of degree m divides $x^{2^m-1}+1$.

An irreducible polynomial $p(x)$ of degree m is said to be primitive if the smallest positive integer n, such that $p(x)$ divides x^n+1, is $n = 2^m - 1$.

Next, a method for constructing the Galois field of 2^m elements ($m>1$) from the binary field $GF(2)$ is presented. We begin with the two elements 0 and 1, from $GF(2)$ and a new symbol; call it α. Define a multiplication " $\bullet$ " by introducing a sequence of powers of α as follows:

$0 \cdot 0 = 0,$
$0 \cdot 1 = 1 \cdot 0 = 0,$
$1 \cdot 1 = 1,$
$0 \cdot \alpha = \alpha \cdot 0 = 0,$
$1 \cdot \alpha = \alpha \cdot 1 = \alpha,$
$\alpha^2 = \alpha \cdot \alpha,$
$\alpha^3 = \alpha \cdot \alpha \cdot \alpha,$
$\vdots$
$\alpha^j = \alpha \cdot \alpha \cdots \alpha \quad (j \text{ times})$
$\vdots$

It follows from the above definition of multiplication that
$0 \cdot \alpha^j = \alpha^j \cdot 0 = 0,$
$1 \cdot \alpha^j = \alpha^j \cdot 1 = \alpha^j,$
$\alpha^i \cdot \alpha^j = \alpha^j \cdot \alpha^i = \alpha^{i+j}.$

By this means, one obtains the following set of elements on which a multiplication operation is defined:
$$F = \{0, 1, \alpha, \alpha^2, \cdots, \alpha^j, \cdots\},$$
The element 1 is defined sometimes by α^0.

Next a condition is imposed on the element α in such a manner that the set F contains only 2^m elements and is closed under multiplications. Let $p(x)$ be a primitive polynomial of degree m over $GF(2)$. Also assume that $p(\alpha) = 0$, i.e. α is a root of $p(x)$. But by Proposition 2.17, $p(x)$ divides $x^{2^m - 1} + 1$. Hence one has
$$x^{2^m - 1} + 1 = q(x) p(x)$$
for same polynomial q(x).
If x is replaced by α in the above equation, one obtains
$$\alpha^{2^m - 1} + 1 = q(\alpha) p(\alpha),$$
where $q(x)$ is some polynomial over $GF(2)$. Thus, since $p(\alpha) = 0$, one has
$$\alpha^{2^m - 1} + 1 = q(\alpha) \cdot 0 = 0.$$
As a result, the following equality is obtained:
$$\alpha^{2^m - 1} + 1 = 0.$$
The addition of 1 to both sides of $\alpha^{2^m - 1} + 1 = 0$ (using modulo-2 addition) results in the following equality:
$$\alpha^{2^m - 1} = 1.$$
Therefore, under the condition that $p(\alpha) = 0$, the set F becomes finite and contains the following elements:

$$F = \{0, 1, \alpha, \alpha^2, \cdots, \alpha^{2^m-2}\}.$$

Now it is shown that the set F^* of nonzero elements of F are closed under the operation of multiplication. To see this, let i and j be two integers such that $0 \leq i, j < 2^m - 1$. If $i + j < 2^m - 1$, then $\alpha^i \cdot \alpha^j = \alpha^{i+j}$, which evidently is a nonzero element in F. If $i + j \geq 2^m - 1$, $i + j$ can be expressed as follows: $i + j = (2^m - 1) + r$, where $0 \leq r < 2^m - 1$. Then

$$\alpha^i \cdot \alpha^j = \alpha^{i+j} = \alpha^{(2^m-1)+r} = \alpha^{2^m-1} \cdot \alpha^r = 1 \cdot \alpha^r = \alpha^r,$$

which is also a nonzero element in F. Hence, one concludes that the nonzero elements of F are closed under the operation of multiplication.

The nonzero elements in F form a commutative group F^* under multiplication. First, element 1 is the unit element. Also the multiplication operation is clearly commutative and associative. For $0 < i < 2^m - 1$, the power α^{2^m-i-1} is the multiplicative inverse of α^i since $\alpha^{2^m-i-1} \cdot \alpha^i = \alpha^{2^m-1} = 1$. (Note that $\alpha^0 = \alpha^{2^m-1} = 1$.) It is clear that the elements $1, \alpha, \alpha^2, \cdots, \alpha^{2^m-2}$ represent $2^m - 1$ distinct elements. Therefore, the nonzero elements of F form a group of order $2^m - 1$ under the operation of multiplication.

The next step is to define an addition operation "+" on F in such a manner that F forms a commutative group under "+." For $0 \leq i < 2^m - 1$ divide the polynomial x^i by $p(x)$ and obtain the following:

$$x^i = q_i(x) p(x) + a_i(x),$$

where $q_i(x)$ and $a_i(x)$ are the quotient and the remainder polynomials, respectively. The remainder $a_i(x)$ is a polynomial of degree $m-1$ or less over $GF(2)$ and has the form,

$$a_i(x) = a_{i0} + a_{i1}x + a_{i2}x^2 + \cdots + a_{i,m-1}x^{m-1}.$$

Since x and $p(x)$ are relatively prime, i.e., they have no common non-constant polynomial factor except 1, the polynomial x^i is not divisible by $p(x)$. Thus, for any $i \geq 0$, one has $a_i(x) \neq 0$. For $0 \leq i, j < 2^m - 1$, and $i \neq j$, it is shown next that $a_i(x) \neq a_j(x)$. Suppose to the contrary that $a_i(x) = a_j(x)$. Then it follows that

$$x^i + x^j = [q_i(x) + q_j(x)]p(x) + a_i(x) + a_j(x)$$
$$= [q_i(x) + q_j(x)]p(x).$$

Elements of Algebra

This implies that $p(x)$ divides $x^i + x^j = x^i(1+x^{j-i})$ (assuming that $j > i$). Since x^i and $p(x)$ are relatively prime, $p(x)$ must divide $x^{j-i} + 1$. However, this is impossible since $j - 1 < 2^m - 1$ and $p(x)$ is a primitive polynomial of degree m which does not divide $x^n + 1$ for $n < 2^m - 1$. Therefore, the hypothesis that $a_i(x) = a_j(x)$ is not valid. As a result, for $0 \le i, j < 2^m - 1$ and $i \ne j$, one has that $a_i(x) \ne a_j(x)$. Hence, for $i = 0,1,2,\cdots,2^m - 2$, one obtains $2^m - 1$ distinct nonzero polynomials $a_i(x)$ of degree $m - 1$ or less. Now, replacing x by α and using the equality that $q_i(\alpha) \cdot 0 = 0$, one has the following polynomial expression for α^i ($i = 0,1,2,\cdots,2^m - 2$):

$$\alpha^i = a_i(\alpha) = a_{i0} + a_{i1}\alpha + a_{i2}\alpha^2 + \cdots + a_{i,m-1}\alpha^{m-1}.$$

Therefore, the $2^m - 1$ nonzero elements, $\alpha^0, \alpha^1, \cdots, \alpha^{2^m - 2}$ in F^*, are represented by the $2^m - 1$ distinct nonzero polynomials of α over $GF(2)$ of degree $m - 1$ or less. The zero element 0 in F is represented by the zero polynomial. As a result, the 2^m elements in F are represented by 2^m distinct polynomials of α over $GF(2)$ of degree $m - 1$ or less. These polynomial are the 2^m distinct elements of F.

Now define an addition operation on F as follows:
$$0 + 0 = 0,$$
and for $0 \le i, j < 2^m - 1$ let
$$0 + \alpha^i = \alpha^i + 0 = \alpha^i,$$
$$\alpha^i + \alpha^j = (a_{i0} + a_{i1}\alpha + \cdots + a_{i,m-1}\alpha^{m-1}) + (a_{j0} + a_{j1}\alpha + \cdots + a_{j,m-1}\alpha^{m-1})$$
$$= (a_{i0} + a_{j0}) + (a_{i1} + a_{j1})\alpha + \cdots + (a_{i,m-1} + a_{j,m-1})\alpha^{m-1},$$
where $a_{i,j} + a_{j,i}$ is carried out in modulo-2 addition. For $i = j$,
$$\alpha^i + \alpha^i = 0$$
and for $i \ne j$,
$$(a_{i0} + a_{j0}) + (a_{i1} + a_{j1})\alpha + \cdots + (a_{i,m-1} + a_{j,m-1})\alpha^{m-1}$$
is nonzero and must be the polynomial expression for some α^k in F. Hence, the set F is closed under the addition "+" defined by Definition 2.7. It can be verified immediately that F is a commutative group under "+". First, 0 is the additive identity. Using the fact that modulo-2 addition is commutative and associative, the addition defined on F is also commutative and associative. The additive inverse of any element in F is itself.

Up to this point, it has been shown that the set $F = \{0, 1, \alpha, \alpha^2, \cdots, \alpha^{2^m-2}\}$ is a commutative group under an addition operation "+" and the nonzero elements of F form a commutative group under a multiplication operation ".". Using the polynomial representation for the elements in F and distributive law of polynomial multiplication, one readily sees that the multiplication on F is distributive over the addition on F. Therefore, the set $F = \{0, 1, \alpha, \alpha^2, \cdots, \alpha^{2^m-1}\}$ is the Galois or finite field of 2^m elements, denoted by $GF(2^m)$. Also, it follows from the polynomial expressions for the elements of $GF(2^m)$ and Proposition 2.16 that $GF(2^m)$ is a vector space over $GF(2)$. The set $\{1, \alpha, \cdots, \alpha^{m-1}\}$ is called the *canonical* basis of $GF(2^m)$ over $GF(2)$.

Notice that the above addition and multiplication operations, defined on $F = GF(2^m)$, require the use of modulo-2 additions and multiplications. Hence, the subset $\{0,1\}$ forms a subfield of $GF(2^m)$ [i.e., $GF(2)$ is a subfield of $GF(2^m)$]. The binary field $GF(2)$ usually is called the ground field of $GF(2^m)$. The characteristic of $GF(2^m)$ is 2.

In the process of constructing $GF(2^m)$ from $GF(2)$, two representations for the nonzero elements of $GF(2^m)$ were developed : The power representation and the polynomial representation. The power representation is the most convenient for multiplications and the polynomial representation is best used for additions.

There is also a very useful function which maps elements of $GF(2^m)$ onto its subfield $GF(2)$. Such a function is called the *trace* function which is defined by

$$Tr(x) = x + x^2 + \cdots + x^{2^{m-1}}. \tag{2.8}$$

It is easy to prove the following proposition for the trace operation :

Proposition 2.20. (1) $Tr(x) \in GF(2)$ for all x in $GF(2^m)$. (2) $Tr(x+y) = Tr(x) + Tr(y)$ for all x,y in $GF(2^m)$.

Let β be any field element of $GF(2^m)$. The monic polynomial $m(x)$ of smallest degree with coefficients in $GF(2)$ such that $m(\beta) = 0$ is called the minimal polynomial of β.

Elements of Algebra

2.8 Arithmetic Operations in $GF(q)$

The theory of finite fields is a useful tool in switching theory and digital signal processing. In channel-coding theory finite-field arithmetic is used extensively in the encoding and decoding processes. These procedures are at the heart of many electronic devices, i.e. compact-disk circuitry, digital television, and magnetic storage (computer memories, tapes, disks) etc. Some communication systems, such as fiber-optic transmissions and satellite transmissions, etc. also make use of computations over a finite field.

Encoding and decoding are the essential operations needed in the application of error-correcting codes. These procedures make possible the addition of redundancy to the information message that is to be transmitted, and the use of it to reconstruct or decode the message in the presence of transmission errors. Frequently in the computation of the remainder after a division by a polynomial one is able to introduce redundancy. Also after reception, by solving certain equations, one is able to reconstruct the source information.

The last section presented the construction of the finite fields, $GF(2^m)$. In general, it can be shown that there exists one and only one finite field of cardinality p^m. This field, $GF(p^m)$, can be represented in the polynomial form, $GF(p)[x]/(p(x))$, where $p(x)$ is irreducible over $GF(p)$ and has degree m. An element a of this field can be represented either by the polynomial $a_0 + \cdots + a_{m-1} x^{m-1}$ ($a_i \in GF(p)$), or by the sequence (vector) $(a_0, \cdots, a_{m-1})$.

Most elementary digital electronic devices are two-state components, and these components are used to implement encoders and decoders. Thus this constraint entails for the most commonly used codes that their coefficients be in the finite field $GF(2^m)$, e.g. the binary codes (over $GF(2)$) or the non-binary codes such as the Reed-Solomon codes (over $GF(2^m)$).

At this point it should be noted that the best basis for encoding information is the real number e. This shows that the closest characteristic should be 3. However, tri-state electronic components are more difficult to implement and less reliable than binary devices and are usually not fast enough to compete with the bi-state components.

As a consequence, algorithms used in coding theory employ calculations which involve the elements of $GF(2^m)$. Thus if u and v are two such elements, it is necessary that one be able to calculate $u+v$, uv and u^{-1} by the use of electronic circuits.

In the last section it is demonstrated that the arithmetic operations in a finite field $GF(2^m)$ are quite distinct from their counterparts in the binary number system. The elements of $GF(2^m)$ can be represented in terms of a basis. With a basis representation, the addition and subtraction operations over $GF(2^m)$ are simple, but the multiplication and division operations are not. However, since multiplications and divisions are used in many decoding algorithms the digital realization of these operations is very important.

It is demonstrated in the last section that any element u of $GF(2^m)$ is representable as a linear combination of elements of the canonical basis $\{1, \alpha, \ldots, \alpha^{m-1}\}$. In general, any element u of $GF(2^m)$ can be expressed as the sum of m linearly independent elements of the field, i.e.,

$$u = u_0 \gamma_0 + u_1 \gamma_1 + \cdots + u_{m-1} \gamma_{m-1},$$

where $\{\gamma_0, \gamma_1, \ldots, \gamma_{m-1}\}$ is any basis of $GF(2^m)$ over $GF(2)$ and $u_i \in GF(2^m)$ for $i = 1, 2, \ldots, m-1$ are the coordinates of u with respect to this basis. Another example of a basis is the so-called *normal* basis : $\{\alpha, \alpha^2, \ldots, \alpha^{2^{m-1}}\}$. Both canonical and normal bases are used widely to represent field elements.

Let $\{\tau_0, \tau_1, \ldots, \tau_{m-1}\}$ be any basis of $GF(2^m)$ over $GF(2)$. Then this basis is said to be dual to the primal basis $\{\gamma_0, \gamma_1, \ldots, \gamma_{m-1}\}$ if

$$Tr(\tau_i \gamma_j) = \delta_{i,j}, \qquad (2.9)$$

where $Tr(\cdot)$ is the trace function defined in Eq. (2.3) as $Tr(x) = \sum_{l=0}^{m-1} x^{2^l}$ and

$$\delta_{i,j} = \begin{cases} 1, & i = j, \\ 0, & i \neq j. \end{cases}$$

If the canonical basis $\{1, \alpha, \ldots, \alpha^{m-1}\}$ is chosen to be the primal basis, the dual-basis coordinates of a field element can be obtained easily -- in fact, they can be obtained by the use of a simple linear feedback-shift register. However, the reverse transformation is not always simple [8]. The combination of the canonical basis and its corresponding dual basis plays an

Elements of Algebra

important role in an efficient realization of a finite-field multiplication operation [2]. However, there does not seem to be an efficient realization which uses both a combination of a normal basis and its dual basis. What follows next is the development of, what is called the triangular basis.

For $j = 0, 1, \ldots, m-1$. let $\beta_j = \sum_{i=0}^{m-1-j} p_{i+j+1} \alpha^i$, where p_i for $i = 0, 1, \ldots, m-1$ are the coefficients of the primitive polynomial $p(x)$ which defines the field $GF(2^m)$. Then $\{\beta_0, \beta_1, \ldots, \beta_{m-1}\}$ is called the triangular basis of $GF(2^m)$ over $GF(2)$ corresponding with the canonical basis $\{1, \alpha, \ldots, \alpha^{m-1}\}$. This fact can be seen by arranging the basis elements in the following matrix form and then by noting that the $m \times m$ transformation matrix is nonsingular. This transformation is

$$\begin{bmatrix} \beta_0 \\ \beta_1 \\ \vdots \\ \beta_{m-1} \end{bmatrix} = \begin{bmatrix} p_1 & p_2 & \cdots & p_{m-1} & 1 \\ p_2 & p_3 & \cdots & 1 & 0 \\ \vdots & \vdots & \ddots & \vdots & \vdots \\ 1 & 0 & \cdots & 0 & 0 \end{bmatrix} \begin{bmatrix} 1 \\ \alpha \\ \vdots \\ \alpha^{m-1} \end{bmatrix}. \quad (2.10)$$

The element u can be represented in the triangular basis as follows:

$$u = \sum_{i=0}^{m-1} \tilde{u}_i \beta_i,$$

where $\tilde{u}_i \in GF(2)$ for $i = 0, 1, \ldots, m-1$ are the triangular basis coordinates of u.

A basis other than the canonical basis can be chosen as the primal basis, and its corresponding triangular basis can be found in a similar manner. As a simple illustration consider the following example.

Example 2.20

Let $p(x)$ be $1 + x^3 + x^4$. Evidently $p(x)$ is a primitive polynomial of degree 4 over $GF(2)$. Then the canonical basis is $\{1, \alpha, \alpha^2, \alpha^3\}$. By a use of Eq. (2.5) the corresponding triangular basis is found to be as $\{\alpha^{14}, \alpha^{13}, \alpha^{12}, 1\}$.

Any element which is represented by either a canonical basis or a triangular basis can be represented uniquely in term of the other basis. In this connection, one obtains

$$\sum_{i=0}^{m-1} u_i \alpha^i = \sum_{j=0}^{m-1} \tilde{u}_j \sum_{l=0}^{m-1-j} p_{l+j+1} \alpha^l \tag{2.11}$$

for $p(x) = \sum_{i=0}^{m-1} p_i x^i$. Equating the coefficients of α^i for $i = 0,1,\ldots,m-1$ on both sides of Eq. (2.11), one obtains

$$\tilde{u}_i = u_{m-1-i} + (1-\delta_{i,0})\sum_{l=1}^{i} \tilde{u}_{i-l} p_{m-l} \pmod{2} \tag{2.12a}$$

and

$$u_{m-1-i} = \sum_{l=0}^{i} \tilde{u}_{i-l} p_{m-l} \pmod{2} \tag{2.12b}$$

for $i = 0,1,\ldots,m-1$. Eqs. (2.12a) and (2.12b) correspond to a transformation of coordinates from the canonical basis to the triangular basis and vice versa; they can be realized by well-known shift-register configurations.

In a basis representation the addition and subtraction operations are quite simple. In electronics, these operations are computed by the use of the exclusive-OR(XOR) or its equivalent, the modulo 2 adder. The multiplication uv and the inverse of w (or division operation) are more complicated. Therefore, two fundamental problems with the arithmetic in $GF(2^m)$ are the multiplication of the two elements u and v and the calculation of the inverse of an element. The implementation of these operations is discussed next.

A direct implementation of the multiplication is suggested in [1] by combinational logic. In such an implementation, a canonical basis is used to represent the elements of the field. Depending on the primitive polynomial $p(x)$, this implementation requires as many as $m^3 - m$ two-input adders over $GF(2)$. Because of its circuit complexity and lack of regularity, it is often advantageous to use other hardware realizations to implement the operations of multiplication.

Two commonly used multiplication algorithms are the dual-basis multiplier invented by Berlekamp[2] and the normal basis multiplier found by Massey and Omura [6]. The representation of field elements with respect to a normal basis is unconventional, but it results in a very simple squaring operation. This is advantageous for the design of inversion circuitry. Multiplications using the Massey-Omura algorithm require the same logic circuitry for all

Elements of Algebra

product coordinates. In the following discussion these two multipliers are presented briefly in the following discussion.

Berlekamp's dual-basis multiplier is described first as follows: Let $\{\tau_0, \tau_1, \ldots, \tau_{m-1}\}$ be the dual of the canonical basis $\{1, \alpha, \ldots, \alpha^{m-1}\}$. Also let w_i^τ for $i = 0, 1, \ldots, m-1$ be the coordinates of w with respect to the dual basis, i.e.

$$w = \sum_{i=0}^{m-1} w_i \alpha^i = \sum_{i=0}^{m-1} w_i^\tau \tau_i.$$

Let $w = \alpha^l$ be the power representation. Then w can be represented by its dual basis: $w = \alpha^l = \sum_{i=0}^{m-1} w_i^\tau \tau_i$. Thus, if w is the product of any two elements u and v, it follows from Eq. (2.4) and Proposition 2.20 that for $l = 0, 1, \ldots, m-1$

$$w_l^\tau = Tr(\alpha^l w) = Tr(\alpha^l uv)$$

$$= Tr(\alpha^l u \sum_{j=0}^{m-1} v_j \alpha^j) = \sum_{j=0}^{m-1} v_j Tr(\alpha^{l+j} u) \qquad (2.13)$$

$$= \sum_{j=0}^{m-1} v_j Tr(\alpha^j (\alpha^l u)) = \sum_{j=0}^{m-1} v_j (\alpha^l u)_j^\tau.$$

Eq. (2.13) requires the calculation of an inner product over $GF(2)$ that involves the canonical basis coordinates of b and the dual basis coordinates $(\alpha^l u)$. The latter can be obtained recursively as follows:

$$\left(\alpha^l u\right)_j^\tau = Tr\left(\alpha^j \alpha^l u\right)$$

$$= \begin{cases} Tr\left(\alpha^{j+1} \alpha^{l-1} u\right), & j = 0, 1, \ldots, m-2 \\ Tr\left(\alpha^m \alpha^{l-1} u\right), & j = m-1, \end{cases} \qquad (2.14)$$

$$= \begin{cases} \left(\alpha^{l-1} u\right)_{j+1}^\tau, & j = 0, 1, \ldots, m-2 \\ \sum_{i=0}^{m-1} p_i \left(\alpha^{l-1} u\right)_i^\tau, & j = m-1. \end{cases}$$

where p_i for $i=0,1,\ldots,m-1$ satisfy $\alpha^m = \sum_{i=0}^{m-1} p_i \alpha^i$.

The realization of Berlekamp's dual-basis multiplier in bit-serial form is shown in Figure 2.3. In general, the circuit of Berlekamp's bit-serial multiplier consists of a linear feedback shift register(LFSR), and it requires $O(m)$ gates. The registers are initially loaded with the dual-basis coordinates of u, and the inner-product output is w_0^τ. After the first shift, the register contains the dual basis coordinates of αu, and the output is w_1^τ. After the second shift, the register contains the dual basis coordinates of $\alpha^2 u$, and the output is w_2^τ, and so on. The module-2 adder and multiplier can be implemented by XOR and AND gates, respectively.

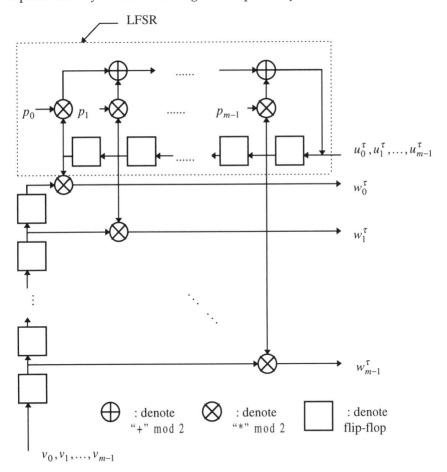

Figure 2.3 Berlekamp's dual basis bit-serial multiplier

Elements of Algebra

The advantage of Berlekamp's bit-serial multiplier is that it requires minimum circuitry when the multiplier v is a constant. However, note that one factor (the multiplier v) is represented in the canonical basis and the other factor (the multiplier u) is represented in the corresponding dual-basis. The product is obtained in the dual basis.

Example 2.21

Consider the finite field $GF(2^4)$ generated by the root α of the primitive irreducible polynomial $p(x) = x^4 + x + 1$ over $GF(2)$. Let $\{\tau_0, \tau_1, \tau_2, \tau_3\}$ be the dual to the canonical basis $\{1, \alpha, \alpha^2, \alpha^3\}$. The elements of $GF(2^4)$ in canonical basis and its dual basis are given in Table 2.5. Assume that the multiplicand $u \in GF(2^4)$ is represented by

$$u = u_0^\tau \tau_0 + u_1^\tau \tau_1 + u_2^\tau \tau_2 + u_3^\tau \tau_3$$

where $\tau_k \in GF(2^4)$ and $u_k^\tau = Tr(a\alpha^k) \in GF(2)$ for $k=0,1,2,3$.

α^i	Coefficients of $\alpha^3\ \alpha^2\ \alpha^1\ \alpha^0$	$Tr(\alpha^i)$	Coefficients of $\tau_0\ \tau_1\ \tau_2\ \tau_4$
0	0 0 0 0	0	0 0 0 0
1	0 0 0 1	0	0 0 0 1
α	0 0 1 0	0	0 0 1 0
α^2	0 1 0 0	0	0 1 0 0
α^3	1 0 0 0	1	1 0 0 1
α^4	0 0 1 1	0	0 0 1 1
α^5	0 1 1 0	0	0 1 1 0
α^6	1 1 0 0	1	1 1 0 1
α^7	1 0 1 1	1	1 0 1 0
α^8	0 1 0 1	0	0 1 0 1
α^9	1 0 1 0	1	1 0 1 1
α^{10}	0 1 1 1	0	0 1 1 1
α^{11}	1 1 1 0	1	1 1 1 1
α^{12}	1 1 1 1	1	1 1 1 0
α^{13}	1 1 0 1	1	1 1 0 0
α^{14}	1 0 0 1	1	1 0 0 0

Table 2.5 Elements of $GF(2^4)$ in canonical and its dual bases

It can be seen from Table 2.1 that $\tau_0 = \alpha^{14}, \tau_1 = \alpha^2, \tau_2 = \alpha, \tau_3 = 1$. For a fixed multiplier $v = \alpha^3 = 0 \cdot 1 + 0 \cdot \alpha + 0 \cdot \alpha^2 + 1 \cdot \alpha^3$, the product uv is also presented in the dual basis $\{\tau_0, \tau_1, \tau_2, \tau_3\}$, i.e.

$$u \cdot v = \sum_{l=0}^{3} w_l^\tau \cdot \tau_{li},$$

where

$$w_l^\tau = \sum_{j=0}^{3} v_j (\alpha^l u)_j^\tau = (\alpha^l u)_j^\tau \tag{2.15}$$

for $l = 0,1,2,3$. In the Berlekamp algorithm the computation of w_l^τ is performed recursively on l for $l = 0,1,2,3$, as follows:

(1) $l=0$; $w_0^\tau = u_3^\tau$ (from Eq. (2.15)).
(2) $l=1$; $w_1^\tau = (\alpha^1 u)_3^\tau = Tr(\alpha^4 \alpha^0 u) = Tr((\alpha+1)u) = Tr(\alpha u) + Tr(u) = u_1^\tau + u_0^\tau$
(from Eqs. (2.12) and (2.15)).
(3) $l=2$;
$w_2^\tau = (\alpha^2 u)_3^\tau = Tr(\alpha^4 \alpha^1 u) = Tr((\alpha^2 + \alpha)u) = Tr(\alpha^2 u) + Tr(\alpha u) = u_2^\tau + u_1^\tau$ (from Eqs. (2.12) and (2.15)).
(4) $l=3$;
$w_3^\tau = (\alpha^3 u)_3^\tau = Tr(\alpha^4 \alpha^2 u) = Tr((\alpha^3 + \alpha^2)u) = Tr(\alpha^3 u) + Tr(\alpha^2 u) = u_3^\tau + u_2^\tau$
(from Eqs. (2.12) and (2.15)).

The flow chart of this Berlekamp bit-serial multiplier is shown in Figure 2.4.

Next consider the Massey-Omura multiplier. Let $u \in GF(2^m)$ be represented in the normal basis $\{\alpha, \alpha^2, \ldots, \alpha^{2^{m-1}}\}$, i.e.,

$$u = \hat{u}_0 \alpha + \hat{u}_1 \alpha^2 + \cdots + \hat{u}_{m-1} \alpha^{2^{m-1}}$$

$$= [\hat{u}_0, \hat{u}_1, \ldots, \hat{u}_{m-1}] \cdot \begin{bmatrix} \alpha \\ \alpha^2 \\ \vdots \\ \alpha^{2^{m-1}} \end{bmatrix}, \tag{2.16}$$

where $\hat{a}_i \in GF(2)$ for $i=1,2,\ldots,m-1$ is the i-th coordinate of a with respect to the normal basis.

Elements of Algebra

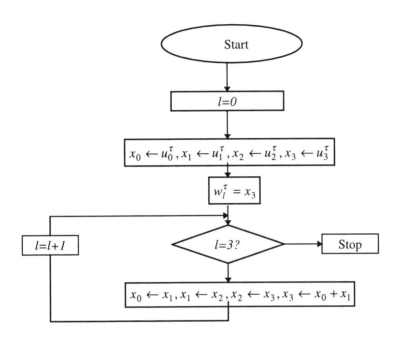

Figure 2.4 Flow chart of Berlekamp bit-serial multiplier

Let w be the product of any two elements u and v of $GF(2^m)$, i.e.,

$$w = [\hat{u}_0, \hat{u}_1, \ldots, \hat{u}_{m-1}] \cdot \begin{bmatrix} \alpha \\ \alpha^2 \\ \vdots \\ \alpha^{2^{m-1}} \end{bmatrix}$$

$$\cdot [\alpha, \alpha^2, \ldots, \alpha^{2^{m-1}}] \cdot \begin{bmatrix} \hat{v}_0 \\ \hat{v}_1 \\ \vdots \\ \hat{v}_{m-1} \end{bmatrix} = \bar{u} M \bar{v}^T ,$$

(2.17)

where "T" denotes matrix transpose, $\bar{u} = [\hat{u}_0, \hat{u}_1, \ldots, \hat{u}_{m-1}]$, $\bar{v} = [\hat{v}_0, \hat{v}_1, \ldots, \hat{v}_{m-1}]$, and the multiplication matrix M is defined as

$$M = \begin{bmatrix} \alpha^{2^0+2^0} & \alpha^{2^0+2^1} & \cdots & \alpha^{2^0+2^{m-1}} \\ \alpha^{2^1+2^0} & \alpha^{2^1+2^1} & \cdots & \alpha^{2^1+2^{m-1}} \\ \vdots & \vdots & \ddots & \vdots \\ \alpha^{2^{m-1}+2^0} & \alpha^{2^{m-1}+2^1} & \cdots & \alpha^{2^{m-1}+2^{m-1}} \end{bmatrix}. \qquad (2.18)$$

The entries $\alpha^{2^i+2^j}$ of the mxm matrix M can be expressed in terms of the normal basis, using Eq. (2.16). That is,

$$\alpha^{2^i+2^j} = \sum_{k=0}^{m} \hat{u}_k(i,j)\alpha^{2^k}.$$

Thus, Eq.(2.18) can be expanded as the expression,

$$M = \sum_{k=0}^{m-1} M^{(k)} \alpha^{2^k}, \qquad (2.19)$$

where $M^{(k)}$ is the mxm Boolean matrix $M^{(k)} = [\hat{u}_k(i,j)]$ for $k=0,1,...,m-1$. From Eqs. (2.17) and (2.19), one obtains the coordinates of the product as

$$\hat{w}_{m-1-l} = \bar{u} M_{m-1-l} \bar{v}^T = \bar{u}^{(l)} M_{m-1} \left(\bar{v}^{(l)}\right)^T \qquad (2.20)$$

for $l=0,1,......,m-1$, where $\bar{u}^{(l)}$ and $\bar{v}^{(l)}$ are the l-step right cyclic shift of $\bar{u}$.

A bit-serial multiplier which is based on Eq. (2.20) is shown in Figure 2.5. This is called a serial-type of Massey-Omura multiplier. Shift registers A and B are initially loaded with the normal- basis coordinates of u and v, respectively. The contents of the registers can be shifted l times to generate the vectors $\bar{u}^{(l)}$ and $\bar{v}^{(l)}$. The contents of the registers are input to an AND-XOR logic array which generates the coordinates of the product by a use of Eq. (2.20). The complexity of this array and consequently that of the multiplier depends on the primitive polynomial $p(x)$, which defines the field. In general, the circuit complexity of the serial-type Massey-Omura multiplier is $O(m^2)$ gates. However, there are certain low-complexity normal bases that require only $O(m)$ gates[9].

Computing inverses is one of the most complicated operations in $GF(2^m)$. Usually, decoders require finite field inversions. A straightforward approach to the computation of the inverse of a nonzero element in $GF(2^m)$ is to use a ROM in which the inverses of the field elements are stored. For $GF(2^m)$, the size of the ROM is on the order of $m2^m$ bits. The coordinates of an element are used as the address of the location in the ROM, where the corresponding inverse is stored. For most of the practically used codes, the size of the ROM

Elements of Algebra

is not very large. As a result, many decoders can be formed that rely on this ROM-based approach to compute inverses. However, the use of a ROM is not advantageous if one wants to design a universal decoder where codes are defined over $GF(2^m)$ for different values of m [10].

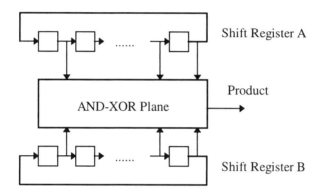

Figure 2.5. Massey-Omura bit-serial multiplier.

Two alternate methods for computing an inverse are possible. One is based on Euclid's algorithm and the other is based on Fermat's theorem. For the first method consider the field elements $u(\alpha)$ to be represented in the canonical basis $\{1, \alpha, \ldots, \alpha^{m-1}\}$ so that $u(\alpha) = \sum_{i=0}^{m-1} u_i \alpha^i$. Since $p(x)$ is a primitive polynomial of degree m over $GF(2)$, $p(x)$ is relatively prime to $u(x)$, and by a use of Euclid's algorithm one can find polynomials $b(x)$ and $d(x)$ such that

$$u(x)b(x) + p(x)d(x) = 1,$$

where $\deg[b(x)] < \deg[p(x)]$. Therefore, $u(\alpha)b(\alpha) = 1$, and $b(\alpha)$ is the canonical basis representation of the inverse of the element a.

An inversion circuit based on Euclid's algorithm requires both polynomial divisions and multiplications. An inverter with a modular structure which is based on Euclid's algorithm is presented in [11].

Since $u \in GF(2^m)$, following Fermat's theorem, $u^{2^m-1} = 1$. Thus one obtains

$$u^{-1} = u^{2^m-2} = u^{2^1} u^{2^2} \cdots u^{2^{m-1}}.$$

Thus, inverters based on Fermat's theorem require recursive squaring and multiplication operations over finite fields. Accordingly, they are suitable when a normal basis is used for the representation of the elements of $GF(2^m)$ [4]. The attractive feature of a normal basis representation is that the squaring can be performed by a simple cyclic shift of a field element's binary digits. A recursive pipeline inverter that uses a Massey-Omura parallel-type multiplier and requires $(m-2)$ multiplications and $(m-1)$ cyclic shifts is discussed in [4].

Problems

2.1. Construct a multiplication table for the set $G=\{0,1,2\}$ *in* such a manner that it forms a multiplicative group.

2.2. A cyclic group is a finite group in which every element can be written as a power of some element *a*. Prove that every cyclic group is Abelian.

2.3. Let **H** be a subgroup of the finite group **G** with $|H|=m$ and $|G|=n$. Prove that $m|n$ (*m* divides *n*) i.e. the order of a subgroup must be a factor of the order of the original group.

2.4. Find the greatest common divisor of the two integers 22471 and 3266 by the use of Euclid's division algorithm.

2.5. Show that the set **R**=$\{0,1,2,...,m-1\}$ is a commutative ring under modulo-*m* addition and multiplication.

2.6. Let *R(x)* be the set of all polynomials with real-number coefficients. Prove that every polynomial ideal *I(x)* is principal.

2.7. Prove that any finite field contains a primitive element.

2.8. Consider the finite filed $GF(2^5)$ generated by $p(x)=1+x^2+x^5$. Let α be a primitive element of $GF(2^5)$,

 (a) Construct a table for $GF(2^5)$ by showing the power polynomial and vector representation of each field element.
 (b) Find all distinct conjugates of α and α^3.

(c) Design a Berlekamp bit-serial multiplier over $GF(2^5)$ and draw a logic diagram of such a multiplier.

2.9. The trace function is defined by $Tr(x) = x + x^2 + \cdots + x^{2^{m-1}}$. Prove
   (a) $Tr(x) \in GF(2)$ for all x in $GF(2^m)$.
   (b) $Tr(x+y) = Tr(x) + Tr(y)$ for all x, y in $GF(2^m)$.

2.10. Prove that $GF(2^m)$ is a vector space over $GF(2)$.

2.11. Let p be the characteristic of the finite field $GF(q)$. Show that if $p \ne q$, then $q = p^n$ for some positive integer n.

2.12. Let β be any field element of $GF(2^m)$ and $f(x)$ be a polynomial with coefficients from $GF(2)$. Show that if β is a root of $f(x)$ over $GF(2)$, then all distinct conjugates β^{2^r} of β are also roots of $f(x)$.

Bibliography

[1] T. C. Bartee and D. I. Schneider, "Computation with finite fields", Information and Computers, Volume 6, pp. 79-98, March 1963.
[2] E. R. Berlekamp, "Bit-serial Reed-Solomon Encoder", IEEE Trans. on Inform. Theory, Vol. IT-28, No. 6, pp.869-874, Nov. 1982.
[3] J. L. Massey and J. K. Omura, "Apparatus for finite fields", Information and computers, Vol. 6, pp.79-98, March 1963.
[4] C. C. Wang, T. K. Truong, H. M. Shao, L. J. Deutsch, J. K. Omura, and I. S. Reed, "VLSI Architecture for computing multiplications and inverses in $GF(2^m)$", IEEE Trans. on Computers, Vol. C-34, pp. 709-717, August 1985.
[5] I. S. Reed, L. J. Deutsch, I. S. Hsu, T. K. Truong, K. Wang, and C. S. Yeh, "The VLSI Implementation of a Reed-Solomon encoder using Berlekamp's bit-serial multiplier algorithm", IEEE Trans. on Computers, Vol. C-33, No. 10, Oct. 1984.
[6] J. K. Omura and J. L.Massey , " Computational method and apparatus for finite-field arithmetic", U.S. Patent, No.4,587,627, May 6, 1986.
[7] E. R. Berlekamp, Algebraic coding theory, McGraw-Hill : New York, 1968.
[8] D. R. Stinson, "On bit-serial multiplication and dual bases in $GF(2^m)$", IEEE Trans. on Inform. Theory, Vol. IT-37, No. 6, pp.1733-1736, Nov. 1991.

[9] D. W. Ash, I. F. Blake, and A. A. Vanstone, "Low complexity normal nases", Discrete Applied Mathematics, Vol. 25, pp. 191-210, 1989.

[10] B. Green and G. Drolet, "A universal Reed-Solomon decoder chip", Proceedings of the 16th Biennial Symposium on Communications, Kingston, Ontario, pp.327-330, may 1992.

[11] K. Araki, I. Fujita, and M. Morisue, " Fast inverter over finite field based on Euclid's algorithm", Transactions of the IEICE, Vol. E 72, No. 11, pp.1230-1234, Nov. 1989.

3 Linear Block Codes

Some elementary concepts of block codes are introduced in Chapter 1. In general, it is known that the encoding and decoding of 2^k codewords of length n can be quite complicated when n and k are large unless the encoder has certain special structures. In this chapter, a class of block codes, called linear block codes, is discussed. Such codes have a linear algebraic structure that provides a significant reduction in the encoding and decoding complexity, relative to that of arbitrary block codes. It might be asked whether restricting our attention to linear codes is limiting in an information-theoretic sense. The Shannon random coding bound shown in Chapter 1 pertains to general block codes. However, it is known [1] that some linear codes can also provide an excellent error-correcting capability. In fact, there is a sequence of linear codes with increasing block length and a fixed rate that is only slightly smaller than channel capacity and has an error probability which approaches zero exponentially as the block length increases.

3.1 Error-Control Block Codes

Let q denote the number of distinct symbols employed on a channel. Here q, in general, can be arbitrary. A block code C is a set of M sequences of channel symbols of length n. These q-ary n-tuples (vectors) are called the *codewords* of the code. In this book, as in all practical systems and most theoretical analyses, the number of code words is taken to be a power of q, i.e. $M = q^k$. The encoding process consists of breaking up the sequence of q-ary message symbols into blocks of length k, and mapping these blocks into codewords of length n in C. It is desirable that this mapping is one-to-one to ensure that the encoded original message data can be recovered at the receiver. The corruption of a codeword by channel noise is modeled usually as the additive process, shown in Figure 3.1.

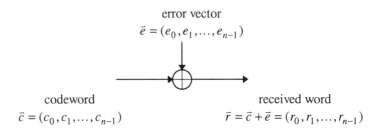

Figure 3.1 A simplified model for an additive noisy channel

Figure 3.1 illustrates a baseband model of a channel which suppresses all of the modulation and demodulation functions, though they are, of course, present in most real communication systems. The modulation format, transmitter power level, and the amount of noise on the channel determines the likelihood that each of the q^n possible error patterns $\vec{e}$ occurs. This model is particularly useful with binary channels in which the error-pattern addition is performed as a component-by-component sum, *module 2*.

The error-control decoder in the receiver performs several functions. In almost all cases the decoder's first task is to examine the received word and to determine whether or not it is a code word. If the received word is found to be invalid, then it is assumed that the channel has caused one or more symbol errors. The determination of whether errors are present in a received word is called error detection.

An error pattern is undetectable if and only if it causes the received word to be a valid code word other than that which was transmitted. Given a transmitted code word $\vec{c}$, there are q^k -1 code words other than $\vec{c}$ that could arrive at the receiver, and thus there are q^k -1 undetectable error patterns.

The decoder can react to a detected error with one of the following three responses:
1. Request a retransmission of the code word.
2. Tag the word as being incorrect and pass it along to the data sink for possible additional processing.

Linear Block Codes

3. Attempt to correct the errors in the received word.

Retransmission requests are used in a very reliable set of error-control strategies which collectively are referred to as automatic-repeat-request (ARQ) protocols. These techniques are examined later in Chapter 9. For now it is noted in passing that this approach is very important for data-transfer applications in which a premium is placed on data reliability.

The second option listed above is commonly referred to as error concealment. It is typical of applications in which delay constraints do not allow for the retransmission of the transmitted word, and it is deemed better to set the received word to a predetermined muted and/or interpolated value than to attempt to correct the errors (e.g., in voice communication and digital audio).

The final option is referred to as forward error correction (FEC). In FEC systems the arithmetic or algebraic structure of the code is used to determine which of the valid code words is most likely to have been sent, given the erroneous received word.

If the decoder performs forward-error correction, it is possible for a detectable error pattern to cause the decoder to select a code word other than that which actually was transmitted. The decoder is then said to have committed a decoder error. Though the initial error pattern, that caused the decoder error is detectable, the decoder error may itself be undetectable. In FEC systems the receiver always assumes, that the code word selected by the decoder, is correct. Usually, the number of error patterns, that cause decoder errors, is greater than the number of undetectable error patterns.

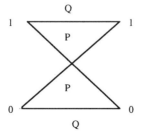

Figure 3.2 The binary symmetric channel

In order to predict the performance of a code, it is necessary to have information about the channel. Most communication channels can be represented approximately by channel models. One of the most commonly used channels is the binary symmetric channel(BSC), shown in Figure 3.2. For the BSC let Q be the probability that the symbol transmitted also is received and $P=1-Q$. It is assumed that $Q>P$, and that each symbol is independent of all others. Such a BSC is called "memoryless".

If the binary symmetric channel is assumed, and if a particular binary codeword of length n is transmitted, the probability that no error occurs is Q^n. The probability that one error occurs in a specified position is PQ^{n-1}. The probability of a particular received word that differs from the transmitted word in i positions is $P^i Q^{n-i}$. Since $Q > P$, the received block with no errors is more likely than any other. Any received word with one error is more likely than any with two or more errors, and so on. In such cases, under the assumption that all codewords are equally likely to be transmitted, the best decision at the receiver would be always to decode a received word into a codeword that differs from the received word in the fewest positions. Thus a received word with one error is more likely than any with two or more errors, and so on. This decision criterion is called maximum-likelihood(ML) decoding. ML decoding can be generalized also to nonbinary memoryless channels.

3.2 Definition of (n,k) linear codes over GF(q)

We are now ready to define a linear code over a general finite-field alphabet. Often it is traditional, especially in engineering texts, to first introduce binary linear codes, and then to generalize to codes over non-binary fields. However, there is no significant conceptual problem in treating at the outset the case of the general non-binary field.

3.2.1 Description of an (n,k) linear code by matrices

Let the message $\bar{m} = (m_0, m_1, \ldots, m_{k-1})$ be an arbitrary k-tuple (k-dimensional row vector) from $GF(q)$. The *linear (n, k) code C over GF(q)* is the set of q^k codewords of row-vector form, $\bar{c} = (c_0, c_1, \ldots, c_{n-1})$, where $c_j \in GF(q)$, which is defined by the following linear transformation :

Linear Block Codes

$$\bar{c} = \bar{m} \cdot G. \tag{3.1}$$

Here G is a $k \times n$ matrix of rank k of elements from $GF(q)$. It is seen in the previous chapter how to add and multiply the elements of a finite field. The matrix multiplication in Eq. (3.1) is performed along conventional lines. Figure 3.3 illustrates the concept of the encoding process and the process of building each codeword vector in terms of adders and multipliers over $GF(q)$.

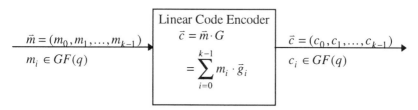

Figure 3.3 Linear (n,k) encoder over $GF(q)$

The matrix G, given in Eq. 3.1, is called the *generator matrix* of the code. It is seen from Eq.(3.1) and Figure 3.3 that a codeword of the code C is comprised of all linear combinations of row vectors of G, i.e.

$$\bar{c} = m_0 \bar{g}_0 + m_1 \bar{g}_1 + \cdots + m_{k-1} \bar{g}_{k-1}, \tag{3.2}$$

where $\bar{g}_i$ is the i-th row vector of G. The rows of G are linearly independent since G is assumed to have rank k. In vector space terminology the code C is called a k-dimensional subspace of the set of all n-tuples. This subspace is spanned by the k linearly independent basis vectors $\bar{g}_i$. The choice of basis vectors is not unique, and thus the same set of codewords can be formed with different generator matrices G, thereby changing only the relationships between the messages and their codewords.

Example 3.1

Now consider the (7, 4) Hamming code over the binary field. The encoding equations for this code are given as follows :

$$c_0 = m_0,$$
$$c_1 = m_1,$$
$$c_2 = m_2,$$
$$c_3 = m_3,$$
$$c_4 = m_0 + m_1 + m_2,$$
$$c_5 = m_1 + m_2 + m_3,$$
$$c_6 = m_0 + m_1 + m_3.$$

By studying the encoding relation in Eq. 3.1 for G, the generator matrix for this example becomes

$$G = \begin{pmatrix} 1 & 0 & 0 & 0 & 1 & 0 & 1 \\ 0 & 1 & 0 & 0 & 1 & 1 & 1 \\ 0 & 0 & 1 & 0 & 1 & 1 & 0 \\ 0 & 0 & 0 & 1 & 0 & 1 & 1 \end{pmatrix}. \tag{3.3}$$

Hence for the message $\bar{m} = (1010)$, one obtains the codeword $\bar{c} = (1010011)$. There are a total of $2^4 = 16$ codewords for this example of a linear code. They are listed in Table 3.1.

$\bar{m}$	$\bar{c}$
0000	0000000
0001	0001011
0010	0010110
0011	0011101
0100	0100111
0101	0101100
0110	0110001
0111	0111010
1000	1000101
1001	1001110
1010	1010011
1011	1011000
1100	1100010
1101	1101001
1110	1110100
1111	1111111

Table 3.1 List of Codewords for the (7,4) linear Hamming code.

Next consider some examples of nonbinary codes.

Example 3.2.
Consider how a simple (3,2) code over GF(4) with symbols labeled $0, 1, \alpha$, and α^2. The arithmetic operations of the finite field GF(4) are given in Table 3.2.

Linear Block Codes

Addition					Multiplication				
+	0	1	α	α^2	*	0	1	α	α^2
0	0	1	α	α^2	0	0	0	0	0
1	1	0	α^2	α	1	0	1	α	α^2
α	α	α^2	0	1	α	0	α	α^2	1
α^2	α^2	α	1	0	α^2	0	α^2	1	α

Table 3.2. Arithmetic table for GF(4).

Thus, if the generator matrix for the code is defined by

$$G = \begin{pmatrix} 1 & 0 & \alpha \\ \alpha & \alpha^2 & 1 \end{pmatrix},$$

then the codeword $\bar{c} = (\alpha, 1, 0)$ for the message $\bar{m} = (1, \alpha)$ is obtained by a simple computation which uses Table 3.2. This code contains $4^2 = 16$ codewords of length 3.

Inspection of the generator matrix in Eq. (3.3) of Example 3.1 shows it has, what is called, the canonical form, namely

$$G = \left(I_{k \times k} \; P_{k \times (n-k)} \right), \tag{3.4}$$

where $I_{k \times k}$ is the $k \times k$ identity matrix and $P_{k \times (n-k)}$ is a $k \times (n-k)$ matrix. The form of this matrix reflects how the n-k additional code symbols are formed from the components of $\bar{m}$. Eqs. (3.1) and (3.4) imply that the first k components of any codeword are precisely the information symbols in their original order. This form of linear encoding is called systematic encoding. Several implementation advantages accrue from the use of systematic codes, not the least of which is the quick-look feature: information can be extracted trivially from the codeword, if desired, without any decoding. Naturally, such a message recovery has a lesser reliability than if the full code vector were used to infer and correct the message content.

Evidently the error-correction properties of a code for any memoryless channel do not change if the symbols are rearranged. Thus a change in the order of the symbols in the codewords does not change the probability of error in the decoding. In general, two such codes are very closely related and are, therefore, called equivalent. More precisely, for linear codes if C is the row space of a matrix G, then C' is the code equivalent to C if and only if

C' is the row space of a matrix G', obtained from G by a rearrangement of the columns of G.

To see this suppose matrix G' is obtained from matrix G by a rearrangement of the columns of G. Clearly, such a rearranging of the columns of G leaves the fundamental coding properties of the codes unchanged; such a re-arrangement only alters the row space. Thus, the permuting of the columns of a generator matrix leads to a generator matrix of an equivalent code.

Also the elementary row operations of any matrix, i.e, an interchange of two rows of G, the multiplication of any row by a scalar in $GF(q)$, or the addition of any one row to any another row, result in a matrix with the same row space. Therefore, given that G is a generator matrix, then an altered matrix G', which is produced by any permutations of the rows or combination of elementary row operations, is also a generator of the same code.

From the above if one matrix can be obtained from another by a combination of row operations (reductions) or column permutations, the two matrices are called combinatorially equivalent. For linear codes, any error-correcting code is equivalent to a code in systematic form. This follows by noting that a succession of elementary row operations and the interchanging of columns of a given G always can transform it into a matrix with the canonical form, given in Eq. (3.4).

Example 3.3
The (3, 2) code over GF(4) is converted into systematic form. Recall the previous code over GF(4), with

$$G = \begin{pmatrix} 1 & 0 & \alpha \\ \alpha & \alpha^2 & 1 \end{pmatrix}.$$

A multiplication of the first row by α and adding it to the second row yields the matrix,

$$G' = \begin{pmatrix} 1 & 0 & \alpha \\ 0 & \alpha^2 & \alpha \end{pmatrix}$$

Then a multiplication of the second row by the reciprocal of the first element in the second row produces

$$G = \begin{pmatrix} 1 & 0 & \alpha \\ 0 & 1 & \alpha^2 \end{pmatrix}$$

Linear Block Codes

which is in the desired systematic form. It is easy to verify that all three of the corresponding sets of codewords are identical, although the association with the messages is changed.

Since every linear block code is equivalent to a systematic code, it is sufficient to focus on the systematic-form generator matrices in the study of linear codes. In systematic form the codeword $\bar{c}$ is comprised of an information segment and a set of $n-k$ symbols that are linear combinations of certain information symbols, determined by the P matrix. That is,

$$c_i = m_i, \text{ for } 0 \leq i < k,$$

$$c_i = \sum_{j=0}^{k-1} m_j P_{j,n-1-i}; \text{ for } k \leq i < n.$$

The second set of equations, given above, is called the set of parity-check equations. These equations show that each redundant symbol is a linear combination of the information digits. These redundant symbols are called *parity-check symbols*, in keeping with the terminology of the already familiar single parity-bit codes that append a single parity bit to a binary k-tuple to make the sum of the symbols in the codeword equal to zero. For this reason linear codes are sometimes referred to as *parity-check codes.*

An encoder of binary systematic linear codes can be implemented by the logic circuit shown in Figure 3.4. In such a circuit, accumulators are used to compute and store the parity-check bits. The parity matrix $P_{k \times (n-k)}$ is stored in the read-only memory(ROM).

It is seen above that the G matrix imposes linear constraints on the code symbols. To gain another important perspective about this set of constraints, another concept of linear algebra is invoked. Any k-dimensional linear subspace of an n-dimensional linear space, for example, a code C, has associated with it a *null space* which is designated by $C^\perp$. This null space $C^\perp$ of code C has the property that every vector in the null space is orthogonal (perpendicular) to every vector in the original space. Let $\bar{c}_i$ and $\bar{h}_j$ be members (vectors) of the linear subspace C and its null space $C^\perp$, respectively. Then by this definition the vector inner product of these vectors is zero, i.e.

$$\bar{c}_i \cdot \bar{h}_j^T = 0 \tag{3.5}$$

is valid for all $\bar{c}_i \in C$ and $\bar{h}_j \in C^\perp$, where the superscript "T" denotes the matrix (vector) transpose operation.

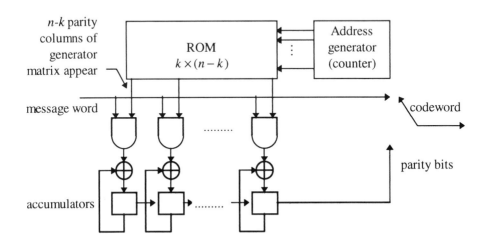

Figure 3.4 An implementation of binary linear systematic encoder

The above null space $C^\perp$ of the space C has dimension $n - k$ and is spanned by a set of $n - k$ linearly independent vectors of length n. This set of $n - k$ vectors when listed in the matrix form,

$$H = \begin{pmatrix} \bar{h}_1 \\ \bar{h}_2 \\ \vdots \\ \bar{h}_{n-k} \end{pmatrix}, \tag{3.6}$$

is called a basis of the null space $C^\perp$. Hence, one has $\bar{h}_i \in C^\perp$ for $i=1,2...,n-k$. This implies by Eq. (3.5) that $\bar{c}_i \bar{h}_i^T = 0$ for $i=1,2...,n-k$. This relation in matrix form is given by

$$\bar{c}_i H^T = 0, \qquad \text{for } \bar{c}_i \in C, \tag{3.7}$$

where 0 represents the vector of the zero elements of $GF(q)$.

Eq. (3.7) constitutes a set of $n-k$ linear equations that the codeword must satisfy, which are called the parity-check equations of the code. Therefore, H is called the **parity-check matrix,** because it describes the total set of parity-

Linear Block Codes

check constraints. If the $n-k$ equations produced by (3.7) are written out, one obtains

$$\sum_{j=0}^{n-1} c_j h_{ij} = 0, \quad i = 1, 2, \ldots, n-k. \tag{3.8a}$$

Example 3.4

For the binary Hamming code, defined in Example 3.1, the last three equations provide the parity check equations, i.e.

$$c_4 = c_0 + c_1 + c_2,$$
$$c_5 = c_1 + c_2 + c_3,$$
$$c_6 = c_0 + c_1 + c_3.$$

Since Eq.(3.7) must hold for any codeword in the space spanned by the rows of 0, one has that G and H must satisfy

$$G \cdot H^T = 0, \tag{3.8b}$$

where "zero" on the right-hand side is a k by $n - k$ matrix of zeros. For any given generator matrix G, many solutions for H are possible.

If the generator matrix G is in systematic form, then finding a parity-check matrix H is simple. For example it is easy to verify by substitution that

$$H = \left(-P_{k \times (n-k)}^T \ I_{(n-k) \times (n-k)} \right) \tag{3.9}$$

satisfies Eq. (3.8).

Since G specifies H (even in the nonsystematic case where H is not found so directly), a completely equivalent but new definition of a linear code is made possible : A linear code C is the set of all words in the null space of the $(n - k) \times n$ matrix H with rank $n - k$. That is, C is the set of vectors $\bar{c}_i$ such that

$$\bar{c}_i \cdot H^T = 0.$$

Furthermore, if G generates an (n, k) linear code C with a null space generated by H, then H, when regarded as a generator matrix, generates an $(n, n-k)$ linear code $C^\perp$ with its null space generated by G. These two codes C and $C^\perp$ are called *dual codes*, with the generator matrix of one being the parity-check matrix of the other. The structure of one code tells a great deal about its dual code. If the code and its dual code are equivalent, the code is called *self-dual*. Such codes have $n-k = k$ or $k = n/2$ and therefore have the rate, $R = 1/2$.

Example 3.5
Recalling the generator matrix for the (7,4) code in Example 3.1, one can write immediately the parity-check matrix for the code, using (3.9), as follows :

$$H = \begin{pmatrix} 1 & 1 & 1 & 0 & 1 & 0 & 0 \\ 0 & 1 & 1 & 1 & 0 & 1 & 0 \\ 1 & 1 & 0 & 1 & 0 & 0 & 1 \end{pmatrix}.$$

Notice again in GF(2) that addition and subtraction are equivalent operations. Thus matrix H also generates a (7, 3) linear code that is a dual code of the (7,4) code, and every codeword in this code is orthogonal to every codeword in the original Hamming code.

3.2.2 Hamming Distance of Linear Codes and Their Error-Protection Properties

The general algebraic properties of linear codes were developed in the last section. Some of these constructs are very useful for describing these codes, but also make possible the realization of their encoders and decoders. We now turn to a study of the performance of a linear code. This focuses on the Hamming distance structure of the linear code and its guarantee to achieve error control.

Consider now a pair of n-symbol vectors $\bar{w} = (w_0, w_1, \ldots, w_{n-1})$ and $\bar{v} = (v_0, v_1, \ldots, v_{n-1})$. There are a variety of ways to measure the "distance" $d(\bar{w}, \bar{v})$ between these two vectors. In the classical Euclidean geometry one encounters the following definition of distance,

$$d_E(\bar{w}, \bar{v}) = \sqrt{\sum_{i=0}^{n-1} (w_i - v_i)^2} \ . \tag{3.10}$$

Euclidean distance is used extensively in the analysis of convolutional and trellis codes in Chapters 9 and 11. In block coding, however, one is more frequently concerned with, what is called, Hamming distance.

The Hamming distance between two vectors $\bar{w}$ and $\bar{v}$ is the number of coordinates in which the two vectors differ, i.e.

$$d_H(\bar{w}, \bar{v}) = |\{i | w_i \neq v_i, i = 0, 1, \ldots, n-1\}| \ . \tag{3.11}$$

Linear Block Codes

For a linear code the Hamming distance between any two codewords is simply described by

$$d_H(\bar{c}_1, \bar{c}_2) = wt(\bar{c}_1 - \bar{c}_2) = wt(\bar{c}_3), \quad (3.12)$$

where $wt(\bar{c})$ denotes the **Hamming weight** of $\bar{c}$ or the number of nonzero positions, and $\bar{c}_1 - \bar{c}_2 = \bar{c}_3$ is the difference between the original code vectors. The difference vector is also a codeword as it is demonstrated next. This fact follows from the definition of the linearity of the encoding operator and the important fact, that the sum of any two codewords is another codeword. Thus, the codeword $\bar{c}_3$ formed by adding $\bar{c}_1$ and $\bar{c}_2$ is

$$\bar{c}_3 = \bar{c}_1 + \bar{c}_2 = \bar{m}_1 \cdot G + \bar{m}_2 \cdot G = (\bar{m}_1 + \bar{m}_2) \cdot G = \bar{m}_3 \cdot G,$$

so that $\bar{c}_3$ is the codeword for the message $\bar{m}_1 + \bar{m}_2$, which is the message k-tuple $\bar{m}_3$. Also, if $\bar{c}_2$ is a codeword, so is $-\bar{c}_2$, corresponding to the message $-\bar{m}_2$, the vector of the additive inverses of the components of vector $\bar{m}_2$. Thus, $\bar{c}_1 - \bar{c}_2$ is also a valid codeword. Furthermore, the n-dimensional all-zeros vector is a codeword of every linear code of length n since it corresponds to the k-dimensional all-zeros message vector.

Certainly, the most important aspect of the Hamming distance of a code is the **minimum Hamming distance**, denoted by d_{min}, between any two codewords. The minimum distance of a block code C is the minimum Hamming distance between all distinct pairs of code words in C, i.e.

$$d_{min} = \min_{\substack{\bar{c}_i, \bar{c}_j \in C \\ \bar{c}_i \neq \bar{c}_j}} \left(d_H(\bar{c}_i, \bar{c}_j) \right). \quad (3.13)$$

By Eq. (3.12) this quantity, d_{min}, is exactly the minimum nonzero Hamming weight, namely $w_{min} = \min_{\bar{c} \in C, \bar{c} \neq 0} \{wt(\bar{c})\}$ of all codewords in a linear code. Thus

$$d_{min} = w_{min}. \quad (3.14)$$

Hamming distance allows for a useful geometric characterization of the error detection and correction capabilities of a block code as a function of the minimum distance d_{min} of the code.

Consider now the use of q-ary codes on the q-ary channel, described in Section 3.1. In this case, the action of the channel can be expressed as the addition of an error pattern $\bar{e}$ to the transmitted codeword $\bar{c}$. Also to achieve maximum-likelihood decoding one selects that codeword $\bar{c}$ which exhibits the fewest discrepancies with the channel output sequence $\bar{r}$ or, what is the same, is closest in Hamming distance to the received word $\bar{r}$.

Suppose that $\bar{c}_0$ is selected for transmission and that the closest codeword is d_{min} in Hamming distance, as shown schematically in Figure 3.5. If the channel-error pattern $\bar{e}$ has

$$t = \left\lfloor \frac{d_{min} - 1}{2} \right\rfloor \qquad (3.15)$$

or fewer errors, one is guaranteed that $\bar{r} = \bar{c}_0 + \bar{e}$ remains closer in Hamming distance to $\bar{c}_0$ than to any other codeword and thus is decoded correctly. As a consequence, $t = \left\lfloor \frac{d_{min} - 1}{2} \right\rfloor$ is called the maximum (random) error-correction capability of the code.

If $t+1$ or more errors occur in a codeword, one may be fortunate to find that $\bar{r}$ is still closer to $\bar{c}_0$ than to any other codeword, but some such error patterns often lead to a decoding error. This geometric argument applies to nonlinear codes as well.

Consider next the problem of error detection. Suppose that the decoder's task is only to detect the presence or absence of errors, and if errors are detected, to label the codeword as unreliable. This detector can be fooled only if $\bar{e}$ takes the transmitted codeword $\bar{c}_0$ into another codeword $\bar{c}_1$, that is $\bar{c}_0 + \bar{e} = \bar{c}_1$. This cannot occur if there are $d_{min} - 1$ or fewer errors in the n positions of the code. Consequently, one knows that $d_{min} - 1$ is the guaranteed error detection capability of the code. Again, if d_{min} or more errors occur, the error pattern may still be detectable. Observe that the guaranteed number of detectable errors is roughly twice the number of correctable errors.

Hybrid modes of error control which employ concurrent error correction and detection are also possible. One such mode of error control is to be described shortly under performance analysis. It is not difficult to show by geometric arguments that one can correct t errors and still detect up to t_d errors provided that $t + t_d < d_{min}$.

Another important channel model is the so-called erasure channel. An example of the binary erasure channel is shown in Figure 3.6. For this channel, the probability of correct transmission of a binary digit (bit) is Q. The probability is $P = 1 - Q$ that a transmitted symbol is erased. (An erasure is indicated by an "X".) The binary symbols are assumed to be affected independently. Note that with this channel the locations of the perturbed symbols are known, and this fact makes the correction of erasures easier than the correction of errors. Generalizations of the erasure channel include a nonbinary erasure channel and a channel with both erasures and errors. The erasure channel is an idealization of a system in which the demodulator is designed to deliver an erasure symbol rather than a 1 or 0 in doubtful cases.

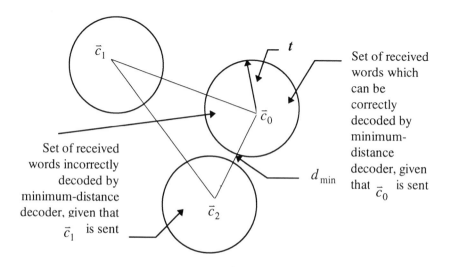

Figure 3.5 Correct and incorrect decoding

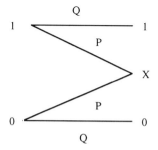

Figure 3.6 The binary erasure channel

An error-and-erasure decoder is provided if the channel occasionally erases code symbols in addition to causing errors. Using similar geometric arguments, one can be assured of simultaneously correcting any combination of t errors and s erasures, provided that $2t + s < d_{min}$.

Other channel models are certainly important, and it must be reiterated that of these types, what is called, hard-decision decoding are often inferior to a true maximum-likelihood decoding which utilizes the actual received physical values of the channel. However. as it is seen later in this chapter, d_{min} plays a key role in many general decoding situations, thereby motivating the search for those linear codes with a large d_{min} for a given n and k.

Another important problem for linear codes is the determination of the minimum distance d_{min}. Recall that a linear code can be specified by the parity-check matrix H. If $\bar{c}$ is a codeword, then $\bar{c} \cdot H^T = 0$, i.e. H produces a zero vector of parity checks. Notice also that $\bar{c} \cdot H^T$ can be written as a linear combination of the columns of H or the rows of H^T. That is,

$$\bar{c} \cdot H^T = \sum_{j=0}^{n-1} c_j \bar{h}_j^T,$$

where $\bar{h}_j$ is the j-th column of H. Thus, the relation

$$\bar{c} \cdot H^T = \sum_{j=0}^{n-1} c_j \bar{h}_j^T = 0$$

means that the columns of H for $c_j \neq 0$ are linearly dependent. Since $d_{min} = w_{min}$, d_{min} is the smallest number of columns that are linearly dependent. Therefore, if for some number d one can show that $d-1$ is the maximum number of columns of the parity check matrix that are linearly independent, then the minimum distance of the code is exactly $d_{min} = d$.

To accurately determine the error-control performance of a linear code, it is important to find the distance structure of the code. The equality $d_{min} = w_{min}$ implies that finding the distance structure of the code one needs only to examine the weight structure of the q^k codewords, rather than the distance between all q^{2k} pairs. Thus the performance analyses of channels in which the input words are uniformly distributed can be performed by focusing only on the transmission of any one codeword. A convenient choice is the all-zeros word which is always a codeword of a linear code.

The *weight distribution* of the code is obtained by listing the Hamming weights i of the codewords, and the number of codewords W_i of Hamming

Linear Block Codes

weight i for $i=0,1,...,n$. This information can be summarized compactly in the form of a polynomial, called the *weight enumerator polynomial*:

$$W(z) = \sum_{i=0}^{n} W_i z^i. \tag{3.16}$$

Clearly, $W_0 = 1$, and $W_i = 0$ for $0 < i < d_{min}$.

Example 3.6
For the (7,4) code discussed in Example 3.1, it is easy to count in Table 3.1 that there is one word of weight 0 (the all-zeros vector), seven words of weight 3, seven of weight 4, and one of weight 7. Hence the weight enumerator polynomial for this code is given by

$$W(z) = \sum_{i=0}^{7} W_i z^i = 1 + 7z^3 + 7z^4 + z^7.$$

For general linear codes, the complete computation of the weight distribution requires the evaluation of the weights of all sums of the rows of G. Most such computations are performed by the use of computers. Even so, this soon becomes impractical as the code size grows. However, for some important codes the weight enumerator is known in analytic form.

The determination of the weight distribution is often simplified by analyzing the dual code and then applying the relations due to MacWilliams [2], which relate the weight distribution of a code to that of its dual code. Specifically, let W_i and $W_i^{\perp}$ denote, respectively, the number of codewords of weight i in an (*n, k*) code C and its (*n,n - k*) dual code $C^{\perp}$. Note that these codes also could be nonlinear codes. The MacWilliams relations or identities, which here are merely stated, are the set of n linear equations:

$$\sum_{i=0}^{n-m} W_i^{\perp} \binom{n-i}{m} = q^{n-k-m} \sum_{j=0}^{n} W_j \binom{n-j}{n-m}, \quad \text{for } m=n,n-1,...,0. \tag{3.17}$$

For linear codes $W_0 = W_0^{\perp} = 1$. The MacWilliams identities are particularly useful in the analysis of very large codes, whose dual codes are, however, small enough to be enumerated directly. Detailed discussions of the weight distribution of block codes can be found in [2][3].

3.2.3 Construction of New Linear Codes from Existing Linear Codes

The practical requirements of electronic systems always impose constraints on the configurations of their components. For an error-control coding subsystem, such constraints determine the allowed codeword length and the code-dimension. Sometimes, the length and dimension of the code one would wish to use is unsuitable so that it needs to be modified in order that the specific requirements of the application can be satisfied. Some of the commonly-used modifications of linear codes are discussed next.

Punctured code A code is punctured by deleting some of its parity symbols (or bits, coordinates). For example, by deleting one parity symbol, an *(n,k)* code becomes a *(n-1,k)* code. In general, a punctured code has an increased information rate and a reduced error-control capability.

Shortened code A code is shortened by deleting some message symbols (bits, or coordinates) from the encoding process. For example, by deleting one message symbol, an *(n,k)* code becomes an *(n-1,k-1)* code.

Extended code A code is extended by adding some additional redundant symbols (bits, or coordinates). For example, by adding one parity symbol, an *(n,k)* code becomes a *(n+1,k)* code. In general, the error-control capability of the extended code can be increased.

For more details on the modification of linear codes the reader is referred to references [2][4].

3.3 Decoding of Linear Block Codes

3.3.1 Maximum-Likelihood Decoding

Suppose that a message $\bar{m}$ is transformed into the codeword $\bar{c}$ by $\bar{c} = \bar{m} \cdot G$ as discussed in Section 3.1. Let this codeword $\bar{c} = (c_0, c_1, \ldots, c_{n-1})$ be transmitted over a noisy channel. In the receiver a real-valued observation vector $\bar{y} = (y_0, y_1, \ldots, y_{n-1})$ is obtained from the demodulator output. For example, when phase-shift-keying(PSK) modulation is employed with binary codes, it is called BPSK. For BPSK each component of the received-vector observation $\bar{y}$ is $y_i = (-1)^{c_i} E_s^{1/2} + n_i$ with n_i being the sampled Gaussian-

Linear Block Codes

noise variables and E_s being the signal energy. If the symbols were in $GF(8)$, then the use of 8-ary orthogonal modulation with noncoherent detection produces eight numbers for each code position. These numbers are the magnitudes (or the squared magnitudes) of the matched-filter outputs.

If the codewords are transmitted "equally-likely", the optimal receiver is the maximum-likelihood(ML) decoder. For a memoryless channel from one symbol to the next symbol, the ML decoder finds that codeword $\bar{c}$ which maximizes the likelihood function, $L(\bar{y}|\bar{c}) = p(\bar{y}|\bar{c}) = \prod_{i=0}^{n-1} p(y_i|c_i)$, where $p(\bar{y}|\bar{c})$, is the conditional probability density function (p.d.f). Instead of maximizing the p.d.f directly it is usually easier to maximize the logarithm of the p.d.f. Then, the optimal choice for the codeword is that codeword $\hat{c}$ which satisfies the identity,

$$p(\bar{y}|\bar{c} = \hat{c}) = \max_{c \in C} \sum_{i=0}^{n-1} \log p(y_i|c_i). \tag{3.18}$$

Example 3.7
For a BPSK signal, which is transmitted over a white Gaussian noise channel, the ML decoder can be represented as follows:

$$\max_{c \in C} \sum_{i=0}^{n-1} \log\left(\frac{1}{\sqrt{2\pi}\sigma} e^{-\frac{(y_i-(-1)^{c_i})^2}{2\sigma^2}}\right), \tag{3.19}$$

where c_i is the i-th bit of the codeword $\bar{c}$. Eq. 3.19 is equivalent to the metric $\min_{c \in C} \sum_{i=0}^{n-1} (y_i - (-1)^{c_i})^2$ which also is equivalent to the quite simple correlation metric $\max_{c \in C} \sum_{i=0}^{n-1} y_i s_i$ where $s_i = (-1)^{c_i}$, this is the optimum matched-filter criterion for binary sequences.

The general ML decoder apparently must compute q^k sums as in Eq. (3.18) and choose that codeword which maximizes the likelihood function. For general linear codes there appears to be no way of surmounting this exponential complexity problem, because the ML decoding problem has been shown to be *NP-complete*. This means that it is equivalent to the class of problems that are known to be computationally difficult in the sense that no solution can been found in polynomial-time. Actually, it is found here that it is possible to decode with a complexity proportional to q^{n-k}, which can be much smaller than q^k. To reduce the complexity of the ML decoder, efficient sub-optimal algorithms, which approximate ML decoding, are developed[3][4].

The general ML decoding problem is discussed further in later chapters, following the development of cyclic codes. The procedures for efficiently organizing approximate ML-decoding schemes also are investigated in these chapters. For the present, however, what are called, algebraic decoding algorithms are discussed in some detail.

3.3.2 Elements of Algebraic Decoding

In the discussions of algebraic decoding algorithms the channel is often viewed as a q-ary input, q-ary output, discrete-memoryless channel. Assume a uniform channel such that when an error is made in transmission of a code symbol with probability P, the error is equally likely to be any of the q-1 possibilities. A possible extension is to let the demodulator produce one of q symbols or an *erasure* when the demodulator has a low confidence in its ability to decide which symbol occurred. Consider now the log-likelihood function for such a channel. The conditional probability that $r_i = c_i$ is received, given that c_i was sent, satisfies:

$$p(r_i|c_i) = \begin{cases} 1-P, & r_i = c_i, \\ \dfrac{P}{q-1}, & r_i \neq c_i, \end{cases}$$

where r_i is the i-th component of the received word $\bar{r} = (r_0, r_1, \ldots, r_{n-1})$.

$$\text{Thus, } \log p(r_i|c_i) = \begin{cases} \log(1-P), & r_i = c_i, \\ \log\left(\dfrac{P}{q-1}\right), & r_i \neq c_i. \end{cases}$$

Hence, the log-likelihood function has only two values so that when the observed demodulator output agrees with the corresponding symbol of a test codeword, a slightly negative metric, namely log(1-P), is assigned, whereas when a discrepancy is found, a more negative value of the metric log(P/(q-1)) is attached. Actually, this may be simplified by noting that the two values of the metric $\log p(r_i|c_i)$ can be translated so that the largest value of the metric is zero and the other scaled so that its smallest value is -1. Then ML

Linear Block Codes

decoding for this channel is equivalent to minimum Hamming-distance decoding. This further emphasizes for the q-ary uniform channel that ML decoding is equivalent to finding that codeword $\bar{c}$ which is closest in Hamming distance to the received word $\bar{r} = (r_0, r_1, \ldots, r_{n-1})$. This type of minimum Hamming-distance decoding is valid and useful in many real communication channels. However, often it is necessary to change the channel symmetry slightly so that the channel output is q-ary, where $1-P > P/(q-1)$.

In general, Hamming distance is the basis for two important possible types of decoders. First, the complete error-correcting decoder is that decoder which, given a received word $\bar{r}$, selects as the transmitted codeword $\bar{c}$ that codeword which minimizes $d_H(\bar{r}, \bar{c})$. The complete decoder is thus, for most channels, the maximum-likelihood decoder. It selects that code word which is closest to the received word regardless of the distance between the two. It is possible for a given received word $\bar{r}$ that there is more than one codeword $\bar{c}$ that minimizes $d_H(\bar{r}, \bar{c})$. In this case the complete decoder chooses randomly from among the closest code words (assuming that there is no additional information that might help to distinguish them).

For many codes such as the Reed-Solomon codes no efficient complete decoding algorithms are known. If such a error-correction requirement is relaxed, however, quite efficient algorithms are readily available.

The second important class of decoders is called the bounded-distance decoder. Given a received word $\bar{r}$, a t-error correcting, bounded-distance, decoder selects that code word $\bar{c}$ which minimizes $d_H(\bar{r}, \bar{c})$ if and only if there exists $\bar{c}$ such that $d_H(\bar{r}, \bar{c}) \leq t$. If no such $\bar{c}$ exists, then a decoder failure is declared.

Clearly the error-correction capability t of a bounded distance decoder must satisfy $t \leq \left\lfloor \dfrac{d_{min} - 1}{2} \right\rfloor$. There is an important distinction to be made between decoder errors and decoder failures in a bounded-distance decoder: The latter are detectable while the former are not. Decoder failures sometimes can be resolved only by means of re-transmission requests.

For both types of decoders, the task is clearly to find the closest codeword in the Hamming-distance sense to $\bar{r}$ among the q^k codewords. Evidently, an exhaustive search of the codewords is possible when q^k is small. However, more efficient methods are required when q^k is large.

The received word may be written as $\bar{r} = \bar{c} + \bar{e}$, where $\bar{e}$ is an *n*-tuple of the channel-error symbols from *GF(q)*, and in addition is over *GF(q)*. Since $d_H(\bar{r},\bar{c}) = wt(\bar{r} - \bar{c}) = wt(\bar{e})$, one needs to find the minimum-weight (or most probable) error pattern that could have produced the given $\bar{r}$ and then subtract this from the received word. If the estimated error pattern $\hat{e}$ agrees with the actual error pattern, the proper codeword is restored, and the errors in the message are corrected. Next, what is called, the standard-array algorithm is discussed. This algorithm shows how the standard array can be used as a decoding table by a simple locating procedure for the minimum-weight error pattern.

3.3.3 Standard Array and Syndrome Decoding

The standard array is a way of tabulating all of the q^n *n*-tuples over *GF(q)*. This $q^{n-k} \times q^k$ rectangular array forms the conceptual basis for decoding linear block codes. The standard array is used as a decoding table by simply locating the minimum-weight *n*-tuple word in the table. This *n*-tuple in the *i*-th column equals the sum of the codeword heading the *i*-th column and the minimum-weight error pattern which appears in the same row of the first column. In group theory the rows of such an array are called cosets, and the element in the first column of each coset is called the coset leader. There are a total of q^{n-k} cosets and coset leaders in a standard array.

Let C be a (*n,k*) linear code over *GF(q)*, and $\bar{c}_1, \bar{c}_2, ..., \bar{c}_{q^k}$ be the q^k codewords of C. The following steps need to be taken to form the standard array, seen in Table 3.3. First, place the q^k codewords of C in the first or top row with the all-zeros word $\bar{c}_1 = (0,0,...,0)$ as the left-most element. From the remaining $q^n - q^k$ *n*-tuples, an error pattern $\bar{e}_2$ which has a minimum-weight is chosen and placed under the all-zeros codeword $\bar{c}_1 = (0,0,...,0)$. The second row is formed by adding $\bar{e}_2$ to each codeword $\bar{c}_i$ for $2 \leq i \leq q^k$ in the top row and placing the sum $\bar{e}_2 + \bar{c}_i$ under $\bar{c}_i$. After having completed the second row, an unused error pattern $\bar{e}_3$, which has a smallest weight

Linear Block Codes

among the remaining error patterns, is chosen and is placed under $\bar{e}_2$. Then a third row is formed by adding $\bar{e}_3$ to each codeword in the top row and placing $\bar{e}_3 + \bar{c}_i$ under $\bar{c}_i$. This process is continued until all of the n-tuples are exhausted so that each possible n-tuple appears somewhere in the array. Thus, we have the rectangular array of rows and columns, shown in Table 3.3.

$\bar{c}_1 = (0,0,...,0)$	$\bar{c}_2$	......	$\bar{c}_i$	......	$\bar{c}_{q^k}$
$\bar{e}_2$	$\bar{e}_2 + \bar{c}_2$	......	$\bar{e}_2 + \bar{c}_i$	......	$\bar{e}_2 + \bar{c}_{q^k}$
$\bar{e}_3$	$\bar{e}_3 + \bar{c}_2$	......	$\bar{e}_3 + \bar{c}_i$	......	$\bar{e}_3 + \bar{c}_{q^k}$
$\vdots$	$\vdots$	......	$\vdots$	......	$\vdots$
$\bar{e}_{q^{n-k}}$	$\bar{e}_{q^{n-k}} + \bar{c}_2$	......	$\bar{e}_{q^{n-k}} + \bar{c}_i$	......	$\bar{e}_{q^{n-k}} + \bar{c}_{q^k}$

Table 3.3 Standard array of an (n,k) linear code

A received word $\bar{r}$ is decoded by looking-up this word $\bar{r}$ in the standard-array, shown in Table 3.3. The error pattern at the left-most column of the row along which $\bar{r}$ is located is the error pattern most likely to have corrupted the received word $\bar{r}$. Thus, if $\bar{r} = \bar{e}_j + \bar{c}_i$, the corresponding error pattern $\bar{e}_j$ is found in the first column of Table 3.3. Finally, the codeword is recovered by computing $\bar{c} = \bar{r} - \bar{e}$.

Example 3.8
Consider the (6,3) linear code over $GF(2)$ specified by
$$G = \begin{pmatrix} 0 & 1 & 1 & 1 & 0 & 0 \\ 1 & 0 & 1 & 0 & 1 & 0 \\ 1 & 1 & 0 & 0 & 0 & 1 \end{pmatrix}.$$
The standard array of this code is given in Table 3.4.

000000	110001	101010	011011	011100	101101	110110	000111
000001	110000	101011	011010	011101	101100	110111	000110
000010	110011	101000	011001	011110	101111	110100	000101
000100	110101	101110	011111	011000	101001	110010	000011
001000	111001	100010	010011	010100	100101	111110	001111
010000	100001	111010	001011	001100	111101	100110	010111
100000	010001	001010	111011	111100	001101	010110	100111
001001	111000	100011	010010	010101	100100	111111	001110

Table 3.4 Standard Array for the (6,3) single-error-correcting code

If the received word is $\vec{r} = (101001)$, the decoded error pattern by the standard array in Table 3.4 is $\vec{e} = (000100)$. Hence, the recovered codeword is $\vec{c} = \vec{r} - \vec{e} = (101101)$.

The standard-array approach is a complete decoding algorithm. The table for q-ary words of length n has q^n words, all of which must be stored in memory. This works well for small codes such as the code in Example 3.8. However, the size q^n of the array grows exponentially with the code length. Thus, for even modestly large values of n and k, the direct use of the standard array becomes impractical. Next, a much simpler decoding method is discussed which reduces the size of the standard array for an (n,k) q-ary code to a q^{n-k} - entry table.

It was shown in Section 3.2 that, given a parity-check matrix H for a code C, a vector $\vec{c}$ in the received space is a codeword if and only if $\vec{c} \cdot H^T = 0$. On the basis of this property a conceptual decoding scheme is outlined in Figure 3.7. Decoding begins by subtracting the low-weight error-pattern vectors $\vec{e}$ from $\vec{r}$, each time checking whether or not $(\vec{r} - \vec{e}) \cdot H^T = 0$. As soon as this parity-check equation is satisfied, a decoded codeword is found. The problem with this algorithm is that a great many error vectors might need to be tested, and a vector subtraction and vector-matrix multiplication must be performed for each test. Fortunately, there is an easier approach to realize this decoding method for linear codes.

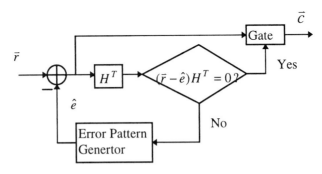

Figure 3.7 A conceptual decoder

Another approach to this decoder is to consider first forming the $(n - k)$-vectors $\vec{s} = \vec{r} \cdot H^T$. Note that

Linear Block Codes

$$\vec{s} = \vec{r} \cdot H^T = (\vec{c} + \vec{e}) \cdot H^T = \vec{e} \cdot H^T, \qquad (3.20)$$

where the last step follows from the parity-check property, $\vec{c} \cdot H^T = 0$, for any code vector $\vec{c}$. Therefore, the elements of $\vec{s}$ are linear combinations only of the errors, e_i, and $\vec{s}$ is called the *syndrome* of the error pattern. It follows from the construction of the standard array that all of the vectors in a given row of a standard array must have the same syndrome. However, the syndromes of the vectors in different rows of the standard array cannot be equal. This can be shown as follows: If two error patterns $\vec{e}_1$ and $\vec{e}_2$ have the same syndrome, i.e. if $\vec{s} = \vec{e}_1 H^T = \vec{e}_2 H^T$, then $(\vec{e}_1 - \vec{e}_2) H^T = 0$. This implies $(\vec{e}_1 - \vec{e}_2) = \vec{c}' \in C$, so that $\vec{e}_1$ and $\vec{e}_2$ would thus be in the same row of the standard array, a contradiction.

This result shows that one needs only to store the coset leaders of the standard array and their syndromes in memory in order to perform ML decoding. The memory table of the coset leaders and their syndrome addresses is shown in Table 3.5.

$\vec{s}_1 = 0$	$\vec{e}_1 = 0$
$\vec{s}_2 = \vec{e}_2 H^T$	$\vec{e}_2$
$\vec{s}_3 = \vec{e}_3 H^T$	$\vec{e}_3$
$\vdots$	$\vdots$
$\vec{s}_{q^{n-k}} = \vec{e}_{q^{n-k}} H^T$	$\vec{e}_{q^{n-k}}$

Table 3.5 Syndrome-decoding table of a (n,k) linear block code

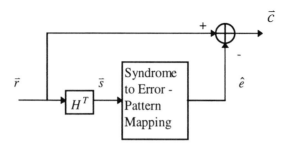

Figure 3.8 Syndrome decoding

The syndrome-decoding table makes possible a complete decoding procedure, called syndrome decoding. The syndrome decoding algorithm is shown in Figure 3.8. Syndrome decoding is performed by the following steps:
1. Compute the syndrome of the received word.
2. Look up the error pattern addressed by this syndrome in the syndrome-decoding table(Table 3.5).
3. Subtract the error pattern from the received word.

Example 3.9
The parity-check matrix of the (6,3) linear code given in Example 3.8 is obtained from Eq. 3.9 as

$$H = \begin{pmatrix} 1 & 0 & 0 & 0 & 1 & 1 \\ 0 & 1 & 0 & 1 & 0 & 1 \\ 0 & 0 & 1 & 1 & 1 & 0 \end{pmatrix}.$$

The syndrome decoding table for this code is constructed by using H. It is shown in Table 3.6.

000	000000
110	000001
101	000010
011	000100
001	001000
010	010000
100	100000
111	001001

Table 3.6 Syndrome decoding table for a (6,3) code

Suppose that the received word is $\bar{r} = (101001)$. The decoder first computes the syndrome $\bar{s} = \bar{r}H^T = (011)$. Looking up this syndrome in the table, the corresponding error pattern is found to be $\bar{e} = (000100)$. The maximum-likelihood transmitted codeword is thus $\bar{r} - \bar{e} = (101101)$, the same corrected codeword obtained in Example 3.8.

The computation of the syndrome $\bar{s}$ is operationally equivalent to that part of an encoding algorithm that involves multiplying a vector by a matrix. For cases where the number of distinct syndromes, q^{n-k}, is reasonably small, such a syndrome decoder represents the most practical decoder. In the syndrome decoding table, one needs only to store the coset leaders for each

Linear Block Codes

syndrome address. In fact, for systematic codes only the k information positions of the coset leaders need to be stored, since normally only estimates of the information are required. Thus, the table size is $q^{n-k} \times k$ symbols. An implementation of such a decoder for a binary linear systematic code is shown in Figure 3.9.

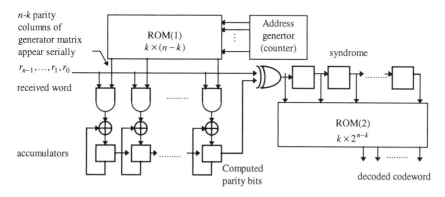

Figure 3.9. An implementation of syndrome decoding

It is seen that the same circuit given in Figure 3.4 for systematic encoding could also be used to compute the syndrome. This is because the computation of the syndrome is equivalent to a computation of the parity bits of the received "message" bits, followed by a comparison of the result with the received "parity" bits. Therefore, the syndrome computation circuit has the same complexity as the encoder.

Notice that the only error patterns that are correctable by the above table look-up decoder are the coset leaders. If an error pattern occurs that is not a coset leader, the decoder retrieves an erroneous error pattern and typically leaves the "corrected" vector with more errors than it originally contained.

Example 3.10
For the (6,3) linear code, specified in Examples 3.8 and 3.9, the non-error event, the six one-error patterns, and the single two-error pattern are corrected. Notice from Table 3.4, however, that three two-error patterns (001001), (010010) and (100100) are equally likely to produce the syndrome $\bar{s} = (111)$. Thus, an attempt to correct the received word with this syndrome is less likely to succeed than an attempt to correct errors when observing other less ambiguous syndromes. It is easy to verify that the minimum

distance of this code is $d_{min} = 3$. Hence, the maximum error-correction capability of the code is $t=1$. The bounded-distance syndrome decoding is performed by the smaller syndrome table, obtained by removing the last row of Table 3.5. If the syndrome of a received word does not equal any syndrome vector in the newly generated syndrome table, an uncorrectable error is detected.

3.4 Performance of Algebraic Decoding

Three types of decision criteria for a decoder are now described. Two of these procedures allow the decoder to avoid decisions by reporting erasures or uncorrectable errors. In the case where the decoder does not produce decisions, it is presumed that some higher-level decoding process can either correct the erasure or request that the same codeword be re-transmitted. The performances of these decoders are measured by their probabilities of errors. The probability expressions are given in terms of the weight distribution $\{W_i\}$ of the code. Thus, they are useful only when the set $\{W_i\}$ is known.

Type 1: Error detection-only decoders

For the detection-only decoder one needs only to test for a zero-syndrome vector since $\bar{s} = \bar{r}H^T = 0$ only if $\bar{r}$ is a valid codeword. If the syndrome is not zero, a detected error is reported. This decision can be wrong only if the error pattern is another codeword.

For the above type of decision device, the performance measure of interest is the probability of an undetected error, denoted by P_{ud}. For the case of the q-ary uniform channel, P_{ud} is computed from the weight distribution $\{W_i\}$ of C as follows:

$$P_{ud} = \Pr ob(\bar{r} = \bar{c}_i + \bar{e}, \bar{e} = \bar{c}_j \neq 0)$$

$$= \sum_{l=d_{min}}^{n} W_l \Pr ob(\bar{e} \text{ is a weight } l \text{ pattern}) \quad (3.21)$$

$$= \sum_{l=d_{min}}^{n} W_l \left(\frac{P}{q-1}\right)^l (1-P)^{n-l}.$$

Thus, knowledge of the complete weight distribution of the code allows for an exact calculation of P_{ud}.

Linear Block Codes

Example 3.11.
For the (6,3) binary code given in Example 3.8, which is transmitted over the BSC, one obtains $P_{ud} = 4P^3(1-P)^3 + 3P^4(1-P)^2$, where P is the bit-error probability of the BSC.

If only the minimum distance of the code and the channel symbol-error rate are known, an upper bound on P_{ud} is given by

$$P_{ud} \leq \sum_{l=d_{min}}^{n} \binom{n}{l} \left(\frac{P}{q-1}\right)^l (1-P)^{n-l},$$

where $\binom{n}{l}$ is the binomial coefficient.

Type 2: Complete decoder
For a complete decoding the decoder is always required to deliver its best estimate of the transmitted codeword that is based on the syndrome measurement. This too can fail, but the probability of a correct decision is given by

$$P_{cd} = P(\bar{e} \text{ is a coset leader}).$$

Then, the probability of decoding error (or an incorrect decoding) for this type of decision is

$$P_{de} = 1 - P_{cd} = P(\bar{e} \text{ is not a coset leader}).$$

An exact calculation of P_{de} requires knowledge of the set of coset leaders. If only the maximum error-correction capability of the code and the channel symbol-error rate are known, an upper-bound for P_{de} is derived next for the BSC. Since the BSC is memoryless, the bit errors occur independently. The number of errors occurring in a codeword n-tuple on a BSC has a binomial distribution with mean nP and variance $nP(1-P)$, where P is the channel transition probability. Hence, the probability that l errors occur in a block of n bits is

$$P_i = \binom{n}{l} P^l (1-P)^{n-l}.$$

If a t-error-correcting block code is used for error correction on such a BSC, one has

$$P_{de} = P(\bar{e} \text{ is not a coset leader}) \leq P(wt(\bar{e}) \geq t+1).$$

That is,

$$P_{de} \le \sum_{l=t+1}^{n} \binom{n}{l} P^l (1-P)^{n-l}. \tag{3.22}$$

Example 3.12.
For the same (6.3) code transmitted over a BSC, one obtains by a use of Table 3.4 that
$$P_{cd} = (1-P)^6 + 6P(1-P)^5 + P^2(1-P)^4$$
and
$$P_{de} = 1 - (1-P)^6 - 6P(1-P)^5 - P^2(1-P)^4.$$
Thus, for the binary (6, 3) code, the probability of error decoding is upper-bounded by
$$P_{de} \le \Pr ob(\bar{e} \text{ has } 2, 3, \ldots, 6 \text{ errors})$$
$$= \sum_{l=2}^{6} \binom{6}{l} P^l (1-P)^{n-l},$$
since all error patterns with two or more errors are not guaranteed to be correctable.

Type 3: Bounded-distance decoder
This type of decoder allows for the correction of $v \le t$ errors, where t is the maximum error-correcting capability of the code. It also makes possible a detection of errors using the remaining syndromes. Thus, a hybrid mode of operation can be adopted which combines both of the 1 and 2 types of the decoder. In essence the standard array is divided into two parts: a set of cosets in which error correction is attempted and another set in which only transmission errors are detected. Thus, three possibilities exist for the so-called Type 3 decoder. One can decode correctly when $wt(\bar{e}) \le t$. One also can decode incorrectly when $\bar{r}$ falls within t units of another codeword. Finally one can detect those errors for which $\bar{r}$ lies in the interstitial space between spheres of radius t drawn about all codewords. The probabilities of these three events clearly are related by the relation,
$$P_{cd} + P_{ud} + P_{de} = 1,$$
where P_{de} denotes the detected-error event.

Example 3.13.
As an example take the (6, 3) code with $v = t = 1$. The Type 3 decoder reports detected errors when the ambiguous syndrome $\bar{s} = (111)$ is obtained. Otherwise, error correction is performed as usual. In this case the probability

of correct decoding is slightly less than that of the Type 2 decoder. This probability is given by

$$P_{cd} = (1-P)^6 + 6P(1-P)^5,$$

However, this smaller probability of error decoding $P_{de} = 1 - P_{cd}$ occurs at the expense of occasionally failing to decode. In general, P_{de} is the probability of an error pattern that is not a coset leader, but that lies in the cosets in which correction is possible. Similarly, P_{de} is the probability that the error pattern falls in the detected-error coset.

Many algebraic decoders are bounded-distance decoders (or, what are called, incomplete decoders), thereby producing that single valid codeword within t Hamming-distance units of the received word if one exists. If no such codeword exists, because more than t errors occur in transmission, the algorithms may simply fail to decode, declaring only detected errors.

The general syndrome decoder, when combined with table look-up, is not practical for long codes since the syndrome table requires q^{n-k} words. This type of decoder rapidly becomes unmanageable, especially for $q > 2$. It is true, however, that the table-look-up decoders, which were not feasible even ten years ago, may well be practical in the future due to the dramatic increase in the sizes and speeds of semiconductor memory.

In addition to the above complexity issue, surrounding general linear codes, it should be noted that both the encoding $\bar{c} = \bar{m}G$ and the syndrome computation $\bar{s} = \bar{r}H^T$ are matrix operations over GF(q). Hence, in general, there are no shortcuts to the avoidance of brute-force matrix multiplication. Hence at least nk multiplications and $n(n-k)$ additions are needed for these usually necessary operations.

In the next chapter cyclic codes are studied. Cyclic codes are linear codes which make possible still simpler encoding and syndrome computations. More importantly, if one wants to avoid table-look-up for the error pattern, algebraic procedures are available for the most important of these codes to solve for the minimum-weight error pattern, or at least to do so with a high probability.

We turn now to some constructive procedures for building certain simple linear codes, the Hamming codes and the Reed-Muller codes. More general constructions are presented later in the context of cyclic codes.

3.5 Hamming Codes

Hamming codes [5] were the first class of binary linear block codes discovered for error correction. These codes have been used widely in error control of computer memories and long-distance telephony. The simplest Hamming code was introduced in Example 3.1. In this section the fundamental structure of this class of codes is discussed further.

The parameters for the class of binary Hamming codes are represented usually as a function of the single integer $m \geq 2$ as follows:

Block code length: $\qquad n = 2^m - 1$
Message length: $\qquad k = 2^m - m - 1$
The minimum Hamming distance: $\qquad d_{\min} = 3$

The construction of binary Hamming codes can be determined by their parity-check matrices. For a Hamming code of length $2^m - 1$, its parity-check matrix is a matrix whose columns consist of the entire set of the non-zero binary m-tuples.

Example 3.14
The parity-check matrix of a (7,4) Hamming code is given already in Example 3.5. An examination of the parity-check matrix in Example 3.5 quickly shows that the *(7,4)* Hamming code is single-error correcting.

In general, it can be shown that the smallest number of distinct nonzero binary m-tuples that can sum to zero is always three. Therefore, by Section 3.2.2 this implies that Hamming codes have the minimum distance three and as a consequence are single-error-correcting.

It was shown in the last section that if a *t*-error-correcting block code is used for error correction on a BSC with a transition probability P that the probability that the decoder fails to decode properly is upper-bounded by the inequality in (3.22). The equality in Eq. (3.22) holds if the linear block code is, what is called, a *perfect code*. Since the code has 2^{n-k} cosets, a perfect code has the property that its standard array has all of the error patterns of t or fewer errors and no others as coset leaders. It is implied from

Linear Block Codes

the parameters $n = 2^{n-k} - 1$ and $t=1$ that a (n,k) Hamming code has $\sum_{l=0}^{t} \binom{n}{l} = 1+n = 2^{n-k}$ cosets. Hence, one has

$$P_{de} = P(\bar{e} \text{ is not a coset leader}) = P(wt(\bar{e}) \geq t+1)$$

$$= \sum_{l=t+1}^{n} \binom{n}{l} P^l (1-P)^{n-l}.$$

Therefore, the Hamming codes form a class of single-error-correcting perfect codes.

Suppose that a small sphere of radius t is placed around each of the possible transmitted codewords. The radii of these spheres can be increased by integer amounts until they cannot be made larger without causing some spheres to intersect. This final value t of the radius is called the packing radius of the code. Such a radius is equal to the number of errors that can be corrected by the code. For a t-error-correcting block code, the number of (received) words in a sphere of radius $t = \lfloor (d_{min} - 1)/2 \rfloor$ is $\sum_{l=0}^{t} \binom{n}{l}$. Since there are 2^k possible transmitted codewords, there are 2^k non-overlapping spheres, each having the radius t. The total number of (received) words enclosed in the 2^k spheres cannot exceed the 2^n possible received words. Thus, a t-error-correcting block code must satisfy the following inequality :

$$2^n \geq 2^k \sum_{l=0}^{t} \binom{n}{l}$$

i.e.

$$2^{n-k} \geq \sum_{l=0}^{t} \binom{n}{l}. \tag{3.23}$$

This is called the Hamming bound or the sphere-packing bound for binary codes. A perfect code satisfies the Hamming bound with equality.

Hamming codes can be decoded by the use of syndrome decoding an discussed in Section 3.3.3. Since Hamming codes are single-error-correcting, the syndrome table has a simple relationship to the parity-check matrix used in the computation of the syndrome. Let the error pattern (vector) be $\bar{e}$ which has a single one in the j-th coordinate position. Let $\{\bar{h}_0, \bar{h}_1, \ldots, \bar{h}_{n-1}\}$ be the set of columns of the parity-check matrix H. When the syndrome is computed from the received word $\vec{r}$, one obtains the transposition of the j-th column of H. That is

$$\vec{s} = \vec{r} \cdot H^T = \vec{e} \cdot H^T = \vec{h}_j^T.$$

Thus, the decoding process for Hamming codes can be simplified as follows.

Hamming-Code Decoding Algorithm
1. Compute the syndrome $\vec{s}$ of the received word $\vec{r}$. If $\vec{s} = \vec{0}$, then STOP.
2. Determine the position j of the column of H that is the transposition of the syndrome.
3. Complement the j-th bit in the received word.

A "C" language program for the encoder and decoder of binary Hamming codes for $3 \leq m \leq 10$ is given in Appendix B.

Non-binary Hamming codes can be constructed by a use of the same basic approach. There are exactly $q^m - 1$ distinct nonzero q-ary m-tuples, but not all pairs of these m-tuples are linearly independent. For each q-ary m-tuple there are $q-1$ distinct nonzero m-tuples that are multiples of that m-tuple, pairs of which are clearly linearly dependent. The q-ary Hamming code parity-check matrix H is constructed by selecting exactly one m-tuple from each set of multiples. This can be done by selecting as columns of H all distinct q-ary m-tuples for which the uppermost nonzero element is 1. H thus has $(q^m - 1)/(q - 1)$ columns and defines a q-ary Hamming code with the following properties.

> Block code length : $n = (q^m - 1)/(q - 1)$
> Message length : $k = (q^m - 1)/(q - 1) - m$
> The minimum Hamming distance : $d_{min} = 3$

3.6 Reed-Muller Codes

Reed-Muller (RM) codes were discovered by Muller [6] in 1954, and shortly thereafter a better algebraic representation was provided by Reed along with an elegant decoding algorithm [7]. The invention of RM codes was an important step beyond the Hamming codes because of their flexibility in correcting multiple numbers of errors per codeword. Since their discovery RM codes have been investigated by researchers with considerable attention. These studies have led to the discovery of other interesting codes [8][9] and

Linear Block Codes

also have resulted in a number of important applications. For example, the first-order RM code of length 32 was used in 1969 for the error-control system of the Mariner and Viking deep-space probes to Mars.

3.6.1 Construction of RM codes

Reed-Muller codes are described most easily through the use of Boolean functions. Let V_m denote the vector space of binary m-tuples $(v_1, v_2, \ldots, v_m)$, i.e. the set of all binary m-tuples. Any function $f(v_1, v_2, \ldots, v_m)$ with values only in $F = \{0,1\}$ is called a Boolean function. Boolean functions are specified completely by, what is called, a "truth" table.

Example 3.15.
Let f be the Boolean function of the three binary variables $\{v_1, v_2, v_3\}$ with the truth table, given in the table 3.7.

A Boolean function f	v_3	v_2	v_1
0	0	0	0
0	0	0	1
0	0	1	0
1	0	1	1
1	1	0	0
0	1	0	1
0	1	1	0
0	1	1	1

Table 3.7. The truth table of a Boolean function

In this example, the 3-tuples (v_3, v_2, v_1) are in the form of natural order, i.e. (0,0,0) corresponds to 0, (0,0,1) corresponds to 1, ..., (1,1,1) corresponds to $2^3 - 1 = 8$. This allows one to unambiguously associate each Boolean function f with a unique binary vector $\vec{f}$. For this example, the binary vector(function) is the first column in the table : $\vec{f} = (00011000)$. Since $\vec{f}$ is a binary vector of length 2^m, there must be 2^{2^m} distinct Boolean functions of m

variables. Under component-wise binary addition of the associated vectors, the Boolean functions form the vector space V_{2^m} over $GF(2)$.

A Boolean function can always be represented in an algebraic formula in term of the operators "+" (XOR) and " $\cdot$ "(AND) by applying the following Boolean-algebra relation :

$$\text{OR} : f \vee g = f + g + f \cdot g, \qquad (3.24)$$
$$\text{NOT} : \bar{f} = 1 + f.$$

It can be verified by the truth table in Table 3.6 that $f = v_1 \cdot v_2 \cdot \bar{v}_3 \vee \bar{v}_1 \cdot \bar{v}_2 \cdot v_3$ in terms of the OR and NOT operations. By a use of Eq. (3.24) and the identity $v_i^2 = v_i$, one obtains finally that $f = v_3 + v_1 \cdot v_2 + v_1 \cdot v_3 + v_2 \cdot v_3$ in terms of the product and sum operators over $GF(2)$.

In general, any Boolean function over V_m can be expressed as a linear combination of the following 2^m monomial product functions :

$$1, v_1, v_2, \ldots, v_m, v_1 \cdot v_2, v_1 \cdot v_3, \ldots, v_{m-1} \cdot v_m, \ldots\ldots, v_1 \cdot v_2 \cdots v_m \qquad (3.25)$$

with coefficients in $F = \{0,1\}$. Since the above 2^m monomial functions are linearly independent, it follows that the vectors with which they are associated are also linearly independent. There is thus a unique Boolean function f for every vector $\bar{f}$ of the form,

$$\bar{f} = m_0 \bar{v}_0 + \sum_{i=1}^{m} m_i \cdot \bar{v}_i + \sum_{\substack{i,j=1 \\ j>i}}^{m} m_{i,j} \cdot \bar{v}_i \cdot \bar{v}_j + \cdots + m_{1,2,\ldots,m} \cdot \bar{v}_1 \cdot \bar{v}_2 \cdots \bar{v}_m, \qquad (3.26)$$

where $\bar{v}_0 = (1,1,\ldots,1)$.

Example 3.16.
The Boolean function f given in Example 3.14 can be represented by
$$\bar{f} = \bar{v}_3 + \bar{v}_1 \cdot \bar{v}_2 + \bar{v}_1 \cdot \bar{v}_3 + \bar{v}_2 \cdot \bar{v}_3,$$
where the four columns in Table 3.6 correspond to the vectors, $\bar{f} = (00011000)$, $\bar{v}_3 = (00001111)$, $\bar{v}_2 = (00110011)$, and $\bar{v}_1 = (01010101)$.

Since there are a total of 2^{2^m} such vectors, the Boolean monomial functions in relation *(3.25)* form a basis for the vector space of Boolean functions in m variables. Reed-Muller codes are defined by the use of these Boolean monomial functions as follows.

Linear Block Codes

Define the degree of a monomial term $v_{i_1} \cdot v_{i_2} \cdots v_{i_r}$ is defined to equal r, and the degree $\deg(f(v_1, v_2, \ldots, v_m))$ of a Boolean function $f(v_1, v_2, \ldots, v_m)$ is set to highest degree monomial which appears. Then, the r-th order binary RM code $R(r,m)$ of length $n = 2^m$, for $0 \leq r \leq m$, is the set of all vectors $\bar{f}$ associated with all Boolean functions $f(v_1, v_2, \ldots, v_m)$ of $\deg(f(v_1, v_2, \ldots, v_m)) \leq r$.

Thus, the r-th order RM code is the set of all linear combinations of monomials of form,

$$\bar{f} = m_0 \bar{v}_0 + \sum_{i=1}^{m} m_i \cdot \bar{v}_i + \sum_{\substack{i,j=1 \\ j>i}}^{m} m_{i,j} \cdot \bar{v}_i \cdot \bar{v}_j + \cdots (\text{up to degree } r)$$

Hence, it is clear that for any $\bar{f}_1, \bar{f}_2 \in R(r,m)$, $\bar{f}_1 + \bar{f}_2 \in R(r,m)$. Therefore, $R(r,m)$ forms a vector space and is a linear code. Since the vectors $\bar{v}_0, \bar{v}_1, \ldots, \bar{v}_m, \bar{v}_1 \cdot \bar{v}_2, \bar{v}_1 \cdot \bar{v}_3, \cdots, \bar{v}_{m-1} \cdot \bar{v}_m, \cdots,$ (up to degree r) form a basis for $R(r,m)$, the generator matrix of $R(r,m)$ is

$$G_{R(r,m)} = \begin{pmatrix} \bar{v}_0 \\ \bar{v}_1 \\ \vdots \\ \bar{v}_m \\ \bar{v}_1 \bar{v}_2 \\ \bar{v}_1 \bar{v}_3 \\ \vdots \\ \bar{v}_{m-1} \bar{v}_m \\ \vdots \end{pmatrix}$$

The dimension of the RM code $R(r,m)$ is $k = 1 + \binom{m}{1} + \cdots + \binom{m}{r}$.

Example 3.17.
Consider the RM code $R(2,4)$ with the codeword length $n = 2^m = 2^4 = 16$ and the dimension $k = 1 + \binom{4}{1} + \binom{4}{2} = 1 + 4 + 6 = 11$. The generator matrix is

$$G_{R(2,4)} = \begin{pmatrix} \bar{v}_0 \\ \bar{v}_4 \\ \bar{v}_3 \\ \bar{v}_2 \\ \bar{v}_1 \\ \bar{v}_4\bar{v}_3 \\ \bar{v}_4\bar{v}_2 \\ \bar{v}_4\bar{v}_1 \\ \bar{v}_3\bar{v}_2 \\ \bar{v}_3\bar{v}_1 \\ \bar{v}_2\bar{v}_1 \end{pmatrix},$$

where

$\bar{v}_0 = (1111111111111111)$,
$\bar{v}_4 = (0000000011111111)$,
$\bar{v}_3 = (0000111100001111)$,
$\bar{v}_2 = (0011001100110011)$,
$\bar{v}_1 = (0101010101010101)$,
$\bar{v}_4\bar{v}_3 = (0000000000001111)$,
$\bar{v}_4\bar{v}_2 = (0000000000110011)$,
$\bar{v}_4\bar{v}_1 = (0000000001010101)$,
$\bar{v}_3\bar{v}_2 = (0000001100000011)$,
$\bar{v}_3\bar{v}_1 = (0000010100000101)$,
$\bar{v}_2\bar{v}_1 = (0001000100010001)$.

In his 1954 paper [7], Reed described a multiple-error-correcting decoding algorithm for RM codes on the basis of a set of parity-check equations. Such a decoding algorithm was called the Reed or majority logic decoding. Majority logic decoding was generalized later by Massey [10] to decode other codes such as the convolutional codes. For RM codes the Reed algorithm provides a maximum-likelihood, hard-decision, decoding in an efficient manner.

An application of the Reed algorithm to the particular RM code $R(2,4)$ given in Example 3.17, is demonstrated next. The non-systematic encoding of RM

Linear Block Codes

codes is accomplished by $\bar{c} = \bar{m} \cdot G_{R(r,m)}$, as discussed in Section 3.2. Let the 11 bits in the message block $\bar{m}$ be written

$$\bar{m} = (m_0, m_4, m_3, m_2, m_1, m_{43}, m_{42}, m_{41}, m_{32}, m_{31}, m_{21}).$$

The subscripts associate each message bit with a row in $G_{R(2,4)}$, $\bar{m}$ is encoded as follows.

$$\bar{c} = (c_0, c_1, \ldots, c_{15})$$
$$= m_0 \bar{v}_0 + m_4 \bar{v}_4 + \cdots + m_1 \bar{v}_1 + m_{43} \bar{v}_4 \bar{v}_3 + \cdots + m_{21} \bar{v}_2 \bar{v}_1$$
$$= [\bar{m}_0 | \bar{m}_1 | \bar{m}_2] \begin{bmatrix} G_0 \\ G_1 \\ G_2 \end{bmatrix}.$$

In this expression the message bits are grouped in accordance with the order or degree of the vectors of which they are coefficients. The block $\bar{m}_0$ consists of the single message bit associated with the zeroth-order term $\bar{v}_0$, the message bits in $\bar{m}_1$ are associated with the first-order terms $\bar{v}_4, \bar{v}_3, \bar{v}_2$, and $\bar{v}_1$, and the bits in $\bar{m}_2$ are associated with the second-order terms $\bar{v}_4 \cdot \bar{v}_3, \ldots, \bar{v}_2 \cdot \bar{v}_1$. For a received word $\bar{r}$, the Reed algorithm begins decoding by determining the highest-order block of message bits, which in this case is $\bar{m}_2$. The estimate for the product $\bar{m}_2 G_2$ is then subtracted from $\bar{r}$, leaving a noise-corrupted, lower-order codeword. This process continues, with the next, lower-order, products $\bar{m}_1 G_1$ being estimated and removed from $\bar{r}$ until only the term $\bar{m}_0 G_0$ remains.

The key to the Reed algorithm lies in the steps by which the message bits are determined. For the RM code $R(2,4)$ given in Example 3.17, one can verify that the minimum distance of the code is $d_{\min} = 4$. Thus this code can correct one random error. The objective of the Reed algorithm is to reconstruct the message $\bar{m} = (\bar{m}_0, \bar{m}_1, \bar{m}_2)$ step-by-step from the received word $\bar{r}$. Assume that $\bar{r} = \bar{c} + \bar{e}$ where $\bar{r} = (r_0, r_1, \ldots, r_{15})$, $\bar{e} = (e_0, e_1, \ldots, e_{15})$ and $\bar{c} = \bar{m} \cdot G_{R(2,4)}$. Then, one has

$$r_0 = m_0 + e_0,$$
$$r_1 = m_0 + m_1 + e_1,$$
$$r_2 = m_0 + m_2 + e_2,$$
$$r_3 = m_0 + m_2 + m_1 + m_{21} + e_3,$$
$$r_4 = m_0 + m_3 + e_4,$$
$$r_5 = m_0 + m_3 + m_1 + m_{31} + e_5,$$
$$r_6 = m_0 + m_3 + m_2 + m_{32} + e_6,$$
$$r_7 = m_0 + m_3 + m_2 + m_1 + m_0 + m_{32} + m_{31} + m_{21} + e_7,$$
$$r_8 = m_0 + m_4 + e_8,$$
$$r_9 = m_0 + m_4 + m_1 + m_{41} + e_9,$$
$$r_{10} = m_0 + m_4 + m_2 + m_{42} + e_{10},$$
$$r_{11} = m_0 + m_4 + m_2 + m_1 + m_{42} + m_{41} + m_{21} + e_{11},$$
$$r_{12} = m_0 + m_4 + m_3 + m_{43} + e_{12},$$
$$r_{13} = m_0 + m_4 + m_3 + m_1 + m_{43} + m_{41} + m_{31} + e_{13},$$
$$r_{14} = m_0 + m_4 + m_3 + m_2 + m_{43} + m_{42} + m_{32} + e_{14},$$
$$r_{15} = m_0 + m_4 + m_3 + m_2 + m_1 + m_{43} + m_{42} + m_{41} + m_{32} + m_{31} + m_{21} + e_{15}.$$

If there are no errors, $\bar{m}_2 = (m_{43}, \cdots, m_{21})$ can be recovered from the following equations.

$$\begin{aligned}
m_{21} &= r_0 + r_1 + r_2 + r_3 \\
&= r_4 + r_5 + r_6 + r_7 \\
&= r_8 + r_9 + r_{10} + r_{11} \\
&= r_{12} + r_{13} + r_{14} + r_{15}, \\
m_{32} &= r_0 + r_2 + r_4 + r_6 \\
&= r_1 + r_3 + r_5 + r_7 \\
&= r_8 + r_{10} + r_{12} + r_{14} \\
&= r_9 + r_{11} + r_{13} + r_{15}, \\
m_{42} &= r_0 + r_2 + r_8 + r_{10} \\
&= r_1 + r_3 + r_9 + r_{11} \\
&= r_4 + r_6 + r_{12} + r_{14} \\
&= r_5 + r_7 + r_{13} + r_{15},
\end{aligned}
\qquad
\begin{aligned}
m_{31} &= r_0 + r_1 + r_4 + r_5 \\
&= r_2 + r_3 + r_6 + r_7 \\
&= r_8 + r_9 + r_{11} + r_{12} \\
&= r_{10} + r_{11} + r_{14} + r_{15}, \\
m_{41} &= r_0 + r_1 + r_8 + r_9 \\
&= r_2 + r_3 + r_{10} + r_{11} \\
&= r_4 + r_5 + r_{12} + r_{13} \\
&= r_6 + r_7 + r_{14} + r_{15}, \\
m_{43} &= r_0 + r_4 + r_8 + r_{12} \\
&= r_1 + r_5 + r_9 + r_{13} \\
&= r_2 + r_6 + r_{10} + r_{14} \\
&= r_3 + r_7 + r_{11} + r_{15}.
\end{aligned} \qquad (3.27)$$

Linear Block Codes

There are four equations (or votes) for each of the values of $m_{43}, \cdots, m_{21}$. It can be seen that if one error occurs, the majority vote is still correct. Therefore, $m_{43}, \cdots, m_{21}$ are decided by the following rule : Give the value "0" to m_{ij} if there exist more than two of its corresponding equations equal to zero and "1" otherwise.

Next, consider the value $\bar{r}^{(1)} = \bar{r} - (m_{43}\bar{v}_4\bar{v}_3 + \cdots + m_{21}\bar{v}_2\bar{v}_1)$.

$r_0^{(1)} = m_0,$
$r_1^{(1)} = m_0 + m_1,$
$r_2^{(1)} = m_0 + m_2,$
$r_3^{(1)} = m_0 + m_2 + m_1,$
$r_4^{(1)} = m_0 + m_3,$
$r_5^{(1)} = m_0 + m_3 + m_1,$
$r_6^{(1)} = m_0 + m_3 + m_2,$
$r_7^{(1)} = m_0 + m_3 + m_2 + m_1,$
$r_8^{(1)} = m_0 + m_4,$
$r_9^{(1)} = m_0 + m_4 + m_1,$
$r_{10}^{(1)} = m_0 + m_4 + m_2,$
$r_{11}^{(1)} = m_0 + m_4 + m_2 + m_1,$
$r_{12}^{(1)} = m_0 + m_4 + m_3,$
$r_{13}^{(1)} = m_0 + m_4 + m_3 + m_1,$
$r_{14}^{(1)} = m_0 + m_4 + m_3 + m_2,$
$r_{15}^{(1)} = m_0 + m_4 + m_3 + m_2 + m_1.$

Notice that $\bar{r}^{(1)}$ is only depends on $\bar{m}_1, \bar{m}_0$, i.e. m_0, m_4, m_3, m_2, m_1. m_4, m_3, m_2, m_1 can be recovered from the following equations.

$$\begin{aligned}
m_1 &= r_0^{(1)} + r_1^{(1)} & m_2 &= r_0^{(1)} + r_2^{(1)} \\
&= r_2^{(1)} + r_3^{(1)} & &= r_1^{(1)} + r_3^{(1)} \\
&= r_4^{(1)} + r_5^{(1)} & &= r_4^{(1)} + r_6^{(1)} \\
&= r_6^{(1)} + r_7^{(1)} & &= r_5^{(1)} + r_7^{(1)} \\
&= r_8^{(1)} + r_9^{(1)} & &= r_8^{(1)} + r_{10}^{(1)} \\
&= r_{10}^{(1)} + r_{11}^{(1)} & &= r_9^{(1)} + r_{11}^{(1)} \\
&= r_{12}^{(1)} + r_{13}^{(1)} & &= r_{12}^{(1)} + r_{14}^{(1)} \\
&= r_{14}^{(1)} + r_{15}^{(1)}, & &= r_{13}^{(1)} + r_{15}^{(1)}, \\
m_3 &= r_0^{(1)} + r_4^{(1)} & m_4 &= r_0^{(1)} + r_8^{(1)} \\
&= r_1^{(1)} + r_5^{(1)} & &= r_1^{(1)} + r_9^{(1)} \\
&= r_2^{(1)} + r_8^{(1)} & &= r_2^{(1)} + r_{10}^{(1)} \\
&= r_3^{(1)} + r_7^{(1)} & &= r_3^{(1)} + r_{11}^{(1)} \\
&= r_8^{(1)} + r_{12}^{(1)} & &= r_4^{(1)} + r_{12}^{(1)} \\
&= r_9^{(1)} + r_{13}^{(1)} & &= r_5^{(1)} + r_{13}^{(1)} \\
&= r_{10}^{(1)} + r_{14}^{(1)} & &= r_6^{(1)} + r_{14}^{(1)} \\
&= r_{11}^{(1)} + r_{15}^{(1)}, & &= r_7^{(1)} + r_{15}^{(1)}.
\end{aligned} \tag{3.28}$$

There are eight equations (or votes) for each m_i for $i=1,2,3,4$. If one error occurs, the majority-logic vote certainly gives each m_i correctly for $i=1,2,3,4$. The decision rule is as follows: For $i=1,2,3,4$ give the value "0" to m_i if there exist more than 5 equations equal to zero, and "1" to m_i otherwise.

Finally, $\bar{r}^{(2)} = \bar{r}^{(1)} - (m_4\bar{v}_4 + m_3\bar{v}_3 + m_2\bar{v}_2 + m_1\bar{v}_1)$. This implies that $\bar{r}_i^{(2)} = m_0$ for $0 \leq i \leq 15$. Therefore, the majority decision rule for m_0 is to give the value "0" to m_0 if $r_i^{(2)}$ for $0 \leq i \leq 15$ has more zeros than ones, and "1" otherwise.

Example 3.18.
Let the received word be $\bar{r} = (0000110100101101)$. The first step is to estimate the message bits $\bar{m}_2 = (m_{43}, \cdots, m_{21})$ by the use of Eq. (3.27). The majority decision yields $\hat{m}_{43} = 1$ and $\hat{m}_{42} = \hat{m}_{41} = \hat{m}_{32} = \hat{m}_{31} = \hat{m}_{21} = 0$. Now subtract $\hat{\bar{m}}_2$ from the received word,

Linear Block Codes

$$\bar{r}^{(1)} = \bar{r} - \hat{\bar{m}}_2 G_2 = (0000110100101101) - (0001000100010001)$$
$$= (0001110000111100).$$

By the use of Eq. (3.28), the estimate of $\bar{m}_1$ is $\hat{\bar{m}}_1 = (0110)$. Finally, one has

$$\bar{r}^{(2)} = \bar{r}^{(1)} - \hat{\bar{m}}_1 G_1 = (0001110000111100) - (0011110000111100)$$
$$= (0010000000000000).$$

The estimate of m_0 is clearly zero. A single bit error in the third bit r_2 has been corrected. The resulting message word is $\bar{m} = (00110000001)$.

The Reed decoding algorithm does not apply if the RM code is encoded systematically. A systematic method for obtaining the majority votes for the Reed algorithm can be found in [10]. It is shown in [8] that the minimum distance of $R(r,m)$ is 2^{m-r}.

3.6.2 Encoding and Decoding of First-Order RM codes

The first order RM code, $R(1,m)$, is a $\left(2^m, m+1, 2^{m-1}\right)$ linear-block code and as a consequence has a low rate and powerful error-correction capability. It is therefore particularly useful for very noisy channels. The best-known application of the first-order codes is probably the use of $R(1,5)$ in the design of the Mariner 9 space probes to transmit pictures from Mars.

The encoding of the first-order RM code is especially simple. The generator matrix for $R(1,m)$ has the form $G = [\bar{1}, \bar{v}_m \bar{v}_{m-1}, \ldots, \bar{v}_1]^T$, where $\bar{1} = (11\ldots\ldots1)$ is the all "1" row vector. The encoding operation for $R(1,m)$ is as follows:

$$\bar{c} = (c_0, c_1, \ldots, c_{15}) = \bar{m} \cdot G = m_0 \bar{1}^T + \sum_{i=1}^{m} m_i \bar{v}_{m+1-i}^T.$$

Example 3.19
Consider the encoding of the $RM(1,3)$ code. A message is encoded into the codeword,

$$\bar{c} = (c_0, c_1, \ldots, c_7) = \bar{m} \cdot \begin{bmatrix} \bar{1} \\ \bar{v}_3 \\ \bar{v}_2 \\ \bar{v}_1 \end{bmatrix} = (m_0, m_1, m_2, m_3) \begin{bmatrix} 1 & 1 & 1 & 1 & 1 & 1 & 1 & 1 \\ 0 & 0 & 0 & 0 & 1 & 1 & 1 & 1 \\ 0 & 0 & 1 & 1 & 0 & 0 & 1 & 1 \\ 0 & 1 & 0 & 1 & 0 & 1 & 0 & 1 \end{bmatrix}$$

Notice that the columns of G have a regular form : they consist of the binary 4-tuples beginning with $(1000)^T$ and running in increasing order to $(1111)^T$. The last 3-bits of each 4-tuple are generated easily by a digital counter. Such a counter processes through the successive states $t_1 t_2 t_3 = 000, 001, 010, 011, 100, 101, 110, 111, 000, 001, \ldots$. This yields a very simple encoding circuit as shown in Figure 3.10.

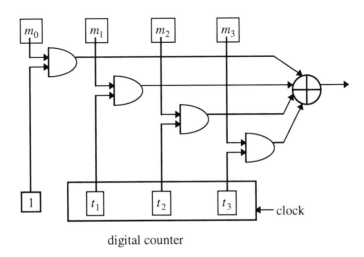

Figure 3.10 Encoder for RM(1,3)

The circuit in Fig. 3.10 forms $m_0 + t_1 m_1 + t_2 m_2 + t_3 m_3$ which is the codeword $\bar{c} = (c_0, c_1, \ldots, c_7)$. This encoder design has the benefit of being very fast and extremely simple.

As it is shown in Section 3.3.1, to transmit signals over a AWGN channel the maximum- likelihood decoding requires comparisons between the received word $\bar{r}$ and each codeword $\bar{c}$. For R(1,m), one must find the distance from $\bar{r}$ to every codeword $\bar{c}$, and then decode $\bar{r}$ as the closest codeword.

To decode R(1,m) first define the transform,
$$T(\bar{x}) = T(x_0, x_1, \ldots, x_{2^m-1}) = \left((-1)^{x_0}, (-1)^{x_1}, \ldots, (-1)^{x_{2^m-1}} \right).$$
Next let
$$\bar{y} = (y_0, y_1, \ldots, y_{2^m-1}) = \left((-1)^{r_0}, (-1)^{r_1}, \ldots, (-1)^{r_{2^m-1}} \right) = T(\bar{r}),$$

Linear Block Codes

$$\bar{s} = (s_0, s_1, \ldots, s_{2^m-1}) = \left((-1)^{c_0}, (-1)^{c_1}, \ldots, (-1)^{c_{2^m-1}}\right) = T(\bar{c}),$$

where $\bar{c}$ and $\bar{r}$ are, respectively, the codeword and received word. Then the metric, $\min_{\bar{c} \in C} d_E^2(\bar{y}, \bar{s}) = \min_{\bar{c} \in C} \sum_{i=0}^{s^m-1}(y_i - s_i)^2$, is equivalent to the metric, $\max_{\bar{c} \in C} \sum_{i=0}^{2^m-1} y_i s_i$. Note that the operations in the expression, $\sum_{i=0}^{2^m-1} y_i s_i$ are real additions and multiplications. Due to the special structure of RM(1,m) codes, a fast algorithm for implementing these operations is developed next.

First, the Hadamard matrix is defined recursively by

$$H_{2^i} = \begin{bmatrix} H_{2^{i-1}}, & H_{2^{i-1}} \\ H_{2^{i-1}}, & -H_{2^{i-1}} \end{bmatrix},$$

where $H_2 = \begin{bmatrix} 1 & 1 \\ 1 & -1 \end{bmatrix}$. Note that $H_{2^i} = H_{2^i}^T$. The Hadamard transform of a row vector $\bar{y}$ is computed by $\hat{y} = \bar{y} \cdot H_{2^m}$.

Recall that the generator matrix for RM(1,m) consists of the row vectors $\bar{1}, \bar{v}_m, \bar{v}_{m-1}, \ldots, \bar{v}_1$. It can be shown that $T(\bar{v}_m), T(\bar{v}_{m-1}), \ldots, T(\bar{v}_1)$ are always rows in the Hadamard matrix H_{2^m}. In general, the $\bar{u}$-th row of H_{2^m}, where $\bar{u}$ has the binary representation $\bar{u} = (u_m, u_{m-1}, \ldots, u_1)$, is the vector $T(\bar{c}_u) = T(u_1 \cdot \bar{v}_1 + \cdots + u_m \bar{v}_m)$.

Example 3.20

For RM(1,3), the row vectors $\bar{v}_3, \bar{v}_2, \bar{v}_1$ are given in Example 3.19. For this example (i=3) the Hadamard Matrix is

$$H_8 = \begin{bmatrix} 1 & 1 & 1 & 1 & 1 & 1 & 1 & 1 \\ 1 & -1 & 1 & -1 & 1 & -1 & 1 & -1 \\ 1 & 1 & -1 & -1 & 1 & 1 & -1 & -1 \\ 1 & -1 & -1 & 1 & 1 & -1 & -1 & 1 \\ 1 & 1 & 1 & 1 & -1 & -1 & -1 & -1 \\ 1 & -1 & 1 & -1 & -1 & 1 & -1 & 1 \\ 1 & 1 & -1 & -1 & -1 & -1 & 1 & 1 \\ 1 & -1 & -1 & 1 & -1 & 1 & 1 & -1 \end{bmatrix}.$$

A close examination of the Hadamard matrix H_8 reveals that the first, second, and fourth rows (where counting begins from 0) are $T(\bar{v}_1), T(\bar{v}_2)$, and $T(\bar{v}_3)$, respectively. The third column of H_8 consists of the vector
$$T(1 \cdot \bar{v}_1 + 1 \cdot \bar{v}_2 + 0 \cdot \bar{v}_3) = (1,-1,-1,1,1,-1,-1,1).$$

Next, the Hadamard transform $\hat{y} = \bar{y} \cdot H_{2^m}$ of vector $\bar{y} = T(\bar{r})$ computes the correlation of all words $T(\bar{c}_u)$, where $\bar{c}_u$ is a linear combination of the vectors $\bar{v}_m, \bar{v}_{m-1}, \ldots, \bar{v}_1$. It follows that each coordinate of the Hadamard transform has the value
$$\hat{y}_{\bar{u}} = 2^m - 2d_H(\bar{r}, \bar{c}_u),$$
where $d_H(\bar{r}, \bar{c}_u)$ denotes the Hamming distance between $\bar{r}$ and $\bar{c}_u$.

Since $-\hat{y}_{\bar{u}}$ is the correlation of $T(\bar{r})$ with $T(\bar{1} + \bar{c}_u) = T(\bar{1} + u_1 \cdot \bar{v}_1 + \cdots + u_m \bar{v}_m)$, one only needs to compute the correlation of $T(\bar{r})$ with half of the codewords in RM(1,m) to determine the maximum-likelihood codeword. The Hadamard transform-decoding algorithm is thus given as follows:

Decoding algorithm for RM(1,m):
Step 1: Compute $\bar{y} = (y_0, y_1, \ldots, y_{2^m-1}) = \left((-1)^{r_0}, (-1)^{r_1}, \ldots, (-1)^{r_{2^m-1}}\right)$ for a given received word $\bar{r}$.
Step 2: Compute the Hadamard transform, $\hat{y} = \bar{y} H_{2^m}$.
Step 3: Find the coordinate $\bar{u} = (u_m, u_{m-1}, \ldots, u_1)$ of the component $\hat{y}_{\bar{u}}$ of $\hat{y}$ which has the greatest magnitude.
Step 4: The codeword is decoded as $\hat{c} = \begin{cases} u_1 \bar{v}_1 + u_2 \bar{v}_2 + \cdots + u_m \bar{v}_m, & \text{if } \hat{y}_{\bar{u}} > 0, \\ \bar{1} + u_1 \bar{v}_1 + u_2 \bar{v}_2 + \cdots + u_m \bar{v}_m, & \text{if } \hat{y}_{\bar{u}} < 0. \end{cases}$

Example 3.21
For RM(1,3) discussed in Examples 3.19 and 3.20, assume that word is $\bar{r} = (01000011)$ is received by the decoder.
Step 1: $\bar{y} = (1,-1,1,1,1,1,-1,-1)$,
Step 2: $\hat{y} = \bar{y} \cdot H_8 = (2,2,2,2,2,2,-6,2)$,
Step 3: The sixth coordinate (110) has the greatest magnitude $6 = |-6|$.
Step 4: Since $\hat{y}_6 = -6 < 0$ the ML codeword is
$$(\bar{1} + 1 \cdot \bar{v}_3 + 1 \cdot \bar{v}_2 + 0 \cdot \bar{v}_1) = (11000011).$$

Note that the Hadamard transform is a Fourier transform that has been generalized to operate over finite abelian groups. Therefore, the Hadamard transform can be implemented by a fast transform.

3.7 Linear Block Codes for Burst-Error Correction

In certain digital storage media, such as magnetic tape, laser disks, and magnetic memories used in computer systems, it is observed that errors occur in the form of both short or long bursts. Examples of error-bursts also occur in wireless communication networks. In such networks messages are transmitted over noisy channels which often are corrupted by large bursts of interference. These noise bursts can occur in both intentional and non-intentional jamming environments. Therefore, it is desirable to find codes and to design coding systems which specifically correct burst errors.

A burst-error environment often uses a random error-correcting code along with an interleaver-de-interleaver. The interleaving device rearranges and spreads out the ordering of a sequence of digits in a deterministic manner. The de-interleaver is a device, which by means of an inverse permutation, restores the bit sequence to its original order. Thus, this approach involves scrambling the coded sequence prior to transmission, that is followed by a descrambling of the sequence after reception. This is done to randomize the burst-error process. The two principal classes of interleavers are the periodic and the pseudo-random interleaver. Such interleaving schemes are discussed further in Chapter 5.

Some linear codes are suitable for the correction of burst errors. For example, as it is discussed later in Chapter 6, the Reed-Solomon codes can be used to correct multiple short bursts or they can be interleaved with the code symbols to correct multiple long bursts. In this section certain interesting properties of the burst-error correction of linear codes are considered.

Unlike random error-correcting codes, in which the most useful codes are constructed by algebraic techniques, the best burst-error-correcting codes are found sometimes by computer-aided search procedures. However, the burst-error-correcting capability of a linear code can be determined by its parameters. Generally, a burst error of length b is defined by the nonzero components of an error-pattern sequence which are in consecutive, symbol/bit, positions, the first and last of which are nonzero.

Example 3.22
An error pattern $\bar{e} = (0,0,\alpha,0,1,0,\alpha^2,0,0,0,0,0,0,0,0)$ of length 15 over GF(4) can be considered to be a burst error of length 5 in symbols.

Now suppose that no burst of length b or less is a codeword of some (n,k) linear code. Then the number of n-vectors, that are zero everywhere except in their first b symbols must be q^b. This follows that any two such n-vectors can not be in the same coset of a standard array. Otherwise, their vector difference, which is a burst of length b or less, would be a codeword. Thus these q^b n-vectors must be in q^b distinct cosets. However, since there are a total of q^{n-k} cosets for any $(n\ k)$ linear code, one must have that

$$q^{n-k} \geq q^b \text{ or } n-k \geq b.$$

Thus the number of parity-check bits $n - k$ must be at least equal to the burst length b of a codeword.

Next assume that a linear code is corrupted by a burst $\bar{u}$ vector of length $2b$ or less where $\bar{u}$ is a codeword. Then this code vector $\bar{u}$ can be expressed as a vector difference of two bursts $\bar{v}$ and $\bar{w}$ of length b or less, so that $\bar{v}$ and $\bar{w}$ are in the same coset of the standard array of this code. If $\bar{v}$ is chosen to be a coset leader, then $\bar{w}$ is, in effect, an uncorrectable error pattern. As a consequence such a code may not be able to correct all burst errors of length b or less. This contradicts the above assumption that the code has a burst of length $2b$ or less as a codeword. Therefore, to correct all bursts of length b or less, no burst of length $2b$ or less can be a code vector.

Reiger[11] showed in 1960 that the number of parity-check symbols of an error-correcting code for a single burst of length b must be at least $2b$. Also, it is shown in [11] that to correct all errors of length b or less and to simultaneously detect all bursts of length l or less, where $l \geq b$, the number of parity-check symbols of this code must be at least $b+l$. These facts are, sometimes, called the Reiger bound for burst-error correction. The Reiger bound is established by the following argument : Suppose that an (n, k) linear code corrects all single bursts of length b or less. Then by the above argument there is no code word that can have a burst of length $2b$ or less. If two such n-vectors were in the same coset, their difference would have to be a code word. Let us choose any two vectors such that each has a burst of length $2b$. If they were in the same coset of a standard array, then their difference would be a code word. However, since it has a burst of length $2b$, no such code word exists. Therefore, two such vectors must be in different cosets, and the number of cosets is at least as large as the number of such vectors. Thus, in order to correct all single burst errors of length b or less, an (n,k) linear code must have at least $2b$ parity-check digits, that is,

$$n-k \geq 2b \text{ or } b \leq (n-k)/2 \qquad (3.29)$$

Eq. (3.29) represents the Reiger upper bound on the capability of an (n, k) code to correct a single burst of length b, which is at most $\lfloor (n-k)/2 \rfloor$ error correcting.

Next in a similar manner every burst of length $b+l$ or less, where $l \geq b$, can be written as the vector sum of a burst of length b or less and a burst of length l or less, where $l \geq b$. If a linear code is to simultaneously correct bursts of length b or less and to detect all bursts of length l or less, then the burst of length b and the burst of length l must be in different cosets, and their sum must not be a code word. Since the number of cosets is at least as large as the number of burst-error patterns of length $b+l$ or less, it follows that

$$q^{n-k} \geq q^{b+l} \text{ or } n-k \geq b+l. \qquad (3.30)$$

Thus, the number of parity-check digits of this code must be at least $b+l$. Eq. (3.30) is the Reiger bound for a code to be able to correct all single bursts of length b or less and simultaneously detect all bursts of length $l \geq b$ or less.

3.8 Product Codes

In many applications the types of channel errors can be both random and burst errors. To illustrate, a digital storage system, e.g. a laser disk store, can have random errors due to thermal energy as well as burst errors due to the surface of the disk being contaminated with fingerprints and scratches. In this subsection, a new code structure is developed which is capable of correcting both random and burst errors.

One way to decode both random and burst errors is to put code words into multidimensional arrays or codes. What is called, a product code is the simplest example of such a multidimensional code. The idea of the product code was introduced first by Elias in 1954 [12].

Product codes are described here in a two-dimensional context; the extensions to higher levels of coding are obvious. Given a $k = k_1 \cdot k_2$ symbol message for transmission, one can imagine storing such a sequence in an array with k_2 rows and k_1 columns. Suppose that each row is encoded in an (n_1, k_1) block code over $GF(q)$, by employing any of the encoding techniques already presented. Let the minimum Hamming distance of this row code be d_1. In this encoding process, an array of k_1 rows and n_1 columns is created. Next to each of these columns an (n_2, k_2) block code is applied over the same finite field. This (n_2, k_2) *code* has the minimum Hamming distance d_2. This process generates a two-dimensional $(n_1 n_2, k_1 k_2)$ code with parity constraints on both the rows and columns. Such a code is called the product code that is constructed from this two component codes.

A systematic product code is shown in Fig. 3.11. In a rectangular array of n_1 columns and n_2 rows, the rows are the code words of the (n_1, k_1) code, and the columns are the code words of the (n_2, k_2) code. The top left subarray contains the $k_1 k_2$ information symbols or bits. The symbols or bits in the top right subarray are computed in accordance with the parity-check rules for the (n_1, k_1) code, and the symbols or bits in the bottom left subarray are computed in accordance with the parity-check rules for the (n_2, k_2) code. The check symbols or bits in the bottom right subarray yield $(n_1 - k_1) \times (n_2 - k_2)$ checks on the check symbols or bits.

It is clear that the product code can be viewed as an $(n_1 n_2, k_1 k_2)$ block code, whose aggregate rate is $R = \dfrac{k_1 k_2}{n_1 n_2} = R_1 R_2$. However, the encoding and decoding process retains the algorithms and complexity of the component codes.

Specifically, decoding takes place in the obvious manner. First, each row is decoded with the corresponding information positions prepared as indicated. Of course, an incorrect row decoding is possible, in which case residual errors are still remain. Next, the columns are decoded separately. This two-step decoding process has a decoding effort, measured by the operation count, which is given approximately by the sum of the individual, row-by-row and column-by-column, decoder complexities.

Linear Block Codes

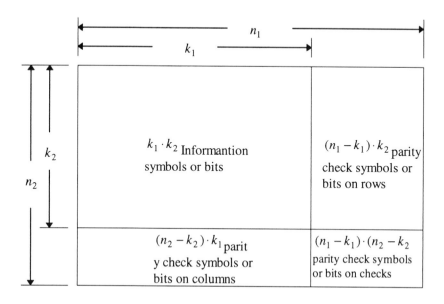

Figure 3.11 A systematic product code

It is demonstrated in Section 3.2.2 that the random error-correction capability of a block code is determined by the minimum distance of the code. Let d_1 and d_2 be the minimum distances of the two codes (n_1, k_1) and (n_2, k_2), respectively. Then a code word of the product code must have at least d_1 nonzero elements in each row and at least d_2 nonzero elements in each nonzero column. Thus, the minimum weight of the product code is $d_1 d_2$. Therefore, at least one code word exists such that the minimum distance is $d = d_1 d_2$. This implies at least for standard codes that the maximum random-error-correction capability of a product code is given by

$$t = \lfloor (d_1 d_2 - 1)/2 \rfloor = 2t_1 t_2 + t_1 + t_2 \qquad (3.31)$$

errors, where $t_1 = \lfloor (d_1 - 1)/2 \rfloor$ and $t_2 = \lfloor (d_2 - 1)/2 \rfloor$.

Unfortunately, this two-step decoding procedure, row decoding followed by column decoding, is, in general, incapable of correcting all error patterns with t or fewer errors in the code array. Nonetheless, such a decoding process is rather efficient and simple, and as it is seen next, many error patterns of more than t errors are correctable by this two-step decoding procedure.

The smallest number of errors which prohibit errors-only decoding in the row/column decoding is now determined. Let t_1 and t_2 be the random error-correcting capability of the row and column codes, respectively. For it to be possible for an array to fail, one must have a certain number of row failures. Row failures can happen when at least $t_1 + 1$ errors occur in any row. Also for the column decoding to fail, it must be true that $t_2 + 1$ row failures have occurred and, even then, these errors must occur in rather special positions. Thus, any error pattern with

$$\tau = (t_1 + 1)(t_2 + 1) - 1 \tag{3.32}$$

errors or less is correctable. Of course, this number τ in Eq. (3.32) is smaller than t in Eq. (3.31) (approximately by one half for large-distance codes).

Example 3.23
Consider a product code generated by two binary codes. Suppose that the row code is a (15, 11) binary code with $d_1 = 3$, and the column code is a (63,51) binary code with $d_2 = 5$. The overall rate is $R = 561/945 \approx 0.6$, and the codeword length is $n = 15 \times 63 = 945$. The minimum distance of the product code is then 15, but by Eq. (3.32) a row/column errors-only decoder can correct $2 \cdot 3 - 1 = 5$ errors in the array, rather than the 7 errors guaranteed by the maximum random error-correction capability in Eq. (3.31). An example of an uncorrectable pattern of 6 errors for a row/column errors-only decoder is illustrated in Figure 3.12(a). Also an 11-error pattern which is correctable is shown in Figure 3.12(b). In Figure 3.12, the transmitted codeword is assumed to be the all-zero codeword. Evidently product codes are powerful because of their high likelihood of correcting errors beyond the guaranteed limit of simple decoders.

As a comparison one might consider for purposes of design other non-product codes of approximately the same block length and rate. A shortened BCH code with $n = 1023$ and $k = 618$ and approximately the same rate $R = 0.6$ can always correct 44 errors, which is far better than the above product code. However, the decoding effort for this non-product code is certainly much larger than that of either of the given component codes. However, the total decoding effort must reflect the need for multiple decodings of both the rows and columns of the product code.

Linear Block Codes

$$
\begin{matrix}
0 & 0 & 0 & \cdots\cdots & 0 & 0 & 0 \\
0 & 1 & 0 & \cdots\cdots & 1 & 0 & 0 \\
\vdots & & & & \vdots & & \\
\vdots & & & & \vdots & & \\
0 & 0 & 1 & \cdots\cdots & 0 & 1 & 0 \\
1 & 0 & 0 & \cdots\cdots & 0 & 0 & 1 \\
0 & 0 & 0 & \cdots\cdots & 0 & 0 & 0 \\
0 & 0 & 0 & \cdots\cdots & 0 & 0 & 0 \\
\end{matrix}
$$

(a) Uncorrectable error pattern $v = 6$.

$$
\begin{matrix}
0 & 0 & 0 & 1 & \cdots\cdots & 0 & 0 & 0 & 0 \\
0 & 1 & 0 & 0 & \cdots\cdots & 0 & 0 & 0 & 0 \\
0 & 0 & 0 & 0 & \cdots\cdots & 1 & 0 & 0 & 0 \\
\vdots & & & & & \vdots & & & \\
\vdots & & & & & \vdots & & & \\
0 & 0 & 0 & 0 & \cdots\cdots & 1 & 0 & 0 & 0 \\
0 & 0 & 0 & 0 & \cdots\cdots & 0 & 1 & 0 & 0 \\
0 & 0 & 0 & 1 & \cdots\cdots & 0 & 0 & 0 & 0 \\
0 & 0 & 0 & 0 & \cdots\cdots & 0 & 0 & 0 & 1 \\
0 & 0 & 0 & 0 & \cdots\cdots & 0 & 1 & 0 & 0 \\
0 & 0 & 0 & 0 & \cdots\cdots & 0 & 0 & 0 & 1 \\
0 & 0 & 0 & 0 & \cdots\cdots & 0 & 0 & 1 & 0 \\
1 & 0 & 0 & 0 & \cdots\cdots & 0 & 0 & 0 & 0 \\
\end{matrix}
$$

(b) Correctable error pattern, $v = 11$.

Figure 3.12. Correctable and uncorrectable error patterns.

A particularly simple application of the product code concept involves single-symbol parity encoding in each dimension, giving each code a minimum distance of 2, and the complete code a minimum distance of 4. Neither row nor column code is capable by itself of correcting any errors, but their combination admits a very simple scheme for correcting one error or detecting three errors. Specifically, any single error is located at the intersection of the row and column wherein the parity checks fail. At the same time any double-error pattern is detectable for this system. In this case, one can attain the maximum random error-correcting capability t with a simple decoder. Consequently, it is concluded that the theoretical capability t of a product code is not particularly useful without a practical decoding algorithm. Related to this are certain decoding algorithms which are applicable to an errors-only decoder for one component of the code and an errors-erasures decoder for the other component of the code.

Blahut [13] describes an intricate procedure whereby the first decoder declares erasures for the entire codeword when decoding fails and, otherwise, passes the suspected error counts to the second decoder. By judiciously processing this side information, the second decoder is able to ensure correction of all error patterns of up to t errors, where t in Eq. (3.31) is the maximum random error-correction capability. A simpler alternative, which under certain approximations achieves the same result, is to perform correction only up to $r_1 < t_1$ errors on rows, and then declaring row erasures when row decoding fails. (Usually, when the error limit r_1 is exceeded, the decoding does not yield an incorrect codeword, but simply a decoding failure, so that one can assume that residual errors, apart from erasures, are negligible). The column decoder can accommodate up to $d_2 - 1$ erasures in a column, assuming that no residual errors are left after row decoding. Thus, error patterns of up to $(d_2)(r_1 + 1) - 1$ errors are guaranteed to be repaired. It should be clear that the reversal of the order of the decoding would guarantee approximately a correction of $(d_1)(r_2 + 1) - 1$ errors, and it is wise to choose the best order.

Example 3.24
Again consider the product code constructed in Example 3.23. Since the Hamming code is a perfect code with $t_1 = 1$, a row decoding failure normally does not occur so that one can choose $r_1 = 0$. That is, the row code is used only for error detection. At least three errors in a row are required to cause incorrect decoding. but single or double errors cause the row to be erased. Column decoding with erasure filling guarantees correction of $d_2(r_1 + 1) - 1 = 2$ errors (3 errors in any single column would cause a product code failure).

Another option is to perform column decoding first, adopting $r_2 = 1$ as the error-correcting radius, instead of the guaranteed limit of two-error correction. As a consequence four errors in a column are required to cause an incorrect decoding. In such a case it may be assumed that whenever two or more errors in a column occur, a column erasure is declared. The row decoder is capable of handling $d_1 - 1$ erasures. Consequently, the guaranteed correction capability with this option is $d_1(r_2 + 1) - 1 = 5$. In this special case the result is no stronger than the simple row/column decoding discussed

Linear Block Codes

previously. However, for more powerful component codes, this method begins to excel and approach the possibilities given by t in Eq. (3.31). Several other techniques that decode a product code up to its maximum random error-correction capability are developed elsewhere[14][15].

Another advantage of product codes is that they have a natural burst-error handling capability. To see how this is possible suppose one forms the complete array shown as in Figure 3.10 and transmits by columns. Decoding proceeds first on the rows after the full array is received and then by columns, as described previously. Notice that any single-error burst, confined to $n_2 t_1$ transmission slots, does not place more than t_1 errors in any row of the received array and is thus correctable. Thus, $n_2 t_1$ is the guaranteed burst-correcting capability. Obviously, an interchange of the order can achieve a burst capability of $n_1 t_2$, and if burst correction is of primary importance, the better choice should be adopted. On this topic the reader is directed to [16] for a related discussion on burst-error correction with "array" codes.

Next consider more generally the burst-error-correction capability of a product code. In Sec. 3.7 a burst length b is defined as a bit string (its first and last bits are nonzero) whose nonzero components are confined to b consecutive bits or symbol positions. Let b_1 and b_2 be the, respectively, burst-error-correcting capabilities of the component codes C_1 and C_2. Suppose that a codeword array of the product code $C_1 \times C_2$ is transmitted row by row. When its received sequence with any burst error of length b_2 or less is re-arranged back into an array in a row-by-row manner, it is clear that $n_1 b_2$ positions affect no more than $b_2 + 1$ consecutive rows. Since each column is affected by a burst of length b_2 or less, the burst is corrected by a column-by-column decoding. Hence, the burst-error-correcting capability of the product code is at least $n_1 b_2$.

Similarly, consider that a word array of the product code is transmitted column-by-column and decoded row-by-row. In this case it can be shown that any error burst of length $n_2 b_1$ or less can be corrected. Thus, the burst-error-correcting capability b of the product code can be expressed by
$$b \geq \max\{n_1 b_2, n_2 b_1\}$$
Thus far, product codes are considered only for either random-error correction or burst-error correction, independently. However, product codes can be used simultaneously for both random and burst-error correction.

In order to simultaneously correct both random and burst errors, a sufficient condition is found in [16] by Burton and Weldon as follows : If the cosets that contain the random-error patterns of weight $t = \lfloor (d_1 d_2 - 1)/2 \rfloor = 2t_1 t_2 + t_1 + t_2$ or less are disjoint from the cosets that contain bursts of length $b = \max\{n_1 t_2, n_2 t_1\}$ or less, the random and burst errors can be corrected simultaneously.

To illustrate this condition, assume first that $n_1 t_2 > n_2 t_1$. Then the maximum burst length is $b = n_1 t_2$. Consider a burst of length $n_1 t_2$ or less. As before, assume that the word array of the product code is arranged on a row-by-row basis. Then each column contains at most t_2 errors in a burst of length $n_1 t_2$ or less. If this burst-error pattern and some random-error patterns of weight not greater than t are in the same coset, then the sum of these two error patterns is a code array(codeword) of the product code. Each column of such a code array has either weight zero or a weight of at least d_2. Since the sum of random and burst-error digits in each column is no greater than d_2, each non-zero column must be composed of at least $d_2 - t_2$ errors from the random-error pattern and at most t_2 errors from the burst-error pattern. Since there are at most t random errors, these errors can be distributed among at most $\lfloor t/(d_2 - t_2) \rfloor$ columns. Hence, the sum array contains at most $\lfloor t/(d_2 - t_2) \rfloor \cdot t_2 + t$ nonzero digits. However,

$$\left\lfloor \frac{t}{(d_2 - t_2)} \right\rfloor t_2 + t \leq t\left(\frac{t_2}{d_2 - t_2} + 1\right) \leq 2t\left(\frac{d_2}{d_2 + 1}\right) < 2t < d.$$

Hence, the sum array contains fewer than $2t < d_1 d_2$ nonzero digits and cannot be a code array in the product code. This contradiction implies that two-error patterns cannot be in the same coset of the standard array of the product code. If $n_2 t_1 > n_1 t_2$, then $b = n_2 t_1$. Thus, the same argument can be applied to rows instead of columns.

Example 3.25
Consider a (36,9,9) binary product code which is constructed from the direct product of a (6, 3, 3) binary code with itself. Such a product code has $t = 4$ and $b = 6$. In order to show that this (36, 9, 9) product code is capable of correcting any combination of $t = 4$ or fewer random errors and simultaneously correcting any error burst of length $b = 6$ or less, it is sufficient to show that an error burst of length $b = 6$ or less and a random error pattern of $t=4$ or fewer errors cannot be in the same coset. For the nine-

bit message, (110,101,101) the encoded arrays of the (36, 9, 9) product code has the form,

$$\begin{array}{c} \text{the } C_1 = (6, 3, 3) \text{ code} \\ \begin{array}{cccccc} 1 & 1 & 0 & 1 & 1 & 0 \\ 1 & 0 & 1 & 0 & 1 & 0 \\ 1 & 0 & 1 & 1 & 0 & 1 \\ 0 & 0 & 0 & 1 & 1 & 1 \\ 0 & 1 & 1 & 0 & 1 & 1 \\ 0 & 1 & 1 & 1 & 0 & 0 \end{array} \quad \text{the } C_2 = (6, 3, 3) \text{ code} \end{array}$$

If this code array of the (36, 9, 9) code is transmitted row by row(which begins from the bottom row), then one sends the sequence,
    011100, 011011, 000111, 101101, 101010, 110110.
Transmission of this code array over a noisy channel can yield a received word with both random and burst errors. Assume that the received word contains four random-bit errors in the marked bit-locations,
    111100, 011011, 000111, 101111, 101110, 110100.
Also assume that the received word containing a burst of length $b = 6$, underlined as follows.
    011100, 011 110, 101 111, 101101, 101010, 110110.
Hence, the random-error pattern of weight $t = 4$ is given by
    100000, 000000, 000000, 000010, 000100, 000010,
and the burst error pattern is
    000000, 000101, 101000, 000000, 000000, 000000.
If these two error patterns are in the same coset, then their sum
    100000, 000101, 101000, 000010, 000100, 000010.
must be a code array in the product code as shown below:

$$\begin{array}{cccccc} 0 & 0 & 0 & 0 & 1 & 0 \\ 0 & 0 & 0 & 1 & 0 & 0 \\ 0 & 0 & 0 & 0 & 1 & 0 \\ 1 & 0 & 1 & 0 & 0 & 0 \\ 0 & 0 & 0 & 1 & 0 & 1 \\ 1 & 0 & 0 & 0 & 0 & 0 \end{array}$$

For this array of errors to be a product-code-word each column of this array should have weight $d_2 = 0$ or weight $d_2 \geq 3$. However, the weights of all columns are less than 3. Hence, this sum array of error cannot be a code array of the product code (36,9,9). There are at most $t = 4$ random errors.

These errors are distributed among two columns. Hence, one concludes that this sum array of errors contains at most $(2) \cdot (1) + 4 = 6$ nonzero bits. Thus, this sum array is not a code array in this product code since the minimum distance of this product code is 9. It can be concluded from this analysis that the received word which contains the random-error pattern of weight $t = 4$ is disjoint from the received word that contains the burst of length $b = 6$.

Problems

3.1. A (4,2) block code is defined by the following encoding table:

(m_0, m_1)	(c_0, c_1, c_2, c_3)
0 0	0 0 0 0
1 0	1 0 0 1
0 1	0 1 1 1
1 1	1 1 1 0

a) Can this code correct a one-bit error?
   If yes, please show how.
   If no, please tell why.
b) Design a decoding rule that can be used to decode a certain bit of error.

3.2. The (7,4) code is given by the following table

Messages	Codewords
0 0 0 0	0 0 0 0 0 0 0
1 0 0 0	1 1 0 1 0 0 0
0 1 0 0	0 1 1 0 1 0 0
1 1 0 0	1 0 1 1 1 0 0
0 0 1 0	0 0 1 1 0 1 0
1 0 1 0	1 1 1 0 0 1 0
0 1 1 0	0 1 0 1 1 1 0
1 1 1 0	1 0 0 0 1 1 0
0 0 0 1	0 0 0 1 1 0 1
1 0 0 1	1 1 0 0 1 0 1
0 1 0 1	0 1 1 1 0 0 1
1 1 0 1	1 0 1 0 0 0 1
0 0 1 1	0 0 1 0 1 1 1
1 0 1 1	1 1 1 1 1 1 1
0 1 1 1	0 1 0 0 0 1 1
1 1 1 1	1 0 0 1 0 1 1

a) Can this code correct a one-bit error?
   If yes, give the decoding rule.
   If not, give the reason.
b) What happens when two errors occur in one codeword?

3.3. Design a maximum-likelihood decoding table for the binary code that consists of the four codewords (0000), (0011), (1100), and (1111).
a) Assume a BSC.
b) Assume a binary-erasure channel.

3.4. The capacity of a BSC is given by
$$C = 1 + p \log p + q \log q,$$
where $q = 1 - p$.

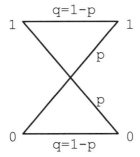

If $p = 10^{-2}$ and $k = 223$, what is the smallest code length n for reliable information transmission?

3.5. The (6,3) code is given by the following table:

Information words	Codewords
0 0 0	0 0 0 0 0 0
1 0 0	1 0 0 0 1 1
0 1 0	0 1 0 1 0 1
1 1 0	1 1 0 1 1 0
0 0 1	0 0 1 1 1 0
1 0 1	1 0 1 1 0 1
0 1 1	0 1 1 0 1 1
1 1 1	1 1 1 0 0 0

The decoding (standard array) table is given below.

000000	100011	010101	110110	001110	101101	011011	111000
000001	100010	010100	110111	001111	101100	011010	111001
000010	100001	010111	110100	001100	101111	011001	111010
000100	100111	010001	110010	001010	101001	011111	111100
001000	101011	011101	111110	000110	100101	010011	110000
010000	110011	000101	100110	011110	111101	001011	101000
100000	000011	110101	010110	101110	001101	111011	011000
001001	101010	011100	111111	000111	100100	010010	110001

a) What is the probability of correctly decoding a received word?
b) What is the probability of an incorrect decoding of an information bit?
c) Show that this code is equivalent to the code given in Example 3.8.

3.6. Consider the systematic (8,4) binary linear code whose parity-check equations are

$$c_4 = c_0 + c_2 + c_3,$$
$$c_5 = c_0 + c_1 + c_3,$$
$$c_6 = c_0 + c_1 + c_2,$$
$$c_7 = c_1 + c_2 + c_3.$$

a) Find the generator matrix $G = (I_4 P)$ and its corresponding parity-check matrix H of the code.
b) Determine the minimum distance of the code.
c) Construct an encoder for this code.

3.7. Consider the binary linear block-code whose generator matrix is

$$G = \begin{pmatrix} 1 & 0 & 1 & 0 & 1 & 1 \\ 0 & 1 & 1 & 1 & 1 & 0 \\ 0 & 0 & 0 & 1 & 1 & 1 \end{pmatrix},$$

a) Find the generator matrix G_s in systematic form for an equivalent code.
b) Find the parity-check matrix H_s for the code in (a).
c) Find the codeword that has (110) as its information word. Show that it is in the row space of G_s and in the null space of H_s.

3.8 Construct a linear block code consisting of eight codewords of length 7 such that any two distinct codewords have a distance of at least 4. Find the generator matrix for the code.

Linear Block Codes

3.9. The design engineer, James, has an encoding device for a binary linear block code. This device has 4 bits of input-data pins and 8-bits of output-data pins. He wants to analyze its code C. After sending the message (information) words

$$\bar{m}_1 = (0001), \bar{m}_2 = (0010), \bar{m}_3 = (0100), \bar{m}_4 = (1000)$$

to the input-data pins, he captured the following corresponding vectors from the output-data pins:

$\bar{c}_1 = (11111111)$,
$\bar{c}_2 = (00011110)$,
$\bar{c}_3 = (01100110)$,
$\bar{c}_4 = (10101010)$.

Help James

(1) Determine a generator matrix G of C from the input and output vectors given above;

(2) Convert this G into the systematic form $G_s = (I_4 P)$ and find the parity-check matrix H_s of the systematic code;

(3) Find the minimum weight of code C.

(4) Construct an encoding device for the dual code $C^\perp$ of C by the use of the encoder for C.

(5) Design a (complete) syndrome-decoding algorithm for code C and decode the following received words:

(a) $\bar{r} = (11100011)$;
(b) $\bar{r} = (11111001)$;

(6) Find the probability that a received word is decoded correctly by the use of the decoding algorithm, designed in part (5) for a BSC with error probability p.

(7) Design a t - error correcting and e - error detecting algorithm for code C for $(t \leq e)$ with $d_{min} = t + e + 1$ and $t > 0, e > 0$. Also decode the received word $\bar{r} = (11001111)$;

(8) Check if the code C has the largest possible minimum distance d_{min} for its codeword length n and dimension k.

3.10. James wants to use code C, analyzed in problem 3.9, to transmit messages through a very noisy channel. First, he measures the channel characteristics and finds that the channel can be approximately modeled as a BSC with $p = 0.15$. James knows for the transmission to be efficient that the

information rate R of code C must be strictly less than the channel capacity C. Help James to
  (1) Decide if code C can be used for an efficient transmission through this channel.
James thought of a method to reduce the information rate R by repeating the last bit m times in each codeword of C, e.g. $\bar{c} = (1110000\,1)$ $\rightarrow \bar{c}' = (1110000\,111)$. He wants to
  (2) Find the smallest m for each codeword of C such that the new generated code C' can perform efficient transmission. Also what is the generator matrix of the code C' ?
  (3) Determine the maximum error-correcting capability of C'.

To further improve the error-control capability, help James
  (4) Construct another binary linear block-code C'' with the information rate $R'' = \frac{1}{2}R$ and the minimum distance $d''_{min} = 2d_{min}$ by using the encoding device of which James has only one.

3.11. James has made a study of both codes and channels. He realizes that he should use a MDS code to design a coding system for the transmission of messages. He ordered an encoder and a decoder for the design. However, because of a shipping error, James only received two identical programmable decoders for a linear systematic MDS $(n, k, n-k+1)$ code and no encoder. These decoders can be set to any possible configuration of $t-$ error-correcting and $e-$ erasure-correcting for the given n and k. Help James to construct an encoder of the code by using one of the decoders.

3.12 Consider the (5,3) 4-ary Hamming code defined over GF(4) by the parity-check matrix
$$H = \begin{bmatrix} 1 & 0 & 1 & 1 & 1 \\ 0 & 1 & 1 & 2 & 3 \end{bmatrix}.$$
The operation tables for this field are given as follows :

+	0	1	2	3
0	0	1	2	3
1	1	0	3	2
2	2	3	0	1
3	3	2	1	0

Linear Block Codes

*	0	1	2	3
0	0	0	0	0
1	0	1	2	3
2	0	2	3	1
3	0	3	1	2

(a) Determine the generator matrix of this code.
(b) Show that this code is a perfect code.
(c) Define a (6,3) 4-ary extended Hamming code.

3.12 For $RM(1,m)$, define the transform $T(\bar{x}) = \left((-1)^{x_0}, (-1)^{x_1}, \ldots, (-1)^{x_{2^m-1}} \right)$.
Show that $T(u_1\bar{v}_1 + u_2\bar{v}_2 + \cdots + u_4\bar{v}_4)$ is the $\bar{u}$–th row of H_{2^m}, where $\bar{u}$ has the binary representation of $\bar{u} = (u_m, u_{m-1}, \ldots, u_1)$. Also show $x_{\bar{u}} = 2^m - 2d_H(\bar{r}, \bar{c}_{\bar{u}})$, where $\bar{x} = \bar{y} \cdot H_{2^m}$, $\bar{y} = T(\bar{r})$, and $\bar{c}_{\bar{u}}$ equals $T^{-1}(\bar{u}-th\ row\ of\ H_{2^m})$.

3.13 Design engineer James uses the Hamming code (7,4,3) to transmit an information sequence. This code has the generator matrix

$$G = \begin{pmatrix} 1 & 1 & 1 & 1 & 1 & 1 & 1 \\ 0 & 0 & 0 & 1 & 1 & 1 & 1 \\ 0 & 1 & 1 & 0 & 0 & 1 & 1 \\ 1 & 0 & 1 & 0 & 1 & 0 & 1 \end{pmatrix}.$$

James designed an encoding device which inserts one parity-check bit in the beginning of each codeword of the Hamming code, i.e. $\bar{c} = \bar{m} \cdot G$ and
$\bar{c}' = (c_0', \bar{c})$ where $\bar{c} = (c_0, c_1, \ldots, c_6)$, $c_0' = \sum_{i=0}^{6} c_i$ and
$\bar{m} = (m_0, m_1, m_2, m_3)$ is a message word. Each word $\bar{c}'$ is mapping to a signal set $\bar{s}' = ((-1)^{c_0'}, (-1)^{c_0}, (-1)^{c_1}, \ldots, (-1)^{c_6})$ and then transmitted over an unquantized discrete memoryless channel with additive white-Gaussian noise.
(a) Please help James to design a soft-decoding algorithm for this code to recover the message words $\bar{m}$.

(b) Decode the received word $\bar{r}' = (0,-2,-1,-1,1,-2,1,-1)$ to find $\bar{c}$ and $\bar{m}$ by using the algorithm you developed in (a).

Bibliography

For books and articles devoted to block codes :

[1] W. W. Peterson and E. J. Weldon, Error Correcting Codes, 2nd ed. Cambridge, MA : MIT Press, 1972.
[2] F. J. MacWilliams and N. J. A. Sloane, The Theory of Error Correcting Codes, Amsterdam North-Holland, 1977.
[3] F. J. MacWilliams, "A theorem on the distribution of weights in a systematic code," Bell System Tech. J., vol. 42, pp.79-94.
[4] S. B. Wicker, Error Control Systems for digital communication and storage, Prentice Hall, Englewood Cliffs, 1995.
[5] R. W. Hamming, "Error detecting and error correcting codes", Bell System Technical Journal, Vol. 29, pp. 147-160, 1950.
[6] D. E. Muller, "Application of Boolean algebra to switching circuit design", IEEE Trans. on Computers, Vol. 3, pp.6-12, Sept. 1954.
[7] I. S. Reed, "A class of multiple-error-correcting codes and a decoding scheme", IEEE Trans. on Inform. Theory, Vol. 4, pp.38-49, Sept. 1954.
[8] A. M. Kerdock, "A class of low-rate nonlinear codes," Info. and Control, No. 20, pp. 182-187, 1972.
[9] F. P. Preparata, "A class of optimum nonlinear double-error correcting codes," Info. and Control, No. 13, pp.378-400, 1968.
[10] J. L. Massey, Threshold decoding, MIT Press, Cambridge, MA, 1963.
[11] S. H. Reiger, "Codes for the correction of 'Clustered' errors," IEEE Trans., on Inform. Theory, IT-6, pp.16-21, 1960.
[12] P. Elias, "Error-free coding", IRE Trans. Info. Theory, Vol. 4, pp.29-37, 1954.
[13] R. Blahut, Theory and Practice of Error Control Codes, Reading, MA : Addison-Wesley, 1983.
[14] G. L. Feng and K. K. Tzeng, "A generalization of the Berlekamp-Massey algorithm for multisequence shift-register synthesis with applications to decoding cyclic codes,", IEEE Trans. on Info. Theory, vol. 37, pp.1274-1287, 1991.
[15] S. Sakata, "Decoding binary 2-D cyclic codes by the 2-D Berlekamp-Massey algorithm," IEEE Trans. on Info. Theory, vol.37, pp.1200-1203, 1991.

[16] R. M. Goodman, R. J. McEliece, and M. Sayano, "Phased burst error-correcting array codes," IEEE Trans. Info. Theory, vol. 39, pp.684-693, March 1993.

4 Linear Cyclic Codes

In general, the linear codes discussed in the last chapter require complicated logic circuitry and a large storage memory for both the encoding and decoding processes when the value of $n-k$ is large. To illustrate for standard-array decoding of an (n,k) binary linear code requires the storage of 2^{n-k} cosets and 2^k words each of n bits, e.g. for the (32,6) binary linear code the storage to $2^{32} \times 32 = 2^{34} \times 8 \approx 16 Gbytes$. To reduce this complexity, syndrome decoding was introduced. But one still has to store 2^{n-k} coset leaders and 2^{n-k} syndromes, a total of $2^{n-k} \times n + 2^{n-k} \times (n-k)$ bits. Again for the (32,6) binary linear code, the storage requirement is $2^{26} \times 32 + 2^{26} \times 26 \approx 500 Mbytes$. Obviously, such coding costs are too high to be acceptable for any application of this simple (32,6) binary linear code. Therefore, for a simpler encoding and decoding some additional properties of a code other than the linear structure need to be used. The linear cyclic codes, discussed in this chapter, can be implemented relatively easily and possess as well a great deal of well-understood mathematical structure.

Linear-cyclic codes form a subclass of linear block-codes. This subclass of linear codes was discovered by E. Prange[1] in 1957. Since then, cyclic codes have been the primary focus of interest for both theoretical coding researchers and practical-minded engineers.

Cyclic codes constitute an important practical error-control coding class for a variety of reasons. Within the class of linear cyclic codes there are certain subclasses of these codes, such as the Reed-Solomon codes and the Golay code, which have excellent error-control performances. The special algebraic and geometric structures of the cyclic codes ensure further simplifications in the encoder and decoder circuitry. A number of efficient encoding and decoding algorithms are derived for cyclic codes by the use of shift-register circuits. These algorithms make it possible to implement long block codes with a large number of codewords in practical communication systems. This is of great interest in many high-speed and broadband applications. Almost all block codes, employed in modern digital practice, are either linear cyclic codes or closely related to them.

The general theory of linear cyclic codes is discussed in this chapter. Also, the shift-register-based algorithms for the encoding and decoding of cyclic codes are investigated in detail. Some important cyclic codes, such as the cyclic redundant-check(CRC) codes, the Golay code, and their applications, are also presented in this chapter.

4.1 Description of Linear Cyclic Codes

4.1.1 Algebraic Structure of Cyclic Codes

An (n, k) linear code C over $GF(q)$ is called a cyclic code if any codeword $\bar{c} = (c_0, c_1, \ldots, c_{n-1}) \in C$, belonging to C also has its cyclic shift, $\bar{c}^{(1)} = (c_{n-1}, c_0, c_1, \ldots, c_{n-2}) \in C$, as a member. As a computer logic operation the codeword $\bar{c}^{(1)}$ is a right cyclic shift of the codeword $\bar{c}$. In discussing cyclic codes, it is often convenient to use a polynomial representation for the codewords, and also for the encoding and decoding operations since the shifting of a codeword is exactly equivalent to a modification of the exponents of a polynomial. Specifically, the codeword $\bar{c}$ is associated with a code polynomial $c(x)$ of degree $n - 1$ or less as follows :

$$c(x) = c_0 + c_1 x + \cdots + c_{n-1} x^{n-1}. \tag{4.1}$$

Evidently $\bar{c}$ as a vector and $c(x)$ as a polynomial are in one-to-one correspondence. Now consider the single right-cyclic shift $\bar{c}^{(1)}$ of $\bar{c}$. The associated, cyclically-shifted, polynomial is

$$c^{(1)}(x) = c_{n-1} + c_0 x + \cdots + c_{n-2} x^{n-1}, \tag{4.2}$$

which again is another polynomial of degree at most $n-1$. Note that the powers found in Eq. (4.1) have merely been increased by one, modulo n. Thus the two polynomials $\bar{c}(x)$ and $\bar{c}^{(1)}(x)$ in Eqs. (4.1) and (4.2) are related, respectively, by the formula,

$$c^{(1)}(x) = x \cdot c(x) \mod (x^n - 1). \tag{4.3}$$

This follows from the fact that

Linear Cyclic Codes

$$x \cdot c(x) = c_0 x + \cdots + c_{n-1} x^n$$
$$= c_{n-1} + c_0 x + \cdots + c_{n-1} x^n - c_{n-1}$$
$$= c_{n-1} + c_0 x + \cdots + c_{n-2} x^{n-1} + c_{n-1}(x^n - 1).$$

Hence, a division of $x \cdot c(x)$ by $x^n - 1$ yields $c^{(1)}(x)$ in Eq. (4.2) as the remainder of this division.

Similarly, if $c(x)$ represents a code word in an (n, k) cyclic code, then the i-th right cyclic shift $x^i c(x) \mod (x^n - 1)$ is also a code word of the cyclic code C which can be expressed by the congruence relation,

$$x^i c(x) = c^{(i)}(x) \mod (x^n - 1). \tag{4.4}$$

Hence the remainder polynomial $c^{(i)}(x)$ represents the n-tuple codeword obtained by cyclically shifting $\bar{c}$ right by i places.

Example 4.1
Consider a $(4,2)$ binary linear code given in Table 4.1

m_0	m_1	$c_0\ c_1\ c_2\ c_3$
0	0	0 0 0 0
1	0	1 0 1 0
0	1	0 1 0 1
1	1	1 1 1 1

Table 4.1 A (4,2) linear code

For the codeword $\bar{c} = (1010)$, one has that the code polynomial $c(x) = 1 + x^2$. The cyclic shift of $\bar{c}$ yields the codeword $\bar{c}^{(1)} = (0101)$ which corresponds to the code polynomial $\bar{c}^{(1)}(x) = x \cdot c(x) = x + x^3$. Thus, this linear code is an example of cyclic code.

Assume there is a non-zero monic polynomial $g(x) = g_0 + g_1 x + \cdots + g_{r-2} x^{r-2} + x^{r-1}$ of *minimum degree* in a cyclic code. Notice that the weight $w(g(x))$, the number of non-zero coefficients of $g(x)$, satisfies $w(g(x)) \geq w_{\min}$ where $w_{\min}$ is the minimum weight of the code. It is shown next that $g(x)$ is unique in a cyclic code for which $g_0 \neq 0$. If $g(x)$ were not unique, then there would exist another such code polynomial of the same degree of the form, $g'(x) = g'_0 + g'_1 x + \cdots + g'_{r-2} x^{r-2} + x^{r-1}$. Thus, one obtains the code polynomial

$g''(x) = g'(x) - g(x) = (g'_0 - g_0) + (g'_1 - g_1)x + \cdots + (g'_{r-2} - g_{r-2})x^{r-2}$. Polynomial $g''(x)$ corresponds to the codeword $(g'_0 - g_0, g'_1 - g_1, \ldots, g'_{r-2} - g_{r-2}, 0, \ldots, 0)$. Since $\deg(g''(x)) < r$, this contradicts the minimality of $g(x)$ so that $g''(x) = 0$. That is, $g'(x) = g(x)$. Next, if $g_0 = 0$, then $g(x) = x(g_1 + g_2 x + \cdots + g_{r-2} x^{r-2} + x^{r-1})$. The n-th right-cyclic-shift of $g(x)$ yields another code polynomial $g'''(x) = g_1 + g_2 x + \cdots + g_{r-1} x^{r-2} + x^{r-1}$. $= x^{n-1} g(x) \mod (x^n - 1)$. However, $\deg(g'''(x)) < r$ again contradicts the fact that $g(x)$ is the non-zero code polynomial of minimum degree. Thus if $g(x) = g_0 + g_1 x + \cdots + g_{r-2} x^{r-2} + x^{r-1}$ with $g_0 \neq 0$ is a minimum weight polynomial in some cyclic code C, then this polynomial is unique.

Another important fact is that a polynomial $c(x)$ of degree $\deg(c(x)) \leq n-1$ is a code polynomial if and only if $c(x) = m(x)g(x)$, where $g(x)$ is the unique polynomial of minimum degree given above, and $m(x)$ is some polynomial of $\deg(m(x)) < n-r$. This fact is proved as follows : If $\deg(c(x)) \leq n-1$ and $c(x) = m(x)g(x)$, then $c(x) = (a_0 + a_1 x + \cdots + a_{n-r-1} x^{n-r-1}) g(x) = a_0 g(x) + a_1 x g(x) + \cdots + a_{n-r-1} x^{n-r-1} g(x)$. This implies that $c(x)$ is a linear combination of $g(x), xg(x), \ldots, x^{n-r-1} g(x)$. Therefore, $c(x)$ is a code polynomial. On the other hand if $c(x)$ is a code polynomial, then by the Euclidean division algorithm $c(x)$ can be represented by $c(x)=m(x)g(x)+q(x)$, where the degree of the remainder polynomial satisfies $\deg(q(x))<\deg(g(x))$. $q(x)$ is a code polynomial since both $c(x)$ and $m(x)g(x)$ are code polynomials. Hence, $\deg(q(x))=0$ since $\deg(q(x))<\deg(g(x))$, a contradiction since $g(x)$ is polynomial of minimum weight in the code. This yields finally $q(x)=0$ so that $c(x)=m(x)g(x)$.

Example 4.2
For the (4,2) cyclic code, discussed in Example 4.1, the code polynomials are
$$c_1(x) = 0,$$
$$c_2(x) = 1 + x^2 = g(x),$$
$$c_3(x) = x + x^3 = xg(x),$$
$$c_4(x) = 1 + x + x^2 + x^3 = (1+x)g(x).$$

The dimension of a cyclic code can be obtained by counting the number of polynomials $c(x)$ of $\deg(c(x)) \leq n-1$ such that $c(x)=m(x)g(x)$. Note that

$deg(m(x)) \leq n-r$ since $deg(g(x))=r$. Thus there are total of q^{n-r} possible polynomials for $m(x)$. This implies that there are total of q^{n-r} code polynomials. Therefore, the dimension of the cyclic code is $k=n-r$.

From the above discussion, it is clear that an (n,k) cyclic code is completely specified by $g(x)$. This suggests that the polynomial $g(x)$ be called the *generator polynomial* of the cyclic code.

The next question is how to select the parameters of a cyclic code with the codeword length n and dimension k. It can be shown that the generator polynomial $g(x)$ of a (n,k) cyclic code is always a polynomial factor of the polynomial $x^n - 1$. This means that any factor of $x^n - 1$ of degree $n-k$ is a possible generator of an (n, k) cyclic code. Since $g(x)$ is a divisor of $x^n - 1$, it follows that

$$x^n - 1 = h(x)g(x), \qquad (4.5)$$

where $h(x)$ is a polynomial of degree k of the form,

$$h(x) = h_0 + h_1 x + \cdots + h_k x^k, \qquad (4.6)$$

where $h_0 = h_k = 1$. $h(x)$ is called the parity polynomial of an (n, k) cyclic code. Since $x^n - 1 = h(x)g(x)$, one has the important relation,

$$h(x)g(x) = 0 \quad \mod(x^n - 1) \qquad (4.7)$$

which implies that the polynomials $g(x)$ and $h(x)$ are orthogonal, modulo the polynomial, $x^n - 1$.

Now, define the message polynomial $m(x)$ to be a polynomial of degree $k - 1$ of the form,

$$m(x) = m_0 + m_1 x + \cdots + m_{k-1} x^{k-1}, \qquad (4.8)$$

with the coefficient set (vector) $\bar{m} = (m_0, m_1, \ldots, m_{k-1})$ representing the k-symbol message sequence. Clearly, the product $m(x)g(x)$ is the polynomial that represents the codeword polynomial of degree $n-1$ or less. Since each code polynomial $c(x)$ is a multiple of $g(x)$, the encoding can be achieved by multiplying the message polynomial $m(x)$ by $g(x)$. Thus an (n, k) cyclic code is completely specified by the monic generator polynomial,

$$g(x) = 1 + g_1 x + \cdots + g_{n-k-1} x^{n-k-1} + x^{n-k}. \qquad (4.9)$$

From Eq. (4.4) it is seen easily that

$$c^{(i)}(x) = x^i m(x) g(x) \quad \mod (x^n - 1) = m_i(x) g(x),$$

This identity proves that if any code word is cyclically shifted i times, that another codeword in the cyclic code C is formed. Thus, any set of q^k codewords which possesses the cyclic property is generated by some generator polynomial $g(x)$.

Example 4.3
Consider the (7, 4) cyclic code over GF(2) whose generator polynomial of degree 3 is a factor of $x^7 - 1$. Since the polynomial $x^7 + 1$ can be factored as $x^7 + 1 = (1+x)(1+x+x^3)(1+x^2+x^3)$, two possible (7,4) cyclic codes can be constructed by the use of the generator polynomials, $g_1(x) = 1+x+x^3$ and $g_2(x) = 1+x^2+x^3$.

Message word $\bar{m}$	$m(x)$	$c(x)=m(x)g(x)$	Codeword $\bar{c}$
0000	0	0	0000000
0001	x^3	$x^3+x^4+x^6$	0001101
0010	x^2	$x^2+x^3+x^5$	0011010
0011	x^2+x^3	$x^2+x^4+x^5+x^6$	0010111
0100	x	$x+x^2+x^4$	0110100
0101	$x+x^3$	$x+x^2+x^3+x^6$	0111001
0110	$x+x^2$	$x+x^3+x^4+x^5$	0101110
0111	$x+x^2+x^3$	$x+x^5+x^6$	0100011
1000	1	$1+x+x^3$	1101000
1001	$1+x^3$	$1+x+x^4+x^6$	1100101
1010	$1+x^2$	$1+x+x^2+x^5$	1110010
1011	$1+x^2+x^3$	$1+x+x^2+x^3$ $+x^4+x^5+x^6$	1111111
1100	$1+x$	$1+x^2+x^3+x^5$	1011010
1101	$1+x+x^3$	$1+x^2+x^6$	1010001
1110	$1+x+x^2$	$x^2+x^3+x^5$	0011010
1111	$1+x+x^2+x^3$	$1+x^3+x^5+x^6$	1001011

Table 4.2 The (7,4) cyclic code generated by $g(x) = 1+x+x^3$.

Let $\bar{m} = (0110)$ be the message sequence to be encoded. Then the corresponding message polynomial is $m(x) = x+x^2$. If one uses $c(x) = m(x)g(x)$ with $g(x) = g_1(x) = 1+x+x^3$ to generate this code, the non-systematic code polynomial is given by

Linear Cyclic Codes

$$c(x) = m(x)g(x) = (x+x^2)(1+x+x^3) = x+x^3+x^4+x^5$$

and the codeword is $\bar{c} = (0101110)$. The (7, 4) cyclic codewords generated by $g(x) = 1+x+x^3$ are shown in Table 4.2 from which it is seen that this code has the minimum distance of 3 and is a single error-correcting code.

4.1.2 Systematic Cyclic-Code Structure

Suppose $g(x)$ is the generator polynomial of an (n, k) cyclic code. This code can be put into the following systematic form:

$$\bar{c} = (-p_0, -p_1, \ldots, -p_{n-k-1}, m_0, m_1, \ldots, m_{k-1}), \tag{4.10}$$

where m_i are the message symbols and p_i are the parity-check symbols. From Eq. (4.10) the message polynomial $m(x)$ and the parity-check polynomial $p(x)$ are expressed by

$$m(x) = m_0 + m_1 x + \cdots + m_{k-1} x^{k-1}, \tag{4.11}$$

$$p(x) = p_0 + p_1 x + \cdots + p_{n-k-1} x^{n-k-1}, \tag{4.12}$$

respectively, where $p(x)$ is the remainder polynomial of degree $n-k-1$ or less which is obtained by dividing $x^{n-k} m(x)$ by $g(x)$. Thus, one has

$$x^{n-k} m(x) = q(x)g(x) + p(x) = m_0 x^{n-k} + \cdots + m_{k-1} x^{n-1} \tag{4.13}$$

which represents the shifted message sequence $\bar{m}$ followed by $n - k$ parity-check symbols. These $n - k$ parity-check symbols are just the coefficients of $p(x)$ as expressed in Eq. (4.12). Thus from Eq. (4.13),

$$c(x) = -p(x) + x^{n-k} m(x) = q(z)g(x) \tag{4.14}$$

is the systematic code polynomial of degree $n - 1$ or less which also is a multiple of $g(x)$. Clearly, this expression corresponds to the codeword in Eq. (4.10) in systematic form.

The systematic encoding algorithm of an (n,k) cyclic code is summarized below.
1. Multiply the message polynomial $m(x)$ by x^{n-k}.
2. Divide the result of Step 1 by the generator polynomial $g(x)$ thereby yielding the remainder polynomial $p(x)$.
3. Set $c(x) = x^{n-k} m(x) - p(x)$.

Example 4.4

Consider the (7, 4) cyclic code with the generator polynomial $g(x) = 1 + x + x^3$. Suppose that the information to be encoded is $m = (0\ 1\ 1\ 0)$. The corresponding information polynomial is $m(x) = x + x^2$. Since $n - k = 3$, one has $x^3 m(x) = x^4 + x^5$. Dividing $x^4 + x^5$ by $g(x) = 1 + x + x^3$ yields the remainder polynomial $p(x) = 1$. Therefore, the code polynomial, expressed by Eq. (4.14), becomes $c(x) = 1 + x^4 + x^5$ or $\bar{c} = (1\ 0\ 0\ 0\ 1\ 1\ 0)$. Similarly, all 16 code words, both in systematic and nonsystematic form, are listed in Table 4.3.

Message sequence	Nonsystematic codeword	Systematic codeword
0000	0000000	0000000
0001	0001101	1010001
0010	0011010	1110010
0011	0010111	0100011
0100	0110100	0110100
0101	0111001	1100101
0110	0101110	1000110
0111	0100011	0010111
1000	1101000	1101000
1001	1100101	0111001
1010	1110010	0011010
1011	1111111	1001011
1100	1011100	1011100
1101	1010001	0001101
1110	1000110	0101110
1111	1001011	1111111

Table 4.3 A (7, 4) Cyclic Code

4.1.3 Generator and Parity-Check Matrices

The generator matrix G in systematic form for an (n, k) cyclic code with the generator polynomial $g(x)$ is found next. Dividing x^{n-k+i} by the generator polynomial $g(x)$ yields the relations,

$$x^{n-k+i} = q_i(x)g(x) + p_i(x) \text{ for } i=0,1,\ldots,k-1, \quad (4.15)$$

where

$$p_i(x) = p_{i,0} + p_{i,1}x + \cdots + p_{i,n-k-1}x^{n-k-1}$$

is the remainder polynomial of degree $n - k - 1$ or less. An evident rearrangement of Eq. (4.15) yields the k identities,

$$c_i(x) = q_i(x)g(x) = -p_i(x) + x^{n-k+i} \quad \text{for } i=0,1,\ldots,k-1,$$

which are multiples of $g(x)$. Evidently these identities constitute the k special codeword polynomials shown below:

Linear Cyclic Codes

$$c_0(x) = -p_0(x) + x^{n-k} \leftrightarrow \bar{c}_0 = (-p_{0,0}, -p_{0,1}, \ldots, -p_{0,n-k-1}, 1, 0, 0, \ldots, 0),$$
$$c_1(x) = -p_1(x) + x^{n-k+1} \leftrightarrow \bar{c}_1 = (-p_{1,0}, -p_{1,1}, \ldots, -p_{1,n-k-1}, 0, 1, 0, \ldots, 0),$$
$$\vdots$$
$$c_{k-1}(x) = -p_{k-1}(x) + x^{n-1} \leftrightarrow \bar{c}_{k-1} = (-p_{k-1,0}, -p_{k-1,1}, \ldots, -p_{k-1,n-k-1}, 0, 0, \ldots, 0, 1).$$

Arranging these k code words as rows of a $k \times n$ matrix produces the generator matrix G in systematic form for the (n, k) cyclic code, namely,

$$G = \begin{pmatrix} -p_{0,0} & -p_{0,1} & \cdots & -p_{0,n-k-1} & 1 & 0 & 0 & \cdots & 0 \\ -p_{1,0} & -p_{1,1} & \cdots & -p_{1,n-k-1} & 0 & 1 & 0 & \cdots & 0 \\ \vdots & & & \vdots & & & & & \\ -p_{k-1,0} & -p_{k-1,1} & \cdots & -p_{k-1,n-k-1} & 0 & 0 & 0 & \cdots & 1 \end{pmatrix}. \quad (4.16)$$

The corresponding parity-check matrix H for this (n, k) cyclic code is

$$H = \begin{pmatrix} 1 & 0 & \cdots & 0 & -p_{0,0} & -p_{1,0} & \cdots & -p_{k-1,0} \\ 0 & 1 & \cdots & 0 & -p_{0,1} & -p_{1,1} & \cdots & -p_{k-1,1} \\ \vdots & & & & \vdots & & & \\ 0 & 0 & \cdots & 1 & -p_{0,n-k-1} & -p_{1,n-k-1} & \cdots & -p_{k-1,k-1} \end{pmatrix}. \quad (4.17)$$

Example 4.5

The generator matrix G of a $(7, 4)$ cyclic code with generator polynomial $g(x) = 1 + x + x^3$ can be constructed from a set of $k = 4$ linearly independent codewords. The basis vectors corresponding with the message words $(1\ 0\ 0\ 0)$, $(0\ 1\ 0\ 0)$, $(0\ 0\ 1\ 0)$ and $(0\ 0\ 0\ 1)$, are found by using Eq. (4.16) as follows:

(1000) ↔ (1101000),
(0100) ↔ (0110100),
(0010) ↔ (1110010),
(0001) ↔ (1010001).

Therefore, the generator matrix G consists of the four linearly independent basis vectors as shown below.

$$G = \begin{pmatrix} 1 & 1 & 0 & 1 & 0 & 0 & 0 \\ 0 & 1 & 1 & 0 & 1 & 0 & 0 \\ 1 & 1 & 1 & 0 & 0 & 1 & 0 \\ 1 & 0 & 1 & 0 & 0 & 0 & 1 \end{pmatrix}.$$

The parity-check matrix H is for this $(7,4)$ code obtained easily by a use of Eq. (4.17) as follows:

$$H = \begin{pmatrix} 1 & 0 & 0 & 1 & 0 & 1 & 1 \\ 0 & 1 & 0 & 1 & 1 & 1 & 0 \\ 0 & 0 & 1 & 0 & 1 & 1 & 1 \end{pmatrix}.$$

Consider now the set of k linearly independent polynomials $g(x)$, $xg(x)$,..., ...$x^{n-k} g(x)$. If the k n-tuples corresponding to these k polynomials are used as the rows of a $k \times n$ matrix, then one obtains the following generator matrix G in nonsystematic form:

$$G(x) = \begin{pmatrix} g_0(x) = g(x) \\ g_1(x) = xg(x) \\ \vdots \\ g_{k-1}(x) = x^{k-1} g(x) \end{pmatrix} \Leftrightarrow G = \begin{pmatrix} g_0 & g_1 & \cdots & g_{n-k} & 0 & \cdots & 0 & 0 \\ 0 & g_0 & g_1 & \cdots & g_{n-k} & 0 & \cdots & 0 \\ \vdots & & & & & & & \\ 0 & 0 & \cdots & 0 & g_0 & g_1 & \cdots & g_{n-k} \end{pmatrix}$$
(4.18)

where $g_0 = g_{n-k} = 1$.

Example 4.6
The generator matrix in nonsystematic form for the (7, 4) cyclic code with $g(x) = 1 + x + x^3$ is obtained from the polynomials $x^i g(x) = x^i + x^{i+1} + x^{i+3}$, $i = 0, 1, 2, 3$. Thus the generator matrix for this (7, 4) cyclic code in nonsystematic form is given by

$$G(x) = \begin{pmatrix} g(x) \\ xg(x) \\ x^2 g(x) \\ x^3 g(x) \end{pmatrix} \Leftrightarrow G = \begin{pmatrix} 1 & 1 & 0 & 1 & 0 & 0 & 0 \\ 0 & 1 & 1 & 0 & 1 & 0 & 0 \\ 0 & 0 & 1 & 1 & 0 & 1 & 0 \\ 0 & 0 & 0 & 1 & 1 & 0 & 1 \end{pmatrix}. \qquad (4.19)$$

In general, a systematic-form generator matrix for a cyclic code is obtained by transforming Eq. (4.18) into Eq. (4.16) as follows:
1. Use $g(x)$ as the first row.
2. To generate the second row, cyclically shift the first row one column to the right. This corresponds with the operation is $xg(x)$ if the $(n - k)$-th column entry is 0. If this entry is 1, add the first row to it. Thus the second row is $xg(x)$ if the coefficient of x^{n-k} in $g(x)$ is 0; otherwise the second row is $xg(x) + g(x)$ if the coefficient is 1.
3. To generate the third row, repeat the same process, i.e., shift the entries in the second row one column to the right. Add $g(x)$ if the $(n - k)$-th column entry is not 0. Repeat this process for all of the rows until one reaches the $(k - 1)$-th row.

Linear Cyclic Codes

It is apparent that the systematic form of G is obtained from the nonsystematic form of G by successive addition of rows using the rules described above. Notice that this operation does not change the code words in the (n,k) cyclic code. It simply re-orders them.

Example 4.7
The systematic form of G for the (n,k) cyclic code, discussed in Example 4.6, is derived by the above algorithm, using Eq. (4.19), to obtain

$$G(x) = \begin{pmatrix} g(x) \\ xg(x) \\ x^2 g(x) + g(x) \\ x^3 g(x) + xg(x) + g(x) \end{pmatrix} \Leftrightarrow G = \begin{pmatrix} 1 & 1 & 0 & 1 & 0 & 0 & 0 \\ 0 & 1 & 1 & 0 & 1 & 0 & 0 \\ 1 & 1 & 1 & 0 & 0 & 1 & 0 \\ 1 & 0 & 1 & 0 & 0 & 0 & 1 \end{pmatrix}.$$

4.1.4 Dual Code of the (n,k) Cyclic Code

An important property of the dual code $C^\perp$ of a cyclic code C is that the (n, n - k) dual code of the (n,k) cyclic code, generated by g(x), is also cyclic and is generated by the polynomial $h^*(x) = x^k h(x^{-1})$, where $h(x) = (x^n - 1)/g(x)$ is the parity-check polynomial of the (n, k) cyclic code. This property is demonstrated briefly as follows: Since $g(x)h(x) = x^n - 1$, this expression also can be re-expressed as $x^n g(x^{-1})h(x^{-1}) = -x^n + 1$ or $g(x^{-1})h(x^{-1}) = x^{-n} - 1$. If one rearranges this relation it becomes $x^{n-k} g(x^{-1}) \cdot x^k h(x^{-1}) = g^*(x)h^*(x) = -x^n + 1$ so that the polynomial $h^*(x) = x^k h(x^{-1})$ is evidently a divisor of $x^n - 1$. Since the degree of h(x) is k, both its reciprocal polynomial $h(x^{-1})$ and $h^*(x)$ have the same degree k. Therefore, it follows that

$$h^*(x) = x^k (h_0 + h_1 x^{-1} + \cdots + h_k x^{-k}) = h_k + h_{k-1} x + \cdots + h_1 x^{k-1} + h_0 x^k \quad (4.20a)$$

Thus it is seen that h*(x) is that divisor of $x^n - 1$ which generates an (n,n-k) cyclic code with the following generator matrix:

$$H = \begin{pmatrix} h^*(x) \\ xh^*(x) \\ \vdots \\ x^{n-k-1} h^*(x) \end{pmatrix} = \begin{pmatrix} h_k & h_{k-1} & \cdots & h_0 & 0 & \cdots & 0 \\ 0 & h_k & h_{k-1} & & h_0 & 0 & \cdots & 0 \\ \vdots & & & & & & \vdots \\ 0 & \cdots & & 0 & h_k & h_{k-1} & \cdots & h_0 \end{pmatrix}. \quad (4.20b)$$

Next consider the multiplication of c(x)=m(x)g(x) by h(x), which yields

$$c(x)h(x) = m(x)g(x)h(x)$$
$$= m(x)(x^n - 1) \tag{4.21}$$
$$= -m_0 - m_1 x - \cdots - m_{k-1} x^{k-1} + m_0 x^n + m_1 x^{n+1} + \cdots + m_{k-1} x^{n+k-1}.$$

One can see that the powers $x^k, x^{k+1}, \ldots, x^{n-1}$ do not appear in Eq. (4.21). Hence, the coefficients of $x^k, x^{k+1}, \ldots, x^{n-1}$ of the expansion of the product $c(x)h(x)$ on the left side of Eq. (4.21) must equal zero. That is,

$$h_0 c_k + h_1 c_{k-1} + \cdots + h_{k-1} c_1 + h_k c_0 = 0,$$
$$h_0 c_{k+1} + h_1 c_{k+2} + \cdots + h_k c_1 = 0,$$
$$\vdots$$
$$h_0 c_{n-1} + h_1 c_{n-2} + \cdots + h_k c_{n-k-1} = 0,$$

which further reduces to the single expression,

$$\sum_{i=0}^{k} h_i c_{n-i-j} = 0 \quad for \quad 0 \le j \le n-k. \tag{4.22}$$

Eq. (4.22) demonstrates the property that any codeword $\bar{c}$ in a cyclic code C is orthogonal to every row of the parity-check matrix H. That is,
$$G \cdot H^T = 0,$$
where G is given in Eq (4.18).

Example 4.8
Consider the (7, 3) dual code of the (7, 4) cyclic code with generator polynomial $g(x) = 1 + x + x^3$. Since the parity-check polynomial is $h(x) = 1 + x + x^2 + x^4$, its (7,3) dual code has by Eq. (4.20a) the generator polynomial $h^*(x) = 1 + x^2 + x^3 + x^4$. The (7, 3) dual code which is generated by $h^*(x) = 1 + x^2 + x^3 + x^4$ is listed in Table 4.4. Since this code has $d_{min} = 4$ for its minimum distance, the maximum random error-correcting capability is $t = \lfloor (d_{min} - 1)/2 \rfloor$, i.e., it is also a single error-correcting code.

Message word $\bar{m}$	$m(x)$	$c(x)=m(x)g(x)$	Codeword $\bar{c}$
000	0	0	0000000
001	x^2	$x^2 + x^4 + x^5 + x^6$	0010111
010	x	$x + x^3 + x^4 + x^5$	0101110
011	$x + x^2$	$x + x^2 + x^3 + x^6$	0111001
100	1	$1 + x^2 + x^3 + x^4$	1011100
101	$1 + x^2$	$1 + x^3 + x^5 + x^6$	1001011
110	$1 + x$	$1 + x + x^2 + x^5$	1110010
111	$1 + x + x^2$	$1 + x + x^4 + x^6$	1100101

Table 4.4 The Dual (7.3) Cyclic Code Generated by $h^*(x) = 1 + x^2 + x^3 + x^4$

Linear Cyclic Codes

By Eq. (4.20b) the row vectors of the generator matrix of the dual code correspond with the polynomials $x^i h^*(x)$, for $i = 0,1, 2$. Hence the generator matrix of the dual code $C^\perp$ has the form,

$$H = \begin{pmatrix} h^*(x) \\ xh^*(x) \\ x^2 h^*(x) \end{pmatrix} = \begin{pmatrix} 1 & 0 & 1 & 1 & 1 & 0 & 0 \\ 0 & 1 & 0 & 1 & 1 & 1 & 0 \\ 0 & 0 & 1 & 0 & 1 & 1 & 1 \end{pmatrix}. \tag{4.23}$$

To verify that Eq. (4.23) is correct, use Eqs. (4.19) and (4.23) to verify $GH^T = 0$.

4.2 Shift-Register Encoders and Decoders of Cyclic Codes

Megabits-per-second to Gigabits-per-second transmission rates are very common in many modern applications such as Digital Television and Optical-Fiber Communications. At the transmission rates, used in these systems, the available error-control techniques are limited by the complexity of the encoding and decoding algorithms. An important property of cyclic codes is that their encoders and decoders can be implemented by the use of simple exclusive-OR gates, switches, shift-registers, and special finite-field adders and multiplier circuits. As it is shown in Chapter 2, shift registers are among the simplest of digital circuits. Shift-registers are a collection of flip-flops (bi-state devices) connected in series. They are thus operable at speeds which are quite close to the fastest speed possible for a single gate in a given device-technology.

4.2.1 Nonsystematic and Systematic Encoders

In Sec. 4.1.1 the non-systematic encoding technique for cyclic codes is given by $c(x)=m(x)g(x)$. In this approach to encoding the product $m(x)g(x)$ is computed as the weighted sum of cyclic shifts of $g(x)$. Such a shift-register circuit is shown in Figure 4.1. At this point considerable confusion can arise over the ordering of both the information and the code sequences. The convention, used in Figure 4.1 that is applied to the message word $\bar{m} = (m_0, m_1, \ldots, m_{k-1})$, is that the right-most symbol of the word is the first symbol to enter the encoder, subscript-order notwithstanding.

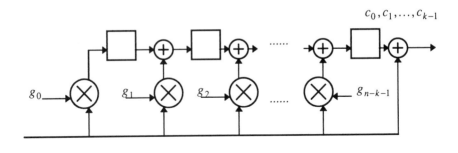

$c_0, c_1, \ldots, c_{k-1}$

$m_0, m_1, \ldots, m_{k-1}$

Figure 4.1 Nonsystematic Shift-Register Encoder for cyclic code. Adders, multipliers, and delay cells are q-ary.

Each time a new message symbol is inputted, the clock of the shift-register is pulsed, and the contents of a shift-register cell are shifted out of that cell and added to the next message symbol, which in turn is sent to the next shift-register cell. At the time that the final message symbol m_0 has been fed into the circuitry of the first shift-register cell, the set of shift-register cells contains the codeword $\vec{c} = (c_0, c_1, \ldots, c_{n-1})$ for the message word $\vec{m} = (m_0, m_1, \ldots, m_{k-1})$.

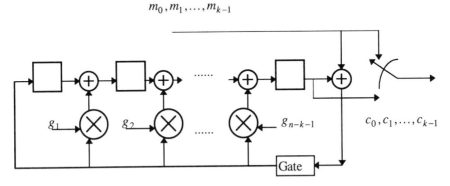

Figure 4.2. Encoder of an (n,k) systematic cyclic code using an $(n-k)$-stage shift register.

It is seen in Sec. 4.1.2 that the encoding of a k-symbol message sequences involves obtaining the parity-check symbols as the remainder $p(x)$, which is obtained by dividing $x^{n-k}m(x)$ by $g(x)$. Thus this encoding procedure is

Linear Cyclic Codes

accomplished by using a division circuit which is an $(n - k)$-stage shift-register with its feedback connections specified by the generator polynomial $g(x)$. In systematic encoding, which is generally used for high-rate codes, the information symbols are transmitted without alteration as shown in Fig. 4.2. Note that the shifting of the information sequence $m(x)$ into the register from the right end of the shift-register is equivalent to a multiplication of $m(x)$ by x^{n-k}. Hence the total number of shifts required to form a codeword is n.

Example 4.9

Consider the (7, 4) cyclic code over $GF(2)$ with $g(x) = 1 + x + x^3$. Figure 4.3 illustrates the encoder specified by $g(x)$, which has as its coefficients $g_0 = g_1 = g_3 = 1$, $g_2 = 0$. Suppose that the input message word is $\bar{m} = (1010)$. When the gate is on, the k-bits of message are shifted into the register and also are sent to the channel. As soon as the information sequence has completely entered the register, the $n - k = 3$ parity-check bits are formed in the register. When the gate is turned off, the parity-check bits are sent into the channel to form the codeword $c = (0\ 0\ 1\ 1\ 0\ 1\ 0)$. The detailed operations of this encoder are shown in Table 4.4.

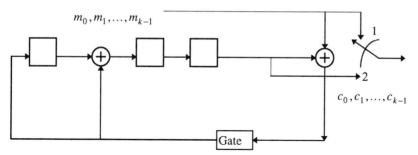

Figure 4.3 Encoder for the (7,4) cyclic code generated by $g(x) = 1 + x + x^3$.

shift number i	register contents	switch	gate	output
0	0 0 0	1	on	0
1	0 0 0	1	on	10
2	1 1 0	1	on	010
3	0 1 1	1	on	1010
4	0 0 1	2	off	11010
5	0 0 0	2	off	011010
6	0 0 0	2	off	0011010

Table 4.5 Encoding operation for the (7, 4) cyclic encoder shown in Fig. 4.3.

The encoding of an (n, k) cyclic code can be accomplished also by using its parity polynomial h(x). It is possible, and sometimes more convenient, to implement the encoder for low-rate codes by the use of a k-stage shift register. The k message symbols are stored initially in the k-stage shift-register, and the register is shifted n - k times to obtain the n - k parity-check symbols as explained below. Since $h_k = 1$, Eq. (4.22) for the parities also can be expressed by

$$c_{n-k-j} = -\sum_{i=0}^{k-1} h_i c_{n-i-j} \quad \text{for } 1 \leq j \leq n-k \tag{4.24}$$

In a systematic code the message symbols in each codeword represent $c_{n-k}, c_{n-k+1}, \ldots, c_{n-1}$, whereas the n - k parity-check symbols $c_0, c_1, \ldots, c_{n-k-1}$ are completely specified by Eq. (4.24). Hence, the n - k parity-check symbols can be determined by a use of Eq. (4.24). An encoder, based on Eq. (4.24) which employs a k-stage shift register, is shown in Fig. 4.4.

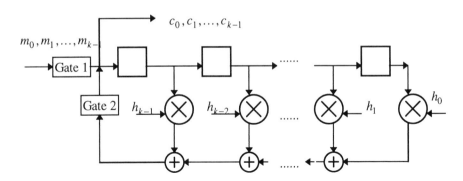

Figure 4.4 A k-stage shift-register encoder for an (n,k) cyclic code generated by the parity-check polynomial $h(x) = h_0 + h_1 x + h_2 x^2 + \cdots + h_k x^k$.

4.2.2 Syndromes and Their Use in Decoding

An important step in decoding most cyclic codes is the ability to compute the syndrome of the received word. Recall that in the decoding of a linear code that the syndromes are defined as the components of the (n - k)-component vector $\bar{s}$ which is produced by multiplying the received word $\bar{r}$ by the parity-check matrix H. That is, $\bar{s} = \bar{r} \cdot H^T$. In a similar manner, the syndromes of a cyclic code can be represented by, what is called, the syndrome polynomial, namely

$$s(x) = s_0 + s_1 x + \cdots + s_{n-k-1} x^{n-k-1}. \tag{4.25}$$

Cyclic codes make possible a highly efficient technique for computing this syndrome polynomial using shift-register circuits.

First consider the systematic case. Recall that a systematic encoding is obtained by dividing the shifted message polynomial $x^{n-k} \cdot m(x)$ by the generator polynomial $g(x)$. The code polynomial $c(x)$ is equal to $x^{n-k} \cdot m(x)$ minus the remainder $p(x)$ obtained from the division operation. The transmitted codeword thus has the form, $\bar{c} = (c_0, c_1, \ldots, c_{n-1}) = (-\bar{p}, \bar{m}) = (-p_0, -p_1, \ldots, -p_{n-k-1}, m_0, m_1, \ldots, m_{k-1})$, where p_i and m_i are the coefficients of the polynomials $p(x)$ and $m(x)$, respectively. Assume that the received word is $\bar{r} = (r_0, r_1, \ldots, r_{n-1}) = (-\bar{p}', \bar{m}') = (-p'_0, -p'_1, \ldots, -p'_{n-k-1}, m'_0, m'_1, \ldots, m'_{k-1})$.

Another vector $\bar{r}_1$ can be derived from the vector $\bar{m}'$ by letting $\bar{r}_1 = \bar{m}' \cdot G = (-\bar{p}'', \bar{m}') = (-p''_0, -p''_1, \ldots, -p''_{n-k-1}, m'_0, m'_1, \ldots, m'_{k-1})$, where G is the systematic generator matrix of this cyclic code. Thus, one has the following vector equation:

$$\bar{r} - \bar{m}' \cdot G = \bar{r} - \bar{r}_1 = (p''_0 - p'_0, p''_1 - p'_1, \ldots, p''_{n-k-1} - p'_{n-k-1}, 0, \ldots, 0). \tag{4.26}$$

Since $G \cdot H^T = 0$, it is clear that the syndrome vector can be expressed by

$$\bar{s} = \bar{r} \cdot H^T - \bar{m}' \cdot G \cdot H^T = (\bar{r} - \bar{r}_1) \cdot H^T$$
$$= (p''_0 - p'_0, p''_1 - p'_1, \ldots, p''_{n-k-1} - p'_{n-k-1}, 0, \ldots, 0) \cdot H^T$$
$$= (p''_0 - p'_0, p''_1 - p'_1, \ldots, p''_{n-k-1} - p'_{n-k-1}).$$

In polynomial form one evidently has $\bar{s}(x) = p''(x) - p'(x)$, where $p'(x) = p'_0 + p'_1 x + \cdots + p'_{n-k-1} x^{n-k-1}$ and $p''(x) = p''_0 + p''_1 x + \cdots + p''_{n-k-1} x^{n-k-1}$. Note also that

$$s(x) = p''(x) - p'(x)$$
$$= x^{n-k} m'(x) - p'(x) - (x^{n-k} m'(x) - p''(x)) = r(x) - q'(x) g(x)$$

where $r(x) = r_0 + r_1 x + \cdots + r_{n-1} x^{n-1}$, $q'(x)$ is a polynomial of degree less than or equal to k, and $g(x)$ is the generator polynomial of the code. Therefore, the syndrome polynomial $s(x)$ in (4.25) of degree $n - k - 1$ or less is the remainder that results from dividing the received polynomial $r(x)$ by the generator polynomial $g(x)$. That is,

$$r(x) = q(x) g(x) + s(x) \tag{4.27a}$$

or

$$s(x) = r(x) \bmod g(x). \tag{4.27b}$$

The syndrome computation in Eqs. (4.27a) or (4.27b) can be accomplished by the use of the (n - k)-stage encoding circuit, shown in Fig. 4.5, except that the received polynomial r(x) can be shifted into the syndrome generator register either from the left or right end. The final content in this register forms the syndrome polynomial $s(x)$.

Syndromes for nonsystematic cyclic codes also can be computed by the use of shift-register techniques. It is shown in Eqs. (4.20b), (4.21) and (4.22) that the syndrome $\bar{s} = \bar{r}H^T$ is the set of coefficients of the following polynomial product,

$$s(x)x^k = s_0 x^k + s_1 x^{k+1} + \cdots + s_{n-k-1} x^{n-1} = r(x)h(x) \bmod (x^n - 1), \tag{4.28}$$

Since $g(x)h(x) = x^n - 1$, the syndrome polynomial s(x) of the non-systematic code also can be computed by a division of $x^{n-k}r(x)$ by the generator polynomial g(x). That is,

$$r(x)x^{n-k} \equiv s(x) \bmod g(x), \tag{4.29}$$

where deg(s(x))<deg(g(x)) with "deg" denoting the "degree" of the polynomial.

Because of their cyclic structure, cyclic codes are especially suitable for error detection and correction. From the above discussion it is clear that the syndrome polynomial is zero, i.e. s(x) = 0 if and only if r(x) = c(x), which is a multiple of g(x). If s(x) ≠ 0, then r(x) is not a code polynomial c(x), and as a consequence r(x) contains the transmission error e(x) due to the channel noise. That is, the received polynomial has the form,

$$r(x)=c(x)+e(x), \tag{4.30}$$

where e(x) denotes the error polynomial. Then Eq. (4.30) coupled with Eq. (4.25), yields e(x)=c(x)+q(x)g(x)+s(x). Since c(x) = m(x)g(x) is a code polynomial which is a multiple of g(x), e(x) becomes

$$e(x) = [m(x) + q(x)]g(x) + s(x) \tag{4.31a}$$

or

$$s(x)=e(x) \bmod g(x). \tag{4.31b}$$

Linear Cyclic Codes

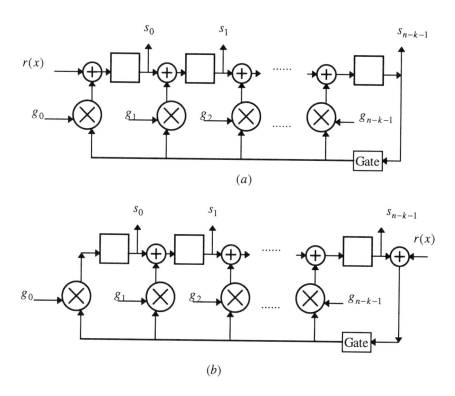

Figure 4.5 An $(n-k)$-stage syndrome generator with inputs from (a), the left end, and from (b), the right end.

Hence the syndrome polynomial $s(x)$ is actually the remainder that results from dividing $e(x)$ by $g(x)$. In fact, $s(x) = 0$ if and only if $e(x) = 0$. Therefore, the error detector tests whether or not $s(x) = 0$. This approach to error detection has a significant advantage over other methods. The encoder of a cyclic code and its error detection circuits are essentially identical so that the design of the encoder/decoder system is simplified. The error-pattern detector is just the syndrome generators shown in Figure 4.5(b) with the syndrome bits as inputs. If the syndrome is not 0, the output is 1, and at least one error is detected.

In Chapter 3 the syndrome values for an arbitrary linear code C are used in a table look-up scheme to implement error correction. The basic idea is that every syndrome value corresponds to a coset of C in all q^{n-k} vectors of length n. The error pattern with the least weight in a given coset is the most likely to occur, and thus is selected as the coset leader.

Maximum-likelihood error correction is performed by first computing the syndrome of the received vector, secondly to use this syndrome to look-up(address) the corresponding coset leader and finally to subtract the selected coset leader from the received vector. The primary drawback to this approach is the size of the syndrome look-up table. For an (n, k) linear code over $GF(q)$ the syndrome table must contain q^{n-k} n-tuples with their n coordinates, taking on any one of q different values. A cyclic code has the interesting property that it allows one to reduce the syndrome table to $\frac{1}{n}$-th of its original size, i.e. from $q^{n-k} \times n$ symbols to q^{n-k} symbols.

Consider the case where the received sequence $r(x)$ is shifted into the syndrome-generating circuit from the left end (see Fig. 4.5a). Let $r^{(1)}(x)$ be the cyclic right shift of $r(x)$. From Eq. (4.3), this is given explicitly by
$$r^{(1)}(x) = x \cdot r(x) - r_{n-1}(x^n - 1). \tag{4.32}$$
By a use of Eqs. (4.27a) and (4.5), Eq. (4.32) can be re-expressed by
$$r^{(1)}(x) = x \cdot (q(x)g(x) + s(x)) - r_{n-1}g(x)h(x) = a(x)g(x) + s^{(1)}(x), \tag{4.33}$$
where $s^{(1)}(x)$ is the remainder that results from dividing $r^{(1)}(x)$ by $g(x)$, i.e. the syndrome polynomial of the cyclic-shifted received word $r^{(1)}(x)$. It follows also from Eq.(4.27a) and (4.33) that
$$x \cdot s(x) = (a(x) - x \cdot q(x) + r_{n-1}h(x))g(x) + s^{(1)}(x). \tag{4.34}$$
Eq. (4.34) tells us that $s^{(1)}(x)$ is also the remainder which results from the division of $xs(x)$ by $g(x)$. That is, if $r^{(1)}(x)$ is the cyclic shift of $r(x)$ one step to the right; the syndrome $s^{(1)}(x)$ which corresponds to $r^{(1)}(x)$ satisfies
$$s^{(1)}(x) = x \cdot s(x) \mod g(x). \tag{4.35}$$
Eq. (4.35) generalizes easily to the i-th cyclical shift case. Thus, if $s(x)$ is the syndrome polynomial, corresponding to a received polynomial $r(x)$, and $r^{(i)}(x)$ is the polynomial obtained by cyclically shifting the coefficients of $r(x)$, i times to the right, then the remainder obtained by dividing $x^i s(x)$ by $g(x)$ is the syndrome $s^{(i)}(x)$ corresponding to $r^{(i)}(x)$.

The above algorithm to obtain the cyclic shifts of the syndrome polynomial greatly reduces the size of syndrome look-up-table decoders for cyclic codes. Given a received vector $\bar{r}$, the corresponding syndrome $\bar{s}$ is obtained by inputting $\bar{r}$ into a shift-register division circuit. As before, when the last

Linear Cyclic Codes

symbol of $\bar{r}$ is shifted into the circuit, the shift-register cells contain the syndrome. The input of an additional zero to this point of the circuit is equivalent to the multiplication of $s(x)$ by x and a division by $g(x)$. The remainder $s^{(1)}(x)$ of this division resides in the shift-register cells. This process, when repeated n times, cycles back to the starting point with the original syndrome in the shift-register cells. Thus one needs to store only one syndrome $\bar{s}$ for an error pattern $\bar{e}$ and all of the cyclic shifts of $\bar{e}$. A syndrome decoder algorithm for a reduced-size look-up table is obtained as follows.

Syndrome decoder for cyclic codes
1. Initialize the counting variable $j = 0$.
2. Calculate the syndrome $\bar{s}$ of the received word $\bar{r}$.
3. Search for the error pattern $\bar{e}$, corresponding to $\bar{s}$ in the syndrome look-up table. If there is such a pattern, go to step 8.
4. Increment counter by $j=j+1$ and input a zero into the shift-register circuit, thereby computing $s^{(j)}$.
5. Search for the error pattern $e^{(j)}$, corresponding to $s^{(j)}$, in the syndrome look-up table. If there is such a pattern, go to step 7.
6. Go to step 4.
7. Determine the error pattern $\bar{e}$, corresponding to $\bar{s}$, by cyclically shifting $e^{(j)}$, j times to the left.
8. Obtain the codeword $\bar{c}$ by $\bar{c} = \bar{r} - \bar{e}$.

Example 4.10
Consider the syndrome circuit for the (7, 4) cyclic code generated by $g(x) = 1 + x + x^3$. This code has $d_{min} = 3$ and is capable of correcting any single error in $e(x)$. There are seven correctable single-error patterns. Suppose that the codeword $\bar{c} = (0110100)$ or $c(x) = x + x^2 + x^4$ is transmitted, and that $\bar{r} = (0110000)$ or $r(x) = x + x^2$ is received. This means that a single error occurred at $e(x) = x^4$. Dividing $r(x)$ by $g(x)$, one obtains its syndrome to be $\bar{s} = (011)$. The received word $\bar{r}$ is shifted into the syndrome register, shown in Fig. 4.6; the contents in that register are given in Table 4.6. At the end of the seventh shift, the register contains the syndrome $\bar{s} = (011)$. If the register is shifted once more with the input gate disconnected, the new contents is $\bar{s} = (111)$, which is the syndrome of $\bar{r} = (0011000)$, the cyclic shift of the received word $\bar{r}$.

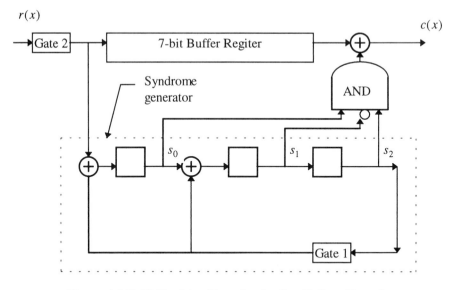

Figure 4.6 Shift-Register Decoder for the (7,4) cyclic code generated by $g(x) = 1 + x + x^3$.

Number of shifts	Input	Shift-Register Contents $S_0\ S_1\ S_2$
0	---	0 0 0
1	0	0 0 0
2	0	0 0 0
3	0	0 0 0
4	0	0 0 0
5	1	1 0 0
6	1	1 1 0
7	0	0 1 1
8	0	1 1 1
9	0	1 0 1

Table 4.6 Contents of the syndrome register.

Since the error patterns consist of all of the cyclic shifts of (0000001), the decoder needs only to be able to recognize one of the seven nonzero syndromes to be able to correct all of the nonzero error patterns. The syndrome $\bar{s} = (101)$, which corresponds to the error pattern $\bar{e} = (0000001)$, is the best choice since it allows one to release the corrected codeword bits before the error location actually is identified.

Linear Cyclic Codes

Figure 4.6 shows the complete decoder. Decoding begins by first setting all of the shift-register cells in the syndrome computation circuit to zero. The received word $\bar{r}$ is then shifted bit-by-bit into the 7-bit received-word buffer and the syndrome-computation circuit, simultaneously. Once the received word is completely shifted into the buffer, the shift-register cells in the syndrome-computation circuit contain the syndrome for the received word. As one continues to shift cyclically the contents of the received-word buffer and the syndrome-computation circuit, the syndrome-computation circuit computes the syndromes for the cyclically shifted versions of the received word, and the corrected bits leave the decoder at the output. If at any point the computed syndrome is $\bar{s} = (101)$, it is "detected" by the AND gate when its output goes to 1. This value is then used to complement and correct the error in the rightmost bit in the buffer as it leaves the buffer.

For the received word $\bar{r} = (0010100)$, when $r(x)$ has been simultaneously shifted into the syndrome register and the buffer register, the syndrome register contains $(0\ 1\ 1)$. In general, when a single error occurs at $e(x) = x^i$, another $6 - i$ shifts are required for the register contents to be $(1\ 0\ 1)$. In this example $e(x) = x^4$, hence two more shifts are needed to obtain $\bar{s} = (1\ 0\ 1)$ after all of the received bits are shifted into the syndrome generator. Thus the next received bit to come out of the buffer register is the erroneous bit. Therefore, only the syndrome $\bar{s} = (1\ 0\ 1)$ needs to be detected. The syndrome $\bar{s} = (1\ 0\ 1)$ is read into the detector of a single AND gate. Since 1 appears at the output of the detector, the erroneous bit r_1 in the rightmost stage of the buffer register needs to be corrected.

For a more detailed exposition on decoders for the general class of cyclic codes, the reader is referred to references[2][3][4]. It should be noted that the complexity of decoders for general cyclic codes increases exponentially with code length n and with the number v of errors to be corrected per received word. In the following chapters specific cyclic codes are examined whose inherent algebraic structure allows for even simpler, though extremely powerful, decoder designs.

4.3 Binary Quadratic Residue Codes and the Golay Code

Let α be a primitive n-th root of unity such that $x^n - 1 = \prod_{i \in Z_n}(x - \alpha^i)$ where Z_n is the integer ring, modulo n, i.e. $Z_n = \{0, 1, \ldots, n-1\}$. It can be shown that there exists a finite field $GF(p^m)$ which contains such a primitive n-th root of unity, α, for a given p, m, and n whenever $n | p^m - 1$. For a cyclic code C it follows from $x^n - 1 = g(x)h(x)$ that the generator polynomial $g(x)$ of C can be represented by $g(x) = \prod_{i \in Q}(x - \alpha^i)$ for $Q \subset Z_n$.

4.3.1 Binary Quadratic Residue Codes

Next consider the family of binary quadratic-residue(QR) codes defined over $GF(2^m)$. Let Q denote the set of quadratic residues, modulo a prime integer n, and let N be the set of nonresidues, i.e. let

$$Q = \{i^2 \pmod{n} \mid i \in Z_n^*\} \tag{4.36a}$$

and

$$N = Z_n^* - Q, \tag{4.36b}$$

where $Z_n^* = Z_n - \{0\}$.

Example 4.11
The quadratic-residue sets Q for $n=7,17,23,31,41$ are given in Table 4.7a.

n	Quadratic residue sets Q
7	{1,2,4}
17	{1,2,4,8,9,13,15,16}
23	{1,2,3,4,6,8,9,12,13,16,18}
31	{1,2,4,5,7,8,9,10,14,16,18,19,20,25,28}
41	{1,2,4,5,8,9,10,16,18,20,21,23,25,31,32,33,36,37,39,40}

Table 4.7a. Quadratic residue sets

Binary QR codes Q, Q', N, N' are defined to be cyclic codes with the following generator polynomials[3][5],

$$q(x) = \prod_{i \in Q}(x - \alpha^i), (x-1)q(x), n(x) = \prod_{i \in N}(x - \alpha^i), (x-1)n(x), \tag{4.36c}$$

Linear Cyclic Codes

respectively, where α is a primitive n-th root of unity in an extension field $GF(2^m)$ of $GF(2)$.

Binary QR codes defined by Eq. (4.36c) have the following properties[3] [6] :

(1) The integer length of the binary QR codes is $n = 8u \pm 1$ where n is a prime and u is some positive integer.

(2) The dimensions of the QR codes satisfy
$$k = \begin{cases} (n+1)/2 & \text{for codes defined by } Q, N, \\ (n-1)/2 & \text{for codes defined by } Q', N'. \end{cases}$$

(3) The minimum distance of the QR codes is lower-bounded by $d_{min} \geq \sqrt{n}$.

The generator polynomials of some binary QR codes are shown in Table 4.7b[6].

(n,k,d)	the generator polynomial q(x)
(7,4,3)	$x^3 + x + 1$
(17,9,5)	$x^8 + x^5 + x^4 + x^3 + 1$
(23,12,7)	$x^{11} + x^{10} + x^6 + x^5 + x^4 + x^2 + 1$
(31,16,7)	$x^{15} + x^{12} + x^7 + x^6 + x^2 + x + 1$
(41,21,9)	$x^{20} + x^{18} + x^{17} + x^{16} + x^{15} + x^{14} + x^{11}$ $+ x^{10} + x^9 + x^6 + x^5 + x^4 + x^3 + x^2 + 1$

Table 4.7b. Gnerator polynomials of binary QR codes.

Two important parameters for characterizing a "good" (n,k, d_{min}) block code are its so-called information rate k/n and error control rate d_{min}/n. For a "good" code both of the "figures-of-merit", k/n and d_{min}/n, should be large under certain constraints, in particular, the Hamming bound. It is seen that the rate k/n for the binary QR codes is approximately 1/2, and it is easy to show that the rate k/n of the extended binary QR codes is exactly 1/2. Also, the minimum distances of many modest size QR codes are quite high for their code length [6].

Also, notice from Table 4.7 that the (7,4,3) Hamming code belongs to the family of binary QR codes. It is shown in Chapter 3 that the (7,4,3) Hamming code is a perfect code and is "optimal" in the sense that its ML decoder has the smallest error probability over a BSC among all binary block codes of length, 7 bits and dimension, 4 bits. In general a (n,k,d_{min}) block code is called *optimum* for a given channel if its error probability is the smallest among all block codes with the same parameters n and k. In the family of binary QR codes, several optimum codes can be found[7][8]. Among these optimum codes the (17,9,5) QR code and the two perfect codes, the (7,4,3) Hamming code and the (23,12,7) Golay code are included. In the next subsection, the (23,12,7) Golay code is discussed in some detail.

4.3.2 The (23,12,7) binary Golay code

The binary Golay code C_{23} is the (23,12,7) quadratic residue code over $GF(2^{11})$ [9][13]. The defining set Q of this code is the set of the quadratic residues, modulo 23 : $Q=\{1,2,3,4,6,8,9,12,13,16,18\}$.

There are two possible generator polynomials for C_{23} :

$$g_1(x) = x^{11} + x^{10} + x^6 + x^5 + x^4 + x^2 + 1, defined\ by\ Q.$$
$$g_2(x) = x^{11} + x^9 + x^7 + x^6 + x^5 + x + 1, defined\ by\ N.$$

Either of these generator polynomials yields a triple error-correcting Golay code. Since C_{23} has dimension 12 and minimum distance 7, it satisfies the Hamming bound with equality:

$$\sum_{i=0}^{t}\binom{n}{i} = \binom{23}{0} + \binom{23}{1} + \binom{23}{2} + \binom{23}{3} = 2^{11} = 2^{n-k}.$$

Thus, the Golay code C_{23} is a perfect code. C_{23} can be extended to form the binary extended Golay code C_{24} by adding one more parity-check bit c_{23} so that $c_{23} = \sum_{i=0}^{22} c_i$. The weight distribution of both C_{23} and C_{24} are given Table 4.6.

The binary Golay code C_{23} was used in the rather revolutional design of an error-control system for the 19.9 Kbits/s PC modem[10][11][12]. Also the

Linear Cyclic Codes

extended binary Golay code C_{24} was used to provide error control on the Voyager spacecraft.

The weight distributions of C_{23} and C_{24} are given in Table 4.8.

i	W_i for C_{23}	W_i for C_{24}
0	1	1
7	253	
8	506	759
11	1288	
12	1288	2576
15	506	
16	253	759
23	1	
24		1

Table 4.8 Weight distributions for C_{23} and C_{24}.

Since C_{23} is a cyclic code, the shift-register syndrome decoding techniques, discussed in Section 4.2.2, can be applied. For C_{23}, one has $n-k=11$ so that approximately only 6k bytes of memory are used to implement the required look-up table for syndrome decoding.

It can be shown (see Problem 4.) for a BSC that the performance of the Golay code C_{23} can be expressed in terms of the probability of an error at the output of the decoder as a function of the probability P of a symbol error of the channel.

$$P_b = \sum_{v=0}^{3}\binom{23}{v}(1-P)^v P^{23-v}$$

$$+ \sum_{j=7}^{16}\frac{j}{23}W_j\left[\sum_{v=0}^{3}\sum_{r=0}^{v}\binom{j}{v-r}\binom{23-j}{r}P^{j-v+2r}(1-P)^{23-j+v-2r}\right]. \quad (4.37a)$$

The performance of a communication system designed to operate at a specified data rate often can be improved by using the appropriate modulation and coding techniques. The efficiency of such systems is usually measured by the received energy per bit-to-noise ratio, denoted by E_b/N_0, where N_0 is the single-sided spectral-noise density required to achieve a specified system bit-error rate (P_b). Without coding, the required E_b/N_0 can be minimized by the selection of an efficient modulation technique. For binary bit-by-bit signaling, the optimum modulation is binary phase-shift keyed (BPSK) modulation.

Suppose the binary symbol output of the Golay encoder is used to modulate an RF carrier sinusoid. Each code symbol results in the transmission of a pulse of the carrier at either of two phase separated by 180^0. Such a sequence of biphase pulses. This signal is transmitted over a channel which adds white Gaussian noise to the signal. After reception and demodulation, a continuous signal-plus-noise output waveform is obtained. To facilitate digital processing by the decoder, this continuous process is next sampled and quantized. If a hard decision (two-level quantization) is used, such a communication channel is a model of BSC for white Gaussian noise with

$$P = Q\left(\sqrt{R \cdot E_b/N_0}\right) \tag{4.37b}$$

where $R = k/n$ is the data rate and where

$$Q(x) = \frac{1}{\sqrt{2\pi}} \int_x^\infty e^{-t^2/2} dt .$$

Note that $R = 12/23$ for the code C_{23}. The performance curves of bit-error rate for the Golay code C_{23} is given in Figure 4.7 which is obtained from the Eqs. (4.37a) and (4.37b) using the weight distributions in Table 4.6 [14]. Series 1 is the bit-error probability with Golay coding and series 2 is the bit-error probability without coding (only with BPSK), which is obtained from Eq.(4.37b) with R=1. It can be seen from Figure 4.7 that, by the use of the Golay code, approximately 2.15dB coding gain is achieved by with $P \approx 10^{-5}$ while 1.33dB coding gain is obtained at $P \approx 10^{-3}$.

Linear Cyclic Codes

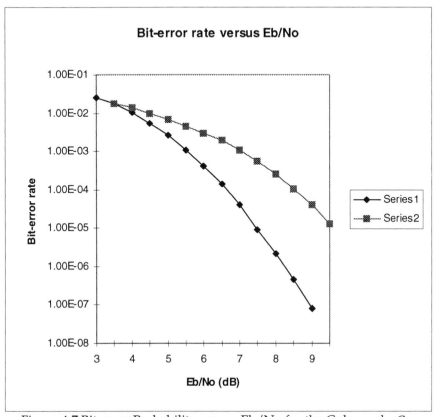

Figure 4.7 Bit-error Probability versus Eb/No for the Golay code C_{23} with hard decision decoding.

4.4 Error Detection with Cyclic and Shortened Cyclic Codes

The reliable detection of transmission errors is required in many data-communication applications. Once such errors are detected, message retransmission is requested. These error-detection schemes are used in the reading and writing of floppy disks, in checking the validity of commands sent to spacecraft (where invalid messages could be disastrous) and in checking the validity of packets in packetized-data communication networks. In the packet-switched network application the term *ARQ* denotes the automatic request of a retransmission upon the detection of message errors. Error detection often is used after powerful error-correction coding takes place in order to detect the presence of residual errors. In such cases error-detection schemes are employed in various data-transmission protocols in the higher layers of the network, and often they are concatenated with another (usually more powerful physical-layer) coding technique.

4.4.1 Error Detection with Cyclic Codes

The popularity of error detection is due to its simple implementation and high reliability for a surprisingly small message overhead or redundancy. As we have seen, the encoder and syndrome generators are simple shift registers with feedback. At the present time integrated circuits are available for the processing of certain international standard codes. Often the same circuit can be configured easily to perform either encoding or decoding functions. The codes for such applications usually are referred to in the literature as *cyclic-redundancy-check* (CRC) codes.

One of the most frequently-used error-control technique over the past 50 years has been the set of CRC error-detecting codes. With the extreme simplicity in implementation, CRC coding techniques owe their predominance to the computer industry and the ubiquity of computer-communication networks. On the other hand, many CRC codes are not as easily understood conceptually; they owe their popularity to their quite simple (and fast) encoder and decoder implementations and their excellent error-detection performance. In this subsection CRC error detection is introduced as shortened cyclic codes. This discussion is then followed by the error-detection performance of cyclic codes in general and the CRC codes in particular.

In the last chapter, several methods for changing the lengths of block codes are discussed. In this subsection shortened systematic cyclic codes are considered. These shortened codes are constructed as follows: Let C be an (n, k) systematic cyclic code. Also let S be the subset of code words in C whose j high-order information coordinates (i.e. the j rightmost coordinates in the codeword) have the value zero. Finally, let C' be the set of words obtained by deleting the j rightmost coordinates from all of the words in S. C' is the $(n - j, k - j)$ shortened systematic code. This procedure is equivalent to the omission of the j bottom-most rows of G or the deletion of the j rightmost columns of H of the original systematic cyclic code.

Example 4.12
Consider the (7,4) cyclic code with $g(x) = 1 + x + x^3$, which is discussed in Examples 4.4 and 4.5. By shortening this code by one bit one obtains the (6,3)

Linear Cyclic Codes

shortened code. The generator matrix G' and the parity-check matrix H' for the (6, 3) shortened code are obtained, respectively, by omitting the last row and column from G of the original code and the last column of H:

$$G' = \begin{pmatrix} 1 & 1 & 0 & 1 & 0 & 0 \\ 0 & 1 & 1 & 0 & 1 & 0 \\ 1 & 1 & 1 & 0 & 0 & 1 \end{pmatrix}, \text{ and } H' = \begin{pmatrix} 1 & 0 & 0 & 1 & 0 & 1 \\ 0 & 1 & 0 & 1 & 1 & 1 \\ 0 & 0 & 1 & 0 & 1 & 1 \end{pmatrix}.$$

The third column in Table 4.2 lists the systematic codewords $\bar{c}$ of the (7,4) cyclic Hamming code. The codewords $\bar{c}$ for the (6, 3) shortened code are found directly from the codewords $\bar{c}$ in the third column of Table 4.3 by taking only those codewords whose rightmost bits are zero and then deleting the rightmost bits of each of these codewords. Thus the code words of the (6, 3) shortened cyclic code are as listed in Table 4.9.

Message words	Shortened codewords
000	000000
001	111001
010	011010
011	100011
100	110100
101	001101
110	101110
111	010111

Table 4.9 The (6,3) shortened code

It is seen from Table 4.9 that the shortened code C' has at least the same minimum distance $d_{min} = 3$ as that of the original code C. In general, since shortening a code decreases the rate of the code, shortened codes have error detection and correction capabilities that are at least as good as the original code.

The (6,3) shortened code is no longer cyclic. In general, codes obtained from cyclic codes by a shortening are almost always non-cyclic. However, it is possible to transform some shortened codes into (n-j,k-j) cyclic codes if they are ideals in the ring of polynomials, module $f(x)$ for some polynomial $f(x)$ of degree n-j [12]. More importantly, this may make it possible to use the same shift-register encoders and decoders that are used for the original cyclic codes.

Consider the systematic cyclic encoder in Figure 4.4. Before encoding begins the shift-register cells are set to zero. If the first j high-order message symbols are equal to zero, then their insertion into the encoding circuit does not change the state of the shift-register cells. This insertion results only in an output of j zeroes. The deletion of the j high-order symbols from the message block thus has no impact on the encoding process, except for the deletion of the j zeroes in the corresponding message positions of the code word. By this same reasoning it can be seen that the shift-register syndrome computation circuit for the original cyclic code (as shown in Figure 4.5) also can be applied to calculate the syndromes for the shortened code as well.

Shortened cyclic codes have been extensively used in computer communications and data networks, under the label of cyclic-redundancy-check (CRC) codes. CRC codes are generally not cyclic, as noted above but are derived from cyclic codes and hence the name. CRC codes are primarily useful in error-detection systems since they take advantage of the considerable burst-error detection capability provided by their cyclic "parent" codes. Common system's responses to detected errors are retransmission request, dismiss the transmitted codeword, or performing error concealment. The actual response is dictated by the specific application. Some of these applications will be discussed in Chapter 9.

CRC codes are determined by a generator polynomial $g(x)$ and may be said to have arbitrary length up to the limit n, the length of the original cyclic code defined by $g(x)$. The literature on CRC codes discusses a variety of "rules-of-thumb" for selecting CRC generator polynomials. The key to the design of a CRC code is to find a generator polynomial of a certain degree that offers optimal or near-optimal performance.

Error-Detection Performance Analysis of CRC Codes
To accurately compute error-detection performance, the weight distributions of CRC codes are needed. However, the weight distributions and even the minimum distances for many cyclic codes are not known. Fortunately, for a given generator polynomial $g(x)$ of a cyclic code C, one often can make meaningful estimation of error-detection capability of this CRC code in a variety of situations. Three situations are considered next that are generally useful in the design of computer-communication systems or data networks.

(a). Error Pattern Coverage Radius.
In communication systems, hardware device failures can result in a total corruption of the transmitted codeword. This is particularly true in high

Linear Cyclic Codes

data-rate systems. In such a situation the received word can appear to be completely independent of the transmitted code word. It is thus desirable to use a code that ensures that the probability, that a received word is a valid codeword, stays below some pre-determined threshold. Let the parameter η denote the ratio of the number of non-codeword n-tuples over $GF(q)$ to the number of n-tuples over $GF(q)$. That is,

$$\eta = \frac{q^n - q^k}{q^n} = 1 - q^{-(n-k)},$$

where k denotes the dimension of the code. This parameter η is, sometimes, called the coverage radius. Note this parameter η is solely a function of the number of redundant symbols in the transmitted codewords.

Next burst-error detection of CRC is discussed. Hardware device faults in computer circuitry often introduce errors over several consecutive transmitted symbols. This creates, from the decoder's perspective, burst errors in the received word. Burst errors are prevalent in mobile communication systems, where the communication channel often suffers from multi-path fading. Recall that a burst-error pattern of length b begins and ends with a nonzero value; the intervening symbols in the pattern may take on any value, including zero. Suppose that the transmission alphabet is $GF(q)$.

It can be shown (see Problem 4.11) for a given length b, that there are $(q-1)^2 \cdot q^{b-2}$ different burst-error patterns. Cyclic codes are quite good at performing burst-error detection. In a low-noise environment that suffers from occasional bursts of noise, a CRC error detection system coupled with retransmission requests is frequently an excellent means of controlling such errors. Using the burst-error pattern model presented above, some useful results, concerning the error detection capability of cyclic codes can be obtained.

Let C denote an (n, k) cyclic code or shortened cyclic code with the generator polynomial $g(x)$ of degree r. (If C is a cyclic code, then $r = n - k$). A cyclic code's burst-error detection capability is best expressed in terms of the ratio η_b, the number of detectable burst-error patterns of length b to the total number of burst-error patterns of length b. Obviously, one has $0 \le \eta_b \le 1$. The case $b = r+1$ is considered in the following.

An error burst $e(x)$ is undetectable if and only if $e(x)$ is a valid code polynomial. Thus an undetectable pattern $e(x)$ must have a degree greater than or equal to $r=n-k$, otherwise the minimality of the degree of $g(x)$ is contradicted. Since the code from which the burst-detecting code is obtained is cyclic, $e(x)$ cannot be a shifted version of a polynomial of degree less than r for the same reason. A degree-r polynomial has $(r + 1)$ coefficients, and thus $\eta_b = 1.0$ for all b less than or equal to r. This implies the following error-detection property :

A q-ary cyclic or shortened cyclic code with generator polynomial g(x) of degree r can detect all burst-error patterns of length r or less.

The only valid code polynomials of degree r are multiples of the generator polynomial, i.e. $a \cdot g(x)$ (otherwise one would be able to generate a nonzero code polynomial of degree less than r by a subtraction). The only undetectable burst-error patterns of length $(r + 1)$ are thus the scalar multiples of $g(x)$ and their $(n - r - 1)$ non-cyclic shifts to the right. With these facts the parameter η_b for $b=r+1$ can be derived in what follows. Notice that the terms, corresponding to shifts of the error patterns, cancel out in the analysis.

$$\eta_{r+1} = \frac{\text{Number of det} ectable\ error\ patterns\ of\ length\ (r+1)}{total\ number\ of\ error\ patterns\ of\ length\ (r+1)}$$

$$= 1 - \frac{(n-r)(q-1)}{(n-r)(q-1)^2 q^{r-1}} = 1 - \frac{q^{-(r-1)}}{q-1}.$$

Therefore, a q-ary cyclic or shortened cyclic code with generator polynomial $g(x)$ of degree r can detect the fraction, $\eta_{r+1} = 1 - \frac{q^{-(r-1)}}{q-1}$, of all burst-error patterns of length $(r + 1)$.

Any code polynomial $c(x)$ of a cyclic or shortened cyclic code is uniquely expressible in the product form, $c(x) = m(x)g(x)$, where the degree of $m(x)$ is less than or equal to $(n - r)$ (notice that $r=n-k$ for a cyclic code). If the code polynomial $c(x) = c_0 + c_1 x + \cdots + c_{n-1} x^{n-1}$ is to be associated with an error burst of length b, starting from c_0, the coefficients c_0 and c_b must be nonzero. Since the coefficients g_0 and g_r of the generator polynomial $g(x)$ must be nonzero, it is necessary and sufficient that the coefficients m_0 and any other $(b-r-1)$ terms of a degree-$(b - r)$ message polynomial be nonzero for the resulting code polynomial to correspond to an undetectable error burst of

Linear Cyclic Codes

length b. Let $f(x) = f_0 + f_1 x + \cdots + f_b x^b$ be any polynomial over $GF(q)$. The procedure to determine η_{r+1} is as follows:

$$\eta_{r+1} = \frac{Number\ of\ detecable\ burst - error\ patterns\ of\ length\ b}{total\ number\ of\ burst - error\ patterns\ of\ length\ b}$$

$$= 1 - \frac{(n-b+1) \cdot \{Number\ of\ polynomials\ of\ degree\ b-r-1\ with\ m_0 \ne 0\}}{(n-b+1)\{Number\ of\ polynomials\ of\ degree\ b-1\ with\ f_0 \ne 0\}}$$

$$= 1 - \frac{(q-1)^2 q^{b-r-2}}{(q-1)^2 q^{b-2}}$$

$$= 1 - q^{-r}.$$

Thus, a q-ary cyclic or shortened cyclic code with generator polynomial $g(x)$ can detect the fraction, $1 - q^{-r}$, of all burst-error patterns of length $b > r+1$.

	CRC-7	CRC-12	CRC-16
Error-detection coverage	99.2188%	99.9756%	99.9985
Detects all single burst of length b or less	b=7	b=12	b=16
Detects η % of all error bursts of length b.	η % =98.44%, b=8.	η % =99.95%, b=13	η % =99.997%, b=17
η_1 % of all error bursts of length b_1.	η_1 %=99.2188%, b_1=8	η_1 %=99.9756%, b_1=13	η_1 %=99.9985%, b_1=17

Table 4.10 Error-detection performance

Example 4.13
If one applies the above results to the binary (n,k) CRC codes, one obtains the following characterization of their error-detection coverage and burst-error detection performances. Note that these performances are independent of the lengths of the encoded message blocks, but depend only on $r=n-k$. From the above discussion, a CRC-r code has an error-detection coverage of $1 - q^{-r}$ and can detect all single bursts of length r or less. Also it can detect

$1 - \dfrac{q^{-(r-1)}}{q-1}$ of all burst-errors of length $r+1$ and detect $1-q^{-r}$ of all burst-errors of length larger than r. Specifically, one has Table 4.10 of error detection performance for $q=2$.

(b) Performance over a Q-ary memoryless symmetric channel
In error-detection applications one is interested only in the probability, denoted P_{ud}, that an error pattern goes undetected at the decoder. As noted earlier, this is exactly the probability that the error pattern corresponds to a nonzero code vector. Thus, knowledge of the complete weight spectrum of a code provides, in principle, the tools for analyzing P_{ud}.

An exact determination of the performance of a CRC code over the binary-symmetric channel requires knowledge of the weight distribution of the code. As this is usually unavailable, the following result can be used as a upper bound on performance under noisy conditions. For general binary (n,k) codes. Korzhik [5] has shown by ensemble-averaging arguments that binary codes exist whose probability of undetected errors on a memoryless binary-symmetric channel is bounded by

$$P_{ud} \leq 2^{-(n-k)}(1-(1-p)^n), \qquad (4.38)$$

where p is the channel-error probability. This implies for *all* p, or no matter how poor the channel, is that

$$P_{ud} \leq 2^{-(n-k)}. \qquad (4.39)$$

For Q-ary codes on uniform channels. the corresponding result is $P_{ud} \leq Q^{-(n-k)}$.

These results suggest that even under very poor channel conditions, where error correction might be prohibitively difficult, that error detection can be quite reliable, being exponentially dependent only on the number of parity symbols that are appended to the message. On typical channels for which the error probability is small, it is expected that the P_{ud} performance would be much better than the $Q^{-(n-k)}$ bound. However to obtain P_{ud} more accurately the dependence of d_{min} and the weight distribution needs to be considered.

These performance bounds do not describe the codes needed to obtain a desired performance, nor does it even hold that easily instrumented cyclic codes behave in this manner. It is rather widely assumed, though incorrectly, that (4.39) holds for any specific (n, k) code with bit-error probability p.

Linear Cyclic Codes

However, it has been shown in [6] that Hamming codes, extended Hamming codes, and double-error-correcting BCH codes satisfy inequality (4.39), and other good codes seem to reflect this fact. But this bound does not hold strictly in many cases of interest, especially under a substantial shortening of a cyclic code.

The performance of any specific (though not necessarily cyclic) code is now analyzed. A Q-ary memoryless symmetric-channel model with error probability p is assumed. If W_i denotes the number of codewords of weight i, then the probability of an undetected error is

$$P_{ud} = \sum_{i=d_{min}}^{n} W_i \left(\frac{p}{Q-1}\right)^i (1-p)^{n-i} . \tag{4.40}$$

This is just the probability that a Q-ary error pattern is produced by the channel which takes an input codeword to another valid codeword. Thus for a small error probability, the probability of undetected errors is given approximately by

$$P_{ud} \approx W_{d_{min}} \left(\frac{p}{Q-1}\right)^{d_{min}} . \tag{4.41}$$

Typically, error-detection schemes have a low redundancy: That is, k is nearly as large as n. Consequently, it is often convenient to find the weight spectrum of the smaller dual code and to use the MacWilliams identity to determine the required weight spectrum of the object code. If $W_i^{\perp}$ denotes the number of weight i words in the dual code, then Eq.(3.17) provides the weight distribution of the desired code, from which (4.39) gives P_{ud}. In the case of binary codes, this reduces to

$$P_{ud} = 2^{-(n-k)} \sum_{i=0}^{n} W_i^{\perp} (1-2p)^i - (1-p)^n . \tag{4.42}$$

For *cyclic codes*, other important error-detection claims can be made. Recall that codewords can be represented (in non-systematic form) by

$$c(x) = m(x)g(x). \tag{4.43}$$

Since $g(x)$ has degree $n - k$, any nonzero code polynomial has exponents that span at least $n-k+1$ positions. Consequently, no nonzero codeword can have all its nonzero symbols confined to $n - k$ or fewer positions, including end-around counting of positions. Since undetectable error patterns are the same set as the nonzero codewords, it is concluded that all error bursts, which are confined to n-k contiguous positions, including end-around bursts, are detectable. Furthermore, among the error patterns of length n-k+1 bits, the fraction of undetectable events is known to be $q^{-(n-k-1)}/(q-1)$, and for still longer error events, the undetectable fraction is $q^{-(n-k)}$ [7] independent of the channel quality. These results also attest to the importance of the number of parity symbols, independent of the message length.

Example 4.14
Consider the (6,3) shortened cyclic code discussed in Example 4.12. From Table 4.9 it is seen that there are 4 codewords of weight 3 and 3 codewords of weight 4. Also the minimum distance of the code is $d_{min} = 3$. When this code is transmitted over a binary symmetric channel with an error probability p between 10^{-6} and 10^{-4}, the probability of undetected error is obtained from Eqs.(4.40)-(4.42) to be
$$P_{ud}(p) = 4p^3(1-p)^6 + 3p^4(1-p)^2 \approx 4p^3.$$
The plot of $P_{ud}(p)$ for various value of p is shown in Figure 4.8.

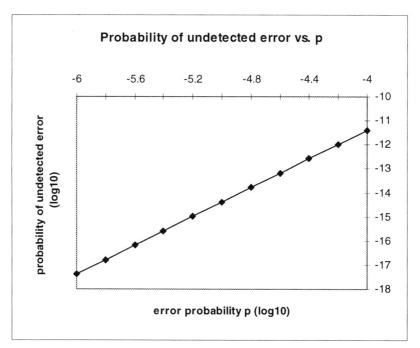

Figure 4.8 Probability of undetected error on BSC for the (6,3) shortened cyclic code.

Linear Cyclic Codes

4.4.2 Applications of CRC in Industry Standards

Many CRC codes are utilized in a variety of industry standards. In the following subsections, some details are given here for several of these applications.

4.4.2.1 The 16-bit CRC-CCITT

CRC codes are used to perform error detection in almost all modern communication systems. For example, in a communication network system, the CRC is a typical method for detecting errors at the data-link layer and the transport layer of the Open Systems Interconnection(OSI) reference model.

The Consultative Committee on International Telephone and Telegraph (CCITT) was an international standards organization and is currently renamed as the International Telecommunication Union (ITU). The 16-bit CRC-CCITT is an error-detecting signature with 16 cyclic-redundancy check bits. Such a CRC technique was recommended by the CCITT for many applications, e.g. data-transfer protocols, computer- interface error detection, etc..

In these applications the data-resending approach is the simplest way to correct transmission errors. These data-resending techniques rely on error-detecting signatures. To detect a large number of errors, it is assumed that the signature values are uniformly distributed; that is , on random data any signature value is about as likely as any other. A properly designed CRC should have this property.

As discussed in the last section, a CRC constructed by an (n,k) cyclic code is capable of detecting any error burst of length n-k or less. Also, the fraction of undetectable error bursts of length n-$k+1$ is $2^{-(n-k+1)}$. For example, a 12-bit CRC misses only one error out of 2^{12}, while a 32-bit CRC will miss only one error in 2^{32}. The most important aspect of a CRC is its length $(n$-$k)$. Long CRCs are more effective than short ones at detecting errors, but also introduce more overhead redundancy and implementation complexity. As a rule of thumb, the 8-bit check-sum error-detection codes are reasonable for packets up to about 128 bytes. Beyond that, the 16-bit CRC-CCITT is adequate for all but the most application-demanding situations.

The 16-bit CRC-CCITT is defined as the remainder obtained by dividing the message polynomial $m(x)$ x^{16} by the generating polynomial $g(x)=x^{16} + x^{12} + x^5 + 1$. The codewords are formed as a k-bit information sequence followed by a check pattern of 16 bits (two 8-bit bytes) at the end of the message. Thus, instead of directly computing the CRC of the message $m(x)$, the CRC of the left-shifted message $m(x)x^{16}$ is computed. That is, two zero bytes are appended to the end of the message. These two zero bytes are the results of this CRC calculation. The CRC is simply a remainder polynomial $q(x)$ of at most 15th degree. Hence, the polynomial $m(x)x^{16} +q(x)$ is dividable by $g(x)$. This suggests that the encoding process adds two zero bytes to the end of the message, which are used when computing the CRC. When a message is sent, these two zero bytes are replaced by the CRC bytes. Systematic encoding is accomplished with the circuit shown in Figure 4.9 and is available commercially in integrated-circuit form. Although this encoder can be used with any packet of length k to produce a $(k + 16, k)$ code (a shortened cyclic code in general), the resulting set of codewords is not cyclic in general since $g(x)$ usually is not a factor of $x^{k+16} -1$. At the receiver the decoder simply computes the CRC of the message and adds the results to the CRC bytes, and then tests to see whether the result equals zero.

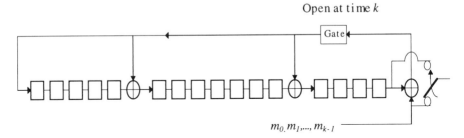

Figure 4.9 Systematic encoder for CCITT polynomial code with $g(x) = x^{16} + x^{12} + x^5 +1$.

Since the 16-bit CRC-CCITT is implemented easily in software or hardware, this error-detection technique is widely used in many communication protocols, such as the XYModem, the Zmodem, and the Kermit file-transfer protocols(FTPs), the X.25 protocol for packet communication, etc. In the following a software implementation of the 16-bit CRC-CCITT is discussed in some detail.

Because the 16-bit CRC uses polynomials, modulo 2, there are only two possibilities at each step of the long-division process: either one divides or one doesn't. This is the key to the bit-wise CRC calculation. In polynomial form if the previous message polynomial is $m(x)$ and the generating polynomial of the CRC is $g(x)$, then $m(x)/g(x)$ can be written in terms of a quotient $q(x)$ and a remainder $r(x)$, i.e. $m(x)/g(x)=q(x)+r(x)/g(x)$. From this decomposition one obtains $r(x)$ as the CRC of the previous message $m(x)$. If the current message bit is b, the new message polynomial can be represented as $m(x)x+b$. (Multiplying m(x) by x raises each power by one, i.e. it performs a left shift). For the previous remainder and the new remainder, the following polynomial equations hold : $(xm(x)+b)/g(x) = xm(x)/g(x)+b/g(x) = xq(x)+(xr(x)+b)/g(x)$. This result demonstrates that the remainder of the divisor $(xm(x)+b)/g(x)$ is given by $(xr(x)+b)/g(x)$. Thus, to determine the new remainder (the new CRC bit), one shifts the new bit b into the previous remainder, and then takes the remainder after a division by $g(x)$. Since the previous remainder is at most of degree 15, the degree of the new remainder cannot be greater than 16th degree. Hence the division is a one-bit test and an "exclusive-or" operation. A straightforward implementation of this 16-bit CRC-CCITT in C language is given as follows :

```
short CRC16(bit, crc)
    int bit;
    short crc;
{
  long longcrc = crc;

  longcrc = (longcrc << 1) ^ bit;   /* compute the next bit */
  if (longcrc&0x10000) longcrc^=0x11021;  /* reduction */
  return(longcrc&0xFFFF);
}
```

In this subroutine, 0x11021 is the binary representation of the generator polynomial of the 16-bit CRC-CCITT. The CRC computations consist of two alternating operations : collect the next bit, and then reduce, modulo, the value obtained so far, by $g(x)$. This subroutine is called once for each bit of data. Consequently this subroutine is not very efficient.

For a faster computation of the 16-bit CRC-CCITT, a table look-up algorithm is developed. Note that if one looks at the operations in the above subroutine over the 8-bit interval, that the eight reductions do not depend on the bits being shifted into the low-order end of the register; these reductions depend only on what already is in the high-order byte. Thus, one can rearrange the calculation to collect 8 bits and then to perform eight reductions. By pre-computing the reduction process in a table, a very efficient algorithm can be generated. The following subroutine (CRC16Table) pre-computes the reductions for all 256 possible values of a byte.

```
static short Crc16Table[256];
void CRC16Table()
{
   short i,j,crc;

   for (i=0; i<256; i++)
     {
        crc = (i<<8);  /* put i into high-order byte */
        for (j=0; j<8; j++)
           crc = (crc<<1)^((crc&0x8000) ? 0x1021 : 0);
        Crc16Table[i] = crc&0xFFFF;
     }
}
```

For most applications memory limitations are not critical. Hence many programs include the above pre-computed table as a list of constants in the source code, rather than having to compute it in real-time. By a use of the pre-computed values in the table, the 16-bit CRC-CCITT can be quickly updated by the following subroutine:

```
CRC16_FAST(ch,crc)
      int ch;
      short crc;
{
   crc = Crc16Table[((crc>>8)&255)]^(crc<<8)^ch;
   return crc&0xFFFF;
}
```

In actual software implementations of the 16-bit CRC-CCITT, such as Zmodem and Kermit, there are many variations of the above algorithms. Other commonly used CRCs, such as a 32-bit CRC, can also be implemented in a similar manner.

4.4.2.2 CRCs in Spread-Spectrum Cellular Systems

In order to improve the reliability of transmission, CRCs are adopted in the standards for cellular mobile telecommunication systems. Next, the CRC's

Linear Cyclic Codes

specified in the cellular system, which use the Code-Division Multiple-Access(CDMA) technology, are discussed as an example.

The currently deployed CDMA cellular technology, referred to as IS-95, is the standard proposed and adopted in North America. Briefly speaking, the signal messages in an IS-95 CDMA system are transmitted over the sync channel, the paging channel, and the access channel. The sync channel is used to provide time and frame synchronization to the mobile station. Only one message, the Sync-Channel Message, is sent on the sync channel. The paging channel is used to send control information to mobile stations that are not assigned yet to a traffic channel. A 30-bit CRC is computed for each Access-Channel signaling message. The CRC includes the message length (MSG_LENGTH) field, which denotes the bit length of the message and the message body. The generator polynomial for the CRC is given as follows:

$$g(x) = x^{30} + x^{29} + x^{21} + x^{20} + x^{15} + x^{13} + x^{12}$$
$$+ x^{11} + x^8 + x^7 + x^6 + x^2 + x + 1.$$

The value of the CRC is computed by the following procedure and the logic structure shown in Figure 4.10:

1. All shift-register elements are initialized to the logical one.
2. The switches are set to the "up" position.
3. The information bit count k is defined as 8 + message body length in bits.
4. The register is clocked k times, with the length and message body of the message as the k input bits.
5. The switches are set to the "down" position.
6. The register is clocked an additional 30 times.
7. The 30 additional output bits are the CRC field.
8. The bits are transmitted in the order in which they appear at the output of the CRC encoder.

The CDMA base station sets the CRC field of a Forward Traffic Channel message to the CRC computed for the message. The CRC computation includes the MSG_LENGTH field and the message body. The CRC field is 16 bits in length.

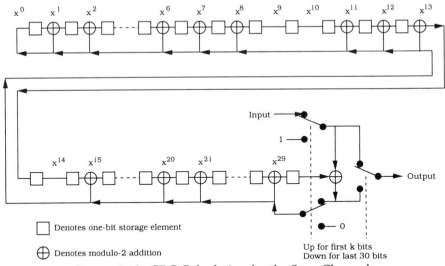

Figure 4.10. CRC Calculation for the Sync Channel, Paging Channel and Access Channel

CRC type	Quotient Polynomial	Applications
CRC-8	$x^8 + x^7 + x^6 + x^4 + x^2 + 1$	
CRC-ATM	$x^8 + x^2 + x + 1$	Employed in the adaptation layer of the ATM protocol.
CRC-12	$x^{12} + x^{11} + x^3 + x^2 + x + 1$	Used when the character length is 6-bits
CRC-16 (or CRC-ANSI[1])	$x^{16} + x^{15} + x^2 + 1$	Used for 8-bit characters
CRC-SDLC[2]	$x^{16} + x^{15} + x^{13} + x^7 + x^4 + x^2 + x + 1$	Used for 8-bit characters. (e.g. IBM Synchronous Data Link Control)
CRC-32	$x^{32} + x^{26} + x^{23} + x^{22} + x^{16} + x^{12} + x^{11} + x^{10} + x^8 + x^7 + x^5 + x^4 + x^2 + x + 1$	IBM Token-Ring, IBM PC Network, CSMA/CD Ethernet LANs.

Table 4.11 A list of CRC codes

[1] ANSI stands for American National Standards Institute.
[2] SDLC stands for IBM Synchronous Data-Link Control.

The generator polynomial for the CRC is the standard CRC-CCITT polynomial discussed in the last section. The CRC is equal to the value computed by the following procedure and logic shown in Figure 4.9.

Linear Cyclic Codes

1. All shift register elements are initialized to the logical one.
2. The switches are set to the "up" position.
3. The information bit-count k is defined as 8 + message body length in bits.
4. The register is clocked k times, with the length and message-body fields of the message as the k input bits.
5. The switches are set to the "down" position.
6. The register is clocked an additional 16 times.
7. The 16 additional output bits are the CRC field.
8. The bits are transmitted in the order in which they appear at the output of the CRC encoder.

The other commonly used CRC polynomials and some of their applications are shown in Table 4.11.

Problems

4.1 Consider the (n,k) binary cyclic code C with the generator polynomial $g(x)$.

(a) Suppose that you have two computer programs which can correctly calculate the minimum distance d_{min} and the minimum weight w_{min} of this code. If you run these programs, should you get $w_{min} < d_{min}$?

(b) Let $w(g(x))$ denote the weight of $g(x)$. Should $w(g(x))$ satisfy $w(g(x)) \geq 2$ for $n > k$?

(c) Should you have $w(g(x)) < w_{min}$?

(d) Should you have $x^n \equiv 1 \mod(g(x))$?

(e) Should d_{min} satisfy $d_{min} > \deg(g(x)) + 1$?

(f) Let $s(x)$ be the syndrome polynomial of a received word polynomial $r(x)$. Does it satisfy $xs(x) \equiv xr(x) \mod(g(x))$?

(g) Should the syndrome polynomial $s(x)$ satisfy $\deg(s(x)) > n - k$? (Here $\deg(s(x))$ denotes the degree of $s(x)$)

(h) If $r(x) = c(x) + e(x)$, where $c(x)$ is a code polynomial and $e(x)$ is a correctable error pattern, then $s(x) \equiv 0 \mod(g(x))$ iff $e(x) = 0$. Is this statement true?

(i) Let $h(x) = \dfrac{x^n + 1}{g(x)}$. Can you guarantee that $h(x)$ is the generator polynomial of the dual code of C?

(j) Does the inequality $w(s(x)) \geq w_{min} - w(r(x))$ hold?

4.2 A cyclic code is generated by $g(x) = x^8 + x^7 + x^6 + x^4 + 1$.
(a) Find the codeword length.
(b) Find the generator matrix and parity-check matrix in systematic form for this code.
(c) Design a linear switching circuit for encoding, one uses $k=7$ stages and the other uses $n-k=8$ stages.
(d) Show that $\alpha, \alpha^2, \alpha^3$ and α^4 are roots of $g(x)$, where α is a root of $x^4 + x + 1$.
(e) Design circuits for calculating $s_i = r(\alpha^i)$ for $i=1,2,3,4$ of any received word.

4.3 Consider a binary cyclic code with odd codeword length n.
(a) Show that the all 1's word is not a codeword if the code has an overall parity check.
(b) Show that it is a codeword if the code does not have an overall parity check.

4.4 For a cyclic code of length n, generated by $g(x)$ over $GF(q)$ and $(n,q)=1$, show that the all 1's word is a codeword if and only if $g(x)$ is not divisible by $x-1$.

4.5 Consider the (15,5) binary linear code C.
(a) What is the maximum burst-error-correction capability of C?
(b) Assume that C is also a cyclic code generated by
$$g(x) = 1 + x + x^2 + x^4 + x^5 + x^8 + x^{10}.$$
Develop an algorithm to decode the received word $\bar{r} = (r_0, r_1, \ldots, r_{14})$ $=$ (111110110010000) to find a burst of length 5 or less.

4.6 In the Ethernet protocol for local-area networks(LANs), a CRC code is employed with the generator polynomial given by
$g(x) = x^{32} + x^{26} + x^{23} + x^{22} + x^{16} + x^{12} + x^{11} + x^{10} + x^8 + x^7 + x^5 + x^4 + x^2 + x + 1$

Linear Cyclic Codes

This polynomial generates a binary Hamming code of length $2^{32}-1$, with $d_{min}=3$. Although this block length is much longer than used in the standard implementation, specify the random and burst-error detecting performance of the system.

(a) If the block is limited to 512 bits or less by shortening, then the minimum Hamming distance increases to 5. In this case, what can be said about the random and burst-error detecting capabilities?

(b) Normally, only error-detection is attempted with such CRC codes, but they could be employed as a combination error-correction and error-detection code. Discuss how the given code could be employed with $n=512$ to correct single errors, while still guaranteeing the detection of any two- or three-error patterns.

4.7 An 8-bit CRC code is employed in the adaptation layer of the ATM protocol by encoding the 4-byte header, containing the addressing and routing information. In effect, one obtains a (40,32) code with $d_{min}=4$. The CRC polynomial is $g(x)=x^8+x^2+x+1$. What claims can be made about the detection of errors in the header field?

4.8 Design an error-detection scheme that operates with 256-bit messages and correctly detects byte(8-bit) burst errors, as well as any four randomly placed errors in a block. (Hint : You should be able to find this code even if the code is not cyclic, but merely produced as if the code were cyclic, for example, $x(x)=m(x)g(x)$, so that the decoder is capable of detecting all error patterns confined to $n-k-1$ bits).

4.9 With the advent of inexpensive, fast semiconductor memories, some encoder and decoder designs can be done effectively by the use of read-only-memory(ROM) tables. For the extended Golay (24,12) code describe a ROM-based implementation of an encoder for the systematic code, and a ROM-assisted syndrome decoder(one still can use a feedback-shift register to compute the syndromes). Repeat for the (48,24) extended quadratic residue code, and comment on the feasibility of this approach.

4.10 Derive the bit-error probability P_b of perfect codes for a BSC by the following steps.
(a) Use of the standard array decoding to show that

$$P_b = \sum_{j=1}^{n} \frac{\mu_j}{k} w_j \Pr(R_j | \bar{c}_0),$$

where $\Pr(R_j | \bar{c}_0)$ is the probability of the event that, when the all-zero codeword $\bar{c}_0 = (0,0,\ldots,0)$ is transmitted, the received word falls in the correctable portion of codewords of weight j in the standard array. μ_j designates the average number of nonzero information symbols associated with codewords of weight j and w_j is the number of codewords of weight j.

(b) Show that
$$\frac{\mu_j}{k} = \frac{j}{n}.$$

(c) Derive Eq. (4.37a) by the use of the results obtained from (a) and (b).

4.11 Suppose that the transmission alphabet is $GF(q)$. Show that there are $\left((q-1)^2 \cdot q^{b-2}\right)$ different burst-error patterns of a length b. (hint : It is known that a burst-error pattern of length b begins and ends with a nonzero value, and the intervening symbols in the pattern may take on any value, including zero.)

4.12 Show that the only cyclic (2k,k), rate- 1/2, codes are those for which $g(x) = x^k - 1$ (which obviously factors $x^{2k} - 1$). These cyclic codes all have $d_{\min} = 2$ for any block length and thus have very poor performance. If one needs to have R=1/2, it is better to modify another code and accept the loss of the cyclic property.

Bibliography

[1] E. Prange, "Cyclic Error-Correcting Codes in Two Symbols", AFCRC-TN-57-103, Air Force Cambridge Research Center, Bedford, Mass., Sept. 1957.
[2] W. W. Peterson and E. J. Weldon, Jr., Error Correcting Codes, 2nd ed. Cambridge, MA. MIT Press, 1972.
[3] F. J. MacWilliam and N. J. A. Sloane, The Theory of Error Correcting Codes, Amsterdam : North-Holland, 1977.
[4] S. Lin and D. J. Costello, Error Control Coding : Fundamentals and Applications, Englewood Cliffs, NJ : Prentice Hall, 1983.
[5] E. Berlekamp, Algebraic Coding Theory, New York : McGraw-Hill, 1968.

[6] Xuemin Chen, <u>Binary Quadratic Residue Codes for Random Error Correction</u>, Ph.D. Dissertation, Communication Sciences Institute(CSI-92-12-06), EE-System Dept., University of Southern California, Los Angeles, 1992.

[7] X. Chen, I. S. Reed, and T. K. Truong, "No binary quadratic code of length 8m-1 is quasi-perfect", IEEE Trans. on Inform. Theory, No. 1994.

[8] M. Elia, "Algebraic decoding of the (23,12,7) Golay code", IEEE Trans. Inform. Theory, Vol. IT-33, pp. 150-151, Jan. 1987.

[9] I. S. Reed, X. Yin, and T.K.Truong, " Decoding the (24,12,8) Golay code", IEE Proc., vol. 137, pt. E. No. 3, pp. 202-206, May 1990.

[10] G. R. Lang and F. M. Longstaff, "A Leech lattice modem", IEEE Journal on selected areas in communications, Vol. 7, No. 6, August, 1989.

[11] J. H. Conway and N. J. A. Sloane, "Decoding techniques for codes and lattices, including the Golay code and the leech lattice," IEEE Trans. Inform. Theory, vol. IT-32, pp.41-50, 1986.

[12] Y. Be'ery and J. Snyders, "Soft decoding of the Golay codes", IEEE Trans. Inform. Theory, 1988.

[13] M.J.E. Golay, "Notes on digital coding", Proc. IRE, 37, p.657, June 1949.

[14] X. Chen, I. S. Reed, and T. K. Truong, "A performance comparison of the binary quadratic residue codes with the 1/2-rate convolutional codes", IEEE Trans. on Inform. Theory, Vol. 40, No. 1, Jan. 1994.

[15] S. B. Wicker, <u>Error Control Systems</u> for digital communication storage, Englewood Cliffs, NJ : Prentice Hall, 1995.

5 BCH Codes

A class of powerful multiple-error-correcting cyclic codes was discovered by Bose and Ray-Chaudhuri in 1960 [1] and independently by Hocquenghem in 1959 [2]. These codes are known as the BCH codes. The BCH codes provide a wide variety of block lengths and corresponding code rates. They are important not only because of their flexibility in the choice of their code parameters, but also because, at block lengths of a few hundred or less, many of these codes are among the most used codes of the same lengths and code rates. Another advantage is that there exists very elegant and powerful algebraic decoding algorithms for the BCH codes. The importance of the BCH codes also stems from the fact that they are capable of correcting all random patterns of t errors by a decoding algorithm that is both simple and easily realized in a reasonable amount of equipment. BCH codes occupy a prominent place in the theory and practice of multiple-error correction.

The original applications of BCH codes were restricted to binary codes of length $2^m - 1$ for some integer m. These were extended later by Gorenstein and Zierler (1961) [3] to the nonbinary codes with symbols from the Galois field $GF(q)$. The most useful and BCH-related nonbinary cyclic codes are the Reed-Solomon codes(1960) which are discussed in the next chapter.

BCH codes and their implementation in the time domain are presented in this chapter. Using discrete Fourier transforms over a finite field, BCH codes also can be defined and implemented in, what is called, the frequency domain. The first decoding algorithm for binary BCH codes was devised by Peterson[4] in 1960. Since then, Peterson's algorithm has been refined by Berlekamp, Massey, Chien, Forney, and many others[5][6][7].

5.1 Definition of the BCH Codes

A BCH code is a cyclic code that usually is defined in terms of the roots of its generator polynomial. Given a finite field $GF(q)$ and a block length $n \geq 3$, which is a divisor of $q^m - 1$ for some positive integer m, an (n,k) BCH code of t-error-correction for $2 \leq 2t \leq n-1$ over $GF(q)$ is generated cyclically by the polynomial,

$$g(x) = LCM\{m_{m_0}(x), m_{m_0+1}(x), \ldots, m_{m_0+2t-1}(x)\}. \qquad (5.1a)$$

Here $m_{m_0+i}(x)$ for $i=0,1,\ldots,2t-1$ are the minimal polynomials of the $2t$ successive powers $\alpha^{m_0}, \alpha^{m_0+1}, \ldots, \alpha^{m_0+2t-1}$ of a field element α whose order is n in some extension field $GF(q^m)$. Also $LCM\{\cdot\}$ denotes the least common multiple polynomial. Normally, the field element α is a primitive element in the extension field $GF(q^m)$ in which case the code is called a primitive BCH code. The order of such an element is $n = q^m - 1$, which is also the length of the code. Furthermore, the codes with $m_0 = 1$ are said to be narrow-sense BCH codes. Notice that the degree of $g(x)$ is less than or equal to $2t \cdot m$, since there are at most $2t$ distinct minimal polynomials involved in the construction of $g(x)$, and each has degree at most m. Thus, for BCH codes over any finite field one has the following relations:

$$n - k = \deg(g(x)) \leq 2t \cdot m \text{ and } n = q^m - 1.$$

In this chapter only the binary BCH codes, i.e. $q=2$, are discussed.

Let $m_i(x)$ be the minimal polynomial of α^i. Also let $c(x) = c_0 + c_1 x + \cdots + c_{n-1} x^{n-1}$ be a code polynomial with coefficients in $GF(2)$. If $c(x)$ has $\alpha, \alpha^2, \ldots, \alpha^{2t}$ as its roots, then $c(x)$ is divisible by the minimal polynomials $m_1(x), m_2(x), \ldots, m_{2t}(x)$ of $\alpha, \alpha^2, \ldots, \alpha^{2t}$, respectively. The generator polynomial $g(x)$ of the binary t-error-correcting BCH code of block length $n = 2^m - 1$ and rate k/n is the lowest degree polynomial over $GF(2)$ with these roots. Thus, the generator polynomial of the code must be the least common multiple of these minimal polynomials. That is,

$$g(x) = LCM\{m_1(x), m_2(x), \ldots, m_{2t}(x)\}, \qquad (5.1b)$$

where $m_i(x)$ is the minimal polynomial of α^i for $i=1,2,\ldots,2t$, and consists of $2t$ successive powers of the field element α, starting at α^i. The order of α is n in the extension field $GF(2^m)$.

BCH Codes

The nonprimitive BCH codes are defined in a similar manner, except that α is replaced by a nonprimitive element β of $GF(q^m)$, and the code length becomes the order of β. It is noted that in the set of the $2t$ minimal polynomials there often will be found repeated polynomials. Since the minimal polynomials are all irreducible, finding the LCM polynomial amounts to a formation of the product of the distinct minimal polynomials in Eq. (5.1).

Example 5.1

Consider the double-error-correcting (15, 7) BCH code. Let α be a primitive element of the finite field $GF(2^4)$ such that $1 + \alpha + \alpha^4 = 0$. If $m_i(x)$ denotes the minimal polynomial of α^i for $i=1,2,3,4$, which are the nonzero elements of $GF(2^4)$, then the list of minimal polynomials for this code is given in Table 5.1.

Field Elements	Conjugates	Minimal Polynomials
α	$\alpha^2, \alpha^4, \alpha^8$	$m_1(x) = 1 + x + x^4$
α^2	$\alpha^4, \alpha^8, \alpha^{16} = \alpha$	$m_2(x) = 1 + x + x^4$
α^3	$\alpha^8, \alpha^{12}, \alpha^{24} = \alpha^9$	$m_3(x) = 1 + x + x^2 + x^3 + x^4$
α^4	$\alpha^8, \alpha^{16} = \alpha, \alpha^{32} = \alpha^2$	$m_4(x) = 1 + x + x^4$
α^5	α^{10}	$m_5(x) = 1 + x + x^2$
α^6	$\alpha^{12}, \alpha^{24} = \alpha^9, \alpha^{48} = \alpha^3$	$m_6(x) = 1 + x + x^2 + x^3 + x^4$
α^7	$\alpha^{14}, \alpha^{28} = \alpha^{13}, \alpha^{56} = \alpha^{11}$	$m_7(x) = 1 + x^3 + x^4$
α^8	$\alpha^{16} = \alpha, \alpha^{32} = \alpha^2, \alpha^{64} = \alpha^4$	$m_8(x) = 1 + x + x^4$
α^9	$\alpha^{18} = \alpha^3, \alpha^{36} = \alpha^6, \alpha^{72} = \alpha^{12}$	$m_9(x) = 1 + x + x^2 + x^3 + x^4$
α^{10}	$\alpha^{20} = \alpha^5$	$m_{10}(x) = 1 + x + x^2$
α^{11}	$\alpha^{22} = \alpha^7, \alpha^{44} = \alpha^{14}, \alpha^{88} = \alpha^{13}$	$m_{11}(x) = 1 + x^3 + x^4$
α^{12}	$\alpha^{24} = \alpha^9, \alpha^{48} = \alpha^3, \alpha^{96} = \alpha^6$	$m_{12}(x) = 1 + x + x^2 + x^3 + x^4$
α^{13}	$\alpha^{26} = \alpha^{11}, \alpha^{52} = \alpha^7, \alpha^{104} = \alpha^{14}$	$m_{13}(x) = 1 + x^3 + x^4$
α^{14}	$\alpha^{28} = \alpha^{13}, \alpha^{56} = \alpha^{11}, \alpha^{112} = \alpha^7$	$m_{14}(x) = 1 + x^3 + x^4$

Table 5.1 Minimal polynomials of the elements in $GF(2^m)$ for $m=4$.

It is easily seen from Table 5.1 that

$$m_1(x) = m_2(z) = m_4(z) = m_8(x),$$
$$m_3(x) = m_6(z) = m_9(z) = m_{12}(x),$$
$$m_5(x) = m_{10}(x),$$
$$m_7(x) = m_{11}(z) = m_{13}(x) = m_{14}(x).$$

Hence, $m_1(x)$, $m_3(x)$, $m_5(x)$, and $m_7(x)$ are the only four distinct irreducible polynomials of $GF(2^4)$. Thus, the generator polynomial of the double-error-correcting binary BCH code of length 15 is given by

$$g(x) = LCM\{m_1(x), m_2(x), m_3(x), m_4(x)\} = LCM\{m_1(x), m_3(x)\}$$
$$= m_1(x)m_3(x) = 1 + x^4 + x^6 + x^7 + x^8.$$

From this example, one sees more generally that every even power of α in the element sequence has the same minimal polynomial as some preceding odd power of α in the sequence. As a consequence, the generator polynomial $g(x)$ of the binary t-error-correcting BCH code of length $2^m - 1$, given by Eq. (5.1), can be reduced to

$$g(x) = LCM\{m_1(x), m_3(x), ..., m_{2t-1}(x)\} \tag{5.2}$$

where the degree of each $m_i(x)$ is no greater than m. Therefore, Eq. (5.2) implies that $c(x)$ is a codeword if and only if $\alpha, \alpha^3, ..., \alpha^{2t-1}$ are the roots of $c(x)$. Since the degree of each minimal polynomial is m or less, the largest possible value for the degree of $g(x)$ is mt. Hence, the code has at most mt parity-check digits, i.e., $n - k \leq mt$.

Example 5.2

Consider the binary primitive BCH codes of length $2^m - 1$ with $m = 4$. The parameters for these codes are shown in the Table 5.2.

n	k	t	n-k	mt
15	11	1	4	4
15	7	2	8	8
15	5	3	10	12
15	1	7	14	28

Table 5.2 Parameters of BCH codes of length $2^4 - 1 = 15$.

Table 5.2 suggests that the number of parity-check digits $n - k$ is exactly equal to mt when t is small, e.g. the (15,11) BCH code and the (15,7) BCH code. However, $n - k < mt$ when t is large, e.g. the (15,5) BCH code and the (15,1) BCH code. In general, the number of parity-check bits for any BCH code is expressed by $n - k \leq mt$, from which it can be computed that the number of information digits satisfies the inequality $k \geq 2^m - mt - 1$.

BCH Codes

In general, for any positive integer pair m,t such that $m \geq 3$ and $t < n/2$, there exists a binary BCH code of block length $n = 2^m - 1$, where the number of parity-check bits satisfies $n - k \leq mt$, and the minimum distance d_{min} satisfies $d_{min} \geq d_0 = 2t + 1$ where d_0 is called the designed distance of the code. The minimum distance d_{min} can be larger than d_0. The following steps are needed to determine the parameters of a BCH code.

1. Choose a primitive polynomial of degree m, and construct $GF(2^m)$.
2. Find the minimal polynomials $m_i(x)$ of α^i for $i = 1, 3, \ldots, 2t-1$.
3. Obtain $g(x) = LCM\{m_1(x), m_3(x),\ldots, m_{2t-1}(x)\}$.
4. Determine k from the relation $\deg(g(x))=n - k$.
5. Find the minimum distance $d_{min} \geq 2t+1$ by referring to the weight of $g(x)$, i.e. $wt(g(x))= d_{min}$.

Example 5.3

Consider the t-error-correcting BCH codes of length $n = 2^5 - 1 = 31$. Let α be a primitive element of $GF(2^5)$ which satisfies $1+\alpha^2 +\alpha^5 = 0$. The extension field $GF(2^5)$ of $GF(2)$ is constructed from the primitive polynomial $p(x) = 1 + x^2 + x^5$ and is given in Table 5.3b.

Elements	Conjugates	Identities	Minimal polynomial
α	$\alpha^2, \alpha^4, \alpha^8, \alpha^{16}$	$m_1(x) = m_2(x) = m_4(x) = m_8(x) = m_{16}(x)$	$m_1(x) = 1+x^2 +x^5$
α^3	$\alpha^6, \alpha^{12}, \alpha^{24}, \alpha^{17}$	$m_3(x) = m_6(x)= m_{12}(x) = m_{24}(x)= m_{17}(x)$	$m_3(x) = 1+x^2 +x^3 +x^4 +x^5$
α^5	$\alpha^{10}, \alpha^{20}, \alpha^9, \alpha^{18}$	$m_5(x) = m_{10}(x) = m_{20}(x)=m_9(x)= m_{18}(x)$	$m_5(x) = 1+x+x^2 +x^4 +x^5$
α^7	$\alpha^{14}, \alpha^{28}, \alpha^{25}, \alpha^{19}$	$m_7(x) = m_{14}(x) = m_{28}(x) = m_{25}(X) = m_{19}(x)$	$m_7(x) = 1+x+x^2 +x^3 +x^5$
α^{11}	$\alpha^{22}, \alpha^{13}, \alpha^{26}, \alpha^{21}$	$m_{11}(x) = m_{22}(x) = m_{13}(x) = m_{26}(x) = m_{21}(x)$	$m_{11}(x) = 1+x+x^3 +x^4 +x^5$

Figure 5.3a Results of Example 5.3.

Finite field elements	Binary representation
0	0 0 0 0 0
1	1 0 0 0 0
α	0 1 0 0 0
α^2	0 0 1 0 0
α^3	0 0 0 1 0
α^4	0 0 0 0 1
$\alpha^5 = 1 + \alpha^2$	1 0 1 0 0
$\alpha^6 = \alpha + \alpha^3$	0 1 0 1 0
$\alpha^7 = \alpha^2 + \alpha^4$	0 0 1 0 1
$\alpha^8 = 1 + \alpha^2 + \alpha^3$	1 0 1 1 0
$\alpha^9 = \alpha + \alpha^3 + \alpha^4$	0 1 0 1 1
$\alpha^{10} = 1 + \alpha^4$	1 0 0 0 1
$\alpha^{11} = 1 + \alpha + \alpha^2$	1 1 1 0 0
$\alpha^{12} = \alpha + \alpha^2 + \alpha^3$	0 1 1 1 0
$\alpha^{13} = \alpha^2 + \alpha^3 + \alpha^4$	0 0 1 1 1
$\alpha^{14} = 1 + \alpha^2 + \alpha^3 + \alpha^4$	1 0 1 1 1
$\alpha^{15} = 1 + \alpha + \alpha^2 + \alpha^3 + \alpha^4$	1 1 1 1 1
$\alpha^{16} = 1 + \alpha + \alpha^3 + \alpha^4$	1 1 0 1 1
$\alpha^{17} = 1 + \alpha + \alpha^4$	1 1 0 0 1
$\alpha^{18} = 1 + \alpha$	1 1 0 0 0
$\alpha^{19} = \alpha + \alpha^2$	0 1 1 0 0
$\alpha^{20} = \alpha^2 + \alpha^3$	0 0 1 1 0
$\alpha^{21} = \alpha^3 + \alpha^4$	0 0 0 1 1
$\alpha^{22} = 1 + \alpha^2 + \alpha^4$	1 0 1 0 1
$\alpha^{23} = 1 + \alpha + \alpha^2 + \alpha^3$	1 1 1 1 0
$\alpha^{24} = \alpha + \alpha^2 + \alpha^3 + \alpha^4$	0 1 1 1 1
$\alpha^{25} = 1 + \alpha^3 + \alpha^4$	1 0 0 1 1
$\alpha^{26} = 1 + \alpha + \alpha^2 + \alpha^4$	1 1 1 0 1
$\alpha^{27} = 1 + \alpha + + \alpha^3$	1 1 0 1 0
$\alpha^{28} = \alpha + \alpha^2 + \alpha^4$	0 1 1 0 1
$\alpha^{29} = 1 + \alpha^3$	1 0 0 1 0
$\alpha^{30} = \alpha + \alpha^4$	0 1 0 0 1

Table 5.3b Finite field *GF(32)* with generated by $p(x) = 1 + x^2 + x^5$.

BCH Codes

Let $m_i(x)$ be the minimal polynomials of α^i, $1 \leq i \leq 2t$, which are to be the roots of the code polynomial $c(x)$. Then $m_i(x)$ is obtained below with the aid of Table 5.3b. These results are also listed in Table 5.3a.

$m_1(x) = (x+\alpha)(x+\alpha^2)(x+\alpha^4)(x+\alpha^8)(x+\alpha^{16}) = 1+x^2+x^5$.

$m_3(x) = (x+\alpha^3)(x+\alpha^6)(x+\alpha^{12})(x+\alpha^{24})(x+\alpha^{17}) = 1+x^2+x^3+x^4+x^5$.

$m_5(x) = (x+\alpha^5)(x+\alpha^{10})(x+\alpha^{20})(x+\alpha^9)(x+\alpha^{18}) = 1+x+x^2+x^4+x^5$.

$m_7(x) = (x+\alpha^7)(x+\alpha^{14})(x+\alpha^{28})(x+\alpha^{25})(x+\alpha^{19}) = 1+x+x^2+x^3+x^5$.

$m_{11}(x) = (x+\alpha^{11})(x+\alpha^{22})(x+\alpha^{13})(x+\alpha^{26})(x+\alpha^{21}) = 1+x+x^3+x^4+x^5$.

Example 5.4

Consider again the t-error-correcting BCH codes of length $n = 2^5 - 1 = 31$. BCH codes are designed from the specification of n and t. The generator polynomials are constructed with the aid of the minimal polynomials found in Example 5.3. Let $g_t(x)$ be the generator polynomials of the length-31 BCH codes for $t = 1, 2, 3, 5, 7$. The generator polynomial for the single-error-correcting BCH code of length $n = 2^5 - 1 = 31$ is obtained as follows:

$g_1(x) = LCM\{m_1(x), m_2(x)\} = m_1(x) = 1 + x^2 + x^5$

because $m_1(x) = m_2(x)$. Since $g_1(x)$ is of degree 5, one has $n - k = 5$. Therefore, $k = 26$ and the (31, 26) BCH code is determined to have $d_{min} \geq 3$, which is identical to a single error-correcting Hamming code. Since $g_1(x)$ is also a code polynomial of weight 3, the minimum distance of this code is exactly 3. Continuing in this manner, one can construct the generator polynomials for other primitive BCH codes of block length 31. $g_t(x)$ and d_{min} for $t = 2, 3, 5, 7$, are determined in a similar manner and are listed in Table 5.4.

Beyond $t = 7$ the BCH code is undefined because the number of nonzero field elements is 31. Thus, it should be noted that BCH codes are designed from the specification of n and t.

The above analysis is based on the code length of the form $n = 2^m - 1 = 31$ for $m = 5$. The first code is a Hamming code, so it has a primitive polynomial $p(x)$ of degree m as its generator polynomial $g_1(x)$. The next generator polynomial $g_2(x)$ of the same length $n = 31$ is found by multiplying $p(x) = g_1(x)$ by $m_3(x)$, the minimal polynomial of α^3. If one continues in this fashion, the previous generator polynomial is multiplied by a new minimal polynomial at each step. This process produces a sequence of BCH-code generator polynomials that increase in degree and error-correcting capability as shown in Table 5.4.

t	(n,k)	$g_t(x)$	d_{min}
1	(31,26)	$g_1(x) = LCM\{m_1(x), m_2(x)\} = m_1(x) = 1 + x^2 + x^5$	3
2	(31,21)	$g_2(x) = LCM\{m_1(x), m_2(x), m_3(x), m_4(x)\} = g_1(x)m_3(x)$ $= m_1(x)m_3(x) = 1 + x^3 + x^5 + x^6 + x^8 + x^9 + x^{10}$	7
3	(31,16)	$g_3(x) = g_2(x)m_5(x) = 1 + x + x^2 + x^3 + x^5 + x^7 + x^8 + x^9 + x^{10} + x^{11} + x^{15}$	11
5	(31,11)	$g_5(x) = LCM\{m_1(x), m_3(x), m_5(x), m_7(x), m_9(x)\} =$ $g_3(x)m_7(x) = 1 + x^2 + x^4 + x^6 + x^7 + x^9 + x^{10} + x^{13} + x^{17} + x^{18} + x^{20}$	11
7	(31,6)	$g_7(x) = m_1(x)m_3(x)m_5(x)m_7(x)m_{11}(x) = g_5(x)m_{11}(x)$ $= 1 + x + x^2 + x^5 + x^9 + x^{11} + x^{13} + x^{14} + x^{15} + x^{16} + x^{18} + x^{19} + x^{21} + x^{24} + x^{25}$	15

Table 5.4 The generator polynomials of
t-error-correcting (31, k) BCH codes

5.2 The BCH Bound on the Minimum Distance d_{min}

Let $c(x) = c_0 + c_1 x + \cdots + c_{n-1} x^{n-1}$ be a code polynomial of a primitive t-error-correcting BCH code of length $n = 2^m - 1$. If $c(x)$ has $\alpha, \alpha^2, \ldots, \alpha^{2t}$ as its roots, then $c(x)$ is divisible by the generator polynomial $g(x)$ which is the least common multiple of the $m_i(x) m_i(x)$ for $1 \le i \le 2t$, and $c(x)$ is the code polynomial of a cyclic code with g(x) as its generator.

Since α^i for $1 \le i \le 2t$ are the roots of $c(x)$, it follows that
$$c(\alpha^i) = c_0 + c_1(\alpha^i) + \cdots + c_{n-1}(\alpha^i)^{n-1} = 0. \qquad (5.3)$$
Eq. (5.3) can be expressed in the following matrix-product form:

$$(c_0, c_1, \ldots, c_{n-1}) \cdot \begin{pmatrix} 1 \\ \alpha^i \\ \vdots \\ (\alpha^i)^{n-1} \end{pmatrix} = 0, \quad 1 \le i \le 2t, \qquad (5.4a)$$

or

$$\bar{c} \cdot H_i^T = 0 \qquad (5.4b)$$

because also $\vec{c} = (c_0, c_1, \ldots, c_{n-1})$ is a code word of the t- error-correcting BCH code. Hence, the code space of the BCH code is the null space of H_i. Thus, the parity-check matrix of the code is any matrix of form,

$$H_i = \left(1, \alpha^i, \ldots, (\alpha^i)^{n-1}\right), \quad \text{for } 1 \leq i \leq 2t, \tag{5.5}$$

where the entries of H_i are the non-zero elements in $GF(2^m)$. If each entry of H_i is replaced by its corresponding m-tuple over $GF(2)$ and arranged in column form, a binary parity-check matrix for the code is produced.

Next, consider the following matrix $\hbar$ which is formed by putting H_i for $1 \leq i \leq 2t$ as its rows, i.e.

$$\hbar = \begin{pmatrix} 1 & \alpha & \cdots & \alpha^{n-1} \\ 1 & \alpha^2 & \cdots & (\alpha^2)^{n-1} \\ \vdots & & & \vdots \\ 1 & \alpha^{2t} & \cdots & (\alpha^{2t})^{n-1} \end{pmatrix}. \tag{5.6}$$

Since the i-th row vector of $\hbar$ is denoted by $H_i = \left(1, \alpha^i, \ldots, (\alpha^i)^{n-1}\right)$, the inner product of codeword $\vec{c}$ and H_i is then

$$\vec{c} \cdot H_i^T = \sum_{j=0}^{n-1} c_j (\alpha^i)^j = 0,$$

if and only if the vectors $\vec{c}$ and H_i^T are orthogonal, i.e., $\vec{c} \cdot \hbar^T = 0$. Now one wants to show that any set of the $d_0 - 1$ or $2t$ columns of $\hbar$ cannot be linearly dependent, which implies that the t-error-correcting BCH code has a minimum distance of at least d_0 or $2t+1$. To prove this, one needs to show that no $2t$ or fewer columns of $\hbar$, given in Eq. (5.6), sums to zero. Let α^j for $j=2i,4i,\ldots$, be the conjugates of α^i for some i. Then $c(\alpha^i) = 0$. Thus, the even-numbered rows of $\hbar$ given in Eq. (5.6) can be omitted.

Assume now that a codeword exists whose components consist only of the nonzero digits $c_{j_u} = 1, 1 \leq u \leq 2t$, i.e., the zero components are omitted. Then, it follows from $\vec{c} \cdot \hbar^T = 0$ in Eq. (5.4b) that

$$\left(c_{j_1}, c_{j_2}, \ldots, c_{j_{2t}}\right) \begin{pmatrix} \alpha^{j_1} & (\alpha^{j_1})^2 & \cdots & (\alpha^{j_1})^{2t} \\ \alpha^{j_2} & (\alpha^{j_2})^2 & \cdots & (\alpha^{j_2})^{2t} \\ \vdots & & & \vdots \\ \alpha^{j_{2t}} & (\alpha^{j_{2t}})^2 & \cdots & (\alpha^{j_{2t}})^{2t} \end{pmatrix} = 0 \qquad (5.7)$$

where $c_{j_1} = c_{j_2} = \cdots = c_{j_{2t}} = 1$.

Recall Sec. 3.2.1 that the minimum weight of a code word $\bar{c}$ is equal to the smallest number of columns of $\hbar$ that sum to zero. If no $2t$ (i.e. $d_o - 1$) or fewer columns of $\hbar$ add to zero, the code has the minimum weight d_o. Now, the second matrix on the left side of Eq. (5.7) is a $2t \times 2t$, or $(d_o - 1) \times (d_o - 1)$ square matrix. It is well known, that to satisfy the matrix system of Eq. (5.7), the determinant of this $(d_o - 1) \times (d_o - 1)$ matrix must be 0, i.e.,

$$D = \begin{vmatrix} \alpha^{j_1} & (\alpha^{j_1})^2 & \cdots & (\alpha^{j_1})^{2t} \\ \alpha^{j_2} & (\alpha^{j_2})^2 & \cdots & (\alpha^{j_2})^{2t} \\ \vdots & & & \vdots \\ \alpha^{j_{2t}} & (\alpha^{j_{2t}})^2 & \cdots & (\alpha^{j_{2t}})^{2t} \end{vmatrix} = 0, \qquad (5.8)$$

for $0 \le j_1 < j_2 < \cdots < j_{2t} \le n-1$.

However, the determinant D of Eq. (5.8) is, which is called, a van der Monde determinant, is shown next to be nonzero. This contradicts Eq. (5.7) and its conclusion in Eq. (5.8).

To show this fact in detail, evaluate D by factoring out α^{j_u}, for $u = 1, 2, \ldots, 2t$ to obtain

$$D = \alpha^{(j_1 + j_2 + \cdots + j_{2t})} \begin{vmatrix} 1 & \alpha^{j_1} & \cdots & (\alpha^{j_1})^{2t-1} \\ 1 & \alpha^{j_2} & \cdots & (\alpha^{j_2})^{2t-1} \\ \vdots & & & \vdots \\ 1 & \alpha^{j_{2t}} & \cdots & (\alpha^{j_{2t}})^{2t-1} \end{vmatrix} \qquad (5.9)$$

$$= \alpha^{(j_1 + j_2 + \cdots + j_{2t})} \prod_{v<u} \left(\alpha^{j_u} - \alpha^{j_v}\right) \ne 0$$

Therefore, the product on the left side of Eq. (5.9) cannot be zero. Here any set of $d_o - 1$ columns is linearly independent. Thus, the assumption that there exists a nonzero codeword $\bar{c}$ of weight $d_o - 1 \le 2t$ is invalid. This implies that the minimum distance of the t-error-correcting BCH code is at least the designed distance $d_o = 2t + 1$. The true minimum distance of a BCH code is

BCH Codes

always equal to or greater than the designed distance d_o. The designed distance d_o is only a lower bound on the true distance of the code. Thus, the minimum distance d_{min} of a BCH code with designed distance d_o is at least d_o. This fact is called the BCH bound.

Example 5.5
Consider the triple error-correcting (31, 16) BCH code with the minimum distance $d_{min} = 11$ (Table 5.4). Let α be a primitive element of $GF(2^5)$ with $1+\alpha^2+\alpha^5 = 0$. Then, by a use of Eq. (5.6) the matrix $\hbar$ of this code is expressed by

$$\hbar = \begin{pmatrix} 1 & \alpha & \alpha^2 & \alpha^3 & \alpha^4 & \alpha^5 & \alpha^6 & \alpha^7 & \alpha^8 & \alpha^9 & \alpha^{10} & \alpha^{11} \\ 1 & \alpha^3 & \alpha^6 & \alpha^9 & \alpha^{12} & \alpha^{15} & \alpha^{18} & \alpha^{21} & \alpha^{24} & \alpha^{27} & \alpha^{30} & \alpha^2 \\ 1 & \alpha^5 & \alpha^{10} & \alpha^{15} & \alpha^{20} & \alpha^{25} & \alpha^{30} & \alpha^4 & \alpha^9 & \alpha^{14} & \alpha^{19} & \alpha^{24} \\ & \alpha^{12} & \alpha^{13} & \alpha^{14} & \alpha^{15} & \alpha^{16} & \alpha^{17} & \alpha^{18} & \alpha^{19} & \alpha^{20} & \alpha^{21} \\ & \alpha^5 & \alpha^8 & \alpha^{11} & \alpha^{14} & \alpha^{17} & \alpha^{20} & \alpha^{23} & \alpha^{26} & \alpha^{29} & \alpha \\ & \alpha^{29} & \alpha^3 & \alpha^8 & \alpha^{13} & \alpha^{18} & \alpha^{23} & \alpha^{28} & \alpha^2 & \alpha^7 & \alpha^{12} \\ & \alpha^{22} & \alpha^{23} & \alpha^{24} & \alpha^{25} & \alpha^{26} & \alpha^{27} & \alpha^{28} & \alpha^{29} & \alpha^{30} \\ & \alpha^4 & \alpha^7 & \alpha^{10} & \alpha^{13} & \alpha^{16} & \alpha^{19} & \alpha^{22} & \alpha^{25} & \alpha^{28} \\ & \alpha^{17} & \alpha^{22} & \alpha^{27} & \alpha & \alpha^6 & \alpha^{11} & \alpha^{16} & \alpha^{21} & \alpha^{26} \end{pmatrix}$$

It is easy to verify for this $\hbar$ that the BCH bound 7 which is less than the minimum distance of the code.

5.3 Decoding Procedures for BCH Codes

There are several algorithms that have been developed specifically to decode BCH codes. The algorithm studied in this chapter was first developed by Peterson for binary BCH codes.

5.3.1 Syndrome computation

Suppose that a code word $c(x)$ is transmitted and because of the channel errors, the received word $r(x)$ is $r(x)=c(x)+e(x)$, where $e(x) = e_0 + e_1 x + \cdots + e_{n-1} x^{n-1}$ is called the error-polynomial pattern. Let there be $v \le t$ non-zero coefficients of $e(x)$. Suppose that these v errors occur in the unknown locations $j_1, j_2, \ldots, j_v$, so that the polynomial $e(x)$ has the form,

$$e(x) = \sum_{i=1}^{v} x^{j_i} \quad \text{for } 0 \le j_i \le n-1. \tag{5.10}$$

Since $\alpha, \alpha^2, \ldots, \alpha^{2t}$ are the roots of each code polynomial, one has $c(\alpha^i) = 0$ for $1 \le i \le 2t$. Therefore, it follows from $r(x) = c(x) + e(x)$ that

$$r(\alpha^i) = e(\alpha^i) \quad \text{for } i = 1, 2, \ldots, 2t. \tag{5.11}$$

As usual, the first step in decoding a cyclic code sequence is to compute the syndrome polynomial $s(x)$ from the received-word polynomial $r(x)$. To decode a t-error-correcting BCH code, which is a cyclic code, by Eq. (4.20) the syndrome vector ($2t$-tuple) is given by

$$\bar{S} = (S_1, S_2, \ldots, S_{2t}) = \bar{r} \cdot \hbar^T \tag{5.12}$$

where $\hbar$ is defined in Eq. (5.6). Here also by Eq. (5.11) the i-th component of the syndrome is

$$S_i = r(\alpha^i)$$
$$= r_0 + r_1 \alpha^i + \cdots + r_{n-1} \alpha^{(n-1)i}, \quad \text{for } 1 \le i \le 2t. \tag{5.13}$$

Note that the i-th syndrome component S_i is a field element of $GF(2^m)$ which corresponds to the syndrome polynomial $s_i(x) = s_0^{(i)} + s_1^{(i)} x + \cdots + s_{n-k-1}^{(i)} x^{n-k-1}$ or the binary syndrome vector $(s_0^{(i)}, s_1^{(i)}, \ldots, s_{n-k-1}^{(i)})$. Thus, each syndrome component of $\bar{S}$ is computed by dividing $r(x)$ by the minimal polynomial $m_i(x)$ for $1 \le i \le 2t$ of α^i such that

$$r(x) = q_i(x) m_i(x) + p_i(x) \tag{5.14}$$

The remainder $p_i(x)$, where $x = \alpha^i$, is the syndrome component S_i since $m_i(\alpha^i) = 0$. Thus, in general, computing $r(\alpha^i)$ is equivalent to computing $p_i(\alpha^i)$. Hence, if Eq. (5.13) is combined with Eq. (5.14), the syndrome components are expressed by

$$S_i = p_i(\alpha^i) = r(\alpha^i) = e(\alpha^i) \quad \text{for } 1 \le i \le 2t \tag{5.15}$$

from which it is seen that the syndrome vector $\bar{S}$ depends only on the error vector $\bar{e}$.

Once the syndrome vector is known the remaining problem is to find the error locations. The error pattern is completely described by the error-location numbers. Substituting $S_i = e(\alpha^i)$ from Eq. (5.15) into Eq. (5.10), one has

BCH Codes

$$S_i = \sum_{u=1}^{v} \left(\alpha^{j_u}\right)^i, \quad 1 \le i \le 2t. \tag{5.16}$$

This set of equations evidently relates the syndrome components with the unknown error-location parameters α^{j_u}, $for\ 1 \le u \le v$. More explicitly these relations are

$$\begin{aligned}
S_1 &= \alpha^{j_1} + \alpha^{j_2} + \cdots + \alpha^{j_v} \\
S_2 &= (\alpha^{j_1})^2 + (\alpha^{j_2})^2 + \cdots + (\alpha^{j_v})^2 \\
&\vdots \quad\quad \vdots \\
S_{2t} &= (\alpha^{j_1})^{2t} + (\alpha^{j_2})^{2t} + \cdots + (\alpha^{j_v})^{2t}
\end{aligned} \tag{5.17}$$

where the unknown parameters α^{j_u} for $u = 1, 2, \ldots, v$, are called the error-location numbers.

When the parameters $\alpha^{j_u}, 1 \le u \le v$, are found, the powers $j_u, u = 1, 2, \ldots, v$, give us the error locations in $e(x)$. These $2t$ equations of Eq. (5.17) are known as power-sum symmetric functions. Consequently, to decode the BCH codes a procedure is needed to solve these power-sum symmetric functions for the unknown numbers $\alpha^{j_u}, u = 1, 2, \ldots, v$. In fact, any method of solving Eq. (5.17) is the basis for an error-correction procedure.

Example 5.6
Consider the single-error-correcting (7, 4) BCH code with minimum distance $d_{\min} = 3$. Let $r(x)$ be the received polynomial. The syndrome vector $\bar{S}$ consists of a binary 2-tuple because $2t = 2\lfloor(d_{\min} - 1)/2\rfloor = 1$. That is, $\bar{S} = (S_1, S_2)$. Let α be the primitive element of $GF(2^3)$. Since the minimal polynomials for α, α^2, and α^4 are equal, we have

$$m_1(x) = m_2(x) = m_4(x) = (x + a)(x + a^2)(x + a^4)$$
$$= 1 + x + x^3$$

To compute the syndrome digits, one divides $r(x)$ by $m_1(x)$, to obtain a remainder $p(x)$ of the form $p(x) = p_0 + p_1 x + p_2 x^2$. Substituting α and α^2 into $p(x)$, one obtains S_1 and S_2, respectively. Thus, a syndrome computation circuit for the (7, 4) BCH code can he devised as the circuit shown in Fig. 5.1.

Now, assume that the all-zeros code word $\bar{c} = (0\ 0\ 0\ 0\ 0\ 0\ 0)$ is transmitted and that $\bar{r} = (0\ 0\ 0\ 0\ 0\ 0\ 1)$ is received. The received polynomial is then
$$r(X) = x^6.$$

After $r(x) = x^6$ is divided by $m_1(x) = 1 + x + x^3$, the remainder is
$p(x) = 1 + x^2$.

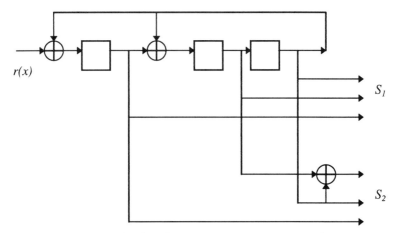

Figure 5.1 Syndrome computation circuit for the single-error-correcting (7, 4) BCH code.

Substituting α and α^2 into $p(x)$ and using the field elements of $GF(2^3)$, one obtains

$$S_1 = 1+\alpha^2 = \alpha^6 = (101),$$
$$S_2 = 1+\alpha+\alpha^2 = \alpha^5 = (111),$$

respectively. Thus, the syndrome vector is given by
$$\bar{S} = (\alpha^6, \alpha^5) = (101,111).$$

Note that it is easy to verify that $S_2 = S_1^2$. Actually, this can be directly proved from Eq. (5.17) by using the fact $(a+b+\cdots+d)^2 = a^2+b^2+\cdots+d^2$ over the finite field $GF(2^m)$.

5.3.2 The error-locator polynomial

Suppose that $v \leq t$ errors actually occur. In order to decode one first computes the syndromes S_i for $i=1,2,\ldots,2t$. These syndromes are then used to find the coefficients of the error-locator polynomial $L(z)$ defined by

$$L(z) = \prod_{i=1}^{v} (1+Z_i z) = \sum_{i=0}^{v} \sigma_i z^i. \tag{5.18}$$

BCH Codes

where Z_i^{-1} for $1 \leq i \leq v$ are the error locators of the v errors, given by $Z_i^{-1} = \alpha^{r_i}$ with the r_i being the locations of the errors. The coefficients of $L(z)$, namely the set $\{\sigma_i\}$, are related by the following elementary symmetric functions [6] of the error locations Z_i^{-1} for $1 \leq i \leq v$:

$$\sigma_0 = 1$$
$$\sigma_1 = Z_1 + Z_2 + \cdots + Z_v$$
$$\sigma_2 = Z_1 Z_2 + Z_2 Z_3 + \cdots + Z_{v-1} Z_v \quad (5.19)$$
$$\vdots$$
$$\sigma_v = Z_1 Z_2 \cdots Z_v.$$

Since α^i is a root of the generator polynomial, the decoder first computes the sum of the kth powers of the error-location numbers by evaluating the received polynomial $r(x)$ at $x = \alpha^i$. This process yields

$$r(\alpha^i) = e(\alpha^i) = S_i = \sum_{j=1}^{v} Z_j^i. \quad (5.20)$$

Given the syndrome components S_i the decoder must then attempt to find for the error-locator polynomial $L(z)$ whose reciprocal roots give the error locations. The relations between $L(z)$ and the S_i's are given by the Newton identities, which are derived by relating the σ_i's, for $1 \leq i \leq v$ with in Eq. (5.19) as demonstrated in the following discussion. If it is possible to determine the coefficients of $L(z)$, i.e., $\sigma_1, \sigma_2, \ldots, \sigma_v$ from the Newton identities, then the error-location numbers Z_j's, $1 \leq j \leq v$, can be found by determining the roots of the error-locator polynomial $L(z)$. The Newton identities are given as follows[6]:

$$S_1 + \sigma_1 = 0$$
$$S_2 + \sigma_1 S_1 + 2\sigma_2 = 0$$
$$S_3 + \sigma_1 S_2 + \sigma_2 S_1 + 3\sigma_3 = 0 \quad (5.21)$$
$$\vdots$$
$$S_v + \sigma_1 S_{v-1} + \sigma_2 S_{v-2} + \cdots + \sigma_{v-1} S_1 + v\sigma_v = 0.$$

Here for the binary case $i\sigma_i = \sigma_i$ when i is odd and $i\sigma_i = 0$ otherwise.

If the σ_i for $1 \leq i \leq v$ are determined from Eq. (5.20), the Z_i for $1 \leq i \leq v$ can be found by evaluating the roots of $L(z)$. The Newton identities in Eq. (5.20) can be expressed also in the following single-equation form:

$$S_i + \sigma_1 S_{i-1} + \cdots + \sigma_{i-1} S_1 + i\sigma_i = 0 \text{ for } i = 1,2,\ldots,v.$$

Consider the case for $i > v$. First a multiplication of $L(x)$ in Eq. (5.18) by x^{-i}, yields

$$x^{-i}L(x) = x^{-i} + \sigma_1 x^{1-i} + \sigma_2 x^{2-k} + \cdots + \sigma_{v-1} x^{v-1-i} + \sigma_v x^{v-i} \quad (5.22)$$

Next substituting the roots of $L(z)$ (i.e., $Z_1^{-1}, Z_2^{-1}, \ldots, Z_v^{-1}$) into Eq. (5.22) produces the following set of equations:

$$Z_1^i + \sigma_1 Z_1^{i-1} + \sigma_2 Z_1^{i-2} + \cdots + \sigma_{v-1} Z_1^{i-v+1} + \sigma_v Z_1^{i-v} = 0$$
$$Z_2^i + \sigma_1 Z_2^{i-1} + \sigma_2 Z_2^{i-2} + \cdots + \sigma_{v-1} Z_2^{i-v+1} + \sigma_v Z_2^{i-v} = 0 \quad (5.23)$$
$$\vdots \qquad \vdots$$
$$Z_v^i + \sigma_1 Z_v^{i-1} + \sigma_2 Z_v^{i-2} + \cdots + \sigma_{v-1} Z_v^{i-v+1} + \sigma_v Z_v^{i-v} = 0$$

Also the summation of these v equations in Eq. (5.23) term-by-term yields the relation,

$$(Z_1^i + Z_2^i + \cdots + Z_v^i) + \sigma_1 (Z_1^{i-1} + Z_2^{i-1} + \cdots + Z_v^{i-1}) + \cdots$$
$$+ \sigma_v (Z_1^{i-v} + Z_2^{i-v} + \cdots + Z_v^{i-v}) = 0 \quad (5.24)$$

Now express Eq. (5.22) in terms of the syndrome components to obtain

$$S_i + \sigma_1 S_{i-1} + \cdots + \sigma_{v-1} S_{i-v+1} + \sigma_v S_{i-v} = 0, \quad \text{for } i > v \quad (5.25)$$

In particular, for $i = v+1$, one gets

$$S_{v+1} + \sigma_1 S_v + \cdots + \sigma_{v-1} S_2 + \sigma_v S_1 = 0. \quad (5.26)$$

Thus, the Newton identities can be extended to the unknown syndromes S_i for $i > v$. Eqs. (5.25) and (5.26) show that the σ_j for $0 \le j \le v$ are closely related to the syndrome components, $S_i, 1 \le i \le v+1$. Thus, σ_j can be determined by solving the set of syndrome equations Eq. (5,25) and (5.26). Then the error-locations $Z_j = \alpha^{i_j}$ for $1 \le j \le v$ can be found by solving for the roots of $L(z)$. This $L(z)$ produces an error pattern $e(x)$ with the minimum number of errors. Hence, if $v \le t$ errors occur, $L(z)$ yields the actual error pattern $e(x)$.

Finally, the error-correcting procedure for the BCH codes is outlined as follows:

1. Using (5.14) compute the syndrome components S_j for $1 \le j \le 2t$ from the received polynomial $r(x)$.
2. Set each $\sigma_j = 0$ for $v+1 \le j \le t$, and solve the first v equations of Eq. (5.21) for the $\sigma_j, 1 \le j \le v$, in terms of the S_j.

BCH Codes

3. Determine the error-locator polynomial $L(z)$ from the these σ_j in terms of syndrome components S_j for $0 \le j \le 2t$.

4. Find the error-location numbers $Z_1, Z_2, \ldots, Z_v$ by solving for the roots of $L(z)$. Use these roots to correct the errors in $r(x)$.

Example 5.7
Continue Example 5.6. Since $t=1$, the syndrome components are computed to be
$$S_1 = 1 + \alpha^2 = \alpha^6 = (101),$$
$$S_2 = 1 + \alpha + \alpha^2 = \alpha^5 = (111).$$
It is known from the Newton's identities given in Eq. (5.20), that $S_1 = \sigma_1$. Hence, the error-locator polynomial $L(z)$ is determined to be
$$L(z) = 1 + \sigma_1 z = 1 + S_1 z.$$
The error location is $Z_1 = \alpha^{-1} = \alpha^6$, i.e. the 2nd bit of the received vector $\bar{r}$.

The above decoding process is, sometimes, called Peterson's direct-solution decoding algorithm. It is evident from Eq. (5.21) that the Newton identities also can be expressed in matrix form as follows:

$$S \cdot \tilde{\Lambda} = \begin{pmatrix} 1 & 0 & 0 & 0 & \cdots & 0 \\ S_2 & S_1 & 1 & 0 & \cdots & 0 \\ \vdots & \vdots & \vdots & \vdots & \ddots & \vdots \\ S_{2v-2} & S_{2v-3} & S_{2v-4} & S_{2v-5} & \cdots & S_{v-1} \end{pmatrix} \begin{pmatrix} \sigma_1 \\ \sigma_2 \\ \vdots \\ \sigma_v \end{pmatrix} = \begin{pmatrix} -S_1 \\ -S_3 \\ \vdots \\ -S_{2v-1} \end{pmatrix}$$
(5.27)

The linear system in Eq. (5.27) has a unique solution if and only if S is nonsingular. If S is nonsingular, it has a nonzero determinant. Given that S is nonsingular, one can solve for $\tilde{\Lambda}$ using the standard techniques of linear algebra. Peterson [4] shows for a binary BCH code that S has a nonzero determinant if there are v or $v-1$ errors in the received word. If fewer than $v-1$ errors occur, one can reduce the size of matrix S by eliminating the two bottom rows and the two rightmost columns, and then check if the remaining matrix is nonsingular. This reduction process continues until the remaining matrix is nonsingular.

It should be noted, that though there may be a solution to Eq. (5.27), it may not lead to the correct error-locator polynomial. Two possible types of errors are the following :

1. If the received word is within the maximum error-correction capability t of an incorrect codeword, then $L(x)$ decodes the received word into an incorrect codeword. This condition is undetectable, and is called a decoder error.
2. If the received word is not within the distance t of any codeword, the error-locator polynomial may have repeated roots, or roots that do not lie in the smallest field containing the primitive n-th root of unity used to construct the code. This condition is detectable and is used to trigger the declaration of, what is called, a decoder failure.

There are some cases in which a valid error-locator polynomial can be constructed that corrects more than t errors, called decoding beyond the maximum error-correction capability of the code. However, such cases are regrettably rare. A summary of Peterson's direct-solution decoding algorithm is given as follows :

Peterson's direct-solution decoding algorithm for a binary t-error-correcting BCH code

1. Compute the syndromes for $\bar{r}$: $\{S_i\} = \{r(\alpha^i)\}$ for $i = 1,2,\ldots,2t$.
2. Construct the syndrome matrix S in Eq. (5.27).
3. Compute the determinant of the syndrome matrix. If the determinant is nonzero, go to step 5.
4. Construct a new syndrome matrix by deleting the two rightmost columns and the two bottom rows from the previous syndrome matrix. Go to step 3.
5. Solve for S and construct $L(x)$.
6. Find the roots of $L(x)$. If the roots are not distinct or $L(x)$ does not have roots in the desired field, go to step 9.
7. Complement the bit positions in $\bar{r}$ indicated by $L(x)$. If the number of errors corrected is less that t, verify that the resulting corrected word satisfies all of the $2t$ syndrome equations. If it does not, go to step 9.
8. Send out the corrected word and then stop.
9. Declare a decoding failure and then stop.

It can be seen that the key procedure for Peterson's algorithm is to construct $L(x)$, i.e. to determine the coefficients σ_i of $L(x)$. For simple cases the σ_i can be solved directly as follows :
- Single-error correction :
$$\sigma_1 = S_1. \tag{5.28}$$
- Double-error correction :

$$\sigma_1 = S_1$$
$$\sigma_2 = S_1^{-1}(S_3 + S_1^3). \tag{5.29}$$

- Triple-error correction :
$$\sigma_1 = S_1$$
$$\sigma_2 = (S_1^3 + S_3)^{-1}(S_1^2 S_3 + S_5) \tag{5.30}$$
$$\sigma_3 = (S_1^3 + S_3) + S_1 \sigma_2.$$

- Quadruple-error correction :
$$\sigma_1 = S_1$$
$$\sigma_2 = ((S_3(S_1^3 + S_3) + S_1(S_1^5 + S_5))^{-1}(S_1(S_7 + S_1^7) + S_3(S_1^5 + S_5))$$
$$\sigma_3 = (S_1^3 + S_3) + S_1 \sigma_2 \tag{5.31}$$
$$\sigma_4 = S_1^{-1}((S_5 + S_1^2 S_3) + (S_1^3 + S_3)\sigma_2).$$

- Quintuple-error correction :
$$\sigma_1 = S_1,$$
$$\sigma_2 = [(S_1^3 + S_3)((S_7 + S_1^7) + S_1 S_3(S_1^3 + S_3)) + (S_5 + S_1^2 S_3)(S_1^5 + S_5)]^{-1}$$
$$\cdot [(S_1^3 + S_3)((S_1^9 + S_9) + S_1^4(S_5 + S_1^2 S_3) + S_3^2(S_1^3 + S_3))$$
$$+ ((S_1^5 + S_5)(S_7 + S_1^7) + S_1(S_3^2 + S_1 S_5))], \tag{5.32}$$
$$\sigma_3 = (S_1^3 + S_3) + S_1 \sigma_2,$$
$$\sigma_4 = (S_1^5 + S_5)^{-1}((S_1^9 + S_9) + S_3^2(S_1^3 + S_3) + S_1^4(S_5 + S_1^2 S_3) + ((S_7 + S_1^7)$$
$$+ S_1 S_3(S_1^3 + S_3))\sigma_2),$$
$$\sigma_5 = (S_5 + S_1^2 S_3) + S_1 \sigma_4 + (S_1^3 + S_3)\sigma_2.$$

Once the error-locator polynomial is known, the roots can be located by the use of, what is called, a Chien search. The Chien search[5] is a systematic means for evaluating the error-locator polynomial at all elements in a field $GF(2^m)$. A Chien-search circuit is shown in Figure 5.2. Each coefficient σ_i of the error locator polynomial is repeatedly multiplied by α^i, where α is primitive in $GF(2^m)$. Each set of products is then summed to obtain $D_i = L(\alpha^i) - 1$.

If α^i is a root of $L(x)$, then $D_i = 1$ and an error is indicated at the coordinate associated with $\alpha^{-i} = \alpha^{n-i}$. The correction bit e_i is then set to 1 and added to the received bit r_{n-i}. If $D_i \neq 1$, then e_i is set to zero, and the corresponding received bit is left unchanged. Note that a verification is performed after error correction to ensure that the resulting word is a codeword. Codeword verification is performed by the use of the same circuit for error detection of cyclic codes, given in Chapter 4. In this circuit, the succeeding bits $c_0, c_1, \ldots, c_{n-1}$ of the corrected word take the place of the error-locator polynomial coefficients. The syndromes $\{s_i'\}$ for the corrected word are computed one at a time. If they are all zero, the corrected word is accepted as a valid codeword; otherwise, a decoder failure is indicated.

The Chien search circuitry in Figure 5.2 is loaded with the coefficients of $L(x)$, to output $L(x)-1$ for $x = \alpha^i$, $i=0,1,\ldots,2^m-1$. From the definition of the error locator polynomial, this sequence $D_i = L(\alpha^i)-1$ determines if an error has occurred in $c_0, c_1, \ldots, c_{n-1}$. Note that if α^{-j} is a root of $L(x)$, so is α^{2^m-1-j}. Thus, in order to check $c_0, c_1, \ldots, c_{n-1}$, $L(x)-1$ is evaluated only at $x = \alpha^{2^m-1-(n-i)}$ for $i=1,2,\ldots,n$.

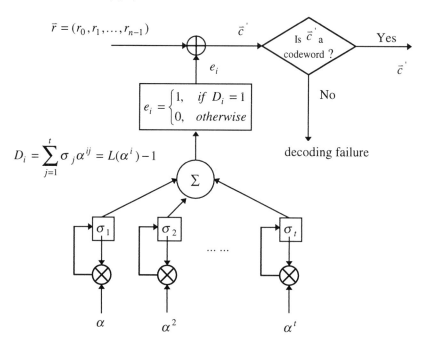

Figure 5.2 Chien search and error-correction for binary code.

BCH Codes

Example 5.8
Illustrate the double-error correction procedure by the use of Peterson's algorithm. By Example 5.4 the double-error-correcting BCH code of length 31 has the generator polynomial $g(x) = 1 + x^3 + x^5 + x^6 + x^8 + x^9 + x^{10}$. The roots of this polynomial include four consecutive powers of α: , namely, the set $\{\alpha, \alpha^2, \alpha^3, \alpha^4\}$, where α is primitive in $GF(32)$. Let the following binary vector; $\bar{r} = (0010000110011000000000000000000)$, be received. This received word corresponds to the received polynomial $r(x) = x^2 + x^7 + x^8 + x^{11} + x^{12}$. The syndrome computations are performed in $GF(32)$ to obtain

$$S_1 = r(\alpha) = \alpha^7,$$
$$S_2 = r(\alpha^2) = \alpha^{14},$$
$$S_3 = r(\alpha^3) = \alpha^8,$$
$$S_4 = r(\alpha^4) = \alpha^{28}.$$

Note here that $S_4 = (S_2)^2 = (S_1)^4$. Then, the error-locator polynomial is obtained by the use of Eq.(5.29) as follows:

$$\sigma_1 = \alpha^7,$$

$$\sigma_2 = \frac{\alpha^8 + (\alpha^7)^3}{\alpha^7} = \alpha^{15},$$

$$L(x) = 1 + \alpha^7 x + \alpha^{15} x^2 = (1 + \alpha^5 x)(1 + \alpha^{10} x).$$

The error locators are $Z_1 = \alpha^5 = \alpha^{31-26}$ and $Z_2 = \alpha^{10} = \alpha^{31-21}$, thereby indicating errors at the 26-th and 21-st coordinates of $\bar{r}$ (counting from the right). The corrected word, with the corrected positions complemented, is

$$\bar{c} = (0010 0\hat{1}0110\hat{1}1 1000000000000000000),$$

i.e. $c(x) = x^2 + x^5 + x^7 + x^8 + x^{10} + x^{11} + x^{12}$. It is easy to verify that $c(x)$ is a code polynomial since $c(x) = x^2 g(x)$.

Now, consider the decoding of nonbinary BCH codes. For such a case one needs to determine not only the locations of the errors, but also their magnitudes. Gorenstein and Zierler(1961) generalized Peterson's decoding algorithm for use with nonbinary codes [3]. Note that the syndromes are now a function of both the magnitudes of the errors and their locations. Assume that the received word is corrupted by v errors. Then its syndromes are given as follows :

$$S_j = e(\alpha^j) = \sum_{k=0}^{n-1} e_k (\alpha^j)^k = \sum_{l=1}^{v} e_{i_l} Z_l^j. \tag{5.33}$$

Eq. (5.33) defines a series of 2t algebraic equations in 2v unknowns as follows:

$$\begin{aligned} S_1 &= e_{i_1} Z_1 + e_{i_2} Z_2 + \cdots + e_{i_v} Z_v, \\ S_2 &= e_{i_1} Z_1^2 + e_{i_2} Z_2^2 + \cdots + e_{i_v} Z_v^2, \\ &\vdots \qquad \vdots \\ S_{2t} &= e_{i_1} Z_1^{2t} + e_{i_2} Z_2^{2t} + \cdots + e_{i_v} Z_v^{2t}. \end{aligned} \tag{5.34}$$

The system in Eq. (5.34) reduces to a set of linear functions in the unknown quantities. It follows from the error-locator polynomial $L(x)$ in Eq. (5.18) that for the error locators $Z_l = e_{i_l}$, one obtains

$$L(Z_l^{-1}) = \sigma_v Z_l^{-v} + \sigma_{v-1} Z_l^{-v+1} + \cdots + \sigma_1 Z_l^{-1} + \sigma_0 = 0 \quad \text{for } l=1,2,\ldots,v. \tag{5.35}$$

Since this expression sums to zero, one can multiply both sides of this equation by the constant e_{i_j} to obtain

$$\begin{aligned} &e_{i_j} Z_l^j (\sigma_v Z_l^{-v} + \sigma_{v-1} Z_l^{-v+1} + \cdots + \sigma_1 Z_l^{-1} + \sigma_0) \\ &= e_{i_j} (\sigma_v Z_l^{j-v} + \sigma_{v-1} Z_l^{j-v+1} + \cdots + \sigma_1 Z_l^{j-1} + \sigma_0 Z_l^j) \\ &= 0 \end{aligned} \tag{5.36}$$

Summing Eq. (5.36) over all indices l, yields an expression from which the Newton identities can be obtained. This expression is

$$\begin{aligned} &\sum_{l=1}^{v} e_{i_l} (\sigma_v Z_l^{j-v} + \sigma_{v-1} Z_l^{j-v+1} + \cdots + \sigma_1 Z_l^{j-1} + \sigma_0 Z_l^j) \\ &= \sigma_v \sum_{l=1}^{v} e_{i_l} Z_l^{j-v} + \sigma_{v-1} \sum_{l=1}^{v} e_{i_l} Z_l^{j-v+1} + \cdots + \sigma_1 \sum_{l=1}^{v} e_{i_l} Z_l^{j-1} + \sigma_0 \sum_{l=1}^{v} e_{i_l} Z_l^j \\ &= \sigma_v S_{j-v} + \sigma_{v-1} S_{j-v+1} + \cdots + \sigma_1 S_{j-1} + \sigma_0 S_j = 0. \end{aligned} \tag{5.37}$$

From Eq. (5.18) it is clear that $\sigma_0 = 1$ is always one. Here Eq. (5.37) can thus be represented by the system,

$$\sigma_v S_{j-v} + \sigma_{v-1} S_{j-v+1} + \cdots + \sigma_1 S_{j-1} = -S_j. \tag{5.38}$$

Assume that $v=t$. Then the following matrix equation is obtained for the symmetric functions σ_j as follows:

$$S' \cdot \bar{\Lambda} = \begin{pmatrix} S_1 & S_2 & S_3 & \cdots & S_{t-1} & S_t \\ S_2 & S_3 & S_4 & \cdots & S_t & S_{t+1} \\ \vdots & \vdots & \vdots & \ddots & \vdots & \vdots \\ S_t & S_{t+1} & S_{t+2} & \cdots & S_{2t-2} & S_{2t-1} \end{pmatrix} \begin{pmatrix} \sigma_t \\ \sigma_{t-1} \\ \vdots \\ \sigma_1 \end{pmatrix} = \begin{pmatrix} -S_{t+1} \\ -S_{t+2} \\ \vdots \\ -S_{2t} \end{pmatrix}. \tag{5.39}$$

BCH Codes

It can be shown that S' is nonsingular if the received word contains exactly t errors. It also can be shown that S' is singular if fewer that t errors occur. If S' is singular, then the rightmost columns and bottom rows can be removed and the determinant of the resulting matrix computed. This process is repeated until one reaches a non-singular matrix. The coefficients of the error-locator polynomial are then found by the use of standard linear-algebraic techniques.

Once the v error locations are known, Eq. (5.34) becomes a system of $2v$ equations in v unknown error magnitudes. This system reduces to the form of the matrix relation given below. Since the $\{Z_i\}$ are nonzero and distinct, the matrix D is a Vandermonde matrix and thus is nonsingular.

$$D \cdot \bar{e} = \begin{pmatrix} Z_1 & Z_2 & \cdots & Z_v \\ Z_1^2 & Z_2^2 & \cdots & Z_v^2 \\ \vdots & \vdots & \ddots & \vdots \\ Z_1^v & Z_2^v & \cdots & Z_v^v \end{pmatrix} \begin{pmatrix} e_{i_1} \\ e_{i_2} \\ \vdots \\ e_{i_v} \end{pmatrix} = \begin{pmatrix} S_1 \\ S_2 \\ \vdots \\ S_v \end{pmatrix}. \qquad (5.40)$$

Decoding is completed by solving for the $\{e_{i_j}\}$.

Peterson-Gorenstein-Zierler decoding algorithm

1. Compute the syndromes for $\bar{r}$: $\{S_i\} = \{r(\alpha^i)\}$, for $i = 1, 2, \ldots, 2t$.
2. Construct the syndrome matrix S' in Eq. (5.39).
3. Compute the determinant of the syndrome matrix. If the determinant is nonzero, goto step 5.
4. Construct a new syndrome matrix by deleting the rightmost columns and the bottom rows from the previous syndrome matrix. Go to step 3.
5. Solve for $\bar{\Lambda}$ and construct $L(x)$.
6. Find the roots of $L(x)$. If the roots are not distinct, or $L(x)$ does not have roots in the desired field, goto step 10.
7. Construct the matrix D in Eq. (5.40) and solve for the error magnitudes.
8. Subtract the error magnitudes from the values at the appropriate coordinates of the received word.
9. Send the decoded codeword and then stop.
10. Declare a decoding failure and then stop.

Example 5.9
Consider a double symbol-error-correcting, non-binary, BCH code (7,3) of length 7 (symbols). This code is constructed over GF(8) with the generator polynomial, $g(x) = (x-\alpha)(x-\alpha^2)(x-\alpha^3)(x-\alpha^4) = x^4 + \alpha^3 x^3 + x^2 + \alpha x + \alpha^3$.

Let the received polynomial be $r(x) = \alpha^2 x^6 + \alpha^2 x^4 + x^3 + \alpha^5 x^2$. Then the syndrome sequence is $S_1 = \alpha^6, S_2 = \alpha^3, S_3 = \alpha^4, S_4 = \alpha^3$. Eq. (5.39) reduces to a system of two equations in two unknowns, given by

$$S' \cdot \bar{\Lambda} = \begin{pmatrix} \alpha^6 & \alpha^3 \\ \alpha^3 & \alpha^4 \end{pmatrix} \begin{pmatrix} \sigma_2 \\ \sigma_1 \end{pmatrix} = \begin{pmatrix} \alpha^4 \\ \alpha^3 \end{pmatrix}.$$

It is easy to verify that matrix S' is nonsingular. Simple substitution then yields the following solution for the coefficients of the error-locator polynomial:

$$\sigma_1 = \alpha^2, \sigma_2 = \alpha,$$
$$L(x) = \alpha x^2 + \alpha^2 x + 1.$$

The error locations are found to be $Z_1 = \alpha^3 = \alpha^{7-4}$ and $Z_2 = \alpha^5 = \alpha^{7-2}$, i.e. the errors are in the fourth and second positions of the received word.

Eq. (5.40) provides the following system of equations for determining the error magnitudes:

$$D \cdot \bar{e} = \begin{pmatrix} \alpha^3 & \alpha^5 \\ \alpha^6 & \alpha^3 \end{pmatrix} \begin{pmatrix} e_3 \\ e_5 \end{pmatrix} = \begin{pmatrix} \alpha^6 \\ \alpha^3 \end{pmatrix}.$$

Solving for the error magnitudes yields the following error polynomial,
$$e(x) = \alpha x^3 + \alpha^5 x^5.$$

The error polynomial is added to the received polynomial to obtain the closest codeword in Hamming distance. Remember that addition and subtraction are the same operations in binary finite fields of the form $GF(2^m)$. Finally
$$c(x) = r(x) + e(x)$$
$$= \alpha^2 x^6 + \alpha^5 x^5 + \alpha^2 x^4 + \alpha^3 x^3 + \alpha^5 x^2,$$
which is a codeword since $c(x) = \alpha^2 x^2 g(x)$.

Both the Peterson and the Peterson-Gorenstein-Zierler decoding algorithms are considered to be too complex to implement when the number of errors is large. More efficient algorithms are required for the correction of a large number of errors. Some faster, less complex, algorithms are discussed in the next chapter.

5.4 Algebraic Decoding of Quadratic Residue Codes

The BCH decoding method discussed in the last section can be extended to decode other cyclic codes. The decoding algorithm presented next, for the quadratic residue codes were developed by Elia[9], and generalized by Bours, et.al.[11], and Reed, et.al [12][13].

To demonstrate the decoding algorithm for the quadratic residue codes, consider the (23,12,7) binary Golay code as an example. Let $m(x) = m_0 + m_1 x + \cdots + m_{11} x^{11}$ be the message polynomial. Also let the codeword $c(x)=m(x)g(x)$ be transmitted through a noisy channel to obtain a received codeword of the form, $r(x)=c(x)+e(x)$, where $e(x)$ is the polynomial of the received error-pattern vector. Thus, the syndromes are defined over $GF(2^{11})$ by $S_i = r(\alpha^i)$ for all $i \in Q$ where Q is the defining set. Consider the generator polynomial $g(x)$ which is associated with the defining set : $Q=\{1,2,3,4,6,8,9,12,13,16,18\}$. Since α, α^3, and α^9 are the roots of $g(x)$ as well as the roots of the code polynomials $c(x)$, the three syndromes are computed by

$$S_1 = r(\alpha), S_3 = r(\alpha^3), S_9 = r(\alpha^9).$$

These syndromes are used next to compute the coefficients of the error-locator polynomial $L(x)$.

There are four cases to discuss.

Case 1 : If no errors occur in the received word, $S_1 = S_3 = S_9 = 0$ and $L(x)=0$.

Case 2 : If a single error occurs, the error-polynomial $e(x)$ has the form x^j, the syndromes have the values,

$$S_1 = \alpha^j, S_3 = (\alpha^j)^3, S_9 = (\alpha^j)^9,$$

i.e. $S_3 = S_1^3, S_9 = S_1^9$. It follows from the Newton identities given in Eq.(5.20) that the error-locator polynomial is

$$L(x) = x + S_1.$$

Case 3 : If two errors occur in the received word, it is a simple matter to verify from the Newton identities (see Eqs. (5.22) and (5.29)) that the coefficients of the error-locator polynomial satisfy the relations,

$$\sigma_1 = S_1,$$
$$\sigma_2 = \frac{S_3 + S_1^3}{S_1}.$$

Therefore, the error-locator polynomial is
$$L(x) = x^2 + S_1 x + \frac{S_3 + S_1^3}{S_1}.$$

Case 4 : Finally, if three errors occur in the received word the coefficients of the error-locator polynomial are derived from Eq. (5.21) in a similar manner to obtain the following identities:
$$S_1 + \sigma_1 = 0,$$
$$S_3 + S_1^3 + \sigma_2 S_1 + \sigma_3 = 0,$$
$$S_5 + S_1^5 + \sigma_2 S_3 + \sigma_3 S_1^2 = 0,$$
$$S_7 + S_1 S_3^2 + \sigma_2 S_5 + \sigma_3 S_1^4 = 0,$$
$$S_9 + S_1^9 + \sigma_2 S_7 + \sigma_3 S_3^2 = 0.$$

An elimination of the unknown syndromes S_5 and S_7 yields the four identities,
$$\sigma_1 = S_1,$$
$$\sigma_2 = S_1^2 + \sqrt[3]{\Delta},$$
$$\sigma_3 = S_3 + S_1 \sqrt[3]{\Delta},$$

where $\Delta = (S_3 + S_1^3) + \frac{(S_9 + S_1^9)}{(S_3 + S_1^3)} = (\sigma_2 + S_1^2)^3$. Since the cube root of Δ exists, the error-locator polynomial can be found. Then, the error locations can be determined by a use of the Chien search. In this case the Chien search only needs to evaluate $L(x)$ at the n-th root of unity α^i for $i=1,2,...,n$ ($n=23$ for the Golay code C_{23}).

Important efforts by Reed et. al. [13][14] developed a number of powerful algebraic decoding algorithms for several specific quadratic residue codes. Some algorithms for solving both quadratic and cubic equations over finite fields also were uncovered [14]. More recently a new approach for decoding entire sets of cyclic codes was found by Chen et. al. [15]-[17]. This was accomplished by a use of, what is called, the Groebner-basis technique, such a new approach is able to decode cyclic codes up to their maximum error-correction abilities.

BCH Codes

5.5 BCH Codes as Industry Standards

Many BCH codes are utilized in a variety of industry standards. In the following subsections, a few of these applications are given. The goal of this section is to show how to define, evaluate, and realize BCH codes for these applications.

5.5.1 Use of the (511,493) BCH Code in the ITU-T Rec. H.261

The ITU-T (the ITU Telecommunication Standardization Sector) is a permanent organ of the International Telecommunication Union (ITU). The ITU-T Recommendation (ITU-T Rec.) H.261, which is entitled, "Video codec for audiovisual services at $p \times 64$ kbit/s," is a video coding standard used primarily for video conferencing and videophone. The core of H.261 is a video compression algorithm that employs a hybrid of motion-compensated interframe-prediction algorithms to reduce the temporal redundancy and also the Discrete Cosine Transform (DCT) to reduce the spatial redundancy. With variable-length encoding even a one-bit error in the bit-stream can make it impossible to recover the remainder of the bitstream, Hence, the encoded data must be structured in such a manner that error propagation can be avoided. The layered structure of the H.261 coded bit-stream is shown in Figure 5.3.

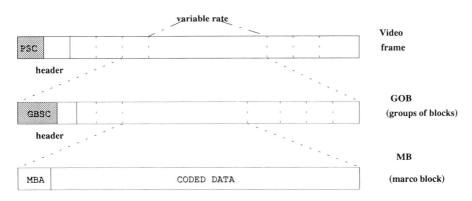

PSC: picture start code, GBSC: group of block start code, MBA:macro block address
Figure 5.3 H.261 Bit-Stream Layer.

The bit-stream of a single video frame comprises a header and several groups of blocks(GOBs). The GOBs comprise a number of macro blocks (MBs). A MB corresponds to six basic data units that are encoded with data from six

8×8 pixel blocks. The number of MBs and the bit-lengths of each MB can vary because the bits in a MB are determined by the coding method, e.g. motion compensation. Bit-stream synchronization is established by using the picture start-code(PSC) in the frame header and the group-of-blocks start-codes(GBSC) in the GOB headers. This structured packing technique is designed to localize the impact of the transmission errors of the bit-stream within a GOBs or a frame.

To meet the customer demand for video-conferencing, video-phone and other audio-visual services, the H.261 has to provide digital-video transmission at B(64kbit/s), H_0 (384kbit/s) rates or their multiples up to the primary rate H_{11} (1536kbit/s) and H_{12} (1920kbit/s). For example, the Integrated-Services Digital Networks(ISDNs) provide a switched transmission service at the B, H_0, or H_{11} and H_{12} bit rates. These transmission media are subject to data errors. Therefore, an error detection and correction scheme is incorporated in H.261. Error correction is especially important in the structured compression bit-stream because a single bit of error typically propagates into an error burst of many bits within a GOB. To select a forward error-correction code for H.261 applications, the following requirements need to be satisfied :
1. High information rate for a low overhead type of compression,
2. low complexity and a low implementation cost,
3. no error propagation,
4. reduction of the bit-error rate from 10^{-4} to 10^{-8}.

When the properties of a binary sequence with random errors are considered, the appropriate code is a binary linear block code. At present the BCH (511,493,5) code is used in H.261 for error correction. The generator polynomial for this cyclic code is specified to be $g(x) = x^{18} + x^{15} + x^{12} + x^{10} + x^8 + x^7 + x^6 + x^3 + 1$. This BCH code is capable of correcting any pattern of, at most, 2 bit errors for each codeword. Conceptually, since there are 511 bit locations, and $2^9 > 511$, nine bits of redundancy are needed to specify the locations of the two-errors. This BCH code provides a reasonably effective way to correct random channel errors. Assume that the channel model is modeled as the BSC, defined in Chapter 3. If the equivalent channel bit-error rate is p, then the decoded bit-error rate can be computed to be

$$p_b = 511^2 p^3.$$

To illustrate its power the code improves an original error rate of 10^{-5} to a decoded error rate of approximately 2.5×10^{-10}. This is achieved with an error-correction overhead of 18/511=3.5%.

BCH Codes

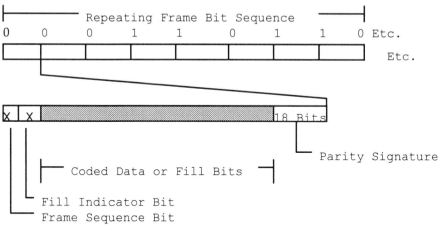

Figure 5.4 Error Correction Frame bits

The error-correction encoding scheme for H.261 involves dividing the compressed-video bit-stream into regular blocks, where for each block a parity signature is generated. The block size is 492 bits and the parity signature is 18 bits. Each block is preceded by a frame-bit sequence and a fill-indicator bit, which is followed by the parity signature. The frame-bit sequence is used to verify and maintain error-correction frame synchronization. These bits are transmitted with a regular repeating frame-alignment pattern of 00011011, so that the sequence bit pattern repeats for each eight error-correction framing blocks. The fill-indicator bit, which is set to one, indicates that the following 492 bits represent encoded image data. If set to zero, the subsequent 492 bits are simply bit-stream filler data. Figure 5.4 illustrates the organization of the error-correction framing bits. The fill-indicator bit is included in the parity-signature calculation. Eighteen BCH check bits are computed from the 493 BCH message bits(1 fill bit and 492 compressed video-data bits) to form a BCH codeword of 511 bits. Note that the synchronization bits are not protected by the BCH code because synchronization is required before BCH decoding can begin.

Since there are no security constraints for a general H.261 bit-stream, the systematic encoding given in Section 4.2, in which the information remains "in the clear", is employed. The BCH encoder computes 18 BCH check bits $p_0, p_1, \ldots, p_{17}$ from the 493 BCH message bits $m_0, m_1, \ldots, m_{492}$ to obtain a BCH codeword of 511 bits. The check polynomial $p(x) = \sum_{i=0}^{17} p_i x^i$ is obtained as the remainder when the shifted message polynomial $x^{18}m(x) = \sum_{i=0}^{492} m_i x^{i+18}$ is divided by any of the binary generator polynomial $g(x)$, specified in the H.261 standard. The algorithm which is used to implement this division for such an encoder is given as follows: Consider the i-th step in this division process is given by

$$m^{(i)}(x) = m(x) - q^{(i)}(x)g(x),$$

where $m^{(i)}(x)$ and $q^{(i)}(x)$ are the remainder and quotient of the division, respectively, at step i. With these definitions the complete algorithm is given as follows:

Step 1: $m^{(0)}(x) = m(x)$ and $q^{(0)}(x) = 0$.

Step 2: Let $s = \deg(m^{(i)}(x))$ and $r = \deg(g(x))$.
do {

$$m^{(i+1)}(x) = m^{(i)}(x) - \frac{m_0}{g_r} \cdot g(x) \cdot x^{s-r};$$

$$q^{(i+1)}(x) = q^{(i)}(x) - \frac{q_0}{g_r} \cdot x^{s-r};$$

$i = i+1;$
} While ($s \geq r$)

Step 3: $r(x) = m(x) \mod g(x) = m^{(i)}(x)$,

$$q(x) = q^{(i)}(x).$$

where m_0 and g_i are the coefficients of $m(x)$ and $g(x)$.

The decoding procedure for the BCH(511,493,5) code follows the four steps given in the Section 5.3. These steps can be accomplished by hardware. Let $\bar{r} = (r_0, r_1, ..., r_{510})$ be the received word and α be the 511-th root of unity. Then, the two syndromes of the decoder are computed by

$$S_1 = \sum_{i=0}^{510} r_i \alpha^i,$$

and

$$S_3 = \sum_{i=0}^{510} r_i \alpha^{3i}.$$

Finally, the decoding algorithm is provided by the following procedure:
1. $S_1 = 0$ and $S_3 = 0$ for the no error case.

BCH Codes

2. If $S_1 \neq 0$ and $S_1^3 - S_3 = 0$, then one error occurs, and the error location is S_1^{-1}.

3. If $S_1 \neq 0$ and $S_1^3 - S_3 \neq 0$, then two errors occur, and the error locations are found by solving the quadratic equation $x^2 + S_1 x + \left(S_3 / S_1 + S_1^2 \right) = 0$.

The implementation of this BCH codec is illustrated next for a 1.0μ CMOS chip. This device has a separate encoder and decoder for the full-duplex operation. The encoder implements the bit-filling, BCH encoding, and the frame-bit generation. The decoder requires both a synchronization acquisition and BCH decoding.

The encoder is implemented by the standard shift-register architecture, shown in Figure 5.5. There are 18 one-bit registers in the encoder, separated by EXCLUSIVE-OR gates at positions that correspond with the nonzero terms of the generator polynomial $g(x)$. The registers are updated every time a new message bit is inputted. After all of the message bits are inputted, the registers contain the check bits which then are shifted serially out of the shift register, with the feedback path disabled by the control signal.

The decoder implements the decoding algorithm in three fully pipelined stages, as shown in Figure 5.6. In general, for a BCH code of length 2^m-1, m bits are required to specify the locations of each error. All internal decoding operations are performed over the m-bit values in the finite field $GF(2^m)$. In H.261 for $m=9$ the sum and the product of any two 9-bit elements of $GF(2^9)$ also belong to $GF(2^m)$. These finite field operations, although unconventional, are less complex than the normal integer operations of a processor. Additions are simply implemented by bit-wise EXCLUSIVE-OR(XOR) operations. A general multiplier, where both the multiplicands and multiplier are variable, consists of 81 AND gates and 83 EXCLUSIVE-OR gates. If the multiplier is fixed, all of the AND gates and some of the EXCLUSIVE-OR gates can be removed. In GF(512), the generator polynomial $g(x)$ has the values α and α^3 among its roots. The first stage of the algorithm is to compute the power sums by evaluating the received polynomial $r(x) = \sum_{i=0}^{510} r_i x^i$ at the roots α and α^3 of $g(x)$, where $r_{510}, r_{509}, \ldots, r_0$ is the 511-bit codeword received from the channel. The evaluation of the syndromes S_i for $i=1,2$, is performed by the use Horner's rule as follows :

$$S_i = (\cdots(r_{510}\alpha^i + r_{509})\alpha^i + \cdots + r_1)\alpha^i + r_0.$$

By Eq.(5.29) the second stage of the decoding process consists of finding the roots β_0 and β_1 of the quadratic equation $x^2 + S_1 x + \left(S_3/S_1 + S_1^2\right) = 0$. To accomplish this the quadratic equation is first transformed into the canonical form $y^2 + y + \left(S_3/S_1^3 + 1\right) = 0$ by the substitution $x = S_1 y$. A root y_0 of this canonical formula is obtained directly from $\left(S_3/S_1^3 + 1\right)$ by means of an EXCLUSIVE-OR network . The roots of the original quadratic equation are $\beta_0 = y_0 S_1$ and $\beta_1 = \beta_0 + S_1$. These computations are accomplished by the use of a general finite-field multiplier. The required division operation is performed as a sequence of multiplications by utilizing the property, $x^{-1} = x^{510}$, for any value x in GF(512). The roots β_0 and β_1 , obtained from the second stage, represent the locations of the errors in the corrupted codeword. These locations are compared with a finite-field counter to generate the error signals in the third stage of the algorithm.

This logic-processing device can be implemented in a array-based technology to minimize the fabrication time. For example, in one implementation[30] it can process data at a throughput rate of 40 Mbit/s , much faster than the H.261 requirement.

A different architecture, with a throughput rate which is more matched to the H.261 requirements , does not lead to any significant reduction in the size of the chip. The needed delay (used to delay the received corrupted bit stream to match the algorithm's processing delay) and the synchronization RAM (for storing the number of matched synchronization bits for the various synchronization phases) already occupies half of the total area of the current chip. In order to save silicon area, the size of the synchronization RAM can be reduced by one half by searching only 256 of the 512 different phases at any one time.

BCH Codes

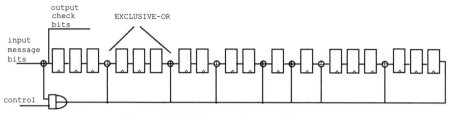

Figure 5.5 BCH encoder architecture.

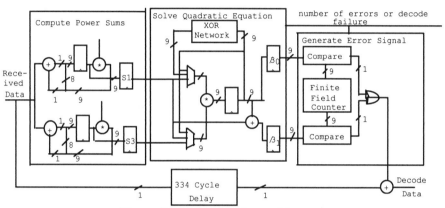

Figure 5.6 BCH decoder architecture

5.5.2 Use of the (40,32) BCH Code in the Header-Error Protection of the cells of the Asynchronous Transfer Mode (ATM)

The next application of BCH codes is to the Asynchronous Transfer Mode (ATM) layer of an optical fiber system. ATM is a high-bandwidth, low-delay, switching and multiplexing technology which is becoming available for both public and private communication networks. ATM principles and ATM-based platforms form the foundation for the delivery of a variety of high-speed digital communication services aimed at corporate users of high-speed data, local-area network interconnections, imaging and multimedia applications. The functions of the ATM technologies can be modeled by the following five layers : the services layer, the adaptation layer, the ATM layer, the convergence layer and the physical layer. Among these layers, the ATM layer takes care of the routing, traffic management and multiplexing of the traffic. The basic unit of the ATM layer is the ATM cell which consists of 53 bytes. The routing and multiplexing of the digital communication channel is controlled by the use of a 5-byte header in the ATM cell. The remaining 48 bytes in the cell are the "payload" which contains information from the upper layers. The ATM layer is concerned with the information contained in the header.

Errors are impossible to eliminate completely in any communication network. As reliable as the optical fiber is, noise and signal dispersion still persist and cause information loss. Random-bit errors occur along the communication route or within the equipment. In an ATM cell these bit errors can occur either in the packet header or in the payload data. The communications circuit and its attendant multiplexing and switching components can give rise to certain characteristic bit errors. When bit errors occur in the header of a cell or packet, the resulting address error is uncorrectable and a misdelivery occurs. This usually results in a loss of the entire packet.

When bit errors occur in the payload data, the damage is limited usually to a degradation of only part of the packet. Of course, if this part of the packet happens to contain important information, the effect can be quite serious, but the rate of such data losses is normally low. In fact, it is extremely low in high-speed optical fiber networks at a bit-error rate which rarely exceeds 10^{-12}. The bit-error rate statistics play a key role in the design of the network.

As the number of errors increases, the error-correction capability must be sufficiently powerful so that the user data would have to be retransmitted. The end effect of such extra-transmission is to create more traffic in the network. Ideally, one would like to design a network that is cost effective in the treatment of its errors. To accomplish this the studies on bit-error rates on an optical fiber and the size of an ATM cell are needed. AT&T conducted a study of the bit-error rates of optical fibers. Such a study is summarized in Figure 5.7.

BCH Codes

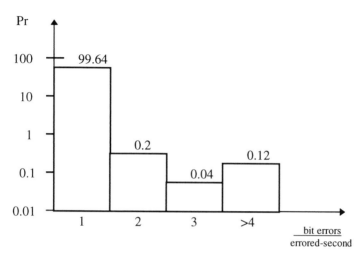

Figure 5.7 Bit-error rate on an optical fiber

From the above facts, in order to minimize packet losses it is found that error-correcting codes need to be applied to the ATM cell header, which contains the address of the packet. In an ATM cell the defined information data are confined to 32 bits. From Figure 5.7 more than 99 percent of the errors are one-bit errors in any errored-second time interval. As a consequence only the simplest code of one error correction needs to be chosen for the ATM cell header. The facts given in Chapter 3 show that 6 parity-check bits are needed to correct a single bit error on a 32 bit word of information data, and 8 parity-check bits are able to correct one bit and detect 84 percent of the protected bits. Therefore, the 40-bit(5-byte) header of an ATM cell, which contains an 8-bit parity check and 32 bits of information pay load, is a reasonable compromise of the error-control rate vs the information rate.

The code selected for the ATM cell header is a shortened cyclic code that corrects 1-bit errors and detects 2-bit errors. Recall that a cyclic code of length $2^m - 1$ with $m > 2$ is generated by a primitive polynomial $g(x)$ of degree m over $GF(2)$. Thus, the encoding of the (40, 32) shortened cyclic code in systematic form consists of three steps:
1. Multiply the message polynomial $m(x)$ by x^8,
2. Divide $x^8 m(x)$ by $g(x)$ to obtain the remainder $b(x)$,

3. Form the code word $b(x) + x^8 m(x)$. All three of these steps can be accomplished with a division circuit which is a linear, $(n-k)$-stage, shift register with feedback connections, which are based on the generator polynomial $g(x) = 1 + x + x^2 + x^8$. Such a circuit is shown in Figure 5.8.

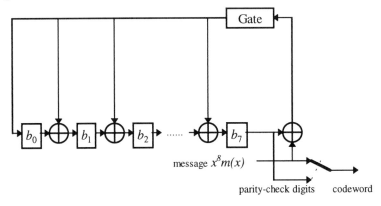

Figure 5.8 Encoding of the (40,32) code

The encoding operation in this digital circuit is carried out as follows:
Step 1. .With the gate turned on, the 32 information digits $m_0, m_1, \ldots, m_{31}$ (or $m(x) = m_0 + m_1 x + \ldots + m_{31} x^{31}$ in polynomial form) are shifted into the circuit and simultaneously into the communication channel. Shifting the message $m(x)$ into the circuit from the front end is equivalent to the pre-multiplication of $m(x)$ by x^8. As soon as the complete message has entered the circuit, the 8 digits in the register form the remainder, and thus they are the parity check bits.
Step 2. Break the feedback connection by turning off the gate.
Step 3. Shift the parity-check digits out and send them into the channel. These 8 parity-check digits $b_0, b_1, \ldots, b_7$, together with the 32 information digits, form a complete codeword.

The decoding circuit for the (40,32) single-error-correcting and double-error-detecting code is shown in Figure 5.9. Let r(x) be the received polynomial. Dividing $x^8 r(x)$ by $g(x)$ and $r(x)$ by $(x+1)$, respectively, one has $x^8 r(x) = a_1(x) g(x) + s_p(x)$ and $r(x) = a_2(x)(x+1) + \sigma$, where $s_p(x)$ is

BCH Codes

of degree 7 or less, and σ is either 0 or 1. If $s_p(x) = 0$ and $\sigma = 0$, $r(x)$ is divisible by $(1+x)g(x)$ and is a codeword polynomial. Otherwise, $r(x)$ is not a codeword polynomial. The syndrome of $r(x)$ is defined as $s(x) = xs_p(x) + \sigma$. If a single error occurs, $s_p(x) \neq 0$ and $\sigma = 1$. On the basis of these facts, a decoder for the (40,32) single-error-correcting and double-error-detecting code can be implemented as follows:

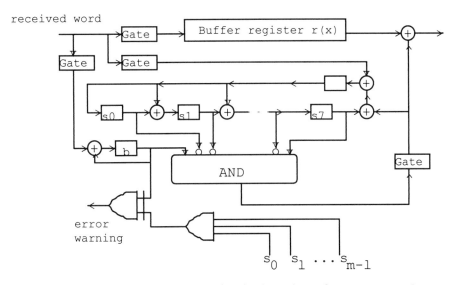

Figure 5.9. Decoding circuit for the (40,32) single-error-correcting and double-error-detecting code

Step 1. For $\sigma = 0$ and $s_p(x) = 0$, the decoder assumes that there is no error in the received polynomial.

Step 2. For $\sigma = 1$ and $s_p(x) \neq 0$, the decoder assumes that a single error occurred. The received word is read out of the buffer register, the syndrome register is shifted cyclically once. Once the syndrome in the register is (0,0,...,0,1), the next digit to come out of the buffer is the erroneous digit, and the output of the 8-input AND gate is 1.

Step 3. For $\sigma = 0$ and $s_p(x) \neq 0$, the decoder assumes that there are two errors and an error-warning bit is turned on.

Step 4. For $\sigma = 1$ and $s_p(x) = 0$, the error-warning bit is also turned on. This happens when an error pattern with an odd number (>1) of errors has occurred, and the error pattern is divisible by $g(x)$.

Using the above error-correction techniques, the ATM cell header is protected by the ITU-T I.432 error-handling procedure given in Figure 5.10.

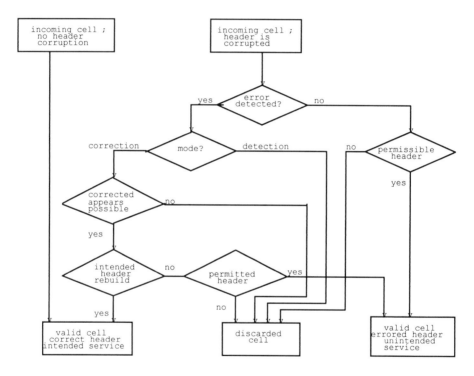

Figure 5.10 ITU-T I-432 view of error handling

The receiving ATM node corrects a 1-bit error, but if an uncorrectable error is detected, the cell is discarded. Cases of undetected multiple-bit errors are rare. Such errors could result in the misdelivery of the cell to a different address. From the standpoint of the receiver, the intruding cell is called a misinsertion and is discarded. The above error-correction technology for headers is effective in reducing the cell loss due to random errors. For optical fibers the probability of discarding cells is given in Figure 5.11 of the standardized ITU-T recommendation for the I.432 method. From the curve

BCH Codes

shown in Figure 5.11, one finds that the cell-loss rate can be minimized to about 10^{-10} for a network bit-error rate of 10^{-7}.

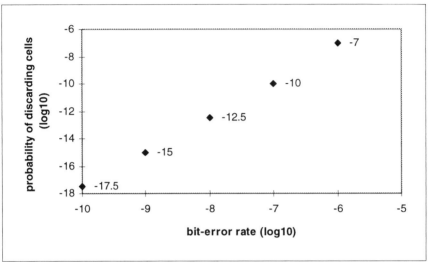

Figure 5.11. Probability of discarding cells for optical fiber.

Problems

5.1 A (15,7) BCH code is generated by
$$g(x) = x^8 + x^7 + x^6 + x^4 + 1.$$
(a) What is the parity-check polynomial $h(x)$, and what are the parameters of the dual cyclic code it generates?
(b) Draw a block diagram of a systematic encoder and syndrome generator for this (15,7) BCH code.
(c) Determine if the received word $\bar{r} = (000110111001000)$ is a valid codeword, where the rightmost bit represents the leading information symbol, by computing the syndrome.
(d) Decode the previous received word by computing the syndromes S_i for $i=0,1,2,3$.

5.2 A (15,5) binary BCH code has $\alpha, \alpha^3, \alpha^5$ as roots of the code-generator polynomial, where α is a primitive element of $GF(2^4)$.
(a) Find the generator polynomial $g(x)$.
(b) Find the parity-check polynomial $h(x)$ of this code.

(c) Draw the block diagrams of the k-stage and (n-k)-stage encoders.
(d) Give the generator and parity check matrices of this code and determine the minimum Hamming distance of the code.
(e) Give a block diagram of a decoder that corrects no less than all two-error patterns.

5.3 A (63,50) BCH code with the minimum Hamming distance $d_{min} = 6$ is generated by
$$g(x) = x^{13} + x^{12} + x^{11} + x^{10} + x^9 + x^8 + x^6 + x^3 + x + 1.$$
Design a 2-error-correction decoder by the use of Peterson's direct-solution algorithm.

5.4 Consider a triple-error-correcting binary BCH code of length 15 with the received words,
(a) $r(x) = x + x^5 + x^8$,
(b) $r(x) = x^5 + x^{10} + x^{12}$
Find the error-locator polynomial $L(x)$ for each case.

5.5 Consider the double-error-correcting binary BCH code with block length 15 which has $\alpha, \alpha^2, \alpha^3$, and α^4 as the roots of its generator polynomial. Assume arithmetic operations over $GF(2^4)$ are defined by the primitive polynomial $\alpha^4 = \alpha + 1$. Draw block diagrams (to the level of individual modulo-2 adders) for the syndrome-computation registers to compute $s_0 = r(\alpha)$ and $s_2 = r(\alpha^3)$.

5.6 It is shown in Sec. 5.4 that the (23,12) Golay code can be decoded for $t=3$ or less errors by modifying the Peterson algorithm. What is the BCH bound of this code and how many errors can be corrected, using the standard Peterson-decoding algorithm.

5.7 Let α be a primitive element of $GF(2^5)$ satisfies $\alpha^5 = \alpha^2 + 1$. A BCH code of length 31 with designed distance 5 is constructed over $GF(2^5)$. If the received word is
$$\bar{r} = (1001011011110000110101010111111),$$
decode this word.

5.8 Construct a ternary BCH code of length 26 and designed distance 5.

BCH Codes

5.9 Let C_1 be a binary cyclic (n,k) code with designed distance $2t_1$ and suppose that $(x+1)$ does not divide $g_1(x)$, the generator of C_1. Now let C_2 be the cyclic $(n,k-1)$ code generated by $(x+1)g_1(x)$ and suppose that this code has designed distance $2t_2 > 2t_1$.
(a) Show that the actual minimum distance of C_1 is at least $2t_1 + 1$.
(b) Prove that the $(17,9)$ and $(65,53)$ BCH codes have distance 5.

5.10 (a) Construct a generator polynomial for a binary BCH code of length 15 and designed distance 4.
(b) Determine the rate of this code.
(c) Compute the generator and parity-check matrices for this code.

5.11 Consider the binary BCH code of length 15 with the generator polynomial, $g(x) = 1 + x^4 + x^6 + x^7 + x^8$.
(a) Determine the minimum distance of this code.
(b) Use Peterson's algorithm to decode the following received words.
    i) $\bar{r} = (001111011101101)$
    ii) $\bar{r} = (100100000000000)$
    iii) $\bar{r} = (011011111001010)$
    iv) $\bar{r} = (001110110000000)$

5.12 Consider the non-binary BCH code of length 7 with the generator polynomial, $g(x) = \alpha^3 + \alpha x + x^2 + \alpha^3 x^3 + x^4$.
(a) Determine how many symbol errors this code can correct.
(b) Use Peterson-Gorenstein-Zierler decoding algorithm to decode the following received words.
    i) $\bar{r} = (010\alpha^4 \alpha^4 01)$,
    ii) $\bar{r} = (0100\alpha^5 00)$,
    iii) $\bar{r} = (000\alpha^4 000)$,
    iv) $\bar{r} = (0100\alpha^5 00)$.

5.13 Draw a Chien search circuit for binary quadratic residue codes and explain why only the n-th root of unity needs to be evaluated in the circuit.

Bibliography

[1] R. C. Bose and D. K. Ray-Chaudhuri, "On a class of error correcting binary group codes", Information and Control, 3, pp.68-79, 1960.
[2] A. Hocquenghem, "Codes correcteurs d'erreurs", Chiffres, 2, pp.147-156, 1959.
[3] D. Gorenstein and N. Zierler, "A class of error correcting code in p^m symbols", Journal of the Society of industrial and applied mathematics, Vol. 9, pp.207-214, June 1961.
[4] W. W. Peterson, "Encoding and error-correction procedures for the Bose-Chaudhuri codes", IRE Transactions on Information Theory, Vol. IT-6, pp.459-470, Sept. 1960.
[5] R. T. Chien, "Cyclic decoding procedure for the Bose-Chaudhuri-Hocquenghem codes", IEEE Transactions on Information Theory, Vol. IT-10, pp.357-363, Oct. 1964.
[6] G. D. Forney, "On decoding BCH codes", IEEE Transactions on Information Theory, Vol. IT-11, pp.549-557, Oct. 1965.
[7] W. W. Peterson and E. J. Weldon, Jr., Error Correcting Codes, 2nd ed. Cambridge, MA. MIT Press, 1972.
[8] F. J. MacWilliam and N. J. A. Sloane, The Theory of Error Correcting Codes, Amsterdam : North-Holland, 1977.
[9] S. Lin and D. J. Costello, Error Control Coding : Fundamentals and Applications, Englewood Cliffs, NJ : Prentice Hall, 1983.
[10] E. Berlekamp, Algebraic Coding Theory, New York : McGraw-Hill, 1968.
[11] M. Elia, "Algebraic decoding of the (23,12,7) Golay code", IEEE Trans. Inform. Theory, Vol. IT-33, pp. 150-151, Jan. 1987.
[12] I. S. Reed, X. Yin, and T.K.Truong, " Decoding the (24,12,8) Golay code", IEE Proc., vol. 137, pt. E. No. 3, pp. 202-206, May 1990.
[13] I. S. Reed, X. Yin, and T. K. Truong, "Algebraic decoding of the (32,16,8) quadratic residue code," IEEE Transaction on Information Theory, Vol. 36, No.4, pp.876-880, July 1990.
[14] I. S. Reed, T. K. Truong, X. Chen, and X. Yin, "Algebraic decoding of the (41,21,9) quadratic residue code," IEEE Transaction on Information Theory, Vol. 38, No. 3, pp.974-976, May 1992.
[15] X. Chen, I.S.Reed, T. Helleseth, and T. K. Truong, "General principles for the algebraic decoding of cyclic codes", IEEE Trans. on Inform. Theory, Vol. 40, No. 5, Sept. 1994.

[16] X. Chen, I.S.Reed, T. Helleseth, and T. K. Truong, "Use of Groebner bases to decode binary cyclic up to the true minimum distance", IEEE Trans. Inform. Theory, No. 5, Sept. 1994.

[17] X. Chen, I.S.Reed, T. Helleseth, and T. K. Truong, "Algebraic decoding of cyclic codes : a polynomial ideal point of view", Contemporary Mathematics, Vol. 168, 1994.

[18] M. Y. Rhee, <u>Error Correcting Coding Theory</u>, McGraw-Hill Publishing Company, New York, 1989.

6 Reed-Solomon Codes

In December of 1958 I. S. Reed and G. Solomon finished the report, entitled "polynomial codes over certain finite fields" at the M.I.T Lincoln Laboratory [1]. In 1960, a slight modification of this report was published as a paper [2] in the Journal of the Society for Industrial and Applied Mathematics(SIAM). This five-page paper described a new class of error-correcting codes that now are called Reed-Solomon(RS) codes. In the decades since their discovery RS codes have enjoyed countless applications from compact discs and digital television in living rooms to spacecraft and satellites in outer space.

Due to the enormous advances in the digital techniques for computers, new digital communication systems are now rapidly replacing the formerly widely used analogue systems. The RS code is one of the most powerful techniques for ensuring the integrity of transmitted or stored digital data against errors or erasures. RS coding has emerged as a key enabling technology in modern digital communications, driven by the ever-increasing demand for higher performance, lower cost, and sustained productivity for daily-life applications.

RS coding involves many modern signal processing techniques, such as the Fast-Fourier Transform (FFT) over finite-fields, iterative algorithms, the linear recursive shift-register, data interleaving and the modern computer architectures for pipeline processing.

In this chapter we provide a detailed, but practical, step-by-step introduction to RS codes. Both the RS encoding and decoding techniques with or without erasure symbols are emphasized. Examples constructed by the primitive

(15,9) RS code are developed in detail in this chapter in order to illustrate the encoding and decoding procedures. These examples include several useful algorithms which are employed to efficiently perform RS coding and decoding. Among these algorithms are the bit-serial encoder, the Berlekamp-Massey decoding algorithm and the Euclidean-decoding algorithm. Also a few examples are introduced in order to further increase our understanding of RS codes. Finally, examples of hardware and software implementations of RS coding are introduced to the interested practical-minded reader.

6.1 The State of RS Coding

Extensive research and development on RS coding has been accomplished over the past forty years. To assess the state-of-the-art of RS coding we first review some historical milestones in the development of RS coding. Following this, a grand tour is taken of the crucial components which are built into an RS coding system. Also an overview of the RS decoding techniques is presented in this chapter.

6.1.1 Milestones in the Development of RS Coding

Roughly speaking, RS coding techniques have gone through two major stages of development : (1) theoretical studies and the algorithm development and (2) the design of architectures for the fast decoding algorithms and their realization for practical applications. Prior to the early 1970s, studies of RS coding focused on questions about the minimum distance, the weight distribution, and the encoding and decoding algorithms. Even then the RS codes were considered to be among the most versatile and powerful error control codes which had the capability of correcting both random and burst errors[3]. Also it was well understood by these early researchers that one of the key obstacles to the practical application of RS codes was the decoding complexity.

The earliest decoding algorithm for RS codes is traceable back to March 1959 in a MIT Lincoln Laboratory report[4]. In this report, I. S. Reed and G. Solomon presented a decoding procedure which used a finite-field version of the expansion of a function in a power series. This made it possible to solve certain sets of simultaneous equations. Though much more efficient than a look-up table, their algorithm was still useful only for the least powerful RS codes.

Reed-Solomon Codes

In 1960 Peterson provided the first explicit description of a decoding algorithm for binary BCH codes[5]. It was found that his algorithm for explicitly solving the syndrome equations was quite useful for decoding RS codes to correct a relatively small number of errors. However, it became computationally intractable as the number of errors became too large. Peterson's algorithm was improved and extended to nonbinary codes by Gorenstein and Zierler (1961)[6], Chien (1964)[7], and Forney (1965)[8]. These efforts were productive, but RS codes capable of correcting, say, more than 6 errors still could not be accomplished in an efficient manner. The decoding complexity of the Peterson-type algorithms was $O(n^3)$. Detractors of coding research in the early 1960s used the lack of an efficient decoding algorithm to claim that RS codes were nothing more than a mathematical curiosity. Fortunately for the modern digital world these detractors' assessments proved to be wrong.

The breakthrough came in 1967 when E. Berlekamp demonstrated an extremely efficient decoding algorithm for both BCH and RS codes[9]. Berlekamp's algorithm allowed for the first time the possibility of a quick and efficient decoding of dozens of symbol errors in some very powerful RS codes. In 1968, Massey showed that the BCH decoding problem is equivalent to the problem of synthesizing the shortest linear-feedback shift register which is capable of generating a given sequence[10]. Massey then demonstrated a fast shift-register-based decoding algorithm for BCH and RS codes that is equivalent to Berlekamp's original version. This shift-register based approach is now commonly referred to as the Berlekamp-Massey(BM) algorithm. The decoding complexity of the BM algorithm is $O(n^2)$.

The Euclidean algorithm for solving the key equation was developed first in 1975 by Sugiyama et al.[11]. It was shown that such an algorithm, which has a complexity order $O(n^2)$, also can be used to efficiently decode RS, BCH and Goppa-type codes. The Euclidean algorithm, named after its discoverer in Greek antiquity, the father of Euclidean geometry, was originally a systematic means for finding the greatest common divisor of two integers. The Euclidean algorithm extended also to more complex collections of algebraic objects, including sets of polynomials which have coefficients in a finite field.

These efficient decoding techniques have made RS codes very popular for quite a large number of different applications. The earliest application of RS codes other than for the military was the use of a one byte error-correction code in a deep-space telecommunication system in about 1970. The first "full-blown" use of RS coding was the Voyager II mission in deep-space exploration, beginning in 1977. Voyager II used a length 255-byte and dimension 223-byte RS code, which corrected 16 byte-errors. The decoder of this specific code successfully used the Euclidean algorithm to solve the decoding equation to find the error-locator polynomial, and finally to correct the errors.

Without doubt it can be claimed that the RS codes are the most frequently used digital error-control codes in the world. This claim rests firmly on the fact that the digital audio disc or compact disc(CD) uses two shortened RS codes for error correction and error concealment. The special properties of these two RS codes make possible the high-quality sound generated by the compact-disc player.

The compact-disc system utilizes a pair of, what are called, *cross-interleaved* Reed-Solomon codes. Since RS codes are nonbinary, each codeword symbol becomes a string of bits when transmitted across a binary channel. If a noise burst corrupts several consecutive bits on the channel, the resulting bit errors are "trapped" within a small number of nonbinary symbols. For each burst of noise the RS decoder needs only to correct a few symbol errors, as opposed to a longer string of bit errors. RS codes can also correct byte-erasures in an efficient manner. The compact-disc error-control system uses the first of the cross-interleaved RS codes to declare byte-erasures, and the second code to correct these erasures. This brilliant piece of engineering allows for the very accurate reproduction of sound despite, sometimes substantial, material imperfections and damage to the surface of the disc.

Finally RS codes, as with any block code used in error control, can be "shortened" to an arbitrary extent. Most error-control codes have a natural length. For example, RS codes defined over the field $GF(2^8)$ have a natural length of 255. The compact-disc system is able to use RS codes of length 32 and 28 eight-bit symbols. while still retaining the byte symbol structure of the length-255 RS code. The shorter code word length keeps the implementation of the interleaver simple, while the 8-bit symbols provide protection against error bursts and also provide a good match for the 16-bit samples obtained from the original analog music source.

It was found in the 1960's [12] that multilevel coding structures, the so-called concatenated codes, could exponentially reduce the probability of error at an information rate less than channel capacity, while the decoding complexity increased only algebraically. In the beginning RS codes were not an obvious choice in a concatenated coding system for correcting random errors in a

Reed-Solomon Codes

transmitted data stream. It was discovered that convolutional codes, which used maximum-likelihood decoding, e.g. the Viterbi algorithm, were an excellent candidate in a concatenated coding system because of their superior random-error correction ability.

However, when uncorrected errors occur, the Viterbi decoder output is very bursty. Thus, the RS code provides the perfect match for, what is called, the "outer" code. Therefore, in most concatenated coding systems, a relatively short convolutional "inner" code is used with maximum-likelihood decoding to achieve a modest error probability, e.g. $P_e \cong 10^{-2}$, at a information rate which is near channel capacity. Then a long high-rate RS "outer code" is used with a powerful algebraic decoding algorithm to correct errors at the output of the convolutional decoder, e.g. the output of the Viterbi decoder, in the concatenated coding system.

Substantial coding gains are achievable with small rate loss in the concatenation of a convolutional with an RS code. Important applications of the concatenated convolutional and RS coding systems include (1) the deep space telecommunication system in the *Voyager* expeditions to Uranus and to Neptune, (2) the Hubble space telescope and more recently (3) the High-Definition TeleVision(HDTV) system (or digital TV system). In all of these systems RS codes are used in the transmission to preserve the reliability of compression-coded photographs or video.

6.1.2 Architectures and Electronics for the Implementation of the RS Codec

To achieve a required performance in error-correction power, throughput, or both, the appropriate choice of computer architecture has to be made. For short codes the decoding methods for cyclic codes, discussed, in Chapters 3 and 4, are able to correct a small number of errors. For example, some simple RS codes can be decoded by means of a syndrome look-up table. The syndromes for a received word are computed, and the corresponding minimum-weight error pattern is found in a table. Then this error pattern is subtracted from the received word, thereby producing a corrected codeword. Unfortunately this approach is usually out of the question for all but the most abbreviated Reed-Solomon codes.

After the discovery of Reed-Solomon codes a search began for efficient encoding and decoding algorithms. Many such algorithms were found through years of work. Among these are the bit-serial encoders, the

Berlekamp-Massey decoding algorithm, the Euclidean decoding algorithm, and the Welch-Berlekamp decoding algorithm, etc. These encoding and decoding algorithms involve computations over finite-fields, such as addition, multiplication, and division. As discussed in Chapter 2, finite-field operations can be realized by logic gates such as the AND, XOR, OR, Flip-Flop, etc.. Each logic gate has a fixed complexity because it is made from elementary transistors. There is a certain delay between the input of the signal received by the logic gate and the availability of the signal at the gate's output. Usually this delay is called the latency of the gate. Also, the energy consumption of a logic gate needs to be considered. To implement encoders and decoders for RS codes, especially for high data rates, the complexity of the logic gates, their latency times, and energy consumption are among the most important factors for choosing architectures and electronic circuits.

By the 1970's the basic concepts and decoding algorithms for RS codes were well understood and developed. But the problems of designing and realizing a low-complexity, high-bit-rate, RS encoder and decoder were still a challenging task. To satisfy the required performance of data rate and error-control, the RS codecs were integrated into the Very Large-Scale Integration(VLSI) chips which contained a great number of logic gates and transistors. One type of such a chip is called an Application-Specific Integrated Circuit(ASIC). At present, an integration scale of millions of transistors can be achieved on a single ASIC chip. The interest in the ASICs resides in their high level of integration and in their performance : they can operate at very high speeds with few support devices, short-connection lengths, high modularity, and complete confidentiality. ASIC devices are divided into Cell-Based designs (or Standard-Cell Designs), and Full Custom Circuits. The disadvantage of using the ASIC technique is that the time required between the design and its fabrication is not as easy to predict as it is in the case of the standard digital circuit technologies.

Many implementations of RS codecs use programmable VLSI devices. These programmable circuits include, what are called, Programmable Logic Arrays(PLAs), Programmable Logic Devices(PLDs), Gate Arrays and Field-Programmable Gate Arrays(FPGAs). These programmable devices are configurable by the user and thus provide a great deal of flexibility. Various techniques for implementing RS codecs on VLSI circuits are discussed in the next chapter.

The electronic implementation of simple RS codecs is performed usually by a use of the classical architectures, discussed in Chapters 4 and 5 for encoding and decoding cyclic and BCH codes. Recently, quite a few decoder designs

Reed-Solomon Codes

and architectures are developed for the long RS codes, e.g., see [13][14][15]. In addition to the search for improved decoding algorithms efforts also were made to perform finite-field arithmetic operations in an efficient manner, e.g., see [16][17][18].

To allow for the direct implementation of complex algorithms in silicon a revolutionary concept, called a systolic architecture or array, was introduced by Kung and his group [29]. The efficient implementation of the solution to a problem using a systolic architecture, required the transformation of the algorithm in such a manner that it can be expressed as a sequence of elementary recurrent operations.

Compared with a classical implementation, the advantages of a systolic architecture are numerous and important as seen below:
(1) The regularity of the basic structures makes possible a low cost VLSI implementation.
(2) The modularity of the architecture makes it easy to extend to solve a large problem of the same type.
(3) The connections between cells are simple.
(4) Data rates can be increased because the large-scale integration has a denser packing.
(5) Power consumption is lower than in more discrete and isolated VLSI circuit designs.
(6) Parallel and pipelined operations require less expensive technologies.
It is clear that the transformation of an algorithm into a systolic structure could lead, for many applications, to a more efficient use of transistors. Usually, a more dense larger scale of integration is possible with systolic-array circuits.

Encoding and decoding of RS codes can be implemented with the following function blocks, using systolic arrays:
(1) The multiplication of two polynomials,
(2) the division of one polynomial by another,
(3) the multiplication of two elements of $GF(2^m)$ (multiplication of two binary polynomials, modulo a third polynomial),
(4) the Computation of the inverse of an element of $GF(2^m)$,
(5) the GCD of two polynomials.

Today's system designer often can find "off-the-shelf" hardware to implement many RS coding algorithms in the form of VLSI integrated circuits and standard macro cells that satisfy a wide range of encoding/decoding needs.

6.1.3 Some Recent Research and Development on RS Coding

RS coding system for space applications

We discuss first certain problems encountered by the *Galileo* mission to Jupiter by means of a space-probe. Difficulties began when the high-gain antenna on board the spacecraft refused to deploy properly so that it became a very low-gain antenna.

As a consequence of this antenna failure all of the data collected by the probe was sent to earth by this low-gain antenna. This resulted in a drastic lowering of the rate at which the spacecraft could reliably transmit data. Engineers from all over the world began to work frantically to find ways to increase the coding gain provided by the concatenated coding system already used by *Galileo*. In [23] Hagenauer, Offer, and Papke discuss several powerful means for attacking this problem. These include the use of, what is called, iterative decoding and the soft-output Viterbi algorithm (SOVA). SOVA provided for the declaration of erasures at the input to the RS decoder, thereby considerably improving performance.

One key result that emerged from such exploratory work on decoding was the demonstration that a concatenated convolutional/RS system could provide reliable data transmission well beyond, what is called, the *cutoff rate.* The cutoff rate is an information-theoretic threshold beyond which many believe that the reliable performance of an error-control system is not possible. The results show for a concatenated error-control system that this supposed barrier can be surpassed. Hagenaner, Offer, and Papke have brought the performance of an RS error-control system within a few decibels of the ultimate barrier, namely the Shannon channel capacity. A series of photos of the surface of a Jubiter moon display the impact of these findings.

RS coding system with feedback

Wicker and Hartz examine various means to use RS codes in applications that allow the receiver to transmit back to the transmitter. Such error-control applications of these procedures include the mobile data-transmission systems and the high-reliability military-communications systems. Along

Reed-Solomon Codes

with their powerful error-correction capabilities, RS codes also can provide a substantial amount of *simultaneous* error detection. The key here lies in the distinction between *a decoder error* and a *decoder failure*.

Consider the decoding of a t error-correcting RS code. If the received word differs from an incorrect code word in t or fewer symbol coordinates, then the closest incorrect code word in Hamming distance is selected by the decoder, thereby resulting in, what is called, a decoder error. This is an undetectable condition that can cause errors in the decoded data.

On the other hand, if a noise-corrupted word differs from all of the code words in $(t + 1)$ or more coordinates, then the decoder declares, what is called, a *decoder failure*. Decoder failures are detectable, so that the receiver is able to request a retransmission of the problematic word.

RS coding in spread-spectrum systems

Spread-spectrum systems can be grouped into two basic types: frequency-hopping spread spectrum (FH/SS) and direct-sequence spread spectrum (DS/SS). An FH/SS system modulates information onto a carrier that is systematically moved or hopped from frequency to frequency. Frequency hopping, sometimes, is used in military communications systems as a powerful means for defeating partial-band jamming, e.g. the Joint Tactical Data System (JTDS) of the U.S. Armed Forces. In general a frequency-hopped system provides protection against any partial-band disturbance. In a peaceful environment. such disturbances may arise from FH, multiple-access interference or from narrow-band noise sources. Partial-band noise can be caused also by frequency-selective fading which occurs on a mobile communication channel. For these reasons and others, frequency hopping is presently receiving serious consideration for use in personal communication systems in the United States. The Global System for Mobile communication(GSM), the European cellular telephone system, has already provided for a limited amount of frequency hopping.

A DS/SS system creates a wide-band signal by means of the phase-shift-keying of its RF carrier along with the sum of the data sequence and the spreading sequence which has a pulse rate that is much larger than that used for the data sequence. When the received signal is "de-spread," the narrow band interfering signals are spread in frequency in such a manner that only a small fraction of their power falls within the bandwidth of the recovered data sequence. As with the FH/SS systems, the DS/SS systems perform well against partial-band disturbances. DS/SS systems have the added benefit of a low probability-of-intercept (LPI) within a wide spectrum. Since the transmitted signal energy is spread across a wide spectrum, only a small amount of signal energy is found in any given narrow band slot in which an enemy might be searching for activity. DS/SS systems are also being considered for use in mobile-radio applications.

In [25] Sarwate begins by describing the design and performance of several DS/SS systems. He then proceeds to discuss how RS codes can be used in the design of frequency-hopping sequences. If these sequences are carefully selected, the interference caused by other users in a multiple-access environment can be reduced greatly. Sarwate shows that some of the hopping sequences, which have been described using a different terminology, can also be viewed as low-rate RS code words. He then proceeds to a description of several DS/SS systems. He also shows that the Gold and Kasami sequences can also be interpreted using the language of RS codes.

In [25] Pursley explores in greater depth the application of RS codes for error correction in FH/SS systems. He focuses on various means by which side information can be obtained from a frequency-hopped system and used to declare erasures at the input to the RS decoder. He begins by considering the use of test symbols and threshold tests, which require very little modification to an existing FH/SS system, but provides a significant improvement in its performance. He then considers more powerful techniques which include the use of binary parity checks on individual code-word symbols and the derivation of the side information from the inner decoder in a concatenated system. Pursley concludes by examining the use of side information in the routing algorithms used in packet-radio networks.

RS coding in computer memory

In [26], Saitoh and Imai describe the error-control problems caused by faults in the integrated circuits used to control data flow in computers. The resulting error patterns in these systems are generally unidirectional so that the erroneous bits have the same value. Saitoh and lmai show that this unidirectional tendency can be exploited to develop codes which are based on RS codes that outperform standard RS codes. Such advanced error-control systems play an integral role in the development of some recent very high-speed super computers.

6.2 Construction of RS Codes

RS codes are constructed and decoded by means of finite-field arithmetic. As noted in Chapter 2 a finite field of q elements is denoted by $GF(q)$. The number of elements in a finite field is of the form, $q=p^m$, where p is a prime number and m is a positive integer. The finite field $GF(q)$ can be completely described by its size q.

Let α be a primitive element in $GF(q)$. Then the $(q-1)$ consecutive powers of α, namely $\{1, \alpha, \alpha^2, \ldots, \alpha^{q-2}\}$, must be distinct. They are the $(q-1)$ nonzero elements of $GF(q)$. The primitive element α is a root of some primitive irreducible polynomial $p(x)$.

The original approach to the construction of RS codes is reasonably straightforward. An information word is of the form $\bar{m} = (m_0, m_1, \ldots, m_{k-1}, m_k)$, where the m_i constitute k information symbols taken from the finite field $GF(q)$. These symbols are used to construct the polynomial $p(x) = m_0 + m_1 x + \cdots + m_{k-2} x^{k-2} + m_{k-1} x^{k-1}$. The RS codeword $\bar{c}$ of the information word $\bar{m}$ is generated by evaluating $p(x)$ at each of the q elements in the finite field $GF(q)$ to obtain

$$\bar{c} = (c_0, c_1, \ldots, c_{q-1}) = [p(0), p(\alpha), p(\alpha^2), \ldots, p(a^{q-1})]$$

as the codeword. A complete set of RS codewords is constructed by allowing the k information symbols to take on all possible values, i.e.

$$C = \left\{ \bar{c} = \begin{bmatrix} 1 & 0 & 0 & \cdots & 0 \\ 1 & \alpha & \alpha^2 & \cdots & \alpha^{k-1} \\ \vdots & \vdots & & & \vdots \\ 1 & \alpha^{q-1} & \alpha^{2(q-1)} & \cdots & \alpha^{(k-1)(q-1)} \end{bmatrix} \cdot \bar{m} : \bar{m} \in [GF(q)]^k \right\}.$$

The RS codes generated in this manner are linear since the vector sum of any two message words of length k is simply another message word of length k. The number of information symbols k is called the dimension of the RS code since the RS codewords form a vector space of dimension k over $GF(q)$. Also, the code length of n symbols equals q since each codeword has q coordinates. When RS codes are discussed, they are usually denoted by their length n and dimension k in symbols as (n,k) codes.

Reed and Solomon's original approach to the construction of their codes fell out of favor following the discovery of the generator polynomial or cyclic-code approach. At the time it was felt that the latter approach led to better decoding algorithms. It was not until 1982 that Tsfasman, Vladut, and Zink, using a technique discovered by Goppa, extended the Reed and Solomon construction to develop a class of codes whose performance exceeded the Gilbert-Varshamov bound [28]. The Gilbert-Varshamov bound is a lower bound on the performance of error-correcting codes that many coding experts were beginning to believe was also an upper bound. Tsfasman, Vladut, and Zink's work uses several powerful tools from algebraic geometry to break open an entirely new field of research that continues at the present time to attract a great deal of interest from coding theorists. Their work yields some new special codes that are extremely powerful.

Today the, so-called, generator-polynomial construction of RS codes is the most commonly used method for generating these codes in the error control literature. Initially, this approach evolved independently of RS codes as a means for describing *cyclic codes* as defined in Chapter 4. In 1961 Gorenstein and Zierler generalized the BCH code to arbitrary finite fields of size p^m, discovering along the way that they had developed a new means for describing RS codes[6].

Cyclic RS codes with codeword symbols from $GF(q)$ have length q-1, one coordinate less than that needed for the original construction. A cyclic RS code can be extended to have length q, or length $q + 1$, but in both cases the resulting codes are usually no longer cyclic. Since the generator-polynomial approach to the construction of RS codes is currently the most popular, one may note that RS codes with symbols in the finite field $GF(q)$ usually have length q-1. For example, the finite field GF(256) is used in many applications because each of the $2^8=256$ field elements can be represented as an 8-bit sequence or a byte. RS codes of length 255 and their shortened codes are thus very popular error-control codes.

The cyclic RS-code construction process proceeds as follows: Suppose that we want to build a t error-correcting RS code of length q - 1 with symbols in $GF(q)$. Recall that the nonzero elements in the finite field $GF(q)$ always can be represented as the $(q - 1)$ powers of some primitive element α. The generator polynomial of a t error-correcting RS code has as roots, the *2t consecutive powers of* α, so that the generator polynomial is of the form,

$$g(x) = \prod_{j=0}^{2t-1}(x-\alpha^{h+j}) = \sum_{j=0}^{2t-1} g_j x^j ,$$

where h is an integer constant which is called, sometimes, the power of the first of the consecutive roots of $g(x)$. Different generator polynomials are formed with different values of h. By carefully choosing the integer h, the circuit complexity of the encoder and decoder often can be reduced[19]. The integer constant h is usually, but not always, chosen to be 1.

Example 6.1.
The generator polynomials for the (7,5) RS codes constructed over $GF(2^3)$ (with the primitive polynomial $p'(x)=x^3+x+1$) has the form,
$$g(x)=(x-\alpha^h)(x-\alpha^{h+1}).$$
The generator polynomials for the two cases, $h=1$ and $h=3$, are given by
$$g(x)=x^2+\alpha^4 x+\alpha^3 \text{ for } h=1,$$
$$g(x)=x^2+\alpha^6 x+1 \text{ for } h=3.$$

Thus a valid code polynomial $c(x)=m(x)g(x)$ can have any degree of $2t$ up to $q-2$. Note here that a code polynomial of degree $(q-2)$ corresponds to a code word of $(q-1)$ coordinates. It follows that the dimension of a code with a generator polynomial of degree $2t$ is $k = q - 2t - 1$. Hence the RS code of length $q-1$ and dimension k can correct up to t errors, where $t = \lfloor (q-k+1)/2 \rfloor$. Here $\lfloor x \rfloor$ denotes the largest integer less than or equal to x. Also $\lfloor x \rfloor$ is called the integer part of the real variable x.

In 1964, Singleton showed that such an ability to correct $t = \lfloor (q-k+1)/2 \rfloor$ random symbol errors is the best possible error-correction capability any block code of the same length and dimension can have[20]. This "optimal" random error-correction capability is named the Singleton bound. Codes which achieve this bound are called *maximum distance separable*(MDS). RS codes are MDS codes. Since RS codes are MDS, they have a number of important properties that lead to several practical consequences.

It is also known from Chapter 3 that the maximum burst-error-correction capability for an (n.k) linear code satisfies *the Rieger Bound* as follows:
$$b \leq \left\lfloor \frac{n-k}{2} \right\rfloor,$$
where b is the length of the burst error in bits or symbols for non-binary codes. Evidently, for RS codes the maximum random error-correction capability equals or coincides with the maximum burst-error-correction capability since $t = \left\lfloor \frac{d-1}{2} \right\rfloor = \left\lfloor \frac{n-k}{2} \right\rfloor = b$ symbols. Therefore, RS codes are the most powerful block codes for both random and burst-error correction.

Also note that RS codes can be shortened by omitting some of the information symbols. But such a shortening does not effect the minimum

Hamming distance of the RS code. Hence, any shortened RS code is also an MDS code. However, a shortened code is not usually a cyclic code.

Any valid code polynomial must be a multiple of the generator polynomial. It follows that any valid code polynomial must have as roots the same $2t$ consecutive powers of α that form the roots of $g(x)$. This provides us with a very convenient means for determining whether or not a received word is a valid code word. One simply makes sure that the corresponding polynomial has the necessary roots. This approach leads to the powerful and efficient decoding algorithms that are introduced later in this chapter.

In almost all of the coding literature RS codes are discussed over the finite fields $GF(2^m)$ because of the simplicity of their implementation using binary devices. Also the power of the first root of the consecutive roots of $g(x)$, namely h, is usually selected to equal one. In this chapter the ensuing discussion satisfies these two conditions.

All RS codes over $GF(2^m)$ for $m \leq 4$ are listed in Table 6.1. In this table r denotes, in percentage, the rate of the code. The rate is the ratio of the number of message symbols k to the block length n, i.e. $r=k/n$. Also $t = \left\lfloor (d-1)/2 \right\rfloor$ denotes the maximum random-symbol error-correction capability.

Table 6.1 RS Codes Over $GF(2^m)$ for $m \leq 4$.

m	n	k	t	r
2	3	1	1	33.3%
3	7	5	1	71.4%
3	7	3	2	42.9%
3	7	1	3	14.3%
4	15	13	1	86.7%
4	15	11	2	72.3%
4	15	9	3	60.0%
4	15	7	4	46.7%
4	15	5	5	33.3%
4	15	3	6	20.0%
4	15	1	7	6.7%

Notice in Table 6.1 that the (3,1) RS code is the shortest RS code which is capable of correcting errors. It should also be noted that as m increases the number of possible RS codes increases in an exponential manner. Thus, once the block length is determined from some fairly large m, an RS code with the

Reed-Solomon Codes

desired pair (r,t), rate versus correction capability, can be selected. It is desirable, but impossible, for the code rate r to approach 100% while the error-correction capability t approaches n. Finally the (15,9) RS code from Table 6.1 is selected as an example throughout this chapter to demonstrate the encoding and decoding processes for RS codes.

block length	n=15 symbols
message length	k=9 symbols
code rate	r=k/n=60%
parity-check symbol length	n-k=6 symbols
minimum distance	d=n-k+1 symbols
error-correction capability	t=(n-k)/2=3 symbols
finite field order	$q=n+1=2^m=2^4=16$ symbols
field generator	$p(x)=x^4+x+1$
codeword-generator polynomial	$g(x) = x^6 + \alpha^{10}x^5 + \alpha^{14}x^4$ $+ \alpha^4 x^3 + \alpha^6 x^2 + \alpha^9 x + \alpha^6$

Table 6.2. The (15,9) RS Code Example Parameters

Example 6.2
The parameters of the (15,9) RS code over $GF(2^4)$ are given in Table 6.2. In this example the code symbols are four binary digits (4-bits) since the code is constructed over $GF(2^4)$ (given by Table 2.1).

6.3 Encoding of RS codes

The encoding procedures stated earlier for cyclic codes are applicable also to RS codes. Let the sequence of k message symbols in $GF(2^m)$ be $\bar{m} =(m_0,m_1,......,m_{k-1})$. Then the message vector is represented in polynomial form as follows :

$m(x)=m_0+m_1x+\cdots+m_{k-1}x^{k-1}$.

One way to generate the code polynomial c(x) is to multiply m(x) by g(x), i.e. c(x)=m(x)g(x). In this encoding procedure the resulting code is not in a systematic form since the k message symbols are not explicitly present in the codeword. As part of the decoding process, one would like to recover the message polynomial m(x) directly from the decoded code polynomial $\hat{c}(x)$.

Example 6.3. Continue the (15,9) RS code in Example 6.2 with the message polynomial $m(x) = \alpha^{11}x$ which represents the message word $\bar{m} = (0\ 0\ 0\ 0\ 0\ 0\ 0\ \alpha^{11}\ 0)$ which in binary form is the sequence (000000000000

0000000000000000011100000). The above non-systematic encoding produces the code polynomial,
$$c(x) = m(x)g(x) = \alpha^{11}x^7 + \alpha^6 x^6 + \alpha^{11}x^5 + x^4 + \alpha^2 x^3 + \alpha^5 x^2 + \alpha^2 x,$$
which represents the codeword $\bar{c} = (0000000\alpha^{11}\alpha^6\alpha^{11}1\alpha^2\alpha^5\alpha^2 0)$. Evidently in this example it is difficult to directly separate the message $m(x)$ from the encoded codeword $c(x)$.

In order to encode so that the message is "in the clear" it is desirable to be able to generate systematic codes. Three common methods to generate systematic RS codes are discussed next.

6.3.1 Encoding by Polynomial Division

To encode RS codes in a systematic manner the codeword polynomial is generated first by a use of the division algorithm. Dividing the shifted message polynomial $x^{d-1}m(x)$ by $g(x)$, one obtains
$$c(x) = x^{d-1}m(x) + p(x), \tag{6.1}$$
where $p(x) = \sum_{j=0}^{d-2} p_j x^j$ is the remainder polynomial of degree less than $d-1$, i.e.
$$p(x) = x^{d-1}m(x) \bmod g(x). \tag{6.2}$$
The polynomial $p(x)$ is called the parity-check polynomial since the sequence $(p_0, p_1, \ldots, p_{d-2})$ is the parity-check word or vector. Evidently, the codeword vector is given by
$$\bar{c} = (p_0, p_1, \ldots, p_{d-2}, m_0, m_1, \ldots, m_{k-1})$$
so that the message is discernible in the encoded word.

Example 6.4
Continue the *(15,9)* RS code in *Example 6.3*. The systematic code polynomial $c(x)$ for the message polynomial $m(x) = \alpha^{11}x$ is computed to be
$$c(x) = x^6 m(x) + (x^6 m(x) \bmod g(x))$$
$$= (x^6)(\alpha^{11}x) + (x^6)(\alpha^{11}x) \bmod g(x)$$
$$= (\alpha^{11}x^7) + (\alpha^{11}x^7) \bmod g(x)$$
$$= \alpha^{11}x^7 + \alpha^8 x^5 + \alpha^{10}x^4 + \alpha^4 x^3 + \alpha^{14}x^2 + \alpha^8 x + \alpha^{12}.$$
Note that the message symbol appears unchanged in the encoded codeword.

To recapitulate, the steps of the above encoding algorithm are given as follows:

Reed-Solomon Codes

Step 1: Premultiply (or shift) the data polynomial $m(x)$ by x^{d-1}.
Step 2: Obtain $p(x)$ by the division algorithm as shown in Eq. (6.2).
Step 3: Combine $p(x)$ and $x^{d-1} m(x)$ to obtain the codeword $c(x)$ as indicated in Eq.(6.1).
The block diagram of such an encoder is given in Figure 6.1.

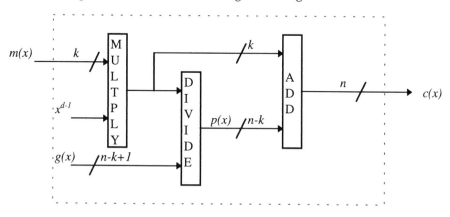

Figure 6.1. Block diagram of a RS encoder

The above encoder depends on the circuit structure needed for the polynomial division. An implementation of the division in Step 2 is shown in Fig. 6.2. It is seen that such a polynomial division requires multiplications in $GF(2^m)$. The complexity of the structure depends on what type of multiplier is used. For the combinational logic multiplier, given in Section 2.3 of Chapter 2, the encoder circuit complexity can be as high as $O(m^3d)$. For the Massey-Omura multiplier discussed in Chapter 2 the circuit complexity is $0(m^2d)$. However, if Berlekamp's bit-serial dual-basis multiplier, discussed in Chapter 2, is used, the complexity reduces to $0(md)$ [19],[21].

In Fig. 6.2, the polynomial $m(x)x^{d-1}$ is created by sequentially sending the message polynomial $m(x)$ into the right side of the division circuit. Therefore, when the input message symbols are received in a bit-serial fashion, a bit-serial multiplication scheme can be used to reduce the computational complexity needed to obtain $p(x)$ at the expense of throughput.

The systematic encoder of an RS code is constructed by a use of the calculator shown in Fig, 6.2. This structure is divided into the three main units as shown in Fig. 6.3.

The fixed-rate pipeline bit-serial encoder, shown in Fig. 6.3, is operated as follows :

1. Row-generation unit: The incoming bit stream is first buffered into the $(m-1)$-stage shift register before a block of m bits is loaded into the linear-feedback shift register (LFSR). Each m-bit data block is considered to be represented in the triangular basis coordinates, defined in Chapter 2 for elements of $GF(2^m)$. As a consequence, the loading operation corresponds to the formation of the 0-th row of the Hankel matrix, described in Chapter 2. The other rows are generated similarly after operations by the LFSR in subsequent clock cycles.

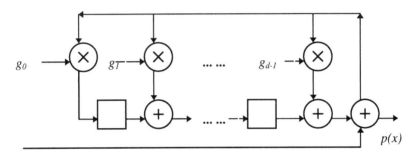

$m_{k-1}, m_{k-2}, \ldots, m_0$

Figure 6.2 Calculator of $p(x)$ by the division in Step 2.

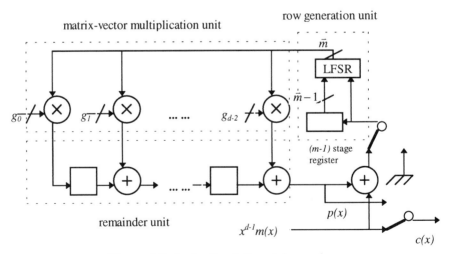

Figure 6.3. A pipeline bit-serial encoder.

2. Matrix-vector multiplication unit: This unit consists of $d-1$ identical multiplication modules. Each module evaluates the inner product of two

Reed-Solomon Codes

input vectors. The common input of these modules is a row of the Hankel matrix. The other input to module i is the vector, consisting of the canonical basis coordinates, defined in Chapter 2 for the g_i. For a fixed-rate encoder g_i can be hard-wired to module i.

3. Remainder unit: This unit stores the coefficients of the remainder polynomial. These coefficients are elements of $GF(2^m)$ and are represented in the basis. Modulo-2 additions are performed serially with respect to the basis.

Example 6.5

The encoder of the (7,5) RS code given in *Example 6.1* with $h=1$, is implemented by the diagram shown in Fig. 6.3. In $GF(2^3)$ every element is expressed by a polynomial in α of degree ≤ 2, i.e. $\alpha^i = b_2\alpha^2 + b_1\alpha + b_0$, where $b_0, b_1, b_2 \in GF(2)$ since $\alpha^3 = \alpha + 1$. Therefore, the two multipliers of the constants are expressed by

$$\alpha^i \cdot \alpha^3 = \alpha^3(b_2\alpha^2 + b_1\alpha + b_0) = (\alpha + 1)(b_2\alpha^2 + b_1\alpha + b_0)$$
$$= (b_1 + b_2)\alpha^2 + (b_0 + b_1 + b_2)\alpha + (b_0 + b_2),$$
$$\alpha^i \cdot \alpha^4 = \alpha^i\alpha^3 \cdot \alpha = (b_1 + b_2)\alpha^3 + (b_0 + b_1 + b_2)\alpha^2 + (b_0 + b_2)\alpha$$
$$= (b_0 + b_1 + b_2)\alpha^2 + (b_0 + b_1)\alpha + (b_1 + b_2).$$

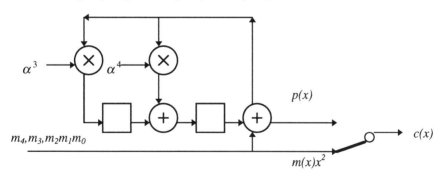

Figure 6.4a A block diagram of the (7,5) RS encoder

In Fig. 6.4, the polynomial division in the encoding process is implemented by a 2-stage, 3-bit, shift register. Between any two consecutive shift registers, there is a 3-bit XOR for the binary finite-field addition. The feedback path which contains the quotient for each division step, is broadcast to the two constant finite-field multipliers. The message bytes are pipelined from the input, and the first 5 bytes of the output are the same as the input, while the last 2 CRC bytes are obtained from the data stored in the two shift registers after all of the message bytes have been computed.

Example 6.6

In this example the bit-serial encoder of the (15,9) RS code given in *Example 6.2*, is developed. Berlekamp's bit-serial multiplier (discussed in Sec.2.8) is used for this encoder, shown in Fig. 6.4b.

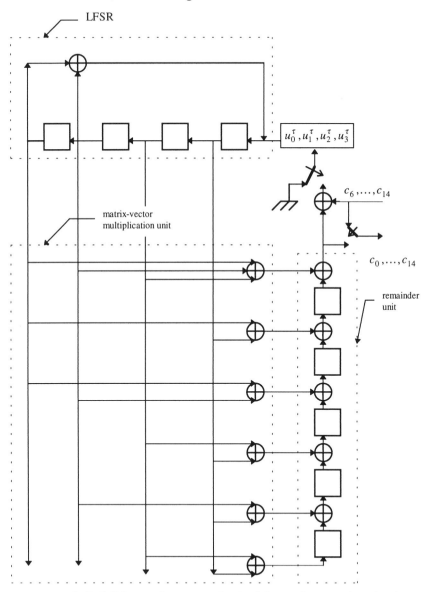

Figure 6.4b (7,5) RS encoder using the Berlekamp bit-serial multiplier

6.3.2 Encoding in the Frequency Domain

An RS code can be generated also in, what is called, the frequency domain [22]. To encode the information symbols in the frequency domain, the first $(n - k)$ symbols, i.e. the C_j for $j = 0,1,...,n - k - 1$, are set to zero. The remaining k symbols in the frequency domain are assigned to be information symbols. The inverse finite-field Fourier transform is given next by the identity,

$$c_i = \sum_{j=0}^{n-1} C_j \alpha^{-ij}, \text{ for } i=0,1,...,n-1. \tag{6.3}$$

Such a finite-field transform also can be used to obtain, what is called, a time domain RS code.

It is seen from Eq. (6.3) that the number of multiplications required for encoding by the use of the Fourier transform is the same number needed to perform polynomial division. However, when the number of information symbols is variable, it could be advantageous to use the Fourier-transform encoding technique since it can accommodate different values of k simply by setting some of the C_j to zero without any change in the hardware. Also, for certain codeword lengths, e.g. $n=2^m$ for $m>0$, it is advantageous to use the Fast Fourier Transform (FFT) with logic multiplies to perform the encoding.

Notice that Eq. (6.3) can be expressed iteratively by
$$c_i = (...(C_{n-1} \alpha^{-i} + C_{n-2}) \alpha^{-i} + ... + C_1) \alpha^{-i} + C_0. \tag{6.4}$$
From this relation a general structure for frequency-domain encoding is derived. Such a structure is called a systolic array. By Eq.(6.4) a code word c_i is generated recursively by the use of one α^{-i} multiplier and one adder.

A structure for computing the iteration in Eq.(6.4) is shown in Fig. 6.5. In this figure, the function of each cell is given by the register-transfer relation, $B_j \leftarrow A_j + B_j \cdot \alpha^{-i}$, where "$\leftarrow$" denotes the operation "is replaced by". The data C_j for $j = n - 1, n - 2,...,0$, are sent to all of the cells simultaneously. Assume initially that all registers are set to zero. After all of the C_j for $j = n - 1, n - 2,...,0$, are entered, the desired c_i are contained in registers B_j. The codeword symbols c_i, computed in this manner, are shifted serially out of the right end of register B_{n-1}. Notice that this structure is independent of the value of t and can be used to generate RS codes of different code rates. It has a circuit complexity of $0(mn)$. Although the structure has a higher circuit complexity than a structure based on polynomial division, it is more suitable as a programmable code-rate encoder since its operation does not depend on t.

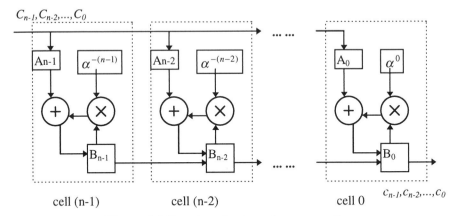

Figure 6.5 Structure for transform encoding

6.3.3 Systematic RS Encoding Using the Generator Matrix

Consider a systematic generator matrix for the (n, k) RS code. This matrix has the form,
$$G = [\ I\ A\ ],$$
where I is the identity matrix of order k and A is a $k \times (n-k)$ matrix.

It is shown in [28] for a Reed-Solomon code that matrix A can be represented by a *Cauchy matrix* which has entries of the form,
$$a_{i,j} = \frac{u_i u_j}{x_i + y_j}, 0 \leq i \leq k-1, 0 \leq j \leq n-k-1,$$
where u_i, v_j, x_i, and y_j are elements of $GF(2^m)$ which are defined by :
$$x_i = \alpha^{n-1-i}, 0 \leq i \leq k-1,$$
$$y_j = \alpha^{n-1-k-j}, 0 \leq j \leq n-k-1,$$
$$u_i = \frac{1}{\prod_{0 \leq l \leq k-1, l \neq i}(\alpha^{n-1-i} - \alpha^{n-1-l})}, 0 \leq i \leq k-1,$$
$$v_j = \prod_{0 \leq l \leq k-1}(\alpha^{n-1-k-j} - \alpha^{n-1-l}), 0 \leq j \leq n-k-1.$$
For the systematic RS codes, the parity-check vector $\bar{p} = [p_0, p_1, \ldots, p_{n-k-1}]$ is

Reed-Solomon Codes

with

$$\bar{p} = \bar{d} \cdot A$$

$$p_j = \sum_{i=0}^{k-1} d_i \cdot a_{i,j}$$

$$= v_j \sum_{i=0}^{k-1} \frac{d_i u_i}{x_i + y_i}, \quad 0 \leq j \leq n-k-1.$$

(6.5)

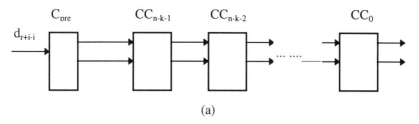

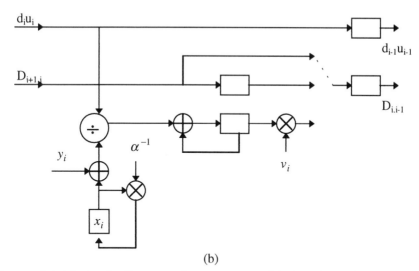

Figure 6.6 (a) A pipeline encoder structure. (b) Cauchy cell CC_j, where $D_j = [D_{j,i}]_{i=0}^{n-1} = [d_0, d_1, \ldots, d_{k-1}, p_j, p_{j+1}, \ldots, p_{r-1}, 0, \ldots, 0]$, $0 \leq j \leq r-1$, and $r = n-k$.

Due to the inversion operations in (6.5), it is evident that an encoding which uses a Cauchy-matrix representation requires more finite field operations than the previous methods. The advantage of using the Cauchy-matrix to represent G is that an encoder structure is now possible which is systolic and does not have any feedback connections. Since interleaving is not required, this type of encoder has a reduced latency delay. This makes it suitable for many applications in which a high throughput is required.

The block diagram of the basic systolic encoder structure is shown in Figure 6.6. The left-most cell C_{pre} computes $d_i u_i$ for $0 < i < k - 1$. In addition to the computation in C_{pre}, the structure has $(n - k)$ identical cells, namely, CC_j for $j = 0, 1,...,n - k - 1$. Cell CC_j computes the component p_j of the parity-check vector during the time steps $n - k - j$ to $n - 1 - j$. Also during these time steps, the switch is set to position A which allows the data symbols to propagate. At the $(n - j)$-th time step the switch is placed at position B, and the parities p_j are *output*ted. During the next k time steps the switch is set to position C so that the parity symbols p_j in the cells on the left side can propagate to the right.

The implementation of each cell in Fig.6.6a requires one divider, two fixed multipliers and two adders, all of which perform operations over $GF(2^m)$ and are of a parallel type. The computational delay in each cell is determined mainly by the divider. Because of the involvement of the divider and the multiplier, the complexity of the encoder can be as high as $O(m^3 t)$ gates. However, bit-serial dividers and multipliers can be used to reduce the complexity at the expense of a small reduction in the throughput.

6.4 Decoding of (n,k) RS Codes

The decoding processes of the (n,k) RS codes are somewhat more difficult to understand and implement than the encoding processes. Today, however, there are many algorithms and commercial devices available which perform RS decoding. The decoding process, because of its increased complexity, usually requires more intensive processing. Decoding needs more shift-register circuits and/or microprocessors than the encoding process. However, the small, less powerful, error-correcting codes are fairly easy to accommodate even at high data rates. The larger, more powerful, codes require considerably more processing, and it is a challenge to efficiently accommodate sustained high-data rates.

First suppose that a code polynomial $c(x)$ of an (n,k) RS code is transmitted over a noisy channel. Consider also receiver systems which have the facility to provide erasure flags that indicate unreliable symbols or, what are called, erasures in the received polynomial $r(x)$. The (n,k) RS decoder needed for

Reed-Solomon Codes

this problem must be able to correct any s erasure symbols and v errors for which $2v+s \leq d-1=n-k$. To accomplish this decoding process let the received polynomials $r(x)$ be represented by

$$r(x)=c(x)+e(x)+e^*(x)$$
$$=c(x)+u(x), \qquad (6.6)$$

where $e(x)$ and $e^*(x)$ are the error and erasure polynomials, respectively, and $u(x)= e(x)+e^*(x)$ is, what is called, the errata polynomial. Assume that there are s *erasures* at positions j_u for $u=1,2,...,s$, and v errors at positions i_w for $w=1,2,...,v$. Note here that $j_u \neq i_w$ for all $v=1,2,...,s$ and $w=1,2,...,v$. To correct an error the decoder must find both its location and its value; to correct an erasure only the value must be found since all positions j_w are known and indicated by erasure flags. In the next section errors-only decoding schemes for RS codes are introduced and discussed.

6.4.1 Errors-only Decoding of RS codes

For errors-only decoding the received polynomial $r(x)$ in Eq. (6.6) has the simplified form, $r(x)=c(x)+e(x)$, where $e(x) = \sum_{i=0}^{n-1} e_i x^i$ represents the error polynomial. The coefficients e_i are elements of $GF(p^m)$. For $e_i \neq 0$ let e_i denote the error value at the i-th position. Suppose that there are v errors, where $0 < 2v \leq (n-k)$ at positions i_w for $w=1,2,...,v$. The objective of the RS decoder is to find the number v of errors, their positions and then their values.

The received polynomial at the input to the decoder is
$$r(x) = c(x)+e(x) = r_0+r_1x+...+r_{n-1}x^{n-1},$$
where the polynomial coefficients are components of the received code-plus-error vector $\bar{r}$. First this polynomial is evaluated at the roots of the generator polynomial $g(x)$ which are $\alpha, \alpha^2, ..., \alpha^{d-1}$. We use the fact that the codeword polynomial $c(x)$ is divisible by $g(x)$, and that $g(\alpha^i)=0$ for $i=1,2,...,d-1$. Thus since $c(\alpha^i) = m(\alpha^i)g(\alpha^i) = 0$ for $i=1,2,...,d-1$, one obtains the relations,

$$r(\alpha^i) = c(\alpha^i)+e(\alpha^i)$$
$$= e(\alpha^i) = \sum_{j=0}^{n-1} e_j \alpha^{ij}, \quad \text{for } i=1,...,d-1.$$

Hence, we find the important fact that this final set of $d-1$ equations involves only components of the error pattern e_i, not those of the code word.

Evidently the above $d-1$ equations can be used to compute the syndromes S_i for $i=1,2,...,d-1$ of the received RS code as follows:

$$S_i = r(\alpha^i) = \sum_{j=0}^{n-1} r_j \alpha^{ij}, \text{ for } i = 1,2,...,d-1. \tag{6.7}$$

Example 6.7.
Continue the *(15,9)* RS code example in *Example 6.4*. Assume that the received code-plus-error polynomial is
$$r(x) = x^8 + \alpha^{11}x^7 + \alpha^8 x^5 + \alpha^{10}x^4 + \alpha^4 x^3 + \alpha^3 x^2 + \alpha^8 x + \alpha^{12}.$$
The error pattern for this received word is $e(x) = r(x) - c(x) = x^8 + x^2$. The error locations are in the second and the eighth symbols (counted from 0) of the received word. Then the syndromes S_j are computed by

$S_1 = r(\alpha) = e(\alpha) = 1$, $S_2 = r(\alpha^2) = e(\alpha^2) = 1$, $S_3 = r(\alpha^3) = e(\alpha^3) = \alpha^5$,
$S_4 = r(\alpha^4) = e(\alpha^4) = 1$, $S_5 = r(\alpha^5) = e(\alpha^5) = 0$, $S_6 = r(\alpha^6) = e(\alpha^6) = \alpha^{10}$.

After the syndromes in Eq. (6.7) are evaluated, the error pattern e_j for $j = 0,1,...,n-1$ can be determined. Suppose that v errors, $0 \le v \le t$, occur in the unknown locations $j_1, j_2, ..., j_v$. Then the error polynomial is expressed by
$$e(x) = e_{j_1} x^{j_1} + e_{j_2} x^{j_2} + \cdots + e_{j_v} x^{j_v},$$
where e_{j_l}, is the value of the l-th error. Neither the error locations $j_1, j_2, ..., j_v$ nor the error values $e_{j_1}, e_{j_2}, ..., e_{j_v}$ are known. In fact, we do not even know the number v of errors. All of these quantities must be computed in order to correct the errors.

In order to accomplish error correction it is convenient first to simplify the terminology by defining the error values to be
$$Y_l = e_{j_l} \text{ for } l=1,2,...,v,$$
and the error locations to be
$$X_l = \alpha^{j_l} \text{ for } l=1,2,...,v.$$
Then the set of $d-1$ *syndrome* equations in Eq. (6.7) in terms of the v unknown error-location numbers $X_1, X_2, ..., X_v$ and the v unknown error values $Y_1, Y_2, ..., Y_v$ are given by

$$S_j = Y_1 X_1^j + Y_2 X_2^j + \cdots + Y_v X_v^j \text{ for } j = 1,2,...,d-1. \tag{6.8}$$

In modern algebra the right sides of the equations in Eq. (6.8) are called power-sum symmetric functions. This set of equations must have at least one

Reed-Solomon Codes

solution for the unknown X_i's and Y_i's because of the way in which the syndromes are defined in Eq. (6.7). It is shown next that such a solution is unique for $0 \le v \le t$, i.e. the functional relations between the syndromes of the code and the error pattern of weight $\le t$ is a one-to-one mapping. Such a fact is proved by the following argument : Suppose that $e_1(x)$ and $e_2(x)$ are the error polynomials of two distinct error patterns of weight $\le t$ corresponding to the same syndrome. That is, suppose

$$S_j = e_1(\alpha^j) = e_2(\alpha^j) \quad for \ j = 1,2,\ldots,d-1.$$

Then, $d(x) = e_1(x) - e_2(x)$ is a codeword with weight $\le 2t = d-1$ since $d(\alpha^j) = 0$ for $j=1,2,\ldots,d-1$. One also has that $d(x)$ is divisible by the generator polynomial $g(x)$. But these facts hold if and only if $e_1(x) = e_2(x)$, and the uniqueness of this mapping is proved.

The decoding process is summarized in the following two steps :
    1. Evaluate the syndromes by a use of Eq. (6.7).
    2. Solve the system of $d-1$ nonlinear equations given by Eq. (6.8) to find the error locations and the error values.

The first step in the above decoding process, namely the syndrome evaluation, can be considered to be the taking of the discrete Fourier transform of the received vector $\bar{r}$, where the $2t$ frequency-domain components are calculated in accordance with Eq. (6.7). In the second step, the direct solution of a system of non-linear equations is difficult except for small values of t. A better approach is to first break the problem into a number of intermediate steps. Towards this end, what is called, the error-locator polynomial $\Lambda(x)$ is introduced next. In a manner similar to that discussed in Chapter 5, the error-locator polynomial is defined by the v-th degree polynomial of the form,

$$\Lambda(x) = 1 + \Lambda_1 x + \cdots + \Lambda_v x^v. \tag{6.9}$$

Next this error-locator polynomial in Eq. (6.9) is defined in such a manner that its roots are the inverses of the error-location numbers i.e. X_l^{-1} for $l=1,2,\ldots,v$. Thus the error-locator polynomial has the form,

$$\Lambda(x) = \prod_{l=1}^{v}(1 - xX_l), \ X_l = \alpha^{i_l}. \tag{6.10}$$

Once the coefficients of the error-locator polynomial $\Lambda(x)$ are determined, the inverses of the error locations X_l^{-1} can be found by successive substitutions.

Next the error values Y_i are obtained by the use of any of the following methods :

(1) Let the newly found error locations $X_l = \alpha^{i_l}$ be substituted into Eq. (6.8). Evidently this set of *d-1* equations is a linear system which can be solved for the error values $Y_j, j = 1,2,...,v$. The straightforward solution involves a matrix inversion for each error pattern. This method is not efficient because of the extensive computations required for matrix inversion.

(2) A more efficient way to find the error values by, what is called, the Forney formula. Once again, this method requires the evaluation of $\Lambda(x)$. First, define the syndrome polynomial by

$$S(x) = \sum_{i=1}^{2t} S_i x^i .$$

Next let the error-evaluator polynomial $\Omega(x)$ be formed in terms of the known polynomials *S(x)* and $\Lambda(x)$ by, what is called, the key equation (see section 6.4.2),

$$\Omega(x) = [1 + S(x)]\Lambda(x) \quad \mod x^{2t+1}. \tag{6.11}$$

By this definition the error-evaluator polynomial $\Omega(x)$ is related evidently to the error locations and error values by the *v* relations,

$$\Omega(X_l^{-1}) = Y_l \prod_{i \neq l} (1 - X_i X_l^{-1}) \text{ for } l=1,2,...,v. \tag{6.12}$$

Thus the error values are given explicitly by the formulae,

$$Y_l = -X_l \frac{\Omega(X_l^{-1})}{\Lambda'(X_l^{-1})} \quad \text{for } l = 1,2,...,v, \tag{6.13}$$

where $\Lambda'(x)$ denotes the formal first derivative of $\Lambda(x)$ with respect to *x*. The above equations in *(6.13)* define the Forney algorithm. This technique constitutes a considerable simplification over the use of matrix inversions. However, it does require divisions of elements in the field $GF(2^m)$.

(3) Another algorithm which is a transformed version of the Forney technique is described next. Such an algorithm is used sometimes in certain decoder implementations. Note first that the formal derivative of $\Lambda(x)$ in Eq. (6.10) can be computed to be $\Lambda'(X_l^{-1}) = X_l \prod_{i \neq l}(1 - X_i X_l^{-1})$. Thus, the error values are given by

$$Y_l = \frac{-X_l \Omega(X_l^{-1})}{X_l \prod_{i \neq l}(1 - X_i X_l^{-1})} \quad \text{for } l=1,2,...,v. \tag{6.14}$$

Combining Eq.(6.14) with Eq.(6.11) yields the explicit formulae,

Reed-Solomon Codes

$$Y_l = \frac{\sum_{i=1}^{v} S_{v-i} \sigma_{l,i}}{X_l \prod_{\substack{i=1 \\ i \neq l}}^{v} (X_l - X_i)} \quad \text{for } l=1,2,\ldots,v,$$

of the v error values, where $\sigma_{l,i}$ is defined as the $(v-1-i)$-th coefficient of the polynomial, $\overline{\Lambda}(x) = \prod_{\substack{i=1 \\ i \neq l}}^{v} (x - X_i) = \sum_{i=0}^{v-1} \sigma_{l,i} x^{v-1-i}$.

Example 6.8

Continue the (15,9) RS code example in *Example 6.4*. Assume that two error locations are already determined, i.e. the second and the eighth symbols, namely $X_1 = \alpha^2$, and $X_2 = \alpha^8$. Then by a use of the first method six linear equations are constructed by substituting $X_1 = \alpha^2$ and $X_2 = \alpha^8$ into Eq. (6.8) to obtain

$$Y_1 \alpha^2 + Y_2 \alpha^8 = 1, Y_1(\alpha^2)^2 + Y_2(\alpha^8)^2 = 1, Y_1(\alpha^2)^3 + Y_2(\alpha^8)^3 = \alpha,$$
$$Y_1(\alpha^2)^4 + Y_2(\alpha^8)^4 = 1, Y_1(\alpha^2)^5 + Y_2(\alpha^8)^5 = 0, Y_1(\alpha^2)^6 + Y_2(\alpha^8)^6 = \alpha^{10}.$$

Only two of these equations are needed since there are only two unknown variables. For example, the first two equations (which are the simplest) yield

$$Y_1 \alpha^2 + Y_2 \alpha^8 = 1,$$
$$Y_1 \alpha^4 + Y_2 \alpha = 1.$$

Now use Cramer's rule to solve for the coefficients Y_i as follows.

$$Y_1 = \frac{\det \begin{vmatrix} \alpha^8 & 1 \\ \alpha & 1 \end{vmatrix}}{\det \begin{vmatrix} \alpha^2 & \alpha^8 \\ \alpha^4 & \alpha \end{vmatrix}} = \frac{\alpha + \alpha^8}{\alpha^{10}} = \frac{\alpha^{10}}{\alpha^{10}} = 1,$$

$$Y_1 = \frac{\det \begin{vmatrix} \alpha^2 & 1 \\ \alpha^4 & 1 \end{vmatrix}}{\det \begin{vmatrix} \alpha^2 & \alpha^8 \\ \alpha^4 & \alpha \end{vmatrix}} = \frac{\alpha^2 + \alpha^4}{\alpha^{10}} = \frac{\alpha^{10}}{\alpha^{10}} = 1.$$

Therefore, the first and second error values are $Y_1 = Y_2 = \alpha^0 = 1$.

The error values are determined next by a use of Forney's formula. First, the error-evaluator polynomial $\Omega(x)$ is determined by a use of Eq. (6.11) to be

$\Omega(x) = 1$.
Then the error values are given directly from Forney's formula by

$$Y_1 = \frac{\Omega(1/X_1)}{X_1(1+X_2/X_1)} = \frac{1}{\alpha^2(1+\alpha^6)} = \frac{1}{\alpha^{15}} = 1,$$

$$Y_2 = \frac{\Omega(1/X_2)}{X_2(1+X_1/X_2)} = \frac{1}{\alpha^8(1+\alpha^9)} = \frac{1}{\alpha^{15}} = 1.$$

The error values also can be computed by Eq. (6.14). For the known error locations, $X_1 = \alpha^2, X_2 = \alpha^8$, the error value Y_1 is

$$Y_1 = \frac{S_2\sigma_{1,0} + S_1\sigma_{1,1}}{X_1(X_1+X_2)}$$

where $\sigma_{1,0}$ and $\sigma_{1,1}$ are determined by $X - X_2 = \sigma_{1,0}X + \sigma_{1,1}$, i.e. $\sigma_{1,0} = 1$ and $\sigma_{1,1} = X_2 = \alpha^8$. Now a substitution of $\sigma_{1,0} = 1, S_1 = 1, S_2 = 1$, and $\sigma_{1,1} = \alpha^8$ into the above equation for Y_1 yields

$$Y_1 = \frac{1 \cdot 1 + 1 \cdot \alpha^8}{\alpha^2(\alpha^2+\alpha^8)} = \frac{1+\alpha^8}{\alpha^4+\alpha^{10}} = \frac{\alpha^2}{\alpha^2} = 1.$$

Similarly, the error value Y_2 is obtained as

$$Y_2 = \frac{S_2\sigma_{2,0} + S_1\sigma_{2,1}}{X_2(X_1+X_2)},$$

where $\sigma_{2,0}$ and $\sigma_{2,1}$ are determined by $X - X_1 = \sigma_{2,0}X + \sigma_{2,1}$ so that $\sigma_{2,0} = 1$ and $\sigma_{2,1} = X_1 = \alpha^2$. Therefore,

$$Y_2 = \frac{1 \cdot 1 + 1 \cdot \alpha^2}{\alpha^8(\alpha^2+\alpha^8)} = \frac{1+\alpha^2}{\alpha+\alpha^{10}} = \frac{\alpha^8}{\alpha^8} = 1.$$

Methods to directly find the error-locator polynomial from the syndromes are explored next.

Determination of the Error-Locator Polynomial

Peterson algorithm The error-locator polynomial $\Lambda(x)$ has v roots X_l^{-1} for $l=1,2,\ldots, v$. The error-locations X_l indicate errors at location i_l for $l=1,2,\ldots, v$. To determine X_l one needs first to obtain the Λ_l, the coefficients of $\Lambda(x)$ for $l=1,2,\ldots, v$. The computations of the Λ_l for $l=1,2,\ldots, v$, that are needed to decode RS codes, center on the solutions of the system of equations,

Reed-Solomon Codes

$$S_j = -\sum_{i=1}^{v} \Lambda_i S_{j-i}, \quad \text{for } j = v+1, v+2, \ldots, 2t, \tag{6.15}$$

using the smallest value of v for which a solution exists. As discussed in Section 5.3 for the decoding of BCH codes, the straightforward way is to directly invert the matrix equation:

$$\begin{bmatrix} S_1 & S_2 & \cdots & S_v \\ S_2 & S_3 & \cdots & S_{v+1} \\ \vdots & \vdots & \ddots & \vdots \\ S_v & S_{v+1} & \cdots & S_{2v-1} \end{bmatrix} \begin{bmatrix} \Lambda_v \\ \Lambda_{v-1} \\ \vdots \\ \Lambda_1 \end{bmatrix} = \begin{bmatrix} -S_{v+1} \\ -S_{v+2} \\ \vdots \\ -S_{2v} \end{bmatrix}. \tag{6.16}$$

However, the value of the number of errors v is unknown at this point. But this can be solved by the use of the Peterson algorithm to determine Λ_l for $l=1,2,\ldots, v$ as follows: As a trial value first set v to the error-correcting capability of the code, namely t, and compute the determinant of the matrix in Eq. (6.16). It is shown in Chapter 5 that if this determinant is nonzero, then this is the correct value of v. If this determinant is zero, then one reduces the trial value of v by 1 and repeats. Continue this process repeatedly until a nonzero determinant is obtained. The actual number v of errors is determined from the size of this final matrix, and the Λ_l for $l=1,2,\ldots, v$ are obtained from inverting of this matrix. Therefore, the original problem is solved by the following Peterson algorithm: If it exists, find the smallest $v \leq t$ for which the system of equations in Eq. (6.15) has a solution. Then find the Λ_l for $l=1,2,\ldots, v$ is the solution. Because of the constraint, $v \leq t$, there need not be a solution for an arbitrary sequence $S_1, \ldots, S_{2t}$. If there is no solution an uncorrectable-error-pattern should be indicated.

For moderate values of v, the above direct method of solution by determinants and a matrix inversion is not unreasonable since the number of computations necessary to invert a v by v matrix is proportional to v^3. However, when a code is applied to correct a large number of errors, one desires for a more efficient method to obtain a solution. The matrix in Eq. (6.16) is not arbitrary in its present form, but is highly structured. This structure is now used to advantage to obtain the polynomial $\Lambda(x)$ by a method that is conceptually more complicated but computationally simpler than the above straightforward procedure of Peterson. There are several more efficient techniques for solving these equations. Two of these methods are examined in this chapter.

Berlekamp-Massey Algorithm. The Berlekamp-Massey algorithm is an efficient algorithm for determining the error-locator polynomial. The structure of the matrix in Eq. (6.16) is used to advantage to obtain Λ_l for $l=1,2,...,v$ by a method that is conceptually more intricate than a direct matrix inversion, but computationally simpler and more efficient.

The Berlekamp-Massey algorithm solves the following modified problem : Find the smallest v for $v \leq 2t$ such that the system of Eqs. (6.15) has a solution and find a vector $(\Lambda_1,...,\Lambda_v)$ that is a solution. It is sufficient to solve the modified problem because, if the original problem has a solution, it is the same as the solution to the modified problem.

For a fixed $\Lambda(x)$ Eq. (6.15) is the equation for, what is called, the classical auto-regressive filter. It is known that such a filter can be implemented as a linear-feedback shift register(LFSR) with taps, given by the coefficients of $\Lambda(x)$. Thus the modified problem can be viewed as the task of designing the LFSR that generates the known sequence of syndromes. Many such shift registers may exist, but we wish to find the smallest LFSR with this property. This shift register would yield the least-weight error pattern that explains the received data. That is, one must find that error-locator polynomial $\Lambda(x)$ of smallest degree. If $\deg \Lambda(x) > t$ or if $\Lambda_v = 0$, then $\Lambda(x)$ is not a legitimate error-locator polynomial. If this occurs, then the syndromes cannot be explained by a correctable error pattern. For this case a pattern of more than t errors has been detected.

Any procedure for designing the above-described LFSR is also a method for solving Eq. (6.16). Such a LFSR design method is obtained next. For the syndrome sequence $S_1,...,S_{2t}$ over the finite field, the procedure of such a LFSR design method yields finally a LFSR with a nonzero rightmost tap. This implies that the length L of the shift register equals the degree of $\Lambda(x)$.

The design procedure is iterative. For each i , starting with $i=1$, a minimum-length LFSR $\Lambda^{(i)}(x)$ is designed for producing the syndromes $S_1,...,S_i$. This shift register for stage i need not be unique, and several choices may exist. The length L_i of this shift register can be greater than the degree of $\Lambda^{(i)}(x)$ since some of the rightmost stages might not have taps in some of the iteration steps. Therefore, a shift register is designated by the pair $\left(\Lambda^{(i)}(x), L_i\right)$ that gives both the feedback-connection polynomial $\Lambda^{(i)}(x)$ and the length L_i of the LFSR. At the start of the i-th iteration, it is assumed that a list of LFSR had been constructed, i.e. the list

Reed-Solomon Codes

$(\Lambda^{(1)}(x), L_1), (\Lambda^{(2)}(x), L_2), \ldots, (\Lambda^{(i-1)}(x), L_{i-1})$ for $L_1 \leq L_2 \leq \cdots \leq L_{i-1}$ already is found.

A key trick of the Berlekamp-Massey algorithm is to use these earlier iterates to compute a new shortest-length LFSR $(\Lambda^{(i)}(x), L_i)$ that generates the sequence $S_1, \ldots, S_{i-1}, S_i$. This is accomplished by the use of a LFSR which generates the syndrome sequence $S_1, \ldots, S_{i-2}, S_{i-1}$ as a trial candidate, and if necessary, to increase the length and to adjust the tap weights.

The design of the shortest LFSR for the Berlekamp-Massey algorithm is iterative and inductive. At the iteration step i, compute the next output of the $(i-1)$-th LFSR to obtain the next estimate of the i-th syndrome as follows:

$$\hat{S}_i = -\sum_{j=1}^{L_{i-1}} \Lambda_j^{(i-1)} S_{i-j}.$$

Next, subtract this "estimated" $\hat{S}_i$ from the desired syndrome output S_i to get an "error" quantity Δ_i, that is called the i-th discrepancy, i.e.

$$\Delta_i = S_i - \hat{S}_i = S_i + \sum_{j=1}^{L_{i-1}} \Lambda_j^{(i-1)} S_{i-j}, \tag{6.17a}$$

or equivalently,

$$\Delta_i = \sum_{j=0}^{L_{i-1}} \Lambda_j^{(i-1)} S_{i-j}. \tag{6.17b}$$

If $\Delta_i = 0$, then set $(\Lambda^{(i)}(x), L_i) = (\Lambda^{(i-1)}(x), L_{i-1})$, and the i-th iteration is complete. Otherwise, the LFSR taps are modified as follows:

$$\Lambda^{(i)}(x) = \Lambda^{(i-1)}(x) + \Delta_i A(x)$$
$$= \sum_{l=0}^{L_i} (\Lambda_l^{(i-1)} + \Delta_i a_l) x^l \tag{6.18}$$

for some polynomial $A(x) = \sum_{l=0}^{L_i} a_l x^l$ yet to be found, where L_i is specified by a mathematical lemma. This lemma, which is demonstrated in [10], shows when $\Delta_i \neq 0$, that one must have $L_i = \max\{L_{i-1}, i - L_{i-1}\}$. This implies that if $\Delta_i \neq 0$ and $L_{i-1} \geq i - L_{i-1}$, then $L_i = L_{i-1}$, which indicates that still another polynomial of the same length L_{i-1} can be found.

Now re-compute a new discrepancy $\hat{\Delta}_i$ from this new polynomial $\Lambda^{(i)}(x)$ as follows:

$$\hat{\Delta}_i = \sum_{l=0}^{L_i} \Lambda_l^{(i)} S_{i-l}$$

$$= \sum_{l=0}^{L_i} \Lambda_l^{(i-1)} S_{i-l} + \Delta_i \sum_{l=0}^{L_i} a_l S_{i-l}. \qquad (6.19)$$

Notice that $\Lambda_l^{(i-1)} = 0$ for $l > L_{i-1}$. To obtain a new discrepancy of zero-value, it is seen from Eq.(6.19) that the coefficients a_j of $A(x)$, given in Eq. (6.18), must satisfy the following relation:

$$\hat{\Delta}_i = \sum_{l=0}^{L_{i-1}} \Lambda_l^{(i-1)} S_{i-l} + \Delta_i \sum_{l=0}^{L_i} a_l S_{i-l} = 0. \qquad (6.20)$$

It follows from Eq.(6.17b) and Eq.(6.20) that the coefficients a_l of $A(x)$ finally must satisfy

$$\sum_{\substack{l=0 \\ l<j}}^{L_i} a_l S_{i-l} = \begin{cases} 1, & \text{for } j = i, \\ 0, & \text{for } j < i. \end{cases} \qquad (6.21)$$

Now, the objective is to find a polynomial $A(x)$ which satisfies the above conditions for the i-th step of the iteration and has a minimum length.

To accomplish this objective concentrate first on the problem of finding $A(x)$ to satisfy Eq. (6.21) in the most computationally efficient manner. Consider some previous step u for $u<i$ at which the condition $\Delta_u \neq 0$, and the polynomial $\Lambda^{(u)}(x)$ is able to generate the first u syndromes, i.e.,

$$\sum_{\substack{l=0 \\ l<j}}^{L_u} \Lambda_l^{(u-1)} S_{j-l} = \begin{cases} \Delta_u, & \text{for } j = u \\ 0, & \text{for } j < u. \end{cases}$$

Thus, if $A(x)$ is selected to be

$$A(x) = \Delta_u^{-1} x^{i-u} \Lambda^{(u-1)}(x), \qquad (6.22)$$

i.e., if

$$a_l = \begin{cases} \Delta_u^{-1} \Lambda_{l-i+u}^{(u-1)} & \text{for } l = (i-u), (i-u+1), \ldots, (L_{u-1}+i-u), \\ 0 & \text{otherwise,} \end{cases}$$

this would yield

Reed-Solomon Codes

$$\sum_{\substack{l=0 \\ l<j}}^{L_i} a_l S_{j-l} = \Delta_u^{-1} \sum_{l=(i-u)}^{(L_{u-1}+i-u)} \Lambda_{l-i+u}^{(u-1)} S_{j-l}$$

$$= \Delta_u^{-1} \sum_{k=0}^{L_{u-1}} \Lambda_k^{(u-1)} S_{j-l-k+u}$$

$$= \begin{cases} 1, & \text{for } j = i, \\ 0, & \text{for } j < i. \end{cases}$$

That is, the $A(x)$ that is selected by Eq. (6.22) satisfies (6.21). Thus the new shift register specified by $\Lambda^{(i)}(x)$ is able to generate the syndrome sequence, $S_1, \ldots, S_{i-1}, S_i$.

However, we do not want just any such shift register. We want that shift register with the minimum length. The last problem is to find that step u for which $A(x)$ has minimum length. $\Lambda^{(u)}(x)$ is the minimum-length polynomial which generates the first u syndromes $S_1, S_2, \ldots, S_u$. Its length is L_u. Therefore, it is observed from Eq.(6.22) that the length of $A(x)$ is $L_u + i - u$. This length is minimum if $(i-u)$ has the smallest value. This happens only if u is the most recent step for $u<i$ such that $\Delta_u \neq 0$ and $L_u > L_{u-1}$.

A (n,k) RS code is able to correct $t = \lfloor (n-k)/2 \rfloor$ errors and the number of errors v ($\leq t$) is unknown. Therefore, the Berlekamp-Massey algorithm must iterate $2t=n-k$ steps in order to determine $\Lambda(x)$. Although the development of the Berlekamp-Massey algorithm is somewhat tedious, the computation of the algorithm is quite straightforward.

Often the inner operations of the Berlekamp-Massey algorithm appear somewhat mysterious. At some iterations, say the i-th, there may be more than one LFSR of minimum length that generates the required sequence. These all produce the same required sequence of r symbols but differ in the future symbols, which were not required at the i-th iteration. During later iterations, whenever possible, other LFSRs of this length are selected to produce the next required symbol. If no such LFSR exists, then the shift register is lengthened. The test to see if a lengthening is required, however, does not involve the next symbol other than to determine if $\Delta_{i+1} \neq 0$. Hence, whenever one alternative shift register is available, then there are at least as many alternative shift registers as the number of values minus one that the $(i+1)$-th symbol can assume.

A flowchart of the Berlekamp-Massey algorithm for the errors-only decoding of RS codes is shown in Figure 6.7.

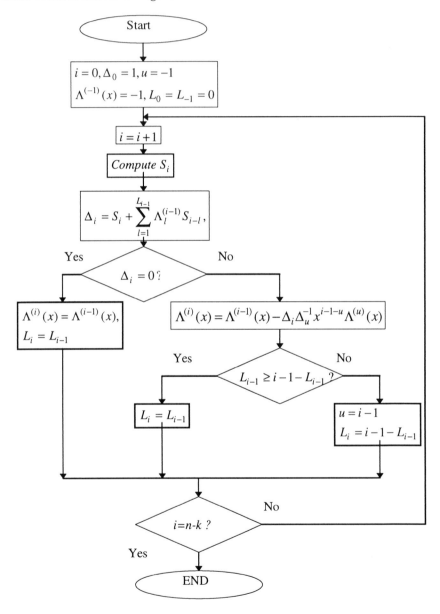

Figure 6.7 Berlekamp-Massey algorithm to find the error-locator polynomial $\Lambda(x)$ for errors-only decoding

Reed-Solomon Codes

Example 6.9
Continue the *(15,9)* RS code example in *Example 6.7*. The Berlekamp-Massey algorithm is demonstrated as follows:
- For $i=0$, $\Lambda^{-1}(x) = 1, \Delta_{-1} = \Delta_0 = 1, L_0 = 0, u = -1$.
- for $i=1$,
  $\Delta_1 = S_1 = 1$,
  $\Lambda^{(1)}(x) = \Lambda^{(0)}(x) + \Delta_1 \Delta_{-1}^{-1} x \Lambda^{-1}(x) = 1 + x, \Delta_1 = S_2 + \Lambda_1^{(1)} S_1 = 1 + 1 \cdot 1 = 1 + 1 = 0$,
  $L_1 = Max(L_0, L_{-1} + i + u_0) = \max(0,1) = 1, u = -1$.
- for $i=2$, $\Lambda^{(2)}(x) = \Lambda^{(1)}(x) = 1 + x, L_2 = L_1 = 1, u = 0$.
- for $i=3$, $\Delta_2 = S_3 + \Lambda_1^{(2)} S_2 = \alpha^5 + 1 \cdot 1 = \alpha^{10}$, ...
- ......
- for $i=2t=6$, $\Lambda(x) = \Lambda^{(6)}(x) = 1 + x + \alpha^{10} x^2$.

The results of this example are summarized in Table 6.3. Notice that the computations of $\Lambda^{(i)}(x)$ and Δ_i are performed over the finite field $GF(16)$ and the calculations for L_i and $i - L_i$ are performed over the real field. From step 6 in Table 6.3 it is evident that the error-locator polynomial is given by $\Lambda(x) = 1 + x + \alpha^{10} x^2$.

i	$\Lambda^{(i)}(x)$	Δ_i	L_i	$i - L_i$	u
-1	1	1	0	-1	--
0	1	1	0	0	--
1	$1+x$	1	1	0	-1
2	$1+x$	0	1	1	-1
3	$1+x+\alpha^{10}x^2$	α^{10}	2	1	0
4	$1+x+\alpha^{10}x^2$	0	2	2	0
5	$1+x+\alpha^{10}x^2$	0	2	3	0
6	$1+x+\alpha^{10}x^2$	0	---	---	---

Table 6.3 Example of the Berlekamp-Massey Algorithm

6.4.2 Error-and-Erasure Decoding of RS codes

In this section, a procedure for correcting both the errors and erasures of RS codes by the use of the Euclidean algorithm is discussed. In this decoding procedure, the Euclidean algorithm is used to directly solve the Berlekamp-Forney key equation for the errata locator polynomial and the errata evaluator polynomial at the same time with a common algorithm.

The Berlekamp-Forney key equation. For error-and-erasure decoding suppose that s errors and v erasures occur in the received word $\bar{r}$, and that $2v + s \le d-1$. Next let α be a primitive element in $GF(2^m)$. Then $\gamma = \alpha^l$ is also a primitive element in $GF(2^m)$, where $(l,n)=1$. To minimize the complexity of an RS encoder, it is usually desirable that the generator polynomial be symmetric. If γ is a root of the generator polynomial of the code, it is shown [19] that the generator polynomial $g(x)$ is symmetric if and only if

$$g(x) = \sum_{j=0}^{d-1} g_j x^j = \prod_{j=b}^{b+(d-2)} (x - \gamma^j),$$

where $g_0 = g_{d-1} = 1$, and b satisfies the equality $2b + d - 2 = 2^m - 1$.

The syndromes of the code are given by

$$S_{(b-1)+w} = r\left(\gamma^{b-1+w}\right) = \sum_{i=0}^{n-1} u_i \gamma^{i \cdot (b-1+w)}$$

$$= \sum_{j=1}^{v+s} Y_j X_j^{(b-1)+w} \qquad (6.23a)$$

for $1 \le w \le d-1$, where u_i for $0 \le i \le n-1$ are the coefficients of the errata polynomial $u(x)$, X_j is either the j-th erasure or error location, and Y_j is either the j-th erasure or error magnitude. It is not difficult to show [3] that,

$$S(x) = \sum_{w=1}^{d-1} S_{(b-1)+w} x^{w-1}$$

$$= \sum_{j=1}^{v+s} \frac{Y_j X_j^b}{(1 - X_j x)} - \sum_{j=1}^{v+s} \frac{Y_j X_j^{b+d-1} x^{d-1}}{(1 - X_j x)}. \qquad (6.23b)$$

In order to illustrate the decoding process, first define the following four polynomials:
The erasure locator:

$$\tau(x) = \prod_{j=1}^{s} (1 - X_j x). \qquad (6.24)$$

The error locator:

$$\lambda(x) = \prod_{j=1}^{v} (1 - X_j x). \qquad (6.25)$$

The errata locator:

Reed-Solomon Codes

$$\Lambda(x) = \tau(x) \cdot \lambda(x) = \prod_{j=1}^{v+s}(1 - X_j x). \tag{6.26}$$

The errata evaluator :

$$A(x) = \sum_{j=1}^{v+s} Y_j X_j^b \left(\prod_{i \neq j}(1 - X_i x) \right). \tag{6.27}$$

In terms of the polynomials defined above, Eq. (6.23) becomes

$$S(x) = \frac{A(x)}{\Lambda(x)} + \frac{x^{d-1} \sum_{j=1}^{v+s} Y_j X_j^{b+d-1} \left(\prod_{i \neq j}(1 - X_i x) \right)}{\Lambda(x)} \tag{6.28a}$$

or

$$S(x)\Lambda(x) = A(x) + x^{d-1} \sum_{j=1}^{v+s} Y_j X_j^{b+d-1} \left(\prod_{i \neq j}(1 - X_i x) \right). \tag{6.28b}$$

From Eq. (6.28b) one obtains finally the congruence relation,

$$S(x)\Lambda(x) \equiv A(x) \bmod x^{d-1}, \tag{6.29}$$

for the key equation. Now define the following set of formal power series :

$$F = \left\{ \sum_{i=0}^{\infty} a_i x^{n-i} \mid a_i \in GF(2^m) \text{ and } n \text{ is an integer} \right\}.$$

Since $\Lambda(x)$ in Eq. (6.26) is a polynomial in x with its leading coefficient not equal to zero, it is not difficult to show that the inverse element $\Lambda^{-1}(x)$ always exists in the set of formal power series. Thus, Eq.(6.29) can be solved for $S(x)$ to yield :

$$S(x) \equiv \frac{A(x)}{\lambda(x)\tau(x)} \bmod x^{d-1}. \tag{6.30}$$

Now define, what is called, the Forney syndrome polynomial $T(x)$ as follows:

$$T(x) \equiv S(x)\tau(x) \bmod x^{d-1}. \tag{6.31}$$

By Eqs. (6.30) and (6.31) one obtains, what is called, the Berlekamp-Forney key equation for $\lambda(x)$ and $A(x)$ is given by,

$$A(x) \equiv T(x)\lambda(x) \bmod x^{d-1}, \tag{6.32}$$

where $\deg\{\lambda(x)\} \leq \lfloor (d-1-s)/2 \rfloor$ since the maximum number of errors in an RS code that can be corrected is $\lfloor (d-1-s)/2 \rfloor$. Here also $\deg\{A(x)\} \leq \lfloor (d+v-3)/2 \rfloor$.

The Forney syndrome polynomial $T(x)$ can be computed from Eq. (6.31) since both the syndrome polynomial $S(x)$ and the error-erasure locator polynomial $\tau(x)$ are known. Therefore, the primary step for decoding RS codes is to solve the key equation Eq.(6.32) for the pair $(\lambda(x), A(x))$, the error-locator polynomial and the errata-locator polynomial.

Euclid's Algorithm.
Euclid's algorithm is a fast method for finding the greatest common divisor(GCD) of a collection of elements in an Euclidean domain. It is shown in Chapter 2 that the GCD of a collection of elements can be expressed as the linear combination of those elements. The extended form of Euclid's algorithm is introduced in Chapter 3 as a means to determine the coefficients for this linear combination.

If one uses the extended form of Euclid's algorithm to determine the GCD of $T(x)$ and x^{d-1}, one generates pairs of solutions $(\lambda_i(x), A_i(x))$ that satisfy

$$Q_i(x)x^{d-1} + \lambda_i(x)T(x) = A_i(x). \tag{6.33}$$

The following theorem, proved in references [22][23], can be used to solve for $(\lambda_i(x), A_i(x))$. Such a theorem is the classical Euclidean algorithm for polynomials in matrix form.

Theorem 6.1 : Assume that the two polynomials x^{d-1} and $T(x)$ in Eq. (6.31) are given, where $\deg\{x^{d-1}\} > \deg\{T(x)\}$. Also let $M_0(x) = x^{d-1}$ and $R_0(x) = T(x)$. Next let $N_s(x)$ be the 2x2 matrix equation which satisfies

$$N_i(x) = \begin{bmatrix} 0 & 1 \\ 1 & -Q_{i-1}(x) \end{bmatrix} N_{i-1}(x), \tag{6.34a}$$

where

$$Q_{i-1}(x) = \begin{bmatrix} M_{i-1}(x) \\ R_{i-1}(x) \end{bmatrix}. \tag{6.34b}$$

Finally, let

$$\begin{bmatrix} M_i(x) \\ R_i(x) \end{bmatrix} = \begin{bmatrix} 0 & 1 \\ 1 & -Q_{i-1}(x) \end{bmatrix} \begin{bmatrix} M_{i-1}(x) \\ R_{i-1}(x) \end{bmatrix} = N_i(x) \begin{bmatrix} x^{d-1} \\ T(x) \end{bmatrix} \tag{6.35a}$$

Then for $i=r$, $M_i(x)$ satisfies

$$M_r(x) = c \cdot GCD(x^{d-1}, T(x)) \tag{6.35b}$$

where c is a scalar, and $R_r(x) = 0$.

Reed-Solomon Codes

Next, we prove the important theorem that the errata evaluator polynomial $A(x)$ and the errata locator polynomial $\tau(x)$ can be obtained simultaneously and simply from the known $T(x)$, defined previously in Eq. (6.31) and the key equation in Eq. (6.32).

Theorem 6.2. Let $T(x)$ in Eq. (6.31) be the Forney syndrome polynomial of a v-error and s-erasure correcting RS code under the condition $s + 2v \le d - 1$. Consider the two polynomials x^{d-1} and $T(x)$ in Eq. (6.31). The Euclidean algorithm for polynomials over $GF(2^m)$ can be used to develop two finite sequences $R_i(x)$ and $\Lambda_i(x)$ from the following two recursive formulas:

$$\Lambda_i(x) = (-Q_{i-1}(x))\Lambda_{i-1}(x) + \Lambda_{i-2}(x) \qquad (6.36a)$$

and

$$R_i(x) = R_{i-2}(x) - Q_{i-1}(x) R_{i-1}(x) \qquad (6.36b)$$

for $i = 1, 2, \ldots, 2t$. where the initial conditions are $\Lambda_0(x) = \tau(x)$, $\Lambda_{-1}(x) = 0$, $R_{-1}(x) = x^{d-1}$ and $R_0(x) = T(x)$. Here $Q_{i-1}(x)$ is obtained as the principal part of $R_{i-2}(x) / R_{i-1}(x)$. The recursion in Eqs. (6.36a) and (6.36b) for $R_i(x)$ and $\Lambda_i(x)$ terminates when $\deg\{R_i(x)\} \le \lfloor (d+v-3)/2 \rfloor$ for the first time for some value $i = i'$. Let

$$A(x) = R_{i'}(x) / \Delta \qquad (6.37a)$$

and

$$\Lambda(x) = \Lambda_{i'}(x) / \Delta. \qquad (6.37b)$$

Also in Eq. (6.37) $\Delta = \Lambda_{i'}(0)$ is a field element in $GF(2^m)$ which is chosen so that $\Lambda_0 = 1$. The pair of polynomials $A(x)$ and $\Lambda(x)$ in Eq. (6.37) constitute the unique solution of $A(x) \equiv T(x)\Lambda(x) \mod x^{d-1}$, where both of the inequalities, $\deg\{\Lambda(x)\} \le \lfloor (d+v-1)/2 \rfloor$ and $\deg\{A(x)\} \le \lfloor (d+v-3)/2 \rfloor$, are satisfied.

Proof: By Theorem 6.1 one observes that

$$N_i(x) = \prod_{j=0}^{i-1} \begin{bmatrix} 0 & 1 \\ 1 & -Q_{i-1}(x) \end{bmatrix} = \begin{bmatrix} A_i(x) & B_i(x) \\ C_i(x) & D_i(x) \end{bmatrix} \qquad (6.38a)$$

where

$$\deg\{D_i(x)\} > \deg\{B_i(x)\}. \qquad (6.38b)$$

It follows from Eq. (3.38a) that the determinant of $N_i(x)$ is $(-1)^i$. Thus, the inverse of $N_i(x)$ is given by

$$N_i^{-1}(x) = \begin{bmatrix} A_i(x) & B_i(x) \\ C_i(x) & D_i(x) \end{bmatrix}^{-1} = (-1)^i \begin{bmatrix} D_i(x) & -B_i(x) \\ -C_i(x) & A_i(x) \end{bmatrix}. \qquad (6.38c)$$

From Eq. (6.38a), $D_i(x)$ satisfies the recursive equations:
$$D_i(x) = D_{i-2}(x) - Q_{i-1}(x) D_{i-1}(x), \qquad (6.39)$$
for $i=1,2,...$, where the initial conditions are $D_{-1}(x) = 0$ and $D_0(x) = 1$. Also from Eq. (6.37c), one obtains the equations,
$$M_i(x) = R_{i-1}(x), \qquad (6.40a)$$
$$R_i(x) = M_{i-1}(x) - Q_{i-1}(x) R_{i-1}(x), \qquad (6.40b)$$
for $i=1,2,...$, where
$$\deg\{M_i(x)\} > \deg\{R_i(x)\}. \qquad (6.40c)$$
Thus, from Eqs.(6.40a) and (6.40b), the recursive formula for $R_i(x)$ is
$$R_i(x) = R_{i-2}(x) - Q_{i-1}(x) R_{i-1}(x) \qquad (6.40d)$$
for $i=1,2,...$, where the initial conditions are $R_{-1}(x) = x^{d-1}$ and $R_0(x) = T(x)$.

A substitution of Eq. (6.38a) into Eq. (6.37c) yields
$$\begin{bmatrix} M_i(x) \\ R_i(x) \end{bmatrix} = \begin{bmatrix} A_i(x) & B_i(x) \\ C_i(x) & D_i(x) \end{bmatrix} \begin{bmatrix} x^{d-1} \\ T(x) \end{bmatrix}. \qquad (6.41)$$

Thus, from Eq. (6.41), one obtains,
$$R_i(x) \equiv T(x) D_i(x) \qquad \mod x^{d-1} \qquad (6.42)$$
for $i=1,2,...$, where the recursive formulas for $D_i(x)$ and $R_i(x)$ are given in Eqs. (6.39) and (6.40d), respectively.

Eq. (6.40d) is the form of Eq. (6.36a) which is used to solve for both $A(x)$ and $\lambda(x)$. To solve for $A(x)$ and $\lambda(x)$, one needs to find by Eq. (6.36b) a unique value $i=i'$ such that:
$$\deg\{D_{i'}(x)\} \leq \lfloor (d-1-v)/2 \rfloor$$
and
$$\deg\{R_{i'}(x)\} \leq \lfloor ((d-1-v)/2) - 1 \rfloor = \lfloor (v+d-3)/2 \rfloor,$$
in Eq. (6.42). Since $\deg\{R_{-1}(x)\} = d-1$ and $\deg\{R_i(x)\}$ is strictly decreasing with an increase in i, there always exists a unique value i' such that
$$\deg\{R_{i'-1}(x)\} \geq \left\lceil \frac{v+d-1}{2} \right\rceil, \qquad (6.43a)$$
and
$$\deg\{R_{i'}(x)\} \leq \left\lceil \frac{v+d-3}{2} \right\rceil \qquad (6.43b)$$

Reed-Solomon Codes

where $\lceil x \rceil$ denotes the least integer greater than x.

Since $\deg\{D_i(x)\}$ is non-decreasing with i, then one needs to show also that
$$\deg\{D_{i'}(x)\} \leq \lfloor (d-1-v)/2 \rfloor. \tag{6.44}$$
To prove Eq. (6.44), it follows from Eq. (6.41) that
$$\begin{bmatrix} x^{d-1} \\ T(x) \end{bmatrix} = N_{i'}^{-1}(x) \begin{bmatrix} M_{i'}(x) \\ R_{i'}(x) \end{bmatrix} \tag{6.45}$$
for $i = i'$, where $N_{i'}^{-1}(x)$ is the inverse of the 2x2 matrix $N_{i'}(x)$ in Eq. (6.38a). The substitution of Eq. (6.38c) into Eq. (6.45) yields:
$$\begin{bmatrix} x^{d-1} \\ T(x) \end{bmatrix} = (-1)^{i'} \begin{bmatrix} D_{i'}(x) & -B_{i'}(x) \\ -C_{i'}(x) & A_{i'}(x) \end{bmatrix} \begin{bmatrix} M_{i'}(x) \\ R_{i'}(x) \end{bmatrix}. \tag{6.46}$$
From Eq.(6.46), one obtains
$$x^{d-1} = (-1)^{i'} \left[D_{i'}(x) M_{i'}(x) - B_{i'}(x) R_{i'}(x) \right]. \tag{6.47a}$$
From Eqs. (6.38b) and (6.40c) one knows that $\deg\{M_{i'}(x)\} > \deg\{R_{i'}(x)\}$ and $\deg\{D_{i'}(x)\} > \deg\{B_{i'}(x)\}$. Thus, by Eq. (6.47a):
$$\deg\{x^{d-1}\} = \deg\{D_{i'}(x)\} + \deg\{M_{i'}(x)\}. \tag{6.47b}$$
Since by Eq. (6.40a) for $i = i'$, $M_{i'}(x) = R_{i'-1}(x)$ and $\deg\{x^{d-1}\} = d-1$, then Eq. (6.47b) becomes:
$$\deg\{D_{i'}(x)\} = d - 1 + \deg\{R_{i'-1}(x)\}. \tag{6.48a}$$
By Eq. (6.43a), $\deg\{D_{i'}(x)\}$ in Eq. (6.48a) is upper bounded by the following inequality:
$$\deg\{D_{i'}(x)\} \leq d - 1 + \left\lceil \frac{v+d-1}{2} \right\rceil = \left\lfloor \frac{d-1-v}{2} \right\rfloor. \tag{6.48b}$$
By Eqs. (6.43b) and (6.48b), the Euclidean algorithm has the following stop conditions:
$$\deg\{R_{i'}(x)\} \leq \lceil (v+d-3)/2 \rceil$$
and
$$\deg\{D_{i'}(x)\} \leq \lfloor (d-1-v)/2 \rfloor.$$
Thus also by Eqs. (6.43a) and (6.48b), polynomials $R_{i'}(x)$ and $D_{i'}(x)$ are solutions of Eq. (6.42). Uniqueness follows from the fact that there is only one solution of Eq. (6.36a) under the conditions of Eq. (6.36b). Therefore, $\Lambda(x) = R_{i'}(x)/\Delta$ and $\lambda(x) = D_{i'}(x)/\Delta$ are the unique solutions of Eq. (6.36), where Δ is chosen to ensure that $\lambda_0 = 1$.

To obtain $\Lambda(x)$, one replaces the recursive equation in Eq. (6.39) by
$$\Lambda_i(x) = -Q_{i-1}(x)\Lambda_{i-1}(x) + \Lambda_{i-2}(x) \tag{6.49}$$
with the changed initial conditions: $\Lambda_0(x) = \tau(x)$ and $\Lambda_{-1}(x) = 0$. To obtain the final errata locator polynomial first let $i=1$. Then Eq. (6.49) becomes, by Eq. (6.39):
$$\Lambda_1(x) = -Q_0(x)\Lambda_0(x) + \Lambda_{-1}(x) = -Q_0(x)\tau(x) = D_1(x) \cdot \tau(x).$$
For $i=2$, Eq. (6.49) becomes, by Eq. (6.39):
$$\Lambda_2(x) = -Q_1(x)\Lambda_1(x) + \Lambda_0(x) = -Q_1(x)D_1(x)\tau(x) + \tau(x)$$
$$= (-Q_1(x)D_1(x) + 1)\tau(x) = D_2(x) \cdot \tau(x),$$
etc. Here, in general, one has $\Lambda_i(x) = D_i(x)\tau(x)$ with $\Lambda_0(x) = \tau(x)$ and $\Lambda_{-1}(x) = 0$.

This concludes the proof of Theorem 6.2. The proof of the Theorem also demonstrates the recursive approach needed to determine the errata locator $\Lambda(x)$ and the errata evaluator $A(x)$ by using the Euclidean algorithm.

The roots of $\Lambda(x)$ are the inverse locations of the v errors and s erasures. These roots are most efficiently found by the Chien search procedure. By Eq. (6.27) it is shown readily that the errata values are

$$Y_j = \frac{A(X_j^{-1})}{\left(X_j^{b-1}\Lambda'(X_j^{-1})\right)} \quad \text{for } j = 1, 2, \ldots, v+s, \tag{6.50}$$

where $\Lambda'(X_j^{-1})$ is the derivative with respect to x of $\Lambda(x)$, evaluated at $x = X_j^{-1}$.

The overall decoding process of the RS codes for correcting errors and erasures, which uses the Euclidean algorithm, is summarized in the following steps:

(1) Compute the transform of the received word over $GF(2^m)$ from Eq. (6.23a). Next calculate the erasure-locator polynomial $\lambda(x)$ from Eq. (6.24) and let $\deg\{\lambda(x)\} = v$.

(2) Compute the Forney syndrome polynomial from $T(x)$ by the use of Eq. (6.31).

(3) To determine the errata-locator polynomial $\Lambda(x)$ and errata evaluator polynomial $A(x)$, where $0 \le v \le d-1$, apply the Euclidean algorithm to x^{d-1} and $T(x)$ as given in Eq. (6.31). The initial values of the Euclidean algorithm

Reed-Solomon Codes

are $\Lambda_0(x) = \tau(x), \Lambda_{-1}(x) = 0, R_{-1}(x) = x^{d-1}$ and $R_0(x) = T(x)$. For $v=d-1$, set $\Lambda(x) = \tau(x)$ and $A(x) = T(x)$.

(4) Compute the errata values from Eq. (6.50).

To illustrate this decoding process for correcting errors and erasures an example for the (15,9) RS code is now presented.

Example 6.10
Consider the *(15,9)* RS code given in Example 6.4. The minimum distance of this code is $d=7$. In this code, s erasures and v errors under the condition $s+2v \leq d-1$ can be corrected. Assume the shifted-message symbols are

$$m(x)x^6 = \alpha^{10}x^{14} + \alpha^{12}x^{13} + \alpha^8 x^{12} + \alpha^5 x^{11} + \alpha^6 x^{10}$$
$$+ \alpha^{14}x^9 + \alpha^{13}x^8 + \alpha^{11}x^7 + \alpha^9 x^6$$

The encoded codeword, which is a multiple of $g(x)$, is

$$c(x) = \alpha^{10}x^{14} + \alpha^{12}x^{13} + \alpha^8 x^{12} + \alpha^5 x^{11} + \alpha^6 x^{10} + \alpha^{14}x^9 + \alpha^{13}x^8$$
$$+ \alpha^{11}x^7 + \alpha^9 x^6 + x^5 + \alpha x^4 + \alpha^2 x^3 + \alpha^6 x^2 + \alpha^{12}x + \alpha^8.$$

Thus its codeword word is given by

$$\bar{c} = (\alpha^{10}, \alpha^{12}, \alpha^8, \alpha^5, \alpha^6, \alpha^{14}, \alpha^{13}, \alpha^{11}, \alpha^9, 1, \alpha, \alpha^2, \alpha^6, \alpha^{12}, \alpha^8).$$

Assume the erasure vector is

$$\bar{e}^* = (0,0,0,0,0,0,0, \alpha^2, 0,0,0,0,0,0,0),$$

and the error vector is

$$\bar{e} = (0,0,0,0,0, \alpha^{11}, 0,0,0,0,0,0, \alpha^7, 0,0,0).$$

Then the errata vector is

$$\bar{u} = (0,0,0,0,0, \alpha^{11}, 0,0, \alpha^2, 0,0,0, \alpha^7, 0,0,0),$$

and the received word is

$$\bar{r} = \bar{c} + \bar{u} = (\alpha^{10}, \alpha^{12}, \alpha^8, \alpha^5, \alpha, \alpha^{14}, \alpha^{13}, \alpha^9, \alpha^9, 1, \alpha, \alpha^{12}, \alpha^6, \alpha^{12}, \alpha^8).$$

The syndromes S_i satisfy

$$S_i = \sum_{j=0}^{14} r_j \alpha^{ji} = \alpha^7 (\alpha^3)^i + \alpha^2 (\alpha^7)^i + \alpha^{11}(\alpha^{10})^i \text{ for } 1 \leq i \leq 6.$$

This yields $S_1 = 1, S_2 = \alpha^{13}, S_3 = \alpha^{14}, S_4 = \alpha^{11}, S_5 = \alpha, S_6 = 0$. Thus, the syndrome polynomial is

$$S(x) = 1 + \alpha^{13}x + \alpha^{14}x^2 + \alpha^{11}x^3 + \alpha x^4.$$

The erasure locator polynomial is $\tau(x) = 1 + \alpha^7 x$. In this example, the maximum erasure correcting capability is $\lfloor (d-1-v)/2 \rfloor = \lfloor (7-1-1)/2 \rfloor = 2$. By Eq. (6.31), one obtains the Forney syndrome polynomial as

$$T(x) \equiv \tau(x)S(x)$$
$$\equiv (1+\alpha^7 x)(1+\alpha^{13}x+\alpha^{14}x^2+\alpha^{11}x^3+\alpha x^4) \mod x^6 \quad (6.51)$$
$$= \alpha^8 x^5 + \alpha^9 x^4 + \alpha x^3 + \alpha^{12}x^2 + \alpha^5 x + 1.$$

In Eq.(6.51) the coefficients of $T(x)$, i.e. $T_0 = 1, T_1 = \alpha^5, T_2 = \alpha^{12}, T_3 = \alpha, T_4 = \alpha^9$ and $T_5 = \alpha^8$, are the Forney syndromes.

The Euclidean algorithm is applied next to polynomials x^{d-1} and $T(x)$, given in Eq. (6.31). The polynomials $\Lambda(x)$ and $A(x)$ are determined by the use of Theorem 6.2. This is accomplished by the recursive formulas Eqs. (6.36a) and (6.36b) illustrated in Table 6.5, where initially $R_{-1}(x) = x^{d-1} = x^6$ and $R_0(x) = T(x) = \alpha^8 x^5 + \alpha^9 x^4 + \alpha x^3 + \alpha^{12}x^2 + \alpha^5 x + 1$. From Table 6.5, one observes that $\deg\{R_{s'}(x)\} = \deg\{R_2(x)\} = 2 \le \lfloor (d+v-3)/2 \rfloor = 2$. Thus, the computation terminates at this point for $s' = 2$, and
$$R_2(x) = \alpha^7 x^2 + \alpha x + \alpha^2,$$
and
$$\Lambda_2(x) = \alpha^7 x^3 + \alpha^{13}x^2 + \alpha^4 x + \alpha^2.$$
By Eqs. (6.36a) and (6.36b) one has finally the solutions,
$$\Lambda(x) = \frac{1}{\alpha^2}\Lambda_2(x) = \alpha^5 x^3 + \alpha^{11}x^2 + \alpha^2 x + 1, \quad (6.52)$$
and
$$A(x) = \frac{1}{\alpha^2}R_2(x) = \alpha^5 x^2 + \alpha^{14}x + 1, \quad (6.53)$$
of the key equation in Eq. (6.32).

i	$Q_{i-1}(x)$	$R_i(x)$	$\Lambda_i(x)$
-1		x^6	0
0		$\alpha^8 x^5 + \alpha^9 x^4 + \alpha x^3$ $+\alpha^{12}x^2 + \alpha^5 x + 1$	$1 + \alpha^7 x$
1	$\dfrac{x}{\alpha^8} + \dfrac{1}{\alpha^7}$	$x^4 + \alpha^{14}x^3 + \alpha^6 x^2$ $+\alpha^2 x + \alpha^8$	$\dfrac{x^2}{\alpha} + \dfrac{x}{\alpha^2} + \dfrac{1}{\alpha^7}$
2	$\alpha^8 x + 1$	$\alpha^7 x^2 + \alpha x + \alpha^2$	$\alpha^7 x^3 + \alpha^{13}x^2 + \alpha^4 x + \alpha^2$

Table 6.5. An example of the Euclidean algorithm used to find $\tau(x)$ and $A(x)$.

Reed-Solomon Codes

By the use of a Chien search, the roots of $\Lambda(x)$ constitute the set $\{\alpha^{-7}, \alpha^{-3}, \alpha^{-10}\}$. The derivative with respect to x of $\Lambda(x)$ in Eq. (6.52) is $\Lambda'(x) = \alpha^5 x^2 + \alpha^2$. Thus, by Eqs. (6.50) and (6.53), the errata values are

$$Y_1 = \frac{A(Z_1^{-1})}{\Lambda'(Z_1^{-1})} = \frac{A(\alpha^{-7})}{\Lambda'(\alpha^{-7})} = \frac{\alpha^5(\alpha^{-7})^2 + \alpha^{14}(\alpha^{-7}) + 1}{\alpha^5(\alpha^{-7})^2 + \alpha^2} = \alpha^2,$$

$$Y_2 = \frac{A(Z_2^{-1})}{\Lambda'(Z_2^{-1})} = \frac{\alpha^5(\alpha^{-3}) + \alpha^{14}(\alpha^{-3}) + 1}{\alpha^5(\alpha^{-3})^2 + \alpha^2} = \alpha^7,$$

and

$$Y_3 = \frac{A(Z_3^{-1})}{\Lambda'(Z_3^{-1})} = \frac{\alpha^5(\alpha^{-10})^2 + \alpha^{14}(\alpha^{-10}) + 1}{\alpha^5(\alpha^{-10})^2 + \alpha^2} = \alpha^{11}.$$

Evidently, the Euclidean algorithm can be used for errors-only decoding by setting the number of erasures to zero in the above-discussed process.

The VLSI architecture of a pipeline RS decoder of this algorithm is presented in Fig. 6.8. In Fig. 6.8 the block diagram can be separated into two parts as indicated by the broken line. The functional units contained in the first part of the decoder architecture in Fig. 6.9 are shown as follows:
(1) The syndrome-computation unit;
(2) The power-calculation unit;
(3) The power-expansion unit;
(4) The polynomial-expansion unit;
(5) The $\lfloor (d+s-3)/2 \rfloor$ generator;
(6) The delay unit.

The syndrome-computation unit accepts received words and computes their syndromes. There are $n-k$ syndrome subcells in an (n,k) RS decoder. The computed syndrome polynomial is labeled as $S(x)$ in Fig. 6.9. The coefficients of $S(x)$ are fed in parallel to the polynomial-expansion unit in order to compute the Forney syndromes.

The power-calculation unit converts the received 1's and 0's into a sequence of α^k 's and 0's, where α is a primitive element of the finite field over which the RS code is defined. These received 1s and 0s indicate the occurrence or nonoccurrence, respectively, of an erasure at a specific location. Since an erasure is at a specific location, the maximum erasure-correcting capability of a *(n,k)* RS decoder is *n-k* so that only *n-k* symbol latches are needed to store all correctable erasure locations.

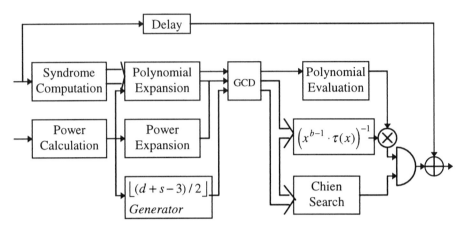

Figure 6.8 Block diagram of pipeline RS error-and-erasure decoder.

A detection circuit for detecting the occurrence of erasures is included in the power-calculation unit. If an erasure occurs at the *k*-th location, a symbol α^k is calculated by the power-calculation unit and latched. The sequence of α^k's is fed first to the polynomial expansion unit, next to the power expansion unit and finally to the $\lfloor (d+s-3)/2 \rfloor$ generator.

The power expansion unit converts the α^k's into an erasure-locator polynomial $\tau(x)$. Therefore, the polynomial $\tau(x)$ has the α^k's as its roots. The erasure-locator polynomial $\tau(x)$ is fed to the modified GCD unit as one of the initial conditions for the modified GCD unit.

The generator unit is used to compute $\lfloor (d+s-3)/2 \rfloor$. The output is sent to the modified GCD unit and used as a stop indicator for Euclid's algorithm. The polynomial-expansion unit is used to compute the required Forney syndromes.

In Fig. 6.9, the erasure-locator polynomial $\tau(x)$ together with the Forney-syndrome polynomial *T(x)* are the inputs to the GCD unit. The outputs of the GCD unit are the errata-locator polynomial, $\tau(x)$, and the errata-evaluator polynomial, *A(x)*. The error-correcting capability of the code is computed as $\lfloor (n-k-s)/2 \rfloor$.

Reed-Solomon Codes

One of the outputs of the GCD unit is the errata-locator polynomial $\tau(x)$. This output is fed to a Chien search unit and to another unit for computing $\left[x^{b-1}\tau'(x)\right]^{-1}$. The other output of the modified GCD is the errata-evaluator polynomial $A(x)$. This is fed to the polynomial-evaluation unit to perform the evaluation of $A(x)$.

The $\left[x^{b-1}\tau'(x)\right]^{-1}$ unit computes part of the data needed for the errata magnitudes. The products of the outputs from the polynomial-evaluation unit and the $\left[x^{b-1}\tau'(x)\right]^{-1}$ unit form the errata magnitudes.

In Fig. 6.9, the Chien search unit is used to search for both the error and erasure locations. The architecture of the Chien-search unit is similar to that of a polynomial evaluation unit, except that there is a zero detector at the end in the Chien-search unit.

The Euclidean-algorithm-based decoding approach is easier to comprehend than the Berlekamp-Massey algorithm but is less efficient in practice because the polynomials that need to be processed have large degrees. However, the difference in computational complexity between the Euclidean and the Berlekamp-Massey approaches is not as dramatic as it is between the Euclidean approach and the direct-solution technique of Peterson, et. al. The principal benefit of the Euclidean approach to the decoding of RS codes (also BCH codes) is the ease with which it can be understood and applied.

Problems

6.1 (a) Compute a generator polynomial for a double-error-correcting RS code of length 31.
(b) Determine the rate of this code.
(c) Construct a parity-check matrix for this code.

6.2. Let the transmitted code be the (7,3) RS code with the generator polynomial $g(x) = x^4 + \alpha^3 x^3 + x^2 + \alpha x + \alpha^3$.
(a) Decode the following received word by a use of the Berlekamp-Massey algorithm:
(1) $\bar{r} = (1001000)$,

(2) $\bar{r} = (1\alpha^5\alpha^4 01\alpha^4 0)$,
(3) $\bar{r} = (\alpha^6 01\alpha^3 100)$,
(4) $\bar{r} = (00\alpha^4 \alpha^4 \alpha^5 10)$.

(b) Decode the following received word by a use of the Euclidean algorithm:
  (1) $\bar{r} = (x00x00\alpha)$,
  (2) $\bar{r} = (0x\alpha 1x10)$,
  (3) $\bar{r} = (x\alpha^5 001x1)$,
  (4) $\bar{r} = (00xx000)$,

where "x" indicates the presence of an erasure.

6.3 The (15,9) RS code over GF(16) is capable of correcting $t=3$ symbol errors. Let the code have roots $\alpha, \alpha^2, \ldots, \alpha^6$, where α is a primitive element in GF(16), so that the generator polynomial is

$$g(x) = \sum_{i=1}^{6}(x - \alpha^i).$$

Suppose that the all-zeros codeword is sent, and that there are two errors α and α^7 in positions 0 and 1 of the codeword.

(a) Compute the syndromes $S_1, S_2, \ldots, S_6$.
(b) Use the Berlekamp-Massey algorithm to correct the received codeword.
(c) Re-check the syndromes to see if the decoded output is a valid codeword.

6.4 Consider a concatenation of the (15,9) Reed-Solomon outer code over GF(16) with a (7,4) binary inner code.

(a) If the binary code symbols are transmitted, using PSK, find the resultant signal bandwidth, normalized to the input bit rate.
(b) Assume that the reception on the coherent AWGN channel has the S/N ratio, $E_b / N_0 = 7$ dB. Determine the probability of error of an inner codeword for both hard-decision decoding and ML decoding of the inner code.
(c) Compute the probability of an outer-code decoding failure for a bounded-distance decoder, assuming that hard decisions are used on the inner codewords.
(d) If one views this scheme as a binary code, what are n and k? What is the overall code rate? Find a shortened BCH code with roughly the same parameters and estimate its minimum distance.

6.5 Derive Eq.(6.12) and Eq. (6.13) by the use of the key equation definition Eq.(6.11).

Bibliography

[1] I.S.Reed and G. Solomon, "Polynomial codes over certain finite fields", M.I.T. Lincoln Laboratory Group Report 47.23, 31 December 1958.
[2] I.S.Reed and G. Solomon, "Polynomial codes over certain finite fields", SIAM Journal of Applied Mathematics, Vol. 8, pp.300-304, 1960.
[3] E. R. Berlekamp, Algebraic Coding Theory, McGraw-Hill : New York, 1968.
[4] I. S. Reed and G. Solomon, " A decoding procedure for polynomial codes", M.I.T. Lincoln Laboratory Group Report 47.24, 6 March 1959.
[5] W. W. Peterson, "Encoding and error-correction procedures for Bose-Chaudhuri Codes", IRE Trans. on Inform. Theory, Vol. IT-6, pp.459-470, Sept. 1960.
[6] D. Gorenstein and N. Zierler, " A class of error correcting codes in p^m symbols", Journal of the Society of Industrial and Applied Mathematics, Vol. 9, pp.207-214, June 1961.
[7] R. T. Chien, "Cyclic decoding procedure for Bose-Chaudhuri-Hocquenghem codes", IEEE Trans. on Inform. Theory, Vol. IT-10, pp.357-363, Oct. 1964.
[8] G. D. Forney, "On decoding BCH codes", IEEE Trans. on Inform. Theory, Vol. IT-11, pp.549-557, Oct. 1965.
[9] E. R. Berlekamp, "Nonbinary BCH decoding", International Symposium on Information Theory, San Remo, Italy, 1967.
[10] J. L. Massey, "Shift register synthesis and BCH decoding", IEEE Trans. on Inform. Theory, Vol. IT-15, No. 1, pp.122-127, Jan. 1969.
[11] Y. Sugiyama, M. Kasahara, S. Hirasawa, and T. Namekawa, "A method for solving key equation for decoding Goppa codes", Inf. and Contr., 1975, 27, pp.87-99.
[12] G. D. Forney, Jr., Concatenated Codes, Cambridge, MA : MIT Press, 1966.
[13] L. R. Welch and E. R. Berlekamp, "Error Correction for Algebraic Block Codes", U. S. Patent No. 4,633,470, issued Dec. 30, 1986.
[14] H. M. Shao, T. K. Truong, L. J. Deutsch, J. H. Yuen, and I. S. Reed, "A VLSI Design of a pipeline Reed-Solomon decoder", IEEE Trans. on Computers, Vol. V-34, pp.393-403, 1985.
[15] H. M. Shao and I. S. Reed, "On the VLSI design of a pipeline Reed-Solomon decoder using systolic arrays", IEEE Trans. on Computers, Vol. C-37, pp.1273-1280, 1988.

[16] C. S. Yeh, I. S. Reed, and T. K. Truong, "Systolic multipliers for finite field ", IEEE Trans. on Computers, Vol. C-33, pp.357-360, 1984.
[17] J. L. Massey and J. K. Omura, "Apparatus for finite field computation", U. S. Patent Application, pp.21-40, 1984.
[18] M. A. Hasan and V. K. Bhargava, "Bit-serial systolic divider and multiplier for $GF(2^m)$", IEEE Trans. on Computers, Vol. 41, pp.972-980, August 1992.
[19] E.R. Berlekamp, "Bit-serial Reed-Solomon encoder", IEEE Trans. on Inform. Theory, Vol. IT-28, No. 6, pp.869-874, Nov. 1982.
[20] R. C. Singleton, "Maximum Distance Q-nary Codes", IEEE Trans. on Inform. Theory, Vol. IT-10, pp.116-118,1964.
[21] I. S. Reed, L.J. Deutsch, I.S. Hsu, T.K. Truong, K. Wang, and C. S. Yeh, "The VLSI implementation of a Reed-Solomon encoder using Berlekamp's bit-serial multiplier algorithm", IEEE Trans. on Computer, Vol. C-33, No.10, Oct. 1984.
[22] R. E. Blahut, Theory and Practice of Error Control Codes, Reading, Mass. : Addison Wesley, 1984.
[23] T. K. Truong, W. L. Eastman, I. S. Reed, and I. S. Hsu, "Simplified procedure for correcting both errors and erasures Reed-Solomon code using Euclidean algorithm", IEE Proceedings, Vol. 135, Pt. E, No. 6, Nov. 1988.

7 Implementation Architectures and Applications of RS Codes

Some basic concepts of the RS codes and various encoding and decoding algorithms are discussed in Chapter 6. In this chapter, certain implementation architectures for RS codes are examined and compared. Also to further demonstrate the power of RS codes certain important applications of are introduced in this chapter.

7.1 Implementation of RS Codes

The symbol-error-correcting capability of the RS codes is used in many practical applications, including magnetic and optical storage systems, space and mobile communications, etc.. However, the design and implementation of the RS codes for certain applications is for many engineers still difficult to treat. Knowledge of RS codes is highly specialized and is usually not within the mainstream of the digital designer's domain of expertise. This is primarily because RS codes operate over the algebra of a finite-field, and the encoding and decoding algorithms often are quite complicated.

In the preceding chapter several available coding algorithms and techniques for carrying out finite-field arithmetic are developed. In this chapter the practical realization of these algorithms is studied in more detail. The main focus is on the decoding algorithms. Examples of microprocessor-based decoder algorithms, Application-Specific Integrated Circuits (ASICs) and Field-Programmable Gate-Array(FPGA) realizations of RS codecs (encoders and decoders) are discussed in the next sections. Many of the arguments for mapping algorithms onto Very Large Scale Integrate(VLSI) architectures can be extended to the implementation of convolutional (and trellis) encoders and decoders introduced in the later chapters.

7.1.1 From Algorithms to VLSI Architectures

The envisaged large number of application of error-control codes to high-speed, layered communication networks calls for channel-coding chips of low manufacturing cost and small size. Manufacturing costs are dominated by the number of integrated chips, the chip packaging and the silicon area per chip. The number of chips can be kept small by the use of a large area of silicon. However, because of the density of defects, production of very large area chips is usually not economic. Thus, high complexity channel-coding systems, such as a multi-chip 80 Mbits/s RS decoder for the deep space application[27], requires, in general, several (or many) chips for implementation. The cost for chip packaging is related to the number of pads. Partly because of this reason the partitioning of a coding system onto a set of chips, one needs to consider carefully the number of interconnections. With the recent advance of semiconductor technologies high-complexity RS coding systems now can be implemented with a fewer number of chips, or, sometimes, even on a single chip.

The ultimate goal is to achieve a VLSI implementation of a RS codec with the smallest silicon area for a specified data rate. As listed in Table 7.1 the data rate for RS coding has a large range that depends on the application. It is obvious that both the hardware and software structures for low data rates such as the audio bit-stream for the compact disk system with 1.41 Mbits/s would be substantially different from those needed for very high data rates such as HDTV of about 19.3 Mbits/s for the terrestrial mode and 38.6 Mbits/s for the high data-rate cable mode.

Data Rate	Type of Application	Notes
44.8 kbits/s	deep-space communication	data rate for Voyager mission
100kbits/s	deep-space communication	(designed) data rate for Galileo mission
1.41Mbits/s	compact disk system	the audio bit-stream
1Mbits/s-- 60Mbits/s	Digital Television	channel bit rate

Table 7.1 A list of data rate for three applications

The silicon area for the VLSI implementation of different algorithms is determined by the required resources such as logic gates, memory, and the interconnection between the modules. The amount of logic depends on the concurrency of the operations. A figure-of-merit for the required concurrency can be specified by the number of concurrent operations. It is evident from the discussions in Chapters 2 and 6 that the number of concurrent operations for RS coding depends on the size of the finite-field, the length of a codeword, the code rate, the specific coding algorithm and the data rate of the application.

The concurrency of operations can be achieved by architectures that utilize either parallel processing or pipelining. Since algorithms can be specified in a hierarchical manner, also parallel processing and pipelining can be defined hierarchically. For example, a pipeline RS error-and-erasure decoder can be specified by tasks as given in Figure 6.9. In summary, these tasks are the syndrome computations, the power calculations, the polynomial expansion, the GCD, and the Chien search, etc. For particular tasks like the GCD and the syndrome computations high computational speed is a requirement. These computational requirements are determined in terms of the number of concurrent operations. These tasks are very regular with a predefined sequence of operations and data accesses. By a refinement of the RS algorithm down to its specific basic operations, one can apply techniques, found in the literature [12][14][17][18], to map the regular algorithms onto architectures which make extensive use of parallel processing and pipelining.

Application of the above discussed mapping strategies leads to a wide variety of architectural solutions for the implementation of RS coding scheme. These solutions include the use of microprocessors or DSP chips, ASICs and FPGA. In order to compare these architectural alternatives an assessment measure, which considers the architectural efficiency, is introduced next.

In general, architectural efficiency can be defined as the ratio of the performance to the cost. There are two commonly-used performance measures. One is called the computational rate R_c which equals the reciprocal of the achieved effective-processing (executive) time T_c for one codeword, i.e. $R_c = 1/T_c$. The other is the throughput rate R_T which is also proportional to $1/T_c$. The determination of costs for a specific architecture is more problematic, since the implementation cost are influenced by a wide range of parameters, such as the silicon area, the design style, the architectural complexity, the semiconductor process, pin count, etc. For a simplified approach, cost can be expressed in terms of the required silicon area $A_{silicon}$ for the well-known Area-Time(AT) product, which is used for architecture assessment. In this approach the architectural efficiency is defined by

$$E = \frac{1}{A_{silicon} \cdot T_c}. \tag{7.1}$$

For an architecture with a specified (fixed) efficiency it follows from Eq. (7.1) that the silicon area is proportional to the computational (or throughput) rate.

$$A_{silicon} = \alpha_c \cdot R_c. \tag{7.2}$$

This relation is evident for the operational parts of a chip, where the increase of the throughput requires an increase of concurrency because of the parallel implementation of the basic processing units.

The required silicon area and the processing time for an implementation of a specific RS coder depend on the applied semiconductor technology. Hence, a realistic architectural assessment needs to consider the gains provided by progress in the semiconductor technology. A sensible approach to achieve a realistic assessment is the normalization of the architecture parameters in accord with the reference technology. In the following subsections a reference process for a grid length of λ_0, e.g. $\lambda_0 = 1.0$, $\lambda_0 = 0.5$, etc. microns (μm), is used. Also, the "constant voltage" model for the scaling of MOS transistors[22] is assumed. In this model the gate delay is proportional to $\left(\lambda/\lambda_0\right)^2$. In order to compare the different alternatives of architectures a normalization by $\left(\lambda/\lambda_0\right)^2$ for the silicon area is used in the following subsections, i.e.

$$A_{silicon} = A_{silicon,0} \cdot \left(\frac{\lambda}{\lambda_0}\right)^2. \tag{7.3}$$

where the index 0 is used for the system with reference length λ_0. This model does not take into account the following three factors : the interconnection delay (which basically does not scale[22]), the short- channel effects and the limitations due to the issue of power dissipation.

For a microprocessor-based implementation of an RS codec, it is useful also to have an empirical measure of the processor speed versus technology. A comprehensive list of microprocessors is given in [21] with clock frequencies and their gate length λ. From the data provided in [21] one can estimate that the cycle time of these processors, which results from a combination of the above-mentioned three factors, scales approximately as $\left(\dfrac{\lambda}{\lambda_0}\right)^{-1.6}$, i.e.

$$R_c = R_{c,0} \cdot \left(\dfrac{\lambda}{\lambda_0}\right)^{-1.6}. \tag{7.4}$$

The combination of Eqs. (7.2), (7.3) and (7.4) yields the estimated silicon area

$$A_{silicon} = \alpha_{c,0} \cdot R_c \cdot \left(\dfrac{\lambda}{\lambda_0}\right)^{3.6}. \tag{7.5}$$

where the index 0 is used for the system with reference length λ_0. As a result it follows that for a given technology there is a linear relation between silicon area and computational (or throughput) rate. Down-scaling effects the silicon area of the operational part by the power 3.6.

For long RS codes the coding processes are performed over large finite-fields. The more accurate model is composed of both an operational part and a memory part which is not effected by the computational (or throughput) rate. The silicon area of the memory part is proportional to the memory capacity C_M. If one considers the effect of the memory part, Eq.(7.5) has to be modified to

$$A_{silicon} = \alpha_{M,0} \cdot C_M \cdot \left(\dfrac{\lambda}{\lambda_0}\right)^2 + \alpha_{c,0} \cdot R_c \cdot \left(\dfrac{\lambda}{\lambda_0}\right)^{3.6}. \tag{7.6}$$

It should be noted that the criteria (7.1) and (7.2) neglect several essential properties of a specific implementation, e.g. design style (full-custom, semi-custom), power dissipation, semiconductor yield and architectural flexibility.

The assessment measures, developed above, are used to analyze several implementations of RS codecs in the next five subsections. These examples of the implementation of RS codecs are introduced only for educational proposes in order to illustrate how to evaluate and compare different architectures.

7.1.2 Implementation of an RS Codec with a Microprocessor

RS codecs can be programmed within a general purpose microprocessor which uses table-lookup algorithms. First, one generates the coefficients needed for the finite-field multiplier. Then these coefficients are stored in memory. Finally the encoding/decoding operations are performed by means of lookup tables.

In the following example, the 100 MHz Sun HyperSparc is used to benchmark a C-program that performs both the encoding and decoding processes. The C-program of this RS codec was developed in the University of Adelaide[23].

This C-program generates the coefficients that are needed for the RS codes with differing parameters. Then it performs the encoding and decoding algorithm via a lookup table. The results of this benchmark of an RS codec are seen in Table 7.2a[24]. The recorded operation times are for the encoding and decoding of the RS code of length 255.

Codec	Execution Time	Throughput
systematic shift-register encoder	0.533ms	478.4 kb/s
Berlekamp-Massey decoding algorithm	5.52ms	46.2kb/s

Table 7.2a RS encoder/decoder executive time and throughput

The above results do not include the processing power used to generate the lookup tables. The lookup tables are assumed to have been previously generated so that they need be initialized only once during the operation of the codec.

The ratio of performance to cost is used as the criteria for characterizing the architectural efficiency of this codec implementation. The simple approach discussed in the preceding subsection to estimate the performance and cost is first to calculate the performance-per-area metric. The HyperSparc 100 MHz processor has a combined die area of 327 mm² and is fabricated in a 0.5 micron process ($\lambda = 0.25\mu$) [1]. Hence, it is known from $327/\lambda^2$ =5232 that

Implementation Architectures and Applications of RS Codes

the area of the chip in units of $10^6 \lambda^2$ is 5232 million λ^2. Evaluating the HyperSparc with a Performance-per-Area metric, one divides the performance, i.e. the computational rate (1/execution time) by the area ($10^6 \lambda^2$). The results of these calculations are given in Table 7.2b.

Codec	Execution Time	Through-put	Area (million λ^2)	Performance/ Area
encoder	0.533ms	478.4 kb/s	5232	0.35
decoder	5.52ms	46.2kb/s	5232	0.035

Table 7.2b Performance per Area

Note that the computational rate is proportional to the throughout rate. The Sparc processor has far more area than is actually needed to perform the encoding and decoding processes. However, one reason for the particularly poor performance could be the need to constantly access the lookup table (external) memory which is necessary to calculate the finite-field multiplications. It is possible to reduce the access rate to an external memory by assigning a local memory to the operative part of the memory, that is, by a pre-fetching of the data in the table. The size of this local data-cache memory depends on the specific access structure of the coding algorithm.

7.1.3 A Digital-Signal-Processor(DSP) Implementation of the RS Codec

The DSP has long been an alternative to the traditional microprocessor. The DSP chips often provide the parallel processing architectures needed for specific applications. RS codecs can be programmed in DSPs for many applications. However, DSPs usually are not designed to implement certain of the specific operations needed for error correction, e.g. the finite-field operations. Fast implementations of finite-field arithmetic are necessary for high-speed codecs. The implementation alternatives, which use lookup tables to provide the fastest finite-field operations, utilize DSP's as well as microprocessors [2].

One such example is the RS codec which uses the TMS320C10 DSP [3]. This particular codec corrects up to 4 errors. The software in this program is written for the *(15,k)* RS code, where *k* ranges from 1 to 5. In this subsection it is shown that specialized hardware can be developed to realize a codec capable of Gbits/s data rates. However, lower data rates can be realized by the TMS320C10 DSP. The resulting throughput for this design ranges between 16 Kb/s to correct 5 symbol errors to 80 Kb/s to correct 1 symbol error.

Another implementation, that uses a DSP for RS error correction, can be accomplished in combination with an FPGA. The TMS320C25 is used to perform the main decoding process which supports 1 symbol error[3], but the finite-field multiplications, which suffer from a low computational speed when performed in software, are accelerated by the use of the FPGA.

Some running-time results for finite-field multiplications over GF(16) are reported in [4] and given in Table 7.3.

Platform	FPGA	TMS320C25 + FPGA	TMS320C 25	Intel 486DX2-66	IBM RS6000 /580
multiplica -tion time	80 ns	700 ns	12 us	6.1 us	4.8us

Table 7.3 Multiplication Times

One interesting thing about Table 7.3 is that the RS code used did not rely on lookup tables to perform the finite-field multiplications. The reason cited were the bottlenecks caused by the memory-access times. In any case, look-up tables are still the most common method of implementing the multiplications in an RS codec.

A more advanced DSP implementation [24] of an RS codec is accomplished by the use of a C-program on a faster DSP chip, the TMS320C30. The TMS320C30 contains only an instruction cache and no data cache so that lookup tables necessitate off-chip memory accesses. However, these could be made faster by using an external Static-Random-Access-Memory(SRAM) cache. On the later TMS family of DSPs (following the TMS320C30), however, data caches are included, so that the lookup-table method becomes a more efficient alternative to perform finite-field arithmetic on-the-fly.

The C-program of the RS codec is compiled by a compiler in the On-Line DSP Lab at Texas Instruments(TI). Such a compiler allows registered users to compile and link the C-code for the TMS320C3x software and the hardware products. These implementations are benchmarked in the evaluation module at TI that runs compiled code with and without optimization on a 33 MHz C30 DSP with *16K* words of SRAM. A break-down of these results is shown in Table 7.4.

Compiler Optimization	Encoder Clock Cycles	Encoder Execution Time	Decoder Clock cycles	Decoder Execution Time	Executable File Size
None	430,762	13.05ms	609,580	18.47ms	9,710 bytes
Level 0	207,246	6.28ms	448,680	13.60ms	4,823 bytes

Table 7.4 Benchmark Results for RS Codec

In order to compare the suitability of a DSP platform for the RS codec it is necessary to apply a common metric. This metric is the measure of the execution time for a message of length 255 bytes divided by the area (scaled by the chip manufacturing process). This metric is in units of seconds/ λ^2, e.g. megabytes per second per unit area often being considered as a metric. However, due to design constraints, the computation of the next message does not commence until the last message is processed. Thus pipelining between messages is not possible, although pipelining and parallelism often are used to process a message.

The die size for the TMS320C30 was estimated to be 6mm x 8.5mm in a 0.5 micron process ($\lambda = 0.25 \mu$). This estimate was obtained from the die size for a 60 ns DSP with 32-bit register operations and an on-chip instruction cache [5]. Scaling the area by feature size yielded 816 million λ^2 as the area of the DSP. For the compiler with the optimization option (see the second row of Table 7.4), its performance-per-area (per million λ^2) is given in Table 7.5.

Codec	Exec. Time	Throughput	Area (million λ^2)	Performance/Area
Encoder	6.28ms	40.6kb/s	816	0.195
Decoder	13.6ms	18.75kb/s	816	0.09

Table 7.5 Performance/Area Results for a DSP Implementation of the RS Codes

7.1.4 Application-Specific Integrated Circuits(ASIC) for RS Codec

The holy grail of a codec implementation with metrics, such as silicon area, clock rate, and power consumption, is the ASIC. One example of such an ASIC chip is the RS codec chip developed by the LSI-Logic Corporation.

The LSI ASIC chip contains an 8-error correcting RS encoder and decoder, both of which are able to independently process 40 Mbytes of data per second. The chip is fabricated in the 1-micron CMOS process with the resulting die size of 9.5 mm x 9.5 mm. The polynomial used to generate the finite-field for the encoder is $x^8 + x^4 + x^3 + x^2 + 1$.

The encoder occupies a small fraction of the chip, only 2000 gates. The total gate count of the chip is 20,000 gates. The chip includes 3Kb of RAM and 4Kb of ROM. Table 7.6 summarizes the characteristics of this LSI-Logic chip.

Technology	1 u CMOS
Die Size	9.5x9.5 mm²
Encoder Clock	40 MHz
Encoder Throughput	40 Mbytes/second
Encoder Latency	3 clock cycles
Decoder Clock	40 MHz
Decoder Throughput	40 Mbytes/second
Decoder Latency	91cycles + codelength
Random Logic-Gate Count	20,000
Encoder-Gate Count	2,000
Memory	3K bits of RAM, 4K bits of ROM

Table 7.6 Summary of the characteristics of the LSI-Logic chip

The performance/Area of this chip is given in Table 7.7.

Codec	Exec. Time	Throughput (Max.)	Area (million λ^2)	Performance /Area
Encoder	6.45us	40MB/s	31.6	4906
Decoder	8.65us	40MB/s	284.4	406

Table 7.7 Analysis of Performance/Area

Implementation Architectures and Applications of RS Codes 295

The execution time of the above-mentioned encoder is calculated by multiplying the cycle time by the code length plus a 3-cycle latency. This encoder occupies 10% of the chip, thereby resulting in an area of 9mm² which is divided by the square of the feature size to yield 31.6 million λ^2. The decoder occupies 81mm², and the execution time is based on a latency of 91 cycles. Note that the metric of performance per area of this chip is much higher than those in the microprocessor and DSP implementations. As a consequence the ASIC is still the best solution to obtain a high bit-rate RS codec realization.

Another example of an ASIC implementation of an RS codec was the chip designed for the Hubble Space Telescope(HST)[24][25]. It was designed on a 1.6 um double-metal CMOS technology. This chip was delivered to the Goddard Space Flight Center in April, 1989. It was installed in the ground communication link for service in the Space Telescope system. The chip was 8.2x8.4mm², contained 200,000 transistors, operated at a sustained data rate of 80 Mbits/s and executes up to 1000 Million Operations Per Second(MOPS) while consuming less than 500 mW of power. The circuit had complete decoder and encoder functions and used a single data-system clock. The error-correction and detection capability of the decoder was selectable by the user. Codeword-length up to 255 bytes were supported by this codec without external buffering. Erasures as well as random errors up to 10 symbols were correctable. Corrected data was outputted at a fixed latency.

Technology	1.6 μm double metal CMOS
Die size	18.2mmx8.4mm
System clock frequency	10MHz
Decoding algorithm	Euclid's algorithm
Decoder through put	80Mbits per second
Decoder latency	2xCode Length+10errors+34clock cycles
Transistor count (including memory)	200,000

Table 7.8 Summary of the characteristics of the Hubble chip

Images transmitted from the HST are encoded with a (255,239) RS code. Note that the minimum distance of this code is $d=255-239+1=17$. Hence, it is known from Chapter 6 that for each message block, the HST code requires the evaluation of 16 equations of order 254 for the syndrome computations, the recursive evaluation of the Euclidean algorithm utilizing the polynomial x^{d-1} of degree 16 and the polynomial $T(x)$ of degree 15, 255 evaluations for each of error-locator and erasure-locator polynomials of degree 8 , 255 evaluations of the errata-evaluator polynomial of degree 7, plus 255 divisions, multiplications and additions of field elements. The total number of calculations per message in the HST (255,239) code is 19386. Operating at 80 Mbits/s, the number of operations per second is 750 million. The full capability of the chip, a (255,235) RS code, requires 1000 MOPS. The computational rate and powerful error correcting features of this chip make it suitable also for use in digital high-definition television applications which require about a 20 Mbits/s data rate.

7.1.5 Field-Programmable Gate-Array(FPGA) designs of RS Codec

To satisfy the requirements of "quick-to-the-market", low volume and real-time operational speeds for some applications, FPGAs often are chosen to prototype a codec. This is due to their fast design time and good performance. One commonly used design approach is to implement an RS codec by the use of a C-program on a microprocessor. Then, in order to reduce the cost, such a codec is transferred onto FPGA hardware.

An example of an FPGA RS-codec design is reported in [6]. In this design, the (31,9) and (15,6) RS decoders are implemented on Xilinx XC3090 FPGA chips. A single XC3090 FPGA chip is needed for a 2-byte error correction of the (15,11) RS codec. However, in order to correct up to 11 errors with the (31,9) RS code, a 4 XC3090 (10MHz) FPGA chip is required.

The FPGA design of an RS codec quickly and cheaply converts circuit descriptions into actual hardware realizations. Such implementations are in-house devices that can be generally quite reliable. Also, design revisions can be made quickly and cheaply. Also, if the error-correction capability for the implemented codec is found not to be sufficient under testing conditions, a more powerful codec design sometimes can be programmed into the FPGA without changing the board layout.

For example, the XC4005 (30MHz) FPGA chip can be used to implement the (31,9) RS decoder in order to reduce the number of chips. Custom ASICs, on the other hand, have a longer design time, a higher fabrication cost, and less flexibility in the choice of the error-correction scheme. A disadvantage in the use of FPGAs is that FPGAs are more suitable for bit-serial operations than for parallel operations due to the chip-routing constraints. Therefore, when

Implementation Architectures and Applications of RS Codes 297

algorithm operations need to be highly parallel, the ASIC design is the better choice.

7.1.6 Comparisons of the Implementations of RS Codecs

The analysis of the implementation approaches, needed to estimate the performance and cost for different realizations, plays an important role in the design and production of RS codecs.

Comparisons of RS codec implementations of the throughputs and performances per area are summarized in Table 7.9 from the data in the above examples. The performance/area metric is used to provide a scaleable measure based on area. As the FPGA chip sizes increase and routing becomes less constrained, FPGAs most likely should be able to support the implementation of more complex designs.

The FPGA encoder design were compiled by the Xilinx tools and simulated in Viewlogic[7]. Such simulations proved the functionality of the encoder design and that the clock delay and area values could be estimated by the use of Xact tools [8].

The values shown for the FPGA decoder are estimates from the designs developed by industry. They are not concise values and should not be used other than to gain a qualitative idea of the order of the magnitude of the performance to area metric to be expected in practice.

Implementation	Throughput	Performance /Area
Sun Microprocessor	478.4 KB/s	.35
TSM320C30 DSP	40.6 KB/s	.195
Xilinx FPGA (programmable length)	29.6 MB/s	585
Xilinx FPGA (fixed length)	33.7 MB/s	685
ASIC	40MB/s	4906

Table 7.9 Comparisons of Implementations of Encoder

The "Performance-per-Area" for different implementations yielded the following comparison of codecs: The microprocessor and DSP RS codec perform similarly, with execution times and areas that were on the same order of magnitude, and hence had a comparable performance/area metric. Since the software used look-up tables, the memory access time was most likely the bottleneck in the processing time. The DSP used in this experiment did not have a data cache. This could account for its performance being somewhat lower than that of the microprocessor. Also the microprocessor ran at a 100 MHz clock rate while the DSP only ran at 33 MHz.

Implementation	Throughput	Performance/Area
Sun Microprocessor	46.2 KB/s	0.035
TSM320C30 DSP	18.75 KB/s	0.09
Xilinx FPGA	10 MB/s(estimate)	20.33(estimate)
ASIC	40 MB/s	406

Table 7.10 Comparisons of Implementations of Decoder

Both the FPGA and ASIC chips were implemented by a network of XOR logic operators for the encoding and achieved similar execution times. However, since the ASIC chip was far more dense than the FPGA, the FPGA suffered in the performance/area metric. It is clear that FPGA's provide an advantage over processors, but how close the FPGAs perform to the ASIC chips is still an area of continuing research. Some results for these two technologies show a performance of the same order of magnitude for the Reed- Solomon encoder.

FPGAs are very useful for fast prototyping. An FPGA prototype can be produced cheaply and quickly and can be modified with low additional cost and effort if the initial approach is not satisfactory. Such designs can be accomplished in-house without having to send them to a fabricator. Since the FPGA design is programmable, this flexibility in the design can lead to the generation of several designs for the same set of parameters. For example, in the design of RS codecs, there are two important parameters: codeword length and message word length. Of course, for different application requirements one might expect to emphasize other design parameters.

In an FPGA implementation the user can quite simply download different designs into a ROM in order to program the FPGA chip. In a full custom ASIC implementation the designer either designs different chips for different requirements or uses more complicated architectures for differing codeword lengths. The ASIC chip is not programmable for a particular error-correction capability because its architecture must be changed for different message word lengths. Overall, the FPGA has many qualitative advantages in using it for an RS codec.

The performance of an FPGA is usually about one order of magnitude lower than that of an ASIC chip, while three orders of magnitude better than that of

Implementation Architectures and Applications of RS Codes

the microprocessor and the DSP. The downside of an FPGA is its cost in large quantities, its typically higher consumption of power than that of an equivalent ASIC, and its limited routing capabilities within the array. Bit-parallel designs are more difficult to implement because of these routing constraints. These downsides clearly constitute areas for future research in re-configurable computing.

7.2 RS codes in industry standards

7.2.1 The CCSDS Standard

At the present time RS codes are a necessity in spacecraft communication systems. As a consequence the Consultative Committee for Space Data Systems(CCSDS), which represent the space agencies for most of the world, created in May 1984 from an official recommendation[9][10][11] for a communication- channel coding standard for spacecraft.

The CCSDS recommended that the RS code for spacecraft be the (255,223) code over the finite-field $GF(2^8)$. This finite-field is generated by the polynomial $p(x) = x^8 + x^7 + x^2 + x + 1$. The generator polynomial for the CCSDS's (255,223) RS code is

$$g(x) = \prod_{j=112}^{143} (x - \alpha^{11j}),$$

where α is a primitive root in the field $GF(2^8)$, i.e. a root of the equation $p(x)=0$.

A VLSI architecture for the (255,223) RS encoder is given in Figure 7.1. In this figure, Q is a 7-bit shift register with reset, and R is a 8-bit shift register with parallel load and reset. S_k for k=0,1,...,30 are 8-bit shift registers with reset. In deep-space communications a low complexity encoder is required for the channel coding.

The choice of parameters is based on the results provided by Belekamp and [12][13] in the early stages of NASA's Galileo project. In particular, Berlekamp discovered a method that used bit-serial, finite-field, arithmetic to simplify the encoding of RS codes. Such a method made use of the dual basis for $GF(2^8)$, relative to a "standard" basis $\{1, \alpha, ..., \alpha^7\}$. The particular choice of field representation and generator polynomial, given above and recommended by the CCSDS standard, was motivated by the desire to minimize the encoder hardware for the bit-serial, dual-basis, encoder. The

Berlekamp bit-serial encoder for the (255,223) RS coder was first implemented by Reed and his research team [14].

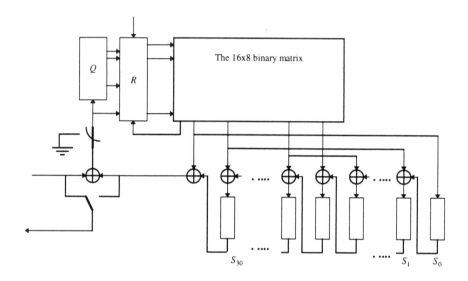

Figure 7.1 A VLSI architecture for the (255,223) RS encoder

The CCSDS standard was adopted for use by numerous planetary missions, including NASA's ill-fated Mars Observer, which was launched in September 1992 and arrived Mars in August 1993. It was used recently in the Pathfinder lander which was launched in June 1996 and arrived Mars in July 1997. Also, it was on Cassini which was launched in 1997 to arrive Saturn in 2004, the joint NASA/ESA(the European Space Agency) Ulysses mission, which was launched in October 1990 to the sun's polar regions, and the ESA missions Giotto, which were launched between 1985 and 1986 to Halley's comet. Finally it is expected to be on Huygens, which was aboard the NASA's Cassini and also on the Cluster and Soho spacecrafts of the International Solar and Terrestrial Physics Program.

7.2.2 The Cross-Interleaved Reed-Solomon Code(CIRC) in the Compact Disc, Digital-Audio, System

The compact-disc(CD), digital-audio, system is the world standard for the storage and reproduction of audio signals. The coding system of the CD also has become the most famous application of error-correction codes. As is usual in a communication system, the various signal operations, shown in Figure. 7.2, are designed into the CD digital-audio system. This system, together with its subsequent modulation and channel encoding, is part of the so-called disc-master process. In this process the information from a digital video-tape recorder system is encoded into the standardized CD format.

The standardized CD format is optically recorded on the surface of a glass disc which is coated with a photo-resist. Following the development and evaporation of the photo-resist, the result is called the master disc. By a further galvanic process the master-disc surface is "transferred" onto a nickel shell (or "father"). From this "father", "sons" or stampers are made, which are suitable for replication. By compression or injection molding the information contained on the surface of the stamper, is transferred to a transparent plastic disc. Finally, after receiving a reflective aluminum coating, over which a protective lacquer is applied, the "Compact Disc" is ready to be played.

With analog storage, such as magnetic tape, there is no opportunity to perform any error-correction processing. Digital-audio storage takes advantage of forward error correction of digital data. In the CD "transmission" system the sources of channel errors are identified as either system or media errors. System errors are mainly caused by signal aliasing, quantization, phase distortion, nonlinearities in the circuit, aperture-error interference and other noise. Dust, scratches, fingerprints, etc, on the magnetic and optical media also cause errors. For example, small unwanted particles or air bubbles in the plastic material or pit inaccuracies due to stamping and stamper errors can be present in the replication process. This can cause errors when the information is optically read. Also, fingerprints or scratches on the disc may occur when the CD is handled. Together with possible surface roughness, these disturbances cause additional channel errors.

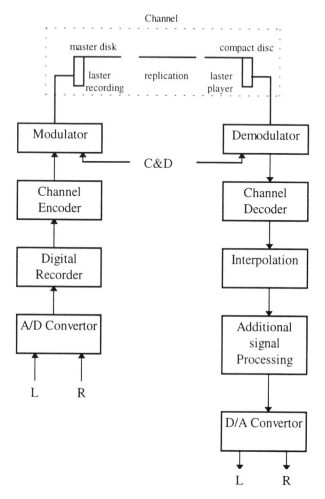

Figure 7.2 The compact disk digital audio system

In the digital audio system, random errors are introduced by modulation and other audio circuits. When a disc is used for laser recording and the read-out of digital signals, there are only a few random errors; most errors occur as burst errors. This is because the dimensions of a pit are small in relation to most of the common mechanical imperfections such as dirt and scratches. Therefore, to determine the requirements for an error-protection codec for a CD the ability to correct both random and burst errors is of great importance. In particular burst errors must be corrected or detected with a high degree of certainty by the error-protection codec. The surface contamination effects areas which are usually quite large compared with the surface used to record a single bit. As a consequence many channel errors occur as bursts when the disc is played. The CD error-control system is designed to handle bursts by means of, what are called, cross-interleaving and the burst error-correcting capability inherent in the Reed-Solomon codes.

The quality characteristics needed for the correction of bursts are the maximum fully-correctable-burst length and the maximum interpolation length. The maximum correctable-burst length is determined by the design of the error-protection codec. Also the maximum interpolation length is the maximum burst length within which all erroneous symbols, that make the error-correction decoder fail, can still be corrected by linear interpolation between adjacent sample values. Finally, random errors also can introduce multiple errors within one codeword.

In general, the greater is the bit-error rate(BER) at the receiving end, the greater is the probability of uncorrectable errors. A measure of the performance for such a system is the number of sample values that have to be reconstituted by interpolation for a given BER. This number of sample-values per unit time is called the sample-interpolation rate. The lower this rate is at a given value of the BER, the better is the quality of the system for random-error correction. An objective assessment of the quality of the error-correcting system requires an indication of the number of errors that pass through unsignalled and, therefore, are not corrected by the system. These unsignalled and uncorrected errors can produce clearly audible 'clicks' in the reproduction.

On the basis of the error and quality characteristics, needed for a CD audio system, the digital information is protected against channel errors by adding parity bytes which are derived separately in two RS encoders. Because the channel has primarily a burst-like error behavior, RS codes were selected for the error-correction. Also, the well-known communications technique of interleaving is used to spread out the errors over a longer time interval. The data streams, which enter the first encoder and leave the second encoder, are scrambled between the two encoders by means of a set of delay lines (see Figure 7.3). The result of this interleaving process is that the bursts as byte errors are spread over a substantially longer time period in order that they are corrected easily with a decoder that corrects only a small number of errors.

At this point the details of the CD error-correction coding system are of primary interest. The CIRC encoder in Figure 2 uses two shortened Reed-

Solomon codes, C_1 and C_2. Both codes use 8-bit symbols(bytes) from the code alphabet GF(256). This provides a nice match to the 16-bit samples that emerge from the A/D converter. The "natural" length of the RS code aver GF(256) is 255, which would lead to a 2040-bit code word and a complex decoder. It should be remembered that the decoders that reside in CD players are for retail sale so that it is extremely important that the decoder costs be minimized. Hence for simplicity the codes are shortened significantly ; C_2 is a (32,28) code, and C_1 is a (28,24) code both over *GF(256)*. Both of these codes have a redundancy of 4 and a minimum distance of 5. The input of the C_1 encoder is a 1.41-Mbps data stream. The CIRC encoder has an effective rate of 3/4, and thus produces a 1.88-Mbps coded data stream.

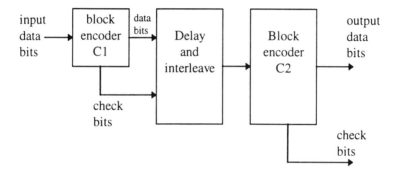

Figure 7.3 A Cross-interleaved block encoder

Each 16-bit sample is treated as a pair of symbols from GF(256). The samples are encoded, 12 at a time, by the C_1 encoder to create a 28-symbol code word. The 28 symbols in each C_1 code word are then passed through a cross-interleaver (of $m = 28$, and $d = 4$ symbols) before being encoded by the C_2 encoder. The resulting 32-symbol C_2 code word is then processed as shown in Figure 7.3.

The present-day CIRC encoding process for the CD system is standard; no matter where you buy a CD (of standard size), the discs can be played on any CD player. However, the CIRC decoding process is not standardized and can vary from player to player [15]. This was done to allow manufacturers to experiment with various designs and to speed the player to market. The basic building blocks of the decoder are shown in Figure 7.4. The C_2 decoder is followed by a cross-deinterleaver and a C_1 decoder. Since both codes have a minimum distance of 5, they can be used to correct all error patterns of weight 2 or less. It is noted in Chapter 6, that even when the full error-correcting capacity of an RS code is used, there remains a significant amount of error-detecting capacity. This is evident in Figures 6-5 and 6-6, which show that a high-weight error pattern is much more likely to cause a decoder failure (which is detectable) than a decoder error (which is not detectable). It is further seen in Chapter 6 that some of the error-correction capacity of an RS code can be exchanged for an increase in the error-detection capacity and a substantial improvement in the reliability of performance.

Most CIRC decoders take advantage of both of these principles. The C_2 decoder is set to correct all single-error patterns. When the C_2 decoder sees a higher-weight error pattern (double-error patterns and any pattern causing a decoder failure), the decoder outputs 28 erasure symbols. Then the cross-deinterleaver spreads these erasures over 28 C_2 code words. C_1 also may be set to correct double-error patterns, or, at the other extreme, it simply may be used to declare erasures (for the least expensive implementation).

The C_1 decoder can correct any combination of e errors and s erasures, where $e + s < 5$. Generally it is designed to decode erasures only (again, an inexpensive solution) owing to the small probability of a C_1 decoder error. Whenever the number of erasures exceeds 4, the C_1 decoder outputs 24 erasures, that correspond to 12 erased music samples. Those errors that cannot be corrected but are still detected, which would give correspondingly unreliable samples, are handled by the error-concealment algorithm. The error-concealment circuitry responds by muting these samples or by interpolating sample values by a use of the correct samples adjacent to the erased samples. A number of additional interleaving and delay operations are included in the encoding and decoding operations in order to enhance the operation of the error-concealment circuitry. For example, samples adjacent in time are separated further by additional interleavers in order to decrease the number of interpolation.

We are now in a position to state a detailed strategy or procedure for decoding the CIRC:
The C_2 decoder:
    If a two-, a single, or a zero-error syndrome occurs,
    then modify at most two symbols accordingly,
    else assign erasure flags to all 28 outgoing symbols.

The C_1 decoder:
    If a zero-error syndrome occurs,
    then no error correction is necessary,

      else if *number of erasures* > 4,
      then   copy C_1 erasure flags from the C_2 erasure flags.
      else begin
            if   *number of erasures* = 4,
            then   try four-erasure decoding,
            if   *number of erasures* = 3,
            then   try three-erasure and no-error decoding,
            if   *number of erasures* = 2,
              begin
                if two-erasure and zero-error syndromes occur,
                then try two-erasure and zero-error decoding;
                if two-erasure and one-error syndromes occur,
                then try two-erasure and one-error decoding;
              end.
          if *number of erasures* = 1,
            begin
              if   one-erasure and zero-error syndromes occur,
              then try one-erasure and zero-error decoding;
              if   one-erasure and one-error syndromes occur,
              then try one-erasure and one-error decoding;
            end.
          if *number of erasures* = 0,
            begin
              if   zero-erasure and one-error syndromes occur,
              then try zero-erasure and one-error decoding;
              if   zero-erasure and two-error syndromes occur,
              then try zero-erasure and two-error decoding;
            end.
          if any of the above decoding processes fail,
            then assign erasure flags to all of the 24 outgoing symbols.
      end.

The error-and-erasure decoding algorithms developed in Chapter 6 also can be used in this decoding strategy. Actually this decoding strategy applies to all RS codes which have a minimum distance of 5. This follows from the fact that the decoding process does not depend on the length of the codeword. Given the above conditions for every error/erasure configuration of the two RS codes, one needs only to check these conditions to decide whether or not a special error/erasure configuration has occurred in the received codeword. Therefore, one expects that the decoding scheme is able to optimize the error and erasure correcting ability of the RS codes while involving no further complexity. These facts make possible, in other applications, the use of two error-correcting RS codes, but with a longer length, which utilize the same error and erasure decoding conditions and algorithm developed above. Performance figures for the CIRC system in a typical CD player are provided in Table 7.11

Longest completely correctable burst.	=4,000 data bits (2.5 mm track length)
Longest interpolation interval. burst length(worst case).	12,300 data bits (7.7 mm track length)
Sample interpolation rate.	One every 10 hours at BER = 10^{-4}, 1000 samples per minute at BER = 10^{-3},
Undetected error rate (clicks in output).	<1 every 750 hours at BER = 10^{-3}, negligible at BER= 10^{-4}
Implementation	1 LSI chip and 2048 bytes RAM

Table 7. 11 CIRC performance in a typical CD player ([15], © 1994 IEEE).

It should be noted that the interleaving used in the CIRC system is accomplished on a symbol-by-symbol basis. The bits, that comprise a given symbol, remain contiguous throughout the entire encoding, transmission, and decoding process. This allows the CIRC system to take advantage of the "burst-trapping" capability of Reed-Solomon codes. The channel itself is binary. When a burst of errors occurs, it may affect many consecutive received bits. Since an RS decoder does not care if a symbol error is caused by 1 to a maximum of 8 bit errors, it has a significant burst-error correcting capability.

7.2.3 RS Product Code in the Digital-Video-Disk(DVD) Specification

In the CD system a concatenation of two codes. namely the EFM (Eight-to-Fourteen Modulation) and the CIRC (Cross Interleaved Reed-Solomon Code) are used. CIRC is used for the correction and detection of erroneously retrieved information while EFM is used to transform the digital audio-bit stream into a sequence of binary symbols, called channel bits, which are suitable for storage on the disc.

Not only are the EFM and CIRC coding schemes useful on the CD for which they were designed, but also they are employed extensively in a large variety of digital audio players and home-storage products such as the CD-ROM, the

CD-I, and the MiniDisc. As in the CD-ROM, the user data is organized into sectors. Sixteen sectors make up one error-correction code (ECC) block.

Under the DVD format rules, sectors of 2048 byte user-data are translated into 2418 bytes (2048 byte user data + 52 sync bytes + 16 header bytes+ 302 parity-check bytes), which, in turn, are translated into 16 x 2418 channel bits. Thus, one user bit is translated into 4836/2048 = 2.36 channel bits. In the conventional CD, an audio bit is translated into 588/192 = 3.05 channel bits, and one concludes that the "format efficiency" of the DVD is improved over the conventional CD by 32%. The new format is more efficient by 47% over the CD-ROM, which uses a "third" error-correction layer. Even though it has a significantly lower data overhead, the error correction code, RS-PC, can cope with longer bursts and burst sequences, and as a consequence with more random errors.

The errors found in both the CD and DVD systems are a combination of both random and burst errors, and in order to simplify the manufacturing process some form of error correction is required. As mentioned earlier. the CD system uses the CIRC ECC. The interleaving scheme is tailored to the specific requirements of the compact-disc audio system. In particular, the adopted, cross-interleaving, technique makes it possible to effectively mask errors if correction is found to be impossible. Depending on the number and magnitudes of the errors to be concealed, this is accomplished by interpolation or by muting the audio signal. Concealment makes errors almost inaudible, and as a result, it allows for a graceful degradation of the sound quality. Specifically, sharp temporary degradation of the audio signal, or the "clicks," are avoided. The judicious positioning of the left and right stereo channels, as well as the audio samples on the even- or odd-number instants within the interleaving scheme, are key parameters to the success of the concealment strategy.

For the DVD a much more powerful error-correction capability needed to be developed. An enhanced error-correction capability in the DVD is required for several reasons. First, the increased physical density implies that physical imperfections affect proportionally more bits. It is further anticipated, since the system margins are much tighter, that the random error rate of the DVD is larger than that of the conventional CD. Secondly, since concealment requirements are no longer the same as the those used in the audio CD, the reliability of the decoded data must be much higher. Since the DVD is a true multimedia disc, its data integrity must be comparable to that used for computer data.

The Reed-Solomon Product Code (RS-PC), like the CIRC, uses a combination of two RS codes denoted by C_1 and C_2. In CIRC, C_1 is (32,28), and C_2 is a (28,24) code. where (n,k) denotes a code with k input and n output bytes. The redundant n-k bytes are generated under the rules used for the RS code. The rate of CRC is $24/32 = 3/4$. In other words, under CIRC rules one redundant byte is added for every three user bytes. The RS-PC code is a classical product code, whereas in the CD, the two codes cooperate. A product code is also part of the format for the digital audio-tape (DAT) recorder.

The two codes of the RS-PC can be imagined to form the two dimensions of a rectangle. In the RS-PC the C_1 and C_2 codes are significantly longer than in CIRC They are the (208,192) and the (182,172) RS codes. As a result the rate of an RS-PC is also much higher than that of CIRC, namely $172*192/(182*208) = 0.872$. Sixteen sectors of the 2,048 bytes of user data make one ECO block.

Figures 7.4 and 7.6, respectively, depict schematically the principles of operation of the CIRC and the RS-PC codes. In the RS-PC, the C_1 and C_2 codes are represented by the rows and columns of a matrix. The codes C_1 and C_2 are combined in such a way in the RS-PC that the parities, which are generated by C_1, are also taken into account in code C_2. The "checks-on-checks" of the RS-PC have the advantage that the error-correcting capability of the codes in tandem is improved over that of the CIRC code, where this "double" check is absent.

The structure of the RS-PC code cures another difficulty of the CIRC code, namely, its convolutional structure. In a cross-interleaved structure the data are formed into an endless array and the codewords are produced on columns and diagonals as shown in Fig. 7.4. Block structures, such as in the RS-PC, are much better adapted to that part of the data world which operates on small segments of information. The CIRC structure of interleaving is designed specifically for very long, non-blocked data segments. such as digital audio or video.

The cross-interleaved structures are found most often in the professional audio and video recorder formats. The disadvantage of the product code structure relative to the CIRC structure is its requirement of twice the memory capacity. However, the falling prices of solid-state devices makes this is a less important requirement than in 1979 when the CD was designed.

The maximum correctable-burst length is approximately 500 bytes (2.4 mm) for the CIRC, while it is 2200 bytes (4.6 mm) for the RS-PC. The RS-PC is capable of reducing a random input-error rate of 2×10^{-2} to a data-error rate of 10^{-15}[16], which is a factor of 10 better than that for the CD.

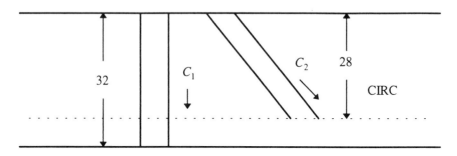

Figure 7.4 Block diagram of CIRC encoder

The "data sector" on a DVD consists of 2048 bytes of Main Data, 12 bytes of identification data (ID) and 2 bytes for the ID Error-Detection code(IED), 6 bytes of Copyrighted Management Information (CPR_MAI), and 4 bytes for the Error-Detection Code (EDC). As shown in Figure 7.5, the data sector consists of 2064 bytes, i.e. 172 bytes x 12 lines. The Main Data are scrambled by the use of a feedback bit-shift-register.

4 bytes ID	2 bytes IED	6 bytes CPR_MAI	Main Data 160 bytes ($D_0 \sim D_{159}$)
	Main Data 172 bytes ($D_{160} \sim D_{331}$)
(12 rows)	:
	Main Data 172 bytes ($D_{1708} \sim D_{1879}$)
	Main Data 168 bytes ($D_{1880} \sim D_{2047}$) 4 bytes EDC

Figure 7.5 Data sector configuration

The message block needed for RS coding is formed with the scrambled 16 data sectors as shown in Figure 7.6. 172 bytes x 192 rows, which is equivalent to 172 bytes x 12 rows x 16 data sectors, is the total message field. The RS encoding process is performed as follows : The 16 bytes of the outer-code parity (PO) are attached first to each column of the 172 columns to form the (208,192) RS outer-code over GF(256). Next, 10 bytes of the inner-code parity (PI) are appended to all 208 rows which include the PO to form the (182,172) RS inner-code over GF(256).

Implementation Architectures and Applications of RS Codes

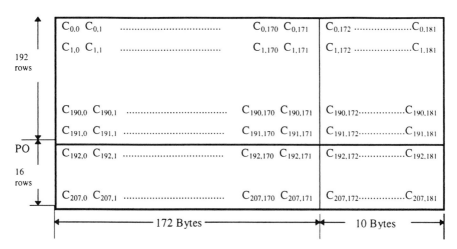

Figure 7.6 DVD RS Code Configuration

To generate the PO and PI of Figure 7.6, the following encoding procedure is performed : First, in column j (for j=0 to 171) attach 16 bytes of $b_{i,j}$ (for i=192 to 207). These are defined by the following remainder polynomial, $r_j(x)$ to form 172 columns of the outer-code RS (208,192), namely

$$r_j(x) = \sum_{i=192}^{207} b_{i,j} x^{207-i}$$
$$= q_j(x) \cdot x^{16} \mod g_{PO}(x),$$

where

$$q_j(x) = \sum_{i=0}^{191} b_{i,j} x^{191-i}, \quad g_{PO}(x) = \prod_{k=0}^{15}(x+\alpha^k).$$

Next, in row i ($i = 0$ to 207), attach 10 bytes of $b_{i,j}$ ($j = 172$ to 181). These are defined by the following remainder polynomial $r_i(x)$ to form 208 rows of the inner-code RS (182,172),

$$r_i(x) = \sum_{j=172}^{181} b_{i,j} x^{181-j} = q_i(x) x^{10} \mod g_{PI}(x)$$

where

$$q_i(x) = \sum_{j=0}^{171} b_{i,j} x^{171-j}, \quad g_{PI}(x) = \prod_{k=0}^{9}(x+\alpha^k),$$

Here α represents the primitive root of the primitive polynomial,
$p(x) = x^8 + x^4 + x^3 + x^2 + 1.$

We know from Chapter 3 that these two RS inner-and outer- codes form a direct-product code. The outer-code RS (208,192) has a minimum distance 17 and is able to correct up to 8 bytes of errors, while the inner-code RS (182,172) has a minimum distance 11 and is able to correct up to 5 bytes of errors. The resulting direct-product code is a two-dimensional (208x182,192x172) code whose minimum distance is 17x11=187 bytes. Therefore, this product code is capable of correcting (17x11-1)/2=93 byte random errors. Notice also that 8 bytes and 5 bytes are, respectively, the burst-error-correcting capabilities of the RS codes (208,192) and (182,172). It is shown in Chapter 3 that the burst-error-correcting capability b of this product code satisfies : $b \geq \max\{208 \times 5, 182 \times 8\} = 1465$ bytes = 11720 bits.

Bibliography

[1] Sun HyperSparc technical specification, 1993.
[2] W. B. Weeks, "Comparison of the Motorola DSP56000 and the Texas Instruments TMS320C25 digital signal processors for implementing a four error correcting (127,99) BCH error control code decoder", IEEE Pacific Rim Conference on communications, computers, and signal processing, June 1st-2nd, pp.345-346, 1989.
[3] M. D. Yucel, "Implementation of a real time Reed Solomon encoder and decoder using TMS32010", IEEE Pacific Rim Conference on communications, computers, and signal processing, June 1st-2nd, pp.341-344, 1989.
[4] C. Parr, "Implementation of a re-programmable Reed Solomon decoder over $GF(2^{16})$ on a digital signal processor with external arithmetic unit", Fourth Int'l European Space Agency Workshop on Digital Signal Processing Techniques Applied to Space Communication, London, Sept. 1994.
[5] C. Caren, b. Benjamin, J. Boddie, and M. Fuccio, "A 60 ns CMOS DSP with On-Chip Instruction Cache", 1987 ISSCC Digest of technical papers, pp. 156-157, Feb. 1987.
[6] S.T.J. Fenn, N. Benaissa, and D. Taylor. "Rapid Prototyping of Reed-Solomon Codecs using Field Programmable Gate Arrays" . Proceedings of the 2nd International Conference on Concurrent Engineering and Electronic Design Automation, pp. 309-312, April 1994,
[7] Viewlogic user manual
[8] Xact manual, 1995.
[9] Recommendation for Space Data System Standards : Telemetry Channel Coding, Issue 1, Washington, D.C., May 1984.

[10] Consultative Committee for Space Data Systems, Blue Book CCSDS 101.0-B-2, Washington, D.C., NASA, 1987.
[11] Consultative Committee for Space Data Systems, Blue Book CCSDS 101.0-B-3, Washington, D.C., NASA, 1992.
[12] E. R. Berlekamp, "Bit serial Reed-Solomon encoder", IEEE Trans. on Inform. Theory, Vol. IT-28, pp.869-874, 1982.
[13] M. Perman and J. -J. Lee, Reed-Solomon encoders -- conventional vs. Berlekamp's architecture, NASA JPL Publication 82-71, Nov. 1982.
[14] I. S. Reed, L. J. Deutsch, I. S. Hsu, T. K. Truong, K. Wang, and C. S. Yeh, "The VLSI implementation of a Reed-Solomon encoder using Berlekamp's bit-serial multiplier algorithm", IEEE Trans. on Computers, Vol. C-33, No. 10, Oct. 1984.
[15] L. B. Vries and K. Odaka, "CIRC---The Error-Correcting Code for the Compact Disk Digital Audio System," Collected Papers AES Premiere Conference, Rye, New York, pp.178-188, New York : Audio Engineering Society, 1982.
[16] K. A. S. Immink, "The digital versatile disc (DVD) : system requirements and channel coding", SMPTE Journal, pp. 483-489, August 1996.
[17] H. M. Shao and I. S. Reed, "On the VLSI design of a pipeline Reed-Solomon decoder using systolic arrays", IEEE Trans. on computers, vol. C-37, pp.1273-1280, 1988.
[18] H. M. Shao, T. K. Truong, L. J. Deutsch, J. H. Yuen, and I. S. Reed, "A VLSI design of a pipeline Reed-Solomon decoder", IEEE Trans. on computers, vol. C-34, pp.393-403, 1985.
[20] H. B. Bakoglu, Circuit Interconnections and Packaging for VLSI, Reading, MA : Addison Wesley, 1987.
[22] N. Weste and K. Eshragion, Principles of CMOS VLSI Design, Reading, MA : Addison Wesley, 1988.
[23] Reed Solomon C Code Web Page, "http://hideki.iis.u-tokyo.ac.jp /robert/rs.c.
[24] Heather Bowers and Hui Zhang, "Comparison of Reed Solomon Codec Implementations" http://infopad.eecs.berkeley.edu /~hbowers /cs252 /rs.html.
[25] S. Whitaker, K. Cameron, G. Maki, J. Canaris, and P. Owsley, "VLSI Reed-Solomon processor for the Hubble space telescope," VLSI Signal Processing IV, Chapter. 35, IEEE Press, 1991.
[26] S. Whitaker, J. Canaris, and K. Cameron, "Reed Solomon VLSI codec for advanced television", IEEE Trans. on Circuits and Systems for Video Technology, vol.1, No.2, June 1991.
[27] G. Maki, P. Owsley, K. Cameron, and J. Venbrux, "VLSI Reed-Solomon decoder design", IEEE Military Communications Conf. Rec., pp.46.5.1-46.5.6, Oct. 1986.

8 Fundamentals of Convolutional Codes

Block codes were the earliest type of codes to be investigated and still remain the subject of the overwhelming bulk of modern coding research. On the other hand, convolutional codes have proved to be equal or superior to block codes in performance in many practical applications and are generally simpler than comparable block codes to implement. In the coding of block codes, each of the $n-k$ redundancy (or parity-check) symbols in a codeword depends only on the corresponding k information symbols of the codeword and not on any other information symbols. This means that the encoder of block codes is memoryless.

Convolutional codes differ from block codes in that the encoder contains memory and the n_0 outputs at any given time unit depend not only on the k_0 inputs of that time unit but also on m previous input words, in fact, on $m \cdot k_0$ information bits. The encoding of an (n_0, k_0, m) convolutional code can be implemented with a k_0 – input and n_0 – output linear sequential circuit with an input memory of m words. Also, the decoding of k_0 information words of an (n_0, k_0, m) convolutional code often needs to use $\hat{m}(\geq m)$ previously decoded information words. Thus the decoder needs digital devices with memory. Typically, k_0 and n_0 satisfy $k_0 < n_0$ and are smaller than the dimension k and length n of the most commonly used block codes. But the memory order m is made sufficiently large in order to obtain a small probability of error. For many applications convolutional codes can achieve the required performance for some desired information rate and also have a low complexity decoder.

Convolutional codes were introduced first by Elias [1] in 1955 as an alternative to block codes. The advances in the various decoding algorithms for convolutional codes proposed by Wozencraft, Fano, Viterbi, etc., spawned many practical applications of convolutional codes to digital transmissions over both wire and wireless channels. Several architectures for convolutional codes used for practical communication channels are developed in Chapter 10.

In this chapter binary convolutional codes are introduced as a subclass of the linear codes. The aim here is to provide some of the background material needed to understand the forthcoming subjects covered in the rest of this book. Starting with the basic definitions, several different encoder structures for convolutional codes are studied. Following this, the distance structure of convolutional codes is developed. To make this possible, generating functions are introduced in order to be able to determine the distance properties of these codes. Then the Viterbi algorithm, a maximum likelihood decoding technique, is derived for the convolutional codes. Also the soft decision version of Viterbi decoding is explained. Finally, upper bounds on the bit-error probability of the maximum-likelihood decoding of convolutional codes are obtained by a use of generating functions.

8.1 Convolutional Encoder

Time-Domain Description

A simple systematic, rate-$\frac{1}{2}$, binary convolutional encoder of constraint length 2 is shown in Figure 8.1(a). The input to this encoder is some binary message sequence, $\bar{m} = (\cdots, m_{-1}, m_0, m_1, \cdots)$. The output is an interleaved sequence $\bar{c} = (\cdots, c_{-1}^{(1)}, c_{-1}^{(2)}, c_0^{(1)}, c_0^{(2)}, c_1^{(1)}, c_1^{(2)}, \cdots)$ of the two binary sequences $\bar{c}^{(1)}$ and $\bar{c}^{(2)}$. Hence the rate is 1/2. The first output sequence $\bar{c}^{(1)}$ is equal simply to the input sequence $\bar{m}$. Thus in the output sequence the message is in the "clear" so that the code is called systematic. The elements $c_i^{(2)}$ of the second sequence $\bar{c}^{(2)}$ are given by the encoding equation,

$$c_i^{(2)} = m_i \oplus m_{i-1} \oplus m_{i-2}, \tag{8.1}$$

where $\oplus$ denotes modulo 2 addition (or the XOR operation). Therefore, the output bits $c_i^{(2)}$ depend not only on the present input bit m_i, but also on the two previous input bits m_{i-1} and m_{i-2}, stored in the shift register. As a consequence we say here for this encoder that the constraint length is 2.

Fundamentals of Convolutional Codes

Finally as defined above the output bits are multiplexed as the sequence, $\bar{c} = (\ldots\ldots,c_0^{(1)}c_0^{(2)},c_1^{(1)}c_1^{(2)},\ldots\ldots)$, called the codeword (or code sequence).

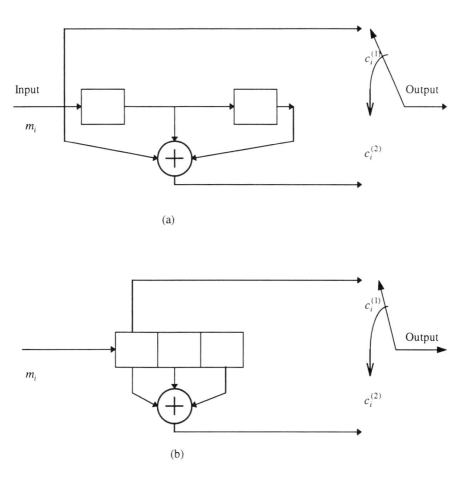

Figure 8.1 (a) A 1/2-rate convolutional code with m=2.
(b) An alternate representation.

In line with automata theory, some authors draw the encoder as in Figure 8.1(b) with the output a function of the memory contents only. Therefore, this code sometimes is said to have a constraint length 3, since the outputs are a function of 3 consecutive input bits. Another measure of constraint length, which is used even less frequently, is to assign a constraint length of $3 \times 2 = 6$ in terms of the output bits of this code.

Evidently, the encoding process for this code is described by a convolution operation in the time domain of the information bits with certain generator sequences over the finite field GF(2), i.e. $\bar{c}^{(1)} = \bar{m} * \bar{g}_1$, $\bar{c}^{(2)} = \bar{m} * \bar{g}_2$, where * denotes the discrete convolution operation over $GF(2)$ and $\bar{g}_1 = (g_{1,0} \ g_{1,1} \ g_{1,2}) = (1 \ 0 \ 0)$, $\bar{g}_2 = (g_{2,0} \ g_{2,1} \ g_{2,2}) = (1 \ 1 \ 1)$ are called the generator sequences of the code. The term, convolutional code, comes from such an observation.

Example 8.1
Let the input sequence be specified by $m_i = 0, i < 0 \text{ and } i > 4$, and $m_0 = m_2 = m_3 = m_4 = 1, m_1 = 0$. Then the output sequences are

$$\bar{c}^{(1)} = (\cdots\cdots \underset{m_{-1}}{0} \ \underset{m_0}{1} \ \underset{m_1}{0} \ \underset{m_2}{1} \ \underset{m_3}{1} \ \underset{m_4}{1} \ \underset{m_5}{0} \cdots\cdots) * (\underset{g_{1,0}}{1} \ \underset{g_{1,1}}{0} \ \underset{g_{1,2}}{0})$$

$$= (\cdots\cdots \underset{c^{(1)}_{-1}}{0} \ \underset{c^{(1)}_0}{1} \ \underset{c^{(1)}_1}{0} \ \underset{c^{(1)}_2}{1} \ \underset{c^{(1)}_3}{1} \ \underset{c^{(1)}_4}{1} \ \underset{c^{(1)}_5}{0} \cdots\cdots)$$

$$\bar{c}^{(2)} = (\cdots\cdots \underset{m_{-1}}{0} \ \underset{m_0}{1} \ \underset{m_1}{0} \ \underset{m_2}{1} \ \underset{m_3}{1} \ \underset{m_4}{1} \ \underset{m_5}{0} \cdots\cdots) * (\underset{g_{2,0}}{1} \ \underset{g_{2,1}}{1} \ \underset{g_{2,2}}{1})$$

$$= (\cdots\cdots \underset{c^{(2)}_{-1}}{0} \ \underset{c^{(2)}_0}{1} \ \underset{c^{(2)}_1}{1} \ \underset{c^{(2)}_2}{0} \ \underset{c^{(2)}_3}{0} \ \underset{c^{(2)}_4}{1} \ \underset{c^{(2)}_5}{0} \ \underset{c^{(2)}_6}{1} \ \underset{c^{(2)}_7}{0} \cdots\cdots),$$

and the final codeword sequence or more simply the codeword is

$$\bar{c} = (\cdots, \underset{c^{(1)}_{-1}}{0} \ \underset{c^{(2)}_{-1}}{0}, \underset{c^{(1)}_0}{1} \ \underset{c^{(2)}_0}{1}, \underset{c^{(1)}_1}{0} \ \underset{c^{(2)}_1}{1}, \underset{c^{(1)}_2}{1} \ \underset{c^{(2)}_2}{0}, \underset{c^{(1)}_3}{1} \ \underset{c^{(2)}_3}{0},$$
$$\underset{c^{(1)}_4}{1} \ \underset{c^{(2)}_4}{1}, \underset{c^{(1)}_5}{0} \ \underset{c^{(2)}_5}{0}, \underset{c^{(1)}_6}{0} \ \underset{c^{(2)}_6}{1}, \underset{c^{(1)}_7}{0} \ \underset{c^{(2)}_7}{0}, \cdots).$$

Transform-Domain Description
The encoder of a convolutional code can be described also by relating the input and output sequences in a transform-domain over GF(2). To accomplish this let the input and output sequences be expressed by

$$m(D) = \sum_{i=-\infty}^{\infty} m_i D^i, \quad c^{(1)}(D) = \sum_{i=-\infty}^{\infty} c^{(1)}_i D^i, \quad c^{(2)}(D) = \sum_{i=-\infty}^{\infty} c^{(2)}_i D^i, \quad (8.2)$$

where the indeterminate D can be interpreted as the unit delay operator. Then the input/output relationships are expressed concisely by

$$c^{(1)}(D) = m(D) \cdot G_1(D), \quad c^{(2)}(D) = m(D) \cdot G_2(D), \quad (8.3)$$

where

$$G_1(D) = \sum_{i=-\infty}^{\infty} g_{1,i} D^i = 1, \quad G_2(D) = \sum_{i=-\infty}^{\infty} g_{2,i} D^i = 1 + D + D^2$$

Fundamentals of Convolutional Codes

are called the *generator polynomials* of the convolutional code. After a multiplexing of the outputs $c^{(1)}(D)$ and $c^{(2)}(D)$, the output codeword polynomial becomes

$$c(D) = c^{(1)}(D^2) + D \cdot c^{(2)}(D^2). \tag{8.4}$$

The generator polynomial of a convolutional code can be determined directly from its encoder structure. On the assumption that each shift-register stage represents a unit-time delay the coefficients of the generator polynomial, associated with a particular output, are obtained simply by looking at the connections in the outputs of the shift-register stages. A coefficient equals one if there exists a connection and zero otherwise. This fact is shown in Figure 8.1.

Example 8.2
For the (2,1,2) code of Figure 8.1, the generator polynomials are $G_1(D) = 1$ and $G_2(D) = 1 + D + D^2$. For the input sequence $m(D) = 1 + D^2 + D^3 + D^4$, the encoding equations in the transform domain are

$$c^{(1)}(D) = (1 + D^2 + D^3 + D^4) \cdot 1 = m(D)$$

$$c^{(2)}(D) = (1 + D^2 + D^3 + D^4) \cdot (1 + D + D^2) = 1 + D + D^4 + D^6,$$

and the multiplexed codeword polynomial is

$$c(D) = 1 + D + D^3 + D^4 + D^6 + D^8 + D^9 + D^{13}.$$

Note that these results are the same as those previously computed using the convolutions in Example 8.1.

The convolutional code specified in Figure 8.1 is called a binary (2, 1, 2) code since the encoder has two output sequences, i.e. $n_0 = 2$, one input sequence ($k_0 = 1$) and an $m = 2$ stage shift register. The rate of this code is $\frac{k_0}{n_0} = \frac{1}{2}$.

More generally a general (n_0, k_0, m) convolutional encoder can be defined by a set of generator polynomials with coefficients in some finite field F. These polynomials can be represented in the form of a $k_0 \times n_0$ matrix as follows:

$$G(D) = \begin{pmatrix} G_{11}(D) & G_{12}(D) & \cdots & G_{1n_0}(D) \\ G_{21}(D) & G_{22}(D) & \cdots & G_{2n_0}(D) \\ \cdots & \cdots & \cdots & \cdots \\ G_{k_01}(D) & G_{k_02}(D) & \cdots & G_{k_0n_0}(D) \end{pmatrix}. \tag{8.5}$$

There are k_0 – input sequences, $\bar{m}(D) = (m^{(1)}(D), m^{(2)}(D), \ldots, m^{(k_0)}(D))$, and n_0 – output sequences, $\bar{c}(D) = (c^{(1)}(D), c^{(2)}(D), \ldots, c^{(n_0)}(D))$. Hence, the rate of the code is $\frac{k_0}{n_0}$. The encoding process is described very compactly by the vector-matrix product,

$$\bar{c}(D) = \bar{m}(D) \cdot G(D), \qquad (8.6)$$

again with all operations over the finite field F. If the constraint length for the i-th input sequence is defined by

$$v_i = \max_{1 \leq j \leq n_0} \left[\deg G_{ij}(D) \right], \qquad (8.7)$$

then the general convolutional encoder can be realized by k_0 shift registers with the i-th shift register of length v_i. The outputs are linear combinations over F of the appropriate shift-register outputs. If the first k_0 columns of the generating matrix $G(D)$ form an identity matrix, then the specified code is a systematic code. Note that the required number of memory elements is equal to the overall constraint length, defined by the sum of the k_0 constraint lengths, i.e.

$$v = \sum_{i=1}^{k_0} v_i. \qquad (8.8)$$

8.2 State and Trellis-Diagram Description of Convolutional Codes

State Diagram

The encoders discussed in the previous section are linear sequential circuits. Actually, all finite-state time-invariant linear causal sequential circuits are candidates for convolutional encoders. Thus, a convolutional encoder can be specified completely by the state diagram in the same manner that any finite-state, time-invariant, linear sequential circuit is defined.

Example 8.3
Consider the encoder of the (2,1,2) convolutional code given in Figure 8.1(a). The state of this encoder is the content of the 2-stage shift register. At the time unit i the state is expressed by (m_{i-1}, m_{i-2}) for input m_i. The next state for the time unit $i+1$ is (m_i, m_{i-1}). Therefore, the input, output and state relationships of this encoder are specified by the entries in the state-behavior Table 8.1.

Fundamentals of Convolutional Codes

Original State	Current Input m_i	Output $c_i^{(1)}$ $c_i^{(2)}$	New State
(0 0)	0	0 0	(0 0)
(1 0)	0	0 1	(0 1)
(0 1)	0	0 1	(0 0)
(1 1)	0	0 0	(0 1)
(0 0)	1	1 1	(1 0)
(1 0)	1	1 0	(1 1)
(0 1)	1	1 0	(1 0)
(1 1)	1	1 1	(1 1)

Table 8.1 Input, output and state table of the (2,1,2) convolutional encoder

In Table 9.1 an output and a new state are uniquely determined by the original state and the input. If the four states are denoted by $S_0 = (0\ 0), S_1 = (1\ 0), S_2 = (0\ 1)$ and $S_3 = (1\ 1)$, the operation of this (2,1,2) convolutional encoder can be described by the state diagram shown in Figure 8.2. In such a state diagram, the labels on the branches denote *input / output*. In this example of a convolutional encoder, there are 2 branches leaving and entering each state.

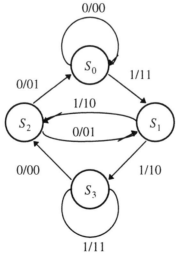

Figure 8.2. State diagram of a (2,1,2) convolutional encoder

For a general binary convolutional code, the state of the encoder is defined above to be the ordered set of binary coefficients of the shift register. Assume that such a convolutional code has an overall constraint length v. Then, there are 2^v states in the state diagram. Each state is assigned a number from 0 to 2^{v-1} which coincides with the binary representation of the content of the shift register. The connecting lines in the state diagram show the state transitions and their associated outputs for all possible inputs. For a rate k_0/n_0 convolutional encoder, there are 2^{k_0} branches (transitions) leaving a state, and 2^{k_0} branches entering each state. Each new sequence of k input bits causes a transition from an initial state to a new state. The encoder output is determined uniquely by the original state and the current input.

Trellis Diagram

To show the time evolution of the coded sequence, it is traditional and instructive to describe all possible transitions of a convolutional code by means of, what is called, a trellis diagram. Such a diagram is obtained directly from the state diagram by including, from left-to-right, the dimension of time.

Example 8.4.

The trellis diagram, corresponding with the state diagram in Figure 8.2, is shown in Figure 8.3.

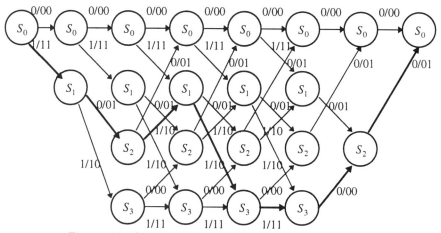

Figure 8.3. diagram of a (2,1,2) convolutional encoder

Fundamentals of Convolutional Codes

In drawing the trellis diagram of Figure 8.3 the same convention used in the state diagram is used also to specify the branches introduced by an input bit "0" and the branches introduced by an input bit "1". Once again the output sequence for the input sequence 10111 (with initial state S_0) is obtained by traveling from state S_0 through the trellis diagram in accordance with the input bits. Such a traversed path is shown as a bold line in Figure 8.3. The output sequence corresponding to this path is 11011010110001 which agrees with the previous example.

One observation is that, for an input sequence of length $L=5$, there are 2^L possible codewords which correspond with the 2^L distinct trellis paths. In the general case of an (n_0, k_0, m) code and an information sequence of length $k_0 L$, there are 2^{k_0} branches leaving and entering each state, and $2^{k_0 L}$ distinct paths through the trellis corresponding with the $2^{k_0 L}$ codewords.

8.3 Nonsystematic Encoder and Its Systematic Feedback Encoder

Nonsystematic Feedforward-only Encoders and Error Propagation

The convolutional code discussed in Examples 8.1 to 8.4 is a systematic code with a feedforward encoder structure (no feedback loops in the encoder). This class of codes constitutes a subset of all convolutional codes. Systematic feedforward-convolutional codes usually do not perform as well as the nonsystematic codes. In fact it is shown in [2] by maximum-likelihood decoding on a memoryless channel that the error probability of a random ensemble of feedforward systematic codes is greater than the general ensemble of the convolutional codes with the same constraint length. Hence one does not want to confine one's attention only to feedforward systematic codes, even though these codes provide the simplest encoding structures.

However, non-systematic, feedforward-only, convolutional codes are not used in most applications. The main reason for this seems to be the fear of uncontrollable error propagation. Consider the following example of a non-systematic code with catastrophic error propagation.

Example 8.5
A simple (2,1,2) non-systematic, feedforward, convolutional code is specified by the following polynomials :

$G_1(D) = 1 + D,$

$G_2(D) = 1 + D^2.$

The encoder is implemented by the shift-register structure shown in Figure 8.4.

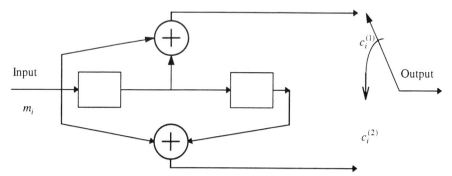

Figure 8. 4. A 1/2 rate non-systematic convolutional code

If the input sequence is an infinitely long sequence of ones, i.e.
$$\bar{m} = (\cdots\cdots \underset{m_{-1}}{0} \ \underset{m_0}{1} \ \underset{m_1}{1} \ \underset{m_2}{1} \ \underset{m_3}{1} \ \underset{m_4}{1} \ \underset{m_5}{1} \cdots\cdots).$$

Then, $\bar{m}(D) = 1 + D + D^2 + D^3 + D^4 + D^5 + \cdots\cdots = \dfrac{1}{1+D}.$ In this case the encoding equations in the transform domain are given by

$$c^{(1)}(D) = \frac{1}{1+D} \cdot (1+D) = 1,$$

$$c^{(2)}(D) = \frac{1}{1+D} \cdot (1+D^2) = 1+D,$$

where the equality, $(1+D)^2 = 1 + D^2$ mod 2, is used. Evidently, for this input sequence the codeword polynomial is given by

$$c(D) = 1 + D + D^3.$$

If this code is transmitted over a BSC and the three first non-zero bits in the codeword are changed, say by errors, to zeroes, the received sequence becomes the all-zero sequence. Any type of decoder would decode such an all "0" received sequence as $\hat{m}(D) = 0$. Therefore, in this case a finite number of channel errors, namely the above change of the 3 non-zero bits to 0's, cause an infinite number of decoded bit errors. Such an effect is called catastrophic error propagation.

Massey and Dain [3] show that a necessary and sufficient condition for a rate k/n convolutional code to be noncatastrophic is that the relation,

Fundamentals of Convolutional Codes

$$\alpha(D) \stackrel{\Delta}{=} GCD\left[\det_1(D), \det_2(D), \ldots, \det_{\binom{n_0}{k_0}}(D)\right] = D^j, \quad (8.9)$$

hold for some $j \geq 0$, where $\det_i(D)$ for $i = 1, 2, \ldots, \binom{n_0}{k_0}$ denotes the determinants of all $\binom{n_0}{k_0}$ distinct $k_0 \times k_0$ submatrices of the generator polynomial matrix of the code and GCD denotes the operation of greatest common divisor. The fact that the feedforward systematic codes are always noncatastrophic follows from the above relation since for such codes $\alpha(D) = 1 = D^0$.

Example 8.6.
It is verified readily by using the generator polynomials in Example 8.2 that $\alpha(D) = GCD(1, 1+D+D^2) = 1 = D^0$. Hence, the (2,1,2) convolutional code specified in Figure 8.1(a) is noncatastrophic. However, Example 8.5 yields $\alpha(D) = GCD(1+D, 1+D^2) = 1+D \neq D^j$ for any j. Therefore, the code specified in Example 8.5 is catastrophic by the above criterion for a convolutional code to be noncatastrophic.

Feedback Systematic Encoders
Forney [4] demonstrated for any noncatastrophic convolutional code that there is an equivalent systematic code with feedback which generates the same set of codewords. The following example represents the construction of such a systematic feedback encoder that is equivalent to a nonsystematic feedforward encoder.

Example 8.7.
Consider the nonsystematic encoder shown in Figure 8.5. The generator polynomial matrix of this code is,

$$G(D) = \begin{bmatrix} 1+D^3+D^4, & 0, & D \\ 1+D^3+D^4, & 1+D^3+D^4, & D+D^2+D^3 \end{bmatrix}.$$

It is easy to verify by the Massey-Dain criterion that this code is non-catastrophic. Also by the use of a sequence of elementary row operations, including the division by polynomials, for matrices of polynomials, this generator polynomial matrix can be put into a systematic-form matrix $G_s(D)$ as follows:

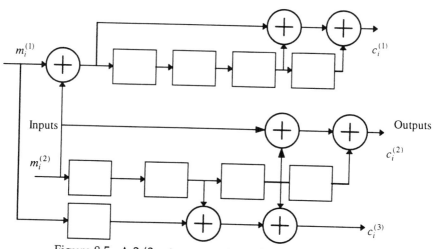

Figure 8.5. A 2/3 rate non-systematic convolutional code.

$$G(D) = \begin{bmatrix} 1+D^3+D^4 & 0 & D \\ 1+D^3+D^4 & 1+D^3+D^4 & D+D^2+D^3 \end{bmatrix}$$

$$\rightarrow \begin{bmatrix} 1+D^3+D^4 & 0 & D \\ 0 & 1+D^3+D^4 & D^2+D^3 \end{bmatrix}$$

$$\rightarrow \begin{bmatrix} 1 & 0 & \dfrac{D}{1+D^3+D^4} \\ 0 & 1 & \dfrac{D^2+D^3}{1+D^3+D^4} \end{bmatrix} \rightarrow G_s(D) = \begin{bmatrix} 1 & 0 & \dfrac{D}{1+D^3+D^4} \\ 0 & 1 & \dfrac{D+D^2}{1+D^3+D^4} \end{bmatrix}.$$

The polynomial divisions in the last column of the final generator polynomial matrix are realized by feedback structures. To accomplish this the output sequences of the encoder for this example are expressed in terms of its input sequences $m^{(1)}(D)$ and $m^{(2)}(D)$ as follows:

$$c^{(1)}(D) = m^{(1)}(D),$$
$$c^{(2)}(D) = m^{(2)}(D),$$
$$c^{(3)}(D) = \frac{m^{(1)}(D)D + m^{(2)}(D) \cdot (D^2+D^3)}{1+D^3+D^4}.$$

The last equation is transformed now into the following form:

$$c^{(3)}(D) = D\Big(m^{(1)}(D) + D\big((m^{(2)}(D) + D\big(m^{(2)}(D) + c^{(3)}(D) + Dc^{(3)}(D)\big)\big)\Big),$$

which is a difference equation which can be implemented by a feedback structure. The feedback realization of this equivalent systematic code is

Fundamentals of Convolutional Codes

shown in Figure 8.6. It is straightforward to verify that this encoder generates the same set of codewords which are generated by the encoders shown in Figures 8.5 and 8.6.

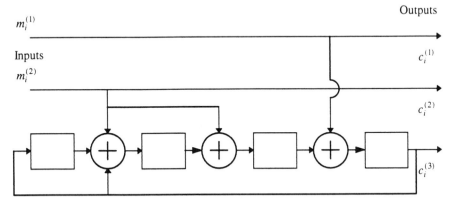

Figure 8.6. A feedback realization of the 2/3 rate systematic encoder of nonsystematic encoder in Figure 8.5.

In general, the generator polynomial matrix of an equivalent systematic feedback encoder can always be obtained by left-multiplying the latter by the appropriate scrambler matrix. That is,

$$G_s(D) = A \cdot G(D),$$

where A is a $k \times k$ matrix of rational functions of the delay operator and $G_s(D)$ denotes the generator polynomial matrix of the equivalent systematic encoder.

Example 8.8.
The generator matrix $G_s(D)$ in Example 8.7 is obtained next by the following multiplication of $G(D)$ and an A-matrix :

$$G_s(D) = \begin{bmatrix} A_{11} & A_{12} \\ A_{21} & A_{22} \end{bmatrix} \cdot G(D),$$

where $G(D)$ is the generator matrix in Example 9.6 and

$$A_{11} = \frac{1}{D+D^2+D^3}, A_{12} = 0, A_{21} = \frac{1}{1+D^3}, A_{22} = \frac{1}{1+D^3}.$$

Systematic codes do possess some advantages over nonsystematic codes. It is seen from the examples given above that the systematic encoders require fewer connections from the shift register to the modulo-2 adders than do the nonsystematic encoders. Also no inverter circuits are needed to recover the information sequence from the code word of a systematic code. The latter property sometimes allows the user to recover the information sequence at the received information sequence prior to any decoding without having to invert the code word. This can be an important advantage in systems where decoding is done off-line, wherein the decoder is subject to temporary failures, or where the channel is known to be "noiseless" during certain time intervals in which decoding becomes unnecessary.

8.4 Distance Properties of Convolutional Codes

The Free Hamming Distance

The performance of a convolutional code depends on the decoding algorithm employed and the distance measures of the code. We begin this section by analyzing the distance properties of convolutional codes. There are several distance measures for convolutional codes. For a BSC the Hamming-distance properties of the code need to be considered. Since convolutional codes are a subclass of all linear codes, the set of Hamming distances between coded sequences equals the set of distances of the coded sequences from the all-zero sequence. Thus, in the following discussion the distance structure of convolutional codes is obtained from the distance of the code to the all-zero sequence.

The set of distances of a convolutional code can be computed by the aid of the code trellis or state diagram.

Example 8.9.
Consider again the trellis diagram of the rate 1/2 convolutional code of Figure 8.3. To find the distance structure of the code, given in Figure 8.3, its trellis diagram is redrawn in Figure 8.7. by labeling the branches with their Hamming weights and distances from the all-zero code sequence.

To find the set of distances for this convolutional code one considers the paths in Figure 8.7 which diverge from S_0 and re-merge into it for the first time at some later time. Examples of such paths are $S_0 S_1 S_2 S_0, S_0 S_1 S_3 S_2 S_0$, and $S_0 S_1 S_3 S_3 S_2 S_0$ with Hamming distances of 4, 4, and 7 from the all-zero sequence, respectively. The input sequence with these three paths are $\cdots 100 \cdots$ (Hamming weight 1), $\cdots 1100 \cdots$ (Hamming weight 2), and $\cdots 11100 \cdots$ (Hamming weight 3), respectively. With more searching one could determine, in principle, the minimum distance of all paths directly from the trellis diagram. For example, such an examination would show that the path

Fundamentals of Convolutional Codes

$S_0 S_1 S_2 S_0$ has the minimum Hamming distance of 4 among all paths of the trellis diagram.

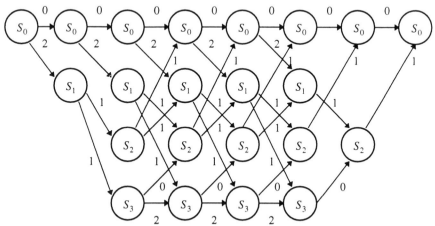

Figure 8.7. Trellis diagram of the encoder shown in Figure 8.3 with branches labeled by their Hamming weights.

The minimum Hamming distance of convolutional codes sometimes is called the free Hamming distance and denoted by d_{free}, i.e.

$$d_{free} \stackrel{\Delta}{=} \min\{d(\bar{c}',\bar{c}''): \bar{c}' \neq \bar{c}''\} = \min\{d(\bar{c},\bar{0}): \bar{c} \neq \bar{0}\} = \min\{w(\bar{c}): \bar{c} \neq \bar{0}\}, \quad (8.10)$$

where $\bar{c}'$ and $\bar{c}''$ are the code words corresponding to the information sequences $\bar{m}'$ and $\bar{m}''$, respectively. Notice that, in the above equation, it is assumed that if $\bar{m}'$ and $\bar{m}''$ are of different lengths, zeros can be appended to the shorter sequence so that their corresponding code words have the same length.

The Generating Function
The above technique, although instructive, is a very time consuming method for determining the distance structure of a convolutional code, particularly for a code of long constraint length. A similar method, based on the state diagram, can be used to obtain a closed-form expression, called the modified generating function, with coefficients that directly provide all of the desired distance information. This method is described next for the same 1/2-rate convolutional code, given previously in Figure 8.1.

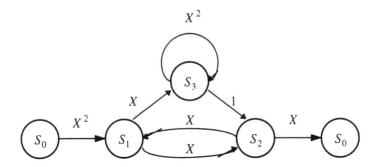

Figure 8.8.The modified state diagram of
the convolutional encoder given in Figure 8.2.

Example 8.10.
Figure 8.8 illustrates the state diagram of Figure 8.2 in a modified form that is convenient for the study of the distance structure of a convolutional code. The branches of this diagram are labeled by, what are called, the branch gains X^i for $i = 0,1,2$, where X is an indeterminant variable and where the exponent i of X denotes the Hamming weight of the branch. For example, the label X^2 in the branch going from S_0 to S_1 indicates that the code segment 11 in the corresponding branch in Figure 8.2 has weight 2. Note that in Figure 8.8 the self-loop at S_0 has been removed since it does not contribute to the distance properties of the code. A continuous circulation around this self-loop simply generates the all-zero sequence of total weight 0. Also, S_0 has been split into two initial and final states.

One can trace all of the paths, obtained from the trellis diagram, onto this modified state diagram. As an example, consider the path $S_0 S_1 S_2 S_0$ which starts from initial state S_0 and terminates at the final state S_0. Multiplying the branch gains of the branches of this path yields the total path gain X^4, whose exponent denotes the Hamming distance of the path from the all-zero sequence. The state equations of the distance profile of the convolutional codes are determined from the modified state diagram. For the modified state diagram of Figure 8.8 the state equations are given by

$$X_1 = X^2 + X \cdot X_2,$$
$$X_2 = X \cdot X_1 + X_3,$$
$$X_3 = X \cdot X_1 + X^2 \cdot X_3,$$

Fundamentals of Convolutional Codes

where X_i for $i = 1,2,3$ are dummy variables with subscripts which show the values of the path gain from the initial state S_0 to S_i as influenced by the other states. The input to the initial state S_0 is unity, and the output $T(X)$ at the final state S_0 is called the generating function for all paths and their distance termination at state S_0. For the present example

$$T(X) = X \cdot X_2$$

since the total gain out of S_0 is the gain X_2 at S_2, followed by X on the branch between S_2 to S_0. Solving the state equations given above and substituting the value of X_2 into the generating function yields finally

$$T(X) = \frac{X^4(2-X^2)}{1-3X+X^4} = 2X^4 - 6X^5 + 17X^6 + \cdots$$

$$= \sum_{d=4}^{\infty} \alpha_d X^d.$$

(8.11)

The coefficients of the power of X of this generating function show that there are only two paths of Hamming distance 4, six paths of Hamming distance 5 from the all-zero path, and seventeen paths of Hamming distance 6, etc..

By introducing new labels or indeterminate variables in the modified state diagram (sometimes also called the error-state diagram) of the convolutional code, additional information about the distance structure of these codes is obtained next from the expansion of, what further is called, an augmented generating function.

Example 8.11.
The new labels in the expansion of the modified state diagram are shown in Figure 8.9. To enumerate the lengths of a given path the label Z is added with exponent 1 for each branch. Thus the exponent of this label is augmented by one every time a branch is passed through. Also the label Y^i is added to each branch to enumerate the Hamming weight of the $i-th$ input bit-sequence associated with each branch of the modified state diagram.

In a manner similar to that used to obtain the modified state diagram, the state equations for the augmented state diagram, shown in Figure 8.9, are

$$X_1 = X^2YZ + XYZ \cdot X_2,$$
$$X_2 = XZ \cdot X_1 + Z \cdot X_3,$$
$$X_3 = XYZ \cdot X_1 + X^2YZ \cdot X_3.$$

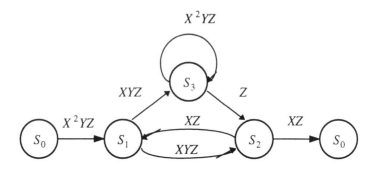

Figure 8.9. The augmented state diagram of the convolutional encoder given in Figure 8.2.

Thus, the augmented generating function at the output terminal state S_0 is given by

$$T(X,Y,Z) = XZ \cdot X_2.$$

The desired augmented generating function is obtained by solving the above state equations and then substituting X_2 into $T(X,Y,Z)$ to obtain finally

$$T(X,Y,Z) = \frac{(XZ - X^3YZ^2 + XYX^2)X^3YZ^2}{1 - X^2YZ - X^2YZ^2 + X^4Y^2Z^3 - X^2Y^2Z^3}$$
$$= X^4YZ^3 + X^4Y^2Z^4 + \cdots\cdots$$

The coefficients of the formal power series of this augmented generating function gives more information about the structure of the paths of the convolutional code. For example, it indicates that one of the paths with the minimum Hamming distance of 4 has length 3 and that its input sequence weight is one, and another sequence has length 4 and that its input sequence has weight 2. These results verify our previous observations about Example 8.3.

It becomes impractical to compute $T(X,Y,Z)$ if the overall constraint length v of the convolutional encoder exceeds 4 or 5. Hence, other means must be used to find d_{free} for most useful convolutional codes. Some examples of efficient algorithms for finding d_{free} which are based on the state diagram of a convolutional code are given in [5] and [6]. These algorithms are capable of computing d_{free} for values of v up to about 20. For larger values of v, when the amount of storage memory required by these algorithms, namely 2^v, becomes unacceptably large, other procedures for finding d_{free}

must be found. It is seen in the next few sections that some decoding algorithms for convolutional codes can, with proper modification, be used to compute the free distance of a code. However, in general no solution to this problem of finding d_{free} for large values of v has yet been discovered.

8.5 Decoding of Convolutional Codes

There are several different algorithms for decoding convolutional codes. These are sequential decoding, majority-logic decoding, and Viterbi decoding. Sequential decoding was the first decoding algorithm for convolutional codes. It was introduced by Wozencraft [7] in 1957. Other versions of this algorithm, most notably, the Fano algorithm and the stack algorithm were developed by Fano [8] and Zigangirov and Jelenik [9]. The majority-logic decoding algorithm is an algebraic decoding method based on the threshold decoding algorithm generalized by Massey [10] from Reed's decoding algorithm for Reed-Muller codes[11]. In 1967, Viterbi [12] introduced a probabilistic nonsequential decoding algorithm for convolutional codes which has since become known as the Viterbi algorithm.

Omura [13] showed later that the Viterbi algorithm is equivalent to a dynamic programming solution to the problem of finding the shortest path through a weighted graph. Finally, Forney [14,15] demonstrated that the Viterbi algorithm is, in fact, a maximum-likelihood decoding algorithm for convolutional codes. In this section only the most powerful Viterbi algorithm is considered. For the other decoding methods the reader is referred further to other references and texts, e.g. see [16].

8.5.1 The Maximum-Likelihood Decoding

The maximum-likelihood principle can be used to decode a particular received sequence $\bar{r}$ into a code sequence $\bar{c}$ which maximizes the likelihood function $P(\bar{r}|\bar{c})$ or the log-likelihood function $\ln P(\bar{r}|\bar{c})$. This type of decoding operation provides the minimum probability of error when all of the transmitted codewords are equally probable.

Assume that a codeword $\bar{c} = (c_0, c_1, \ldots, c_{l-1})$ of length l is transmitted through a memoryless channel and that the received sequence is $\bar{r} = (r_0, r_1, \ldots, r_{l-1})$. Since the noise components are independent in a memoryless channel, the likelihood function $P(\bar{r}|\bar{c})$ is expressed by

$$P(\bar{r}|\bar{c}) = \prod_{i=0}^{l-1} P(r_i|c_i), \qquad (8.12a)$$

where $P(r_i|c_i)$ is a channel transition probability. Also note that the log-likelihood function, in this case, can be written as

$$\ln P(\bar{r}|\bar{c}) = \sum_{i=0}^{l-1} \ln P(r_i|c_i). \qquad (8.12b)$$

Next the case of hard-decision decoding is presented. Maximum-likelihood decoding is obtained by choosing some (there could be more than one) nearest codeword to the received sequence $\bar{r}$ in Hamming distance which maximizes the likelihood function.

Suppose that the codewords are transmitted through a BSC with a channel-transition probability p. Also assume that the received binary sequence $\bar{r}$ differs from codeword $\bar{c}$ in exactly d positions. In such a case the likelihood function $P(\bar{r}|\bar{c})$ is expressed by

$$P(\bar{r}|\bar{c}) = p^d (1-p)^{l-d}.$$

Without loss of generality assume that $p < 1/2$. Then it can be demonstrated easily that a maximization of the likelihood function $P(\bar{r}|\bar{c})$ is, in fact, equivalent to the minimization of the Hamming distance d between $\bar{r}$ and $\bar{c}$.

On the other hand for, what is called, the soft-decision algorithm, the maximum-likelihood decoding is usually obtained by choosing a codeword whose modulated signal sequence is nearest to the received sequence $\bar{r}$ in Euclidean distance.

The likelihood function for soft-decision decoding is found for the special case of a BPSK signal sequence transmitted over an additive white-Gaussian noise (AWGN) channel as follows:

$$P(\bar{r}|\bar{s}) = \prod_{i=0}^{l-1} \frac{1}{\sqrt{\pi N_0}} e^{-\frac{|r_i - s_i|^2}{N_0}}$$

$$= \left(\frac{1}{\pi N_0}\right)^{\frac{l}{2}} \exp\left(-\frac{1}{N_0} \sum_{i=0}^{l-1} |r_i - s_i|^2\right),$$

Fundamentals of Convolutional Codes

where $\bar{s} = (s_0, s_1, \ldots, s_{l-1})$, $s_i = (1-(-1)^{c_i})/2$ and $\bar{c} = (c_0, c_1, \ldots, c_{l-1})$ is the codeword. Also N_0 is the single-sided power-spectral density of the AWGN and

$$d_E^2(\bar{r}, \bar{s}) = \sum_{i=0}^{l-1} |r_i - s_i|^2$$

is the squared Euclidean distance between $\bar{r}$ and $\bar{s}$. Notice that r_i is a number in the real field. The more practical case is when r_i is quantized to have discrete values over a finite field. Such a case is discussed in the next section.

The key observation about the above likelihood function is that the maximum-likelihood decoding rule for the AWGN channel with soft-decision decoding can be formulated as the minimization of the Euclidean distance between the received sequence and the decoded signal sequence (i.e. the modulated codeword).

8.5.2 Viterbi Algorithm

The straightforward implementation of the maximum-likelihood decoder for convolutional codes would require that one examine all code sequences or paths through the trellis and choose any of those code sequences which provide the largest likelihood function (or the smallest distance). Since the number of code paths grows exponentially with sequence length, such a brute force decoding technique would appear to be impracticable.

Consider the trellis diagram drawn in Example 8.4 for the (2,1,2) convolutional code specified in Figure 8.1. Recall that input sequences of length $l+m$, where the last $m=2$ bits are 0's, yield 2^l codewords which correspond with 2^l distinct trellis paths. If a codeword sequence $\bar{c}$ is transmitted through a Discrete Memoryless Channel (DMC), and the received sequence is $\bar{r}$, the decoding process is to search all encoder paths $\bar{c}$ over the trellis diagram by the use of the maximum-likelihood criteria, i.e. find a codeword $\hat{c}$ such that

$$\ln P(\bar{r}|\hat{c}) = \max_{v} \ln P(\bar{r}|\bar{c}'_v) \text{ for } v = 1, 2, \ldots, 2^l.$$

In general, such a direct search method is very hard to realize. Assume that $l = 100$. Then, there are a total of $2^l = 2^{100} > 10^{30}$ codewords or trellis paths. If $l+m = 102$ information bits are transmitted in one second, the transmission rate is only 204bits/s (which is quite low). For this case the maximum-likelihood decoder has to compute and compare more than 10^{30}

likelihood functions within one second. It would seem that the implementation of such a direct search method is impossible.

However, as it is seen in this section, a practical method to realize maximum-likelihood decoding of convolutional codes is made possible, in fact, by the Viterbi algorithm. Assume that an information sequence $\bar{m} = (\bar{m}_0, \bar{m}_1, \ldots, \bar{m}_{l-1})$ of length $k_0 \cdot l$ bits (where k_0 is the length of $\bar{m}_i$) is encoded into a code word $\bar{c} = (\bar{c}_0, \bar{c}_1, \ldots, \bar{c}_{l+m-1})$ of length $N = n_0(l+m)$ bits (where n_0 is the length of $\bar{c}_i$), and that a Q-ary sequence $\bar{r} = (\bar{r}_0, \bar{r}_1, \ldots, \bar{r}_{l+m-1})$ is received over a binary input, Q-ary output, discrete memoryless channel (DMC). Alternately, these sequences can be written as $\bar{m} = (m_0, m_1, \ldots, m_{k_0 l-1})$, $\bar{c} = (c_0, c_1, \ldots, c_{N-1})$ and $\bar{r} = (r_0, r_1, \ldots, r_{N-1})$, where the subscripts now represent more simply the ordering of the symbols in each sequence. The decoder must produce an estimate $\hat{c}$ of the codeword $\bar{c}$ which is based on the received sequence $\bar{r}$. A maximum-likelihood decoder (MLD) for a DMC chooses $\hat{c}$ as that codeword $\bar{c}$ which maximizes the log-likelihood function $\ln P(\bar{r}|\bar{c})$. Since

$$P(\bar{r}|\bar{c}) = \prod_{p=0}^{l+m-1} P(\bar{r}_p|\bar{c}_p) = \prod_{j=0}^{N-1} P(r_j|c_j), \qquad (8.13)$$

holds for a DMC, it follows that

$$\ln P(\bar{r}|\bar{c}) = \sum_{p=0}^{l+m-1} \ln P(\bar{r}_p|\bar{c}_p) = \sum_{j=0}^{N-1} \ln P(r_j|c_j), \qquad (8.14)$$

where $P(r_j|c_j)$ is a channel transition probability. This fact yields a minimum error-probability decoding rule when all codewords are assumed to be equally likely.

The log-likelihood function $\ln P(\bar{r}|\bar{c})$ is called the metric associated with the trellis path $\bar{c}$, and is denoted by $M(\bar{r}|\bar{c})$. The terms $\ln P(\bar{r}_p, \bar{c}_p)$ in the above equation are called the branch metrics, and are labeled more simply by $M(\bar{r}_p|\bar{c}_p)$, i.e. $M(\bar{r}_p, \bar{c}_p) = \ln P(\bar{r}_p, \bar{c}_p)$. Also, the terms $\ln P(r_j, c_j)$ are called bit metrics, and are denoted similarly by $M(r_j|c_j)$. Hence, by these definitions and Eq.(8.14) the path metric $M(\bar{r}|\bar{c})$ is expressed by

$$M(\bar{r}|\bar{c}) = \sum_{p=0}^{l+m-1} M(\bar{r}_p|\bar{c}_p) = \sum_{j=0}^{N-1} M(r_j|c_j) \qquad (8.15)$$

Fundamentals of Convolutional Codes 337

as a sum of branch metrics and as a sum of bit metrics. A partial path metric for the first j branches of a path is expressed similarly in term of branch metrics by

$$M([\bar{r}|\bar{c}]_j) = \sum_{p=0}^{j-1} M(\bar{r}_p|\bar{c}_p). \tag{8.16}$$

The following iterative algorithm, when applied to the received sequence $\bar{r}$ from a DMC, finds the trellis path with the largest metric (i.e. the maximum likelihood path). At each step, it compares the metrics of all paths entering each state, and stores only the path with the largest metric, called the survivor, together with its metric.

The Viterbi Algorithm :
1. Beginning at time unit $j = m$, compute the partial metric for the single path entering each state. Store the path (the survivor) and its metric for each state.
2. Increase j by 1. Compute the partial metric for all of the paths entering a state by adding the branch metric entering that state to the metric of the connecting survivor at the preceding time unit. For each state, store the path with the largest metric (called the survivor) along with its metric and eliminate all other paths.
3. If $j < l + m$, repeat step 2. Otherwise, stop.

There are 2^v survivors from time unit m through time unit l, one for each of the 2^v states. After time unit l there are fewer survivors since fewer states remain while the encoder returns to the all-zero state. Finally, at time unit $l+m$ there is only one state, the all-zero state. Hence there is only one survivor, and the algorithm terminates. Following Step 2 of the algorithm, assume now that maximum-likelihood path were eliminated by the algorithm at time unit t, then for the partial path the metric of the survivor would exceed that of the maximum-likelihood path to this point. Thus if the remaining portion of the maximum-likelihood path were appended onto the survivor at time unit t, then the total metric of this path would exceed the total metric of the maximum-likelihood path. But this would contradict the definition of the maximum-likelihood path as the path with the largest metric. Hence, the maximum-likelihood path cannot be eliminated by the algorithm, and as a consequence must be the final survivor.

Therefore, the final survivor $\hat{c}$ of the Viterbi algorithm is the maximum-likelihood path, that is, $M(\bar{r}|\hat{c}) \geq M(\bar{r}|\bar{c})$ for all $\bar{c} \neq \hat{c}$. The Viterbi algorithm is optimum in the sense that it always finds the maximum-likelihood path through the trellis.

Hard-Decision Decoding
In hard-decision decoding, when each signal value is received, a "binary" decision is made to determine whether the signal represents a transmitted zero or a one. Then these decisions form the input sequence to the Viterbi decoder. If the channel is assumed to be memoryless, then from the decoder's perspective it can be modeled as a BSC.

The first step in defining the bit metrics for this channel is the compilation of the likelihood functions in a table. These conditional probabilities are then converted into log-likelihood functions and applied to the decoding algorithm.

Example 8.12.
Viterbi's decoding algorithm is illustrated next by the use of the same code discussed in Example 8.4. Suppose that the codeword sequence $\bar{c} = (11,01,10,10,11,00,01)$, (the path is shown by the bold path in Figure 8.3) is transmitted, by the use of a binary signaling scheme, say BPSK. Assume that the receiver performs hard-decision detection for $Q = 2$ and provides the Viterbi decoder with the received sequence $\bar{r} = (11,01,11,10,11,00,01)$ which has one error in the sixth location. In such a case the channel can be modeled as a BSC with $p < \frac{1}{2}$. Then, by Example 8.12 and the definition of a Hamming metric one has

$$\underset{\bar{c}}{Max}\ M(\bar{r}|\bar{c}) = \underset{\bar{c}}{Max}(\ln p(\bar{r}|\bar{c})) = \underset{\bar{c}}{Max}\left(d(\bar{r},\bar{c}) \cdot \ln \frac{p}{1-p} + l \cdot \ln(1-p) \right).$$

Thus, since $\ln \frac{p}{1-p} < 0$, $\underset{\bar{c}}{Max}\ M(\bar{r}|\bar{c})$ is equivalent to $\underset{\bar{c}}{Min}\ d(\bar{r},\bar{c})$. Therefore, the Hamming distance between the received sequence and the codeword sequence is the appropriate metric to be minimized. The Viterbi decoder chooses that path through the trellis of the code which has a minimum Hamming distance from the received sequence. Define the $i-th$ branch distance by $d(\bar{r}_i,\bar{c}_i)$. Then, clearly $d(\bar{r},\bar{c}) = \sum_{i=0}^{l+m-1} d(\bar{r}_i,\bar{c}_i)$. In a similar manner the partial path metric, a partial path distance for the first j branches of a path, is expressed by

Fundamentals of Convolutional Codes

$$d\left([\bar{r}]_j,[\bar{c}]_j\right) = \sum_{i=0}^{j-1} d(\bar{r}_i,\bar{c}_i). \tag{8.17}$$

Notice that $d\left([\bar{r}]_j,[\bar{c}]_j\right) = d\left([\bar{r}]_{j-1},[\bar{c}]_{j-1}\right) + d(\bar{r}_{j-1},\bar{c}_{j-1})$ for a given $\bar{r}$ and $\bar{c}$. For each state at time unit j, Step 2 of the above Viterbi algorithm always keeps the path which has the minimum partial distance, namely, $d\left([\bar{r}]_j,[\bar{c}]_j\right)$. The Viterbi decoding process is shown next in Figure 8.10 over a set of partial trellis diagrams.

Time unit	Received branch word	Trellis Diagram with survivor paths	d_s	$\hat{m}$
0	11		2 0	0 1
1	01		3 3 0 2	00 01 10 11

2	11	(trellis diagram with states S_0, S_1, S_2, S_3; labels 00, 00, 01, 11, 11, 01, 10, 01, 10, 11)	1 1 4 2	100 101 010 111
3	10	(trellis diagram with states S_0, S_1, S_2, S_3; labels 00, 00, 11, 11, 01, 11, 01, 10, 10, 10, 10, 11, 00)	2 2 3 1	1000 1001 1110 1011
4	11	(trellis diagram with states S_0, S_1, S_2, S_3; labels 00, 11, 01, 11, 01, 10, 01, 10, 11, 00, 00, 11)	4 2 3 1	11100 10001 10110 10111

Fundamentals of Convolutional Codes 341

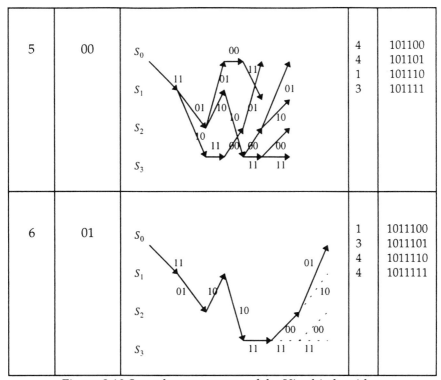

Figure 8.10 Stage-by-stage steps of the Viterbi algorithm

In Figure 8.10, d_s denotes the minimum partial distance for each state at the given time unit and $\hat{m}$ is the estimated "partial" message word for each possible state at a given time unit.

Figure 8.10 illustrates the procedure for the Viterbi algorithm to find the minimum-distance path through the trellis diagram. This process starts at the initial state S_0. Initially the algorithm sets all of the distance metrics, associated with the different states, to zero. At the first stage of the decoding process the algorithm computes the distance for each of the two branches which emanate from state S_0 and arrives at the states S_0 and S_1. These distances are stored together with their surviving paths for states S_0 and S_1.

At the second time unit or stage the surviving paths with their respective metrics are computed and stored for each state. Since there is only one path which arrives at each state at this time unit, no comparisons are needed and only a single path is stored as the survivor for each state.

At the third stage of decoding there are two paths which merge into each state. In this case the decoder computes the partial distance associated with each path, stores only that path with the minimum partial distance and discards the other path. In this way, the decoder proceeds through the trellis at each stage and preserves for each state only one survivor and its distance from the received signal. The algorithm attempts to compute the minimum partial distance for each possible path re-emerging at a specific state (node). At each state the algorithm compares the partial distances of the different paths arriving at that state, preserves the path with the smallest partial distance, and discards the rest.

For example, at time unit 3 there are two branches entering state S_3 : one is the branch (10) from the previous state S_1 and the other is the branch (11) from the previous state S_3 . For the received branch word (10) the partial distance of the path from branch (10) equals 1 which is the summation of the branch distance $d((10),(10)) = 0$ and the minimum partial distance $d_s = 1$ for state S_1 at time unit 2. The partial distance of the path from the branch (11) equals 3 which is the summation of the branch distance $d((11),(10)) = 1$ and the minimum partial distance $d_s = 2$ for the state S_3 at the time unit 2. Therefore, the minimum partial distance rule yields the survivor path which is the path with the branch (10). The surviving path is stored at each node level together with its metric. The algorithm proceeds through the trellis in an iterative manner and preserves a relatively small list of paths that is guaranteed to contain the minimum partial-distance path.

At the 6-th time unit the survivor path is the code sequence closest to the received sequence in Hamming distance. If a decision is made at this stage, such a survivor path is selected as the code path nearest to the received code sequence in the sense of minimum distance. Thus the final survivor path is the decoded codeword $\hat{c} = (11,01,10,10,11,00,01)$ and the final decoded message word is $\hat{m} = (1011100)$. As a consequence the single bit error in $\bar{r}$ is corrected by the Viterbi algorithm, detailed in Figure 8.10.

In the above decoding process a difficulty may arise in a given comparison between the merging paths which have the same metric values. Such ties can be resolved randomly by the flip of a fair coin. The outcome of such a coin tossing does not alter the ultimate performance of the decoder. As an example of a tie consider the paths which arrive at state S_2 at time unit 2.

Fundamentals of Convolutional Codes

The two paths which enter the state S_2 are $(11,10,00)$ and $(00,11,01)$. These two paths have the same partial distance 4. Consequently, one chooses one of these two paths arbitrarily, say the path which goes to S_2 from S_3, and discard the other. Note that this choice has no effect on the final decision since it is eliminated at a later time unit of the decoding algorithm.

The following comments about the Viterbi algorithm are found to arise from the above example and discussion:

1. The Viterbi algorithm consists of $2^{k_0 \cdot m}$ states for a (n_0, k_0, m) convolutional code. A set of path memories for each state is needed to store the survivor path and the estimated message sequence $\hat{m}$. Also needed is the memory of the set of metrics for each state in order to save the partial minimum metric. Therefore, the complexity of the Viterbi algorithm is proportional to $2^{k_0 \cdot m}$, i.e. it increases exponentially with m and k_0. To avoid a high complexity and cost, the value of the memory m is selected usually to satisfy $m \leq 10$.

2. For each state a memory is required to store the complete sequences of the surviving paths. For a sequence of length l the final path length is $n_0 \cdot l$ (or $k_0 \cdot l$). For most practical applications the encoded convolutional-code sequences could require very long lengths l, approaching infinity as $l \to \infty$. In such cases the memory, required to preserve the survivors, becomes large and expensive. Also, such a large memory length $n_0 \cdot l$ (or $k_0 \cdot l$) introduces a long decoding delay which usually is undesirable from a practical point of view.

Note that the first four branches in Figure 8.10 of the surviving paths at the 5th time unit and the 6th time unit of the decoding procedure coincide. Such a fact implies that a decision could be made for the first four time units, and only the survivor branches corresponding with the 4th and 5th time units need to be stored. In practice, for very long coded sequences, it is necessary to truncate the surviving paths at some length τ. In such a case the decoder at any time unit l retains only the survivor branches corresponding to the most recent τ time units back in the trellis. For each new time unit the decoder makes a definite decision on the survivor path for the time units up to the time unit $l - \tau$. If the truncation depth τ is chosen large enough, there is a high probability that all surviving paths at the time unit l contain the same branches at the time units up to $l - \tau$. An analysis by Forney [15] shows that in the small signal-to-noise ratio case, a minimum truncation length of $\tau \approx 5.8m$ is sufficient to ensure that the additional error probability due to truncation is negligible. Extensive simulations and actual experience have verified that $\tau = (4 \text{ or } 5)m$ is usually an acceptable truncation length to use in practice. A real Viterbi decoder, which considers this truncation, is discussed in the next section.

Soft Decision Decoding
It was noted in Chapter 1 that soft demodulator decisions can result in a performance advantage over hard-decision decoding. In soft-decision decoding the receiver takes advantage of the side-information generated by the receiver bit-decision circuitry. Rather than to simply assign a zero or a one to each received bit, a more flexible approach is taken through the use of multiple-level quantization. More than two decision regions are established, ranging from a "one" decision to a "zero" decision. The intermediate quantization values are used to make possible a soft-decision. On an AWGN channel such soft-decision decoding process can provide an increase in coding gain of about 2 to 3 dB over the hard-decision Viterbi decoder.

Consider a binary convolutional code sequence $\bar{c} = (\cdots, c_{-1}, c_0, c_1, \cdots)$ that is transmitted over an AWGN channel. Each component $s_u = (-1)^{c_u}$ of the BPSK-modulated signal sequence $\bar{s}$ is corrupted by the addition of some statistically independent Gaussian-noise variables. Let E_b be the received energy-per-codeword bit and let the one-sided noise spectral density be N_0 Watts/Hz. Then the received signals can be represented as Gaussian random variables with mean $s_u \sqrt{E_b}$ and variance $N_0/2$. The likelihood function then assumes the standard form of a Gaussian probability density function with a nonzero mean. Let $\{a_1, a_2\} = \{-1, 1\}$. Then, when $s_u = a_i$ the u-th component of the (unquantized) received sequence $\bar{r}$ is described by the probability density function,

$$p(r_u | s_u = a_i) = \frac{1}{\sqrt{\pi N_0}} e^{-(r_u - s_u \sqrt{E_b})^2 / N_0} \text{ for } i=1,2. \tag{8.18}$$

The log likelihood function in Eq. (8.14) is simplified, by noting that $(s_u)^2 = 1$, as follows:

Fundamentals of Convolutional Codes

$$\ln p(\bar{r}|\bar{s}) = \sum_{u=0}^{l+m-1} \ln p(r_u|s_u = a_i)$$

$$= \sum_{u=0}^{l+m-1} \left(-\frac{(r_u - s_u\sqrt{E_b})^2}{N_0} - \ln\sqrt{\pi N_0} \right)$$

$$= \frac{-1}{N_0} \sum_{u=0}^{l+m-1} (r_u^2 - 2r_u s_u \sqrt{E_b} + s_u^2 E_b) - \frac{l+m}{2} \ln \pi N_0 \qquad (8.19)$$

$$= C_1 \sum_{u=0}^{l+m-1} (r_u s_u) + C_2$$

$$= C_1 (\bar{r}^T \cdot \bar{s}) + C_2$$

In Eq. (8.19), all terms that are not a function of $\bar{s}$ are collected into the pair of constants C_1 and C_2. The path metrics for the AWGN channel are thus the inner product of the received word and the codewords. The individual bit metrics are

$$M(r_u|s_u) = r_u s_u. \qquad (8.20)$$

The minimization of the path metric in Eq. (8.19) is equivalent to finding the codeword $\bar{s}$ that is closest to $\bar{r}$ in terms of Euclidean distance. This is a different metric than the BSC, which minimizes Hamming distance.

The above analysis assumes that the receiver is capable of processing real numbers with infinite precision. In practical applications of digital circuitry the received signal needs to be quantized, and the Viterbi decoder therefore has some loss in its coding gain. However, the loss of coding gain through quantization to as few as eight levels is surprisingly small. A simulation study conducted in [21] shows that an 8-level quantization resulted in only 0.25 dB loss of coding gain for a number of codes. Some performance points for different levels of quantization are summarized in Table 8.2 [18][21]. It can be seen that coarser quantization levels could result a substantial decline in performance.

Quantization Levels	Loss in Coding Gain (dB)
8	0.25
4	0.9
2	2.2

Table 8.2 Loss in coding gain

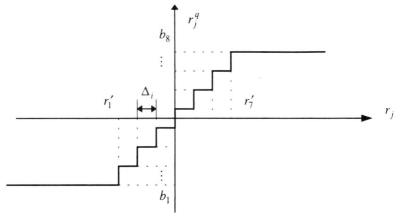

Figure 8.12. Input-output relations of a quantizer.

As illustrated in Figure 8.12, the continuously received quantity r_u is mapped next into a discrete output component r_u^q. That is, r_u^q is restricted to be some discrete set of values of the quantizer output, say $\{b_l\}$, for $l = 1, 2, \ldots, M$. Usually, one selects $M = 2^Q$ for a Q-bit quantization. Given that s_u is a_i, the probability that $r_u^q = b_l$ is denoted by the symbol p_{il}, i.e.

$$p_{il} \stackrel{\Delta}{=} P\left(r_u^q = b_l | s_u = a_i\right). \tag{8.21}$$

The $\{p_{il}\}$, called transition probabilities, are computed by the following integrals of the Gaussian density function of Eq. (8.18) over the quantization intervals $(-\infty, r_1'), (r_1', r_2'), \ldots, (r_{M-2}', r_{M-1}'), (r_{M-1}', \infty)$.

$$p_{2,1} = \int_{-\infty}^{r_1'} \frac{1}{\sqrt{\pi N_0}} e^{-(y-\sqrt{E_b})^2 / N_0} dy = 1 - Q\left(\frac{r_1' - \sqrt{E_b}}{\sqrt{N_0/2}}\right), \tag{8.22a}$$

$$p_{2,l} = \int_{r_l'}^{r_{l+1}'} \frac{1}{\sqrt{\pi N_0}} e^{-(y-\sqrt{E_b})^2 / N_0} dy = Q\left(\frac{r_l' - \sqrt{E_b}}{\sqrt{N_0/2}}\right) - Q\left(\frac{r_{l+1}' - \sqrt{E_b}}{\sqrt{N_0/2}}\right), \tag{8.22b}$$

for $l=2,\ldots,M-1$, and

$$p_{2,M} = 1 - \int_{r_{M-1}'}^{\infty} \frac{1}{\sqrt{\pi N_0}} e^{-(y-\sqrt{E_b})^2 / N_0} dy = 1 - Q\left(\frac{r_{M-1}' - \sqrt{E_b}}{\sqrt{N_0/2}}\right), \tag{8.22c}$$

Fundamentals of Convolutional Codes

where $Q(x) = \int_{x}^{\infty} \frac{1}{\sqrt{2\pi}} e^{-y^2/2} dy$. The symmetry of the channel yields

$$p_{1,l} = p_{2,M+1-l} \quad \text{for } l=1,2,\ldots,M. \tag{8.22d}$$

A C-program for computing the above transition probabilities is given in Appendix B.

input (r_u)	output (b_l)	Transition Probabilities
$-\infty < r_u \leq -1.75$	$b_1 = -1.75$	$p_{1,1} = 0.69159$, $p_{2,1} = 0.00023$
$-1.75 < r_u \leq -1.25$	$b_2 = -1.25$	$p_{1,2} = 0.1498$, $p_{2,2} = 0.00112$
$-1.25 < r_u \leq -7.5$	$b_3 = -0.75$	$p_{1,3} = 0.09189$, $p_{2,3} = 0.00486$
$-0.75 \leq r_u < -0.25$	$b_4 = -0.25$	$p_{1,4} = 0.04406$, $p_{2,4} = 0.01654$
$0.25 \leq r_u < 0.75$	$b_5 = 0.25$	$p_{1,5} = 0.01654$, $p_{2,5} = 0.04406$
$0.75 \leq r_u < 1.25$	$b_6 = 0.75$	$p_{1,6} = 0.00486$, $p_{2,6} = 0.009189$
$1.25 \leq r_u < 1.75$	$b_7 = 1.25$	$p_{1,7} = 0.00112$, $p_{2,7} = 0.1498$
$1.75 \leq r_u < \infty$	$b_8 = 1.75$	$p_{1,8} = 0.00023$, $p_{2,8} = 0.69159$

Table 8.3 The input-output relations of a quantizer.

Example 8.13

Assume that M=8 and b_i ranged from -1.75 to 1.75. Also assume that $\sqrt{E_b} = 1.75$ and $N_0 = 2$. The input-output relations of the quantizer are given in Table 8.3.

For the input $0.75 \leq r_u \leq 1.25$, the transition probability p_{il} is the area between x=0.75 to x=1.25 in Figure 8.12.

In general, the set of probabilities, namely $\{p_{il}\}$ for $i = 1,2,\ldots, A; l = 1,2,\ldots, M$, may be conveniently displayed in a diagram when A and M are small and in a matrix when A and M are large, as it is shown in Figure 8.13. When A=2, such a channel is called a BSC for $M = 2^Q = 2$ and a binary symmetric erasure channel (BSEC) for M=3.

The bit decision in a hard-decision receiver is made by the use of a hard limiter. A similar role in the soft-decision receiver is played by a multiple-bit A/D converter. For example, the model in Fig. 8.12 implies the use of a Q-bit A/D converter in the decision circuitry.

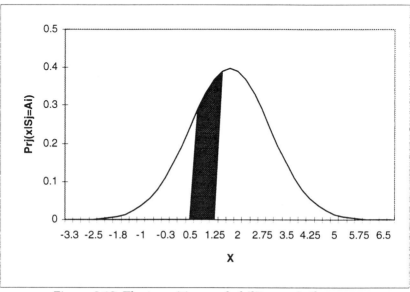

Figure 8.12. The transition probability p_{jl} is the area of the bold-shaded region.

The components of the Gaussian-noise sequence, which the channel adds to the code signal-sequence $\bar{s}$, are statistically independent. If one assumes that each component r_j of the received sequence $\bar{r}$ is subjected sequentially to a quantizer, then each component s_j is affected independently by the same set of transition probabilities $\{p_{il}\}$. Then, the path metric $M(\bar{r}|\bar{s})$ and the partial path metric $M(\lfloor \bar{r}|\bar{s} \rfloor_j)$ for the first j branches of a path for the Viterbi decoding process can be expressed in a manner similar to that in Eqs. (8.15) and (8.16).

Fundamentals of Convolutional Codes

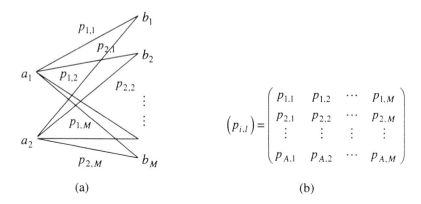

Figure 8.13 Transition probability diagram and matrix (a) A=2, (b) Matrix;

For the quantized received signal the bit metric of Eq. (8.20) is no longer valid. A new metric can be derived directly from Eqs.(8.13) and (8.14) by the use of transition probabilities specified in Eq.(8.21). One commonly-used metric is the log-likelihood function in terms of the normalized transition probabilities, given by

$$M(\bar{r}|\bar{s}) = \sum_{u=0}^{l+m-1} \log\left(\frac{P(r_u^q = b_l | s_u = a_i)}{P(r_u^q = b_l | s_u = a_1) + P(r_u^q = b_l | s_u = a_2)}\right) \quad (8.23a)$$

$$= \sum_{u=0}^{l+m-1} M(r_u^q | s_u),$$

where

$$M(r_u^q | s_u) = \log\left(\frac{P(r_u^q = b_l | s_u = a_i)}{P(r_u^q = b_l | s_u = a_1) + P(r_u^q = b_l | s_u = a_2)}\right). \quad (8.23b)$$

From the point of view of implementation it is usually more convenient to use positive integer numbers as metrics rather than the actual bit metrics. The bit metric $M(r_u^q|s_u)$ can be replaced by $\alpha M(r_u^q|s_u) + \beta$, where α and β are real numbers with $\alpha, \beta > 0$. A path $\bar{c}$ which maximizes $M(\bar{r}|\bar{s}) = \sum_{u=0}^{j-1} M(r_u^q|s_u)$ also maximizes $\alpha \sum_{u=0}^{j-1} M(r_u^q|s_u) + \beta$. Hence such a modified metric can be used without affecting the performance of the Viterbi algorithm. Usually, the constants α and β are chosen so that all metrics

$\alpha \sum_{u=0}^{j-1} M(r_u^q | s_u) + \beta$ can be approximated by their floor

integers $\left\lfloor \alpha \sum_{u=0}^{j-1} M(r_u^q | s_u) + \beta \right\rfloor$. Depending on the choice of α, there are many integer metrics possible for a given DMC. However, the performance of the Viterbi algorithm is now somewhat suboptimum, due to the modified metric approximation by integers, but the degradation is typically slight.

Example 8.14
Consider the transition probabilities given in Table 8.3. Such a table yields the log-likelihood functions of normalized transition probabilities $M(r_u^q | s_u)$ given in Table 8.4a. By choosing $\alpha = 100$ and $\beta = 352$, one obtains the set of bit metrics $\lfloor 100 \cdot M(r_u^q | s_u) + 353 \rfloor$, shown in Table 8.4b. Such metrics can be implemented easily in digital hardware. In order to reduce the implementation complexity, one tries to use a minimum number of bits to represent each metric value.

l	$r_u^q = b_l$	$M(r_u | s_u = a_1)$	$M(r_u | s_u = a_2)$
1	-1.75	-0.0001	-3.5229
2	-1.25	-0.0032	-2.1308
3	-0.75	-0.0224	-1.2984
4	-0.25	-0.1384	-0.5640
5	0.25	-0.5640	-0.1384
6	0.75	-1.2984	-0.0224
7	1.25	-2.1308	-0.0032
8	1.75	-3.5229	-0.0001

Table 8.4a Log likelihood function

l	$r_u^q = b_l$	$100 \, M(r_u | s_u = a_1) + 353$	$100 \, M(r_u | s_u = a_2) + 353$
1	-1.75	352	0
2	-1.25	352	139
3	-0.75	350	223
4	-0.25	339	296
5	0.25	296	339
6	0.75	223	350
7	1.25	139	352
8	1.75	0	352

Table 8.4b Integer metrics.

Fundamentals of Convolutional Codes

For the code discussed in Example 8.4, assume that the codeword,
$$\bar{c} = (11,01,10,10,11,00,01),$$
is transmitted and corrupted by an error pattern. In the first case assume that a hard-decision is used to obtain the received word $\bar{r} = (00,01,10,00,11,00,01)$. Note for this case that the decoder causes a decoding error. However, for the case of a 3-bit soft-decision decoder the following information is made available in the decoder :
$$\bar{r} = (b_5 b_5, b_8 b_3, b_2 b_8, b_5 b_8, b_2 b_2, b_8 b_8, b_8 b_1).$$
It can be shown with this added soft-decision information that the Viterbi algorithm is able to correctly recover the codeword.

Implementation of Viterbi decoder

An implementation architecture of the Viterbi decoding algorithm is shown in Figure 8.14. The key function blocks of this architecture are a block synchronizer, a branch metric computer, path metric updating, storage, etc..

The synchronizer is a device for determining the beginning of a branch in the received symbol sequence and, sometimes, for resolving symbol polarity. In general, convolutional codes have the property that branch synchronization can be resolved by changes of the error rate. The operation of the synchronizer is straightforward and requires an indication that the input sequence contains a significantly larger number of errors than normal.

In the branch metric computer, each time a new branch is received the branch metric computer determines a new set of metric values for each of the different branches that appear in the code trellis. The entire set of metric values are represented by the branch metric table (e.g. Table 8.4).

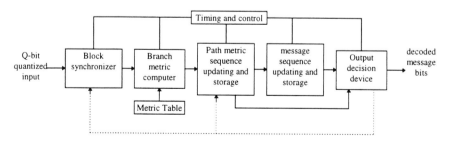

Figure 8.14 Block diagram of a Viterbi decoder

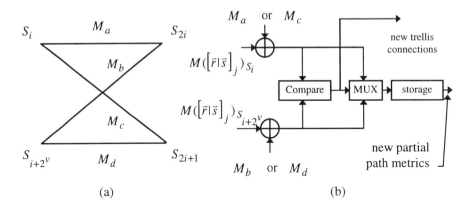

Figure 8.15 Butterfly module and ACS function.

The path metric storage and updating process is illustrated next by a use of a rate-$1/n$ convolutional code. The trellis diagram for any $R = 1/n$ code can be subdivided into basic "butterfly" modules shown in Figure 8.15a. Here M_a, M_b, M_c, and M_d denote the branch metrics. This "butterfly" structure suggests that the key portion of the decoder, concerned with the storage and updating of the path metrics, can be implemented by repeated use of the computational unit shown in Figure 8.15b. This computational unit is often called the add-compare-select (ACS) function. For a constraint-length v convolutional code, this circuit is applied 2^v times. It can be seen that the complexity of this circuit is linearly proportional to the number of bits that are used to represent the path metric.

A C-program of the soft-decision Viterbi algorithm for a additive white-Gaussian-noise channel is provided in [22]. ACS and "butterfly" modules are applied in this program. For a detailed discussion, readers are referred to [23].

8.6 Performance Bounds

The most useful performance measure for convolutional codes is the bit-error probability, P_b, which is defined as the expected number of decoded information-bit errors per information bit. The bit-error probability of a given convolutional code for some channels is upper-bounded by the use of union bounds. We begin the following discussion by evaluating this bound for a BSC.

Fundamentals of Convolutional Codes

From the linear property of a convolutional code one can assume, without loss of generality, that the all-zero code sequence is transmitted. In order to find an upper bound on P_b, an error probability, what is called the first-event error probability, P_e is evaluated. This is, the probability that an error sequence merges with the all-zero (or correct) path for the first time at a given node in the trellis. A first-error sequence decoding at time j is shown in Figure 8.16, where the correct path $\bar{c}$ is eliminated by the incorrect path $\bar{c}'$ and returns for the first time, time j, to the correct path.

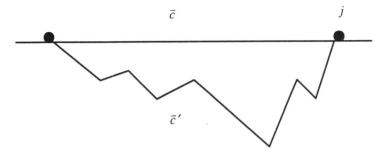

Figure 8.16 The first event error at time j.

Suppose that path sequence $\bar{c}'$ differs from the correct path sequence $\bar{c}$ in d bits. For the BSC the pairwise error probability $P_e^{(d)}$ is the probability that one-half or more of d transmitted bits are in error. Assuming a BSC crossover probability p, $P_e^{(d)}$ is obtained through a simple combinatorial exercise.

$$P_e^{(d)} = \begin{cases} \sum_{j=(d+1)/2}^{d} \binom{d}{j} p^j (1-p)^{n-j}, & d \text{ odd} \\ \frac{1}{2}\binom{d}{d/2} p^{d/2}(1-p)^{d-d/2} + \sum_{j=(d+1)/2}^{d} \binom{d}{j} p^j (1-p)^{n-j}, & d \text{ even} \end{cases}$$

(8.24a)

This expression can be simplified for the case in which d is odd in the following manner (it can be shown that the bound holds for even values of d as well). It is assumed that the BSC crossover probability p is less than or equal to $1/2$.

$$P_e^{(d)} = \sum_{j=(d+1)/2}^{d} \binom{d}{j} p^j (1-p)^{n-j}$$

$$< \sum_{j=(d+1)/2}^{d} \binom{d}{j} p^{d/2} (1-p)^{d/2} \qquad (8.24b)$$

$$= p^{d/2} (1-p)^{d/2} \sum_{j=(d+1)/2}^{d} \binom{d}{j}$$

The bound on $P_e^{(d)}$ in Eq. (8.24b) can be converted into the Bhattacharyya bound by extending the limits of the summation as follows:

$$P_e^{(d)} < p^{d/2} (1-p)^{d/2} \sum_{j=(d+1)/2}^{d} \binom{d}{j}$$

$$< p^{d/2} (1-p)^{d/2} \sum_{j=0}^{d} \binom{d}{j} \qquad (8.24c)$$

$$= 2^d p^{d/2} (1-p)^{d/2}$$

$$= \left(2\sqrt{p(1-p)}\right)^d .$$

Therefore, in a BSC the maximum-likelihood decoder chooses path $\bar{c}'$ in favor of the correct path $\bar{c}$ if more than $\left\lfloor \dfrac{d}{2} \right\rfloor$ errors occur. The probability of error in the pairwise comparison of $\bar{c}$ with $\bar{c}'$ is upper bounded by

$$P_e^{(d)} \leq \left[2\sqrt{p(1-p)}\right]^d .$$

The above inequality is an upper bound of the first-event probability of an error-path segment of distance d from the all-zero (or correct) path. There are many path sequences with different distances that merge with the correct path at a given time j for the first time. The set-union bound for probability yields an upper bound of the first-error probability by summing the error probabilities of all such possible paths. That is,

$$P_e \leq \sum_{d=d_{free}}^{\infty} w_d P_e^{(d)}, \qquad (8.25)$$

where w_d is the number of codewords of Hamming distance d from the all-zero codeword. A comparison of the generating function in Eq. (8.11) of Example 8.10 with Eq. (8.25) yields the inequality,

Fundamentals of Convolutional Codes

$$P_e \le T(X)|_{X=2\sqrt{p(1-p)}},$$

where $T(x)$ is the generating function for all paths which terminate at state S_0.

Next an upper bound of the average bit-error probability P_b is obtained. This probability can be upper-bounded by weighting each pairwise error probability, $P_e^{(d)}$, in Eq.(8.25) by the number n_d of incorrectly decoded information bits for the corresponding incorrect path. For a rate k/n convolutional code this result is divided by k to yield the following bound for the average bit-error probability :

$$P_e \le \frac{1}{k} \sum_{d=d_{free}}^{\infty} w_d n_d P_e^{(d)}.$$

Once again this upper bound can be expressed in terms of a generating function. Recall that the augmented generating function enumerates the number of information bits for each path. In general, this function with $L=1$ (since one is not interested in path length) is expressed by

$$T(X,Y) = \sum_{d=d_{free}}^{\infty} w_d X^d Y^{n_d}.$$

The derivative of this function with respect to Y and evaluated at $Y=1$ is given by

$$\frac{\partial T(X,Y)}{\partial Y}|_{Y=1} = \sum_{d=d_{free}}^{\infty} w_d n_d X^d. \tag{8.26}$$

Comparing Eqs. (8.25) and (8.26) yields

$$P_b \le \frac{1}{k} \frac{\partial T(X,Y)}{\partial Y}|_{Y=1, X=2\sqrt{p(1-p)}}.$$

Next consider the performance of convolutional codes when the receiver performs soft-decisions followed by a maximum-likelihood decoder. As an example of such a decoding algorithm consider binary antipodal transmission through an AWGN channel. Also assume again that the correct path sequence c and the incorrect path sequence c', shown in Figure 8.15, differ in d bit positions. The pairwise error probability for such a path sequence is

$$P_e^{(d)} = \frac{1}{2} \text{erfc}\left(\sqrt{\frac{dE_0}{N_0}}\right), \qquad (8.27)$$

where E_0 represents the average energy per transmitted channel symbol and $\text{erfc}(x) = 1 - 2Q(\sqrt{2}x)$ is called the error function. Upper-bounding the error function $\text{erfc}(x)$ by $\exp(-x^2)$, the pairwise error probability in Eq.(8.27) is bounded by

$$P_d \leq \frac{1}{2} e^{-\frac{dE_0}{N_0}}. \qquad (8.28)$$

A substitution of Eq.(8.28) into Eq.(8.25) and a comparison with Eq.(8.11) yields finally the bound,

$$P_e \leq \frac{1}{2} T(X)|_{X=\exp(-\frac{E_0}{N_0})}.$$

Also, an upper bound on the bit-error probability for a binary input AWGN channel is obtained as follows:

$$P_b \leq \frac{1}{2k} \frac{\partial T(X,Y)}{\partial Y}\Big|_{Y=1, X=\exp(-\frac{E_0}{N_0})}.$$

Tighter bounds can be found by a use of the useful inequality,

$$\text{erfc}(\sqrt{x+y}) \leq \text{erfc}(\sqrt{x})e^{-y}, \quad x \geq 0, y \geq 0. \qquad (8.29)$$

Recall that $d \geq d_{free}$ so that Eq.(8.27) can be written as

$$\begin{aligned} P_e^{(d)} &= \frac{1}{2} \text{erfc}\left(\sqrt{\frac{d_{free}E_0}{N_0} + \frac{(d-d_{free})E_0}{N_0}}\right) \\ &\leq \frac{1}{2} \text{erfc}\left(\sqrt{\frac{d_{free}E_0}{N_0}}\right) e^{-\frac{(d-d_{free})E_0}{N_0}}. \end{aligned} \qquad (8.30)$$

Next a substitution of Eq.(8.27) into Eq.(8.25) and a comparison with Eq.(8.11) yields the bound,

$$P_e \leq \frac{1}{2} \text{erfc}\left(\sqrt{\frac{d_{free}E_0}{N_0}}\right) e^{\frac{d_{free}E_0}{N_0}} T(X)\Big|_{X=\exp(-\frac{E_0}{N_0})}.$$

Similarly, the bit-error probability is bounded by

Fundamentals of Convolutional Codes

$$P_b \leq \frac{1}{2k} erfc\left(\sqrt{\frac{d_{free} E_0}{N_0}}\right) e^{\frac{d_{free} E_0}{N_0}} \frac{\partial T(X,Y)}{\partial Y}\Big|_{Y=1, X=\exp(-\frac{E_0}{N_0})} . \quad (8.31)$$

In order to compare coded systems with uncoded systems, one needs to evaluate the performance of a coded system in terms of energy per information-bit, E_b, rather than the signal energy per channel-symbol, E_0. For a code with rate k/n the energy per information bit, E_b, is defined by

$$E_b \equiv \frac{E_0}{(k/n)}. \quad (8.32)$$

Using Eq.(8.32) the error bounds of convolutional codes, derived in this section, can be re-expressed in terms of energy per information bit.

To compare, using the above upper bounds, the performance between soft-decision decoding and hard-decision decoding note that the difference is about 2.5-3 dB (about 2dB by the results of simulations for $10^{-2} \leq P_b \leq 10^{-6}$.). Although this comparison is based on the performance of a specific code it can be generalized by a random coding argument[20]. Such an analysis shows that about a 2-3 dB gain is obtained using soft-decision decoding over the entire range of signal-to-noise ratios.

8.7 Punctured Convolutional Codes of Rate $(n-1)/n$ with Simplified Maximum-Likelihood Decoding

In many bandwidth-constrained applications, high-rate or low-redundancy, convolutional codes are desirable. However, the implementation of Viterbi decoders for these codes is often quite complicated. For the (n_0, k_0, m) binary convolutional code the principal practical limitation of the Viterbi algorithm is that the complexity of decoding is proportional to $2^{k_0 \cdot m}$, the number of encoder states. As noted in Section 9.6, the decoder memory is proportional to $2^{k_0 \cdot m}$. Also, since $2^{k_0 \cdot m}$ comparisons must be performed per unit time, the sequential decoding time is proportional to $2^{k_0 \cdot m}$. This exponential dependence on $k_0 \cdot m$ is what, at the present time, limits the practical use of the Viterbi algorithm to $k_0 \cdot m \leq 15$.

The speed limitations of the Viterbi algorithm can be alleviated to a considerable extent by employing parallel decoding. Since the $2^{k_0 \cdot m}$

comparisons that must be performed at each time unit are identical, $2^{k_0 \cdot m}$ identical processors can be used to perform the comparisons in parallel rather than having one processor perform all $2^{k_0 \cdot m}$ comparisons in sequence. To accomplish parallel processing each identical processor computes the metric values of the $2^{k_0 \cdot m}$ paths which enter some state, select a path with the best metric, and then store that path and its metric in the memory of the decoder. This parallel implementation of the Viterbi algorithm requires $2^{k_0 \cdot m}$ processors, each doing one computation per unit time in contrast with having only one processor that sequentially computes $2^{k_0 \cdot m}$ operations per unit time. Parallel decoding therefore yields a factor-of-$2^{k_0 \cdot m}$ speed advantage over the standard decoder, but with $2^{k_0 \cdot m}$ times as much hardware. However, for a high-rate code, e.g. an $(n, n-1, m)$ convolutional code, the parallel decoder still might require too much hardware. This complexity is critical especially in high-speed applications. Furthermore, if a change in rate is desired, a fundamentally different decoding structure would be required.

This complexity of the decoder can be overcome in many cases by a process of puncturing, first introduced to the convolutional coding by Cain, Clark and Geist[18]. The idea of a punctured convolutional code is very similar to the concept of puncturing a block code.. One simply deletes certain code symbols from a lower-rate code, while keeping the parameter k_0 fixed, to obtain a convolutional code with an effectively higher rate. This puncturing is performed in a periodic manner, thereby creating a periodically time-varying code.

It is known that the implementation of the Viterbi algorithms for high-rate convolutional codes can be simplified greatly if this code is generated by puncturing a low-rate code. In the standard approach to the decoding of rate-$(n-1)/n$ codes the implementation is complicated by the code structure which has 2^{n-1} paths entering each state rather than only the two paths needed for a rate-$1/n$ code. This makes the resulting comparisons and selection of the path with the best metric much more difficult. In this section some rate-$(n-1)/n$ codes are introduced by examples. These rate-$(n-1)/n$ codes are obtained by deleting or puncturing certain bits from the code words of a code of rate less than or equal to $1/n^*$, where $n^* \leq n$. These codes are decoded using the Viterbi algorithm with roughly the same decoding complexity that a rate-$1/n^*$ code would require. This punctured-code approach is illustrated next by two examples.

Fundamentals of Convolutional Codes

Example 8.15.
Consider first the (3,2,2), 2/3-rate, convolutional code with the generator matrix

$$G(D) = \begin{pmatrix} 1+D & 1+D & 1 \\ D & 0 & 1+D \end{pmatrix}.$$

The encoder block diagram and trellis structure for this code are shown in Figures 8.17a and 8.17b.

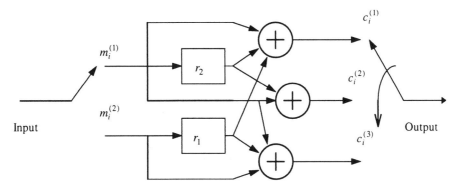

Figure 8.17a. The (3,2,2) 2/3-rate convolutional encoder.

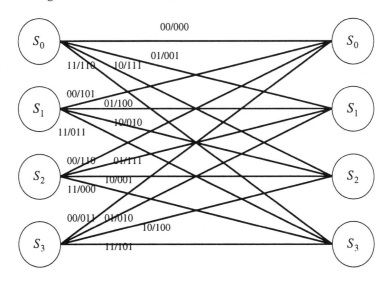

Figure 8.17b. Trellis structure for the (3,2,2) convolutional code

In the trellis diagram in Fig. 8.16b, the state of this encoder in Fig. 8.17a at unit time i is defined as the contents of the two registers r_1 and r_2: $(m_{i-1}^{(2)}, m_{i-1}^{(1)})$, i.e. $S_0 = (0\,0), S_1 = (1\,0), S_2 = (0\,1), S_3 = (1\,1)$. The input and output pairs are denoted by $m_i^{(1)} m_i^{(2)} / c_i^{(1)} c_i^{(2)} c_i^{(3)}$.

When this code is decoded by the Viterbi algorithm in the conventional manner, a 4-ary comparison needs to be made at each state, and one such 4-ary comparison per state needs to be made for every two information bits. The number of such operations contrasts dramatically with the much simpler binary comparisons needed to decode a rate-1/n code.

Consider now how a 2/3-rate punctured code can be obtained from the (2,1,2) convolutional code generated by the generator matrix,

$$G(D) = \left(1 + D + D^2, \quad 1 + D^2\right).$$

If every fourth output bit of this encoder is deleted, this code produces three channel bits for every two information bits, i.e. it is a rate-2/3 code. In fact, if the bit from the second generator polynomial $1 + D^2$ is deleted from every other branch, the resulting punctured code is identical with the (3,2,2) convolutional code discussed above. Formally, this punctured-code encoding process is explained as follows : the outputs of the (2,1,2) convolutional encoder are selected in accordance with the above puncturing matrix, $[P_1, P_2] = [11,10]$, where "0" denotes no- transmission, "1" denotes a transmission. This process produces a single bit-serial stream. The puncturing matrix converts by this means the rate-1/2 encoder into a rate-2/3 encoder. The internal structure of the punctured encoder is illustrated in Figure 8.18a.

This punctured code has the trellis diagram shown in Figure 8.18b where X indicates the deleted bits. In a manner similar to that used in the Example 8.3, a state in the diagram is defined to be the contents of the 2-stage shift register. At time unit i the state is expressed by (m_{i-1}, m_{i-2}). Therefore, the four states are denoted by $S_0 = (0\,0), S_1 = (1\,0), S_2 = (0\,1), S_3 = (1\,1)$.

Notice that the transitions between states and the resulting transmitted bits are identical in Figures 8.17b and 8.18b, but in Figure 8.18b the transitions occur by means of a set of intermediate states since only one bit at a time is shifted into the encoder rather than two bits. Clearly we have succeeded in generating the same code in a different manner.

Fundamentals of Convolutional Codes

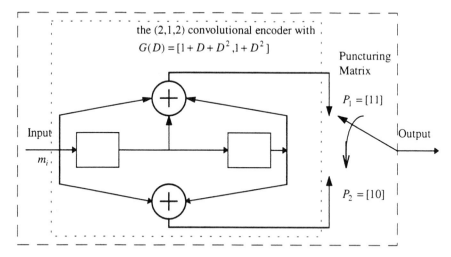

Figure 8.18a. A punctured 2/3 rate convolutional coder : every 2-bit sequential input yields a 3-bit sequential output

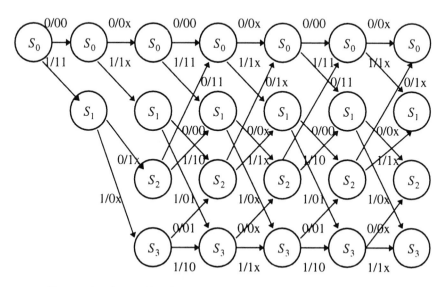

Figure 8.18b. Trellis diagram of the (2,1,2) convolutional code

In this example an (3,2,2) code was constructed by periodically deleting bits from an (2,1,2) code or, in other words, by puncturing the code. In general, puncturing the code reduces its free distance, in this case from 5 to 3. However, this distance is as large as could be achieved with any (3,2,2) convolutional code. A study in [17] shows that there are no free-distance 4 (3,2,2) codes. Thus in this case no further reduction in the free distance can be achieved by the use of a punctured code.

It can be verified that the above-discussed (3,2,2) punctured-code can also be thought of as an (3,1,2) code with the generator matrix

$$G(D) = \left(1+D+D^2 \quad 1+D^2 \quad 1+D+D^2\right),$$

that has the first two bits deleted on the first branch, and then the third bit is deleted on the next branch. Such a fact shows that the generators of a punctured code are in general not unique.

The above discussion suggests that there are two important searches needed to discover efficient punctured-codes : (1) A search for the best punctured-code generators of constraint length v. (2) A hunt for the simplest generator for a selected punctured-code.

An extensive search was performed in [18] to find the best punctured-code generators at each constraint length v. The definition of "best" was in terms of that code which had the best performance over the AWGN channel at a large signal-to-noise ratio. This asymptotic performance probability for a rate-(n-1)/n code was found [19] to be

$$P_e \approx \frac{w_i(d)}{2(n-1)} \, erfc\left(\sqrt{\frac{(n-1) \cdot d}{n} \cdot \frac{E_b}{N_0}}\right),$$

where d is the free distance of the code and $w_i(d)$ is the total input weight of all information sequences which produce weight d paths. All codes with the largest free distance were exhaustively searched in order to find the minimum $w_i(d)$.

While the above criterion does not guarantee that the code selected is the best for any specified error rate, say 10^{-5}, the resulting codes have nearly optimal performance at interesting values of the output bit-error rate. Of course, these codes also must be checked to verify that they are noncatastrophic.

Some of the best punctured codes are listed in Tables 8.5 and 8.6. These codes were found by Cain, Clark, and Geist [18].

Fundamentals of Convolutional Codes

Constraint length K	Code generators (octal)	d_{free}
3	5,7,5,7	3
4	64,74,64,74	4
5	62,56,46,46	4
6	61,47,65,65	5
7	724,534,474,754	6

Table 8.5 Some best rate-3/4 punctured convolutional codes([18], @ 1979 IEEE)

The practical benefits of the punctured-code approach are obvious. This technique makes it possible to implement a rate-2/3 decoder as a rate-1/2 decoder with the additional ability to stuff erasures into the locations of the deleted bits. After these erasures are stuffed into the control sequence, the decoding proceeds just as if the code were a rate-1/2 code. This procedure simplifies the classical Viterbi decoding procedure of the (n-1)/n-rate code by replacing the complex 2^{n-1} – ary comparisons at each state by binary comparisons.

Constraint length K	Code generators (octal)	d_{free}
3	7,5,7	3
4	44,74,64	4
5	62,56,52	5
6	61,53,57	6
7	714,564,714	6
8	676,522,676	8

Table 8.6 Some best rate-2/3 punctured convolutional codes([18], @ 1979 IEEE)

In a high-speed decoder applications, such decoder simplifications have had a major impact on decoder complexity. Even at low data rates, when a serial implementation was used, the standard approach required a factor of $(2^{n-1}-1)/(n-1)$ more binary comparisons per decoded bit than the punctured-code approach.

The punctured-code concept also can be used to provide a high-speed selectable-rate decoder. To do this efficiently one must find two unique generator polynomials which can simultaneously generate good 1/2, 2/3, and 3/4 rate codes. Thus the code rate can be changed simply by changing the manner in which the generators are sampled.

Problems

8.1 A rate 1/2 binary convolutional code is defined by the generator matrix,
$$G(D) = [1, 1 + D + D^2].$$

(a) Draw the encoder block diagram.
(b) What is the encoder constraint length and the memory order?
(c) Is the code produced in systematic form?
(d) How many states does the encoder have?
(e) How many trellis branches enter/exit each state?

8.2 Consider the (2,1,2) convolutional code with
$$G(D) = [1 + D + D^2, 1 + D^2].$$

(a) Draw an encoder block diagram.
(b) Draw the state diagram.
(c) Find the minimum free distance d_{free} of this code.
(d) Draw the trellis diagram for this code with an information sequence of length $L = 5$.
(e) For a BSC with $p < 0.5$ decode $\bar{r} = (10, 10, 00, 01, 11, 01, 11)$ to find both the estimated code sequence $\hat{c}$ and the message sequence $\hat{m}$ by a use of the Viterbi algorithm (show the decoding process).

8.3 Consider the (2,1,3) convolutional code with
$$G(D) = [1 + D^2, 1 + D + D^2 + D^3],$$

(a) Draw a block diagram of the encoder.
(b) Draw the state diagram.
(c) Find the code word $\bar{c}$ corresponding with the information sequence $\bar{m} = (1010)$.
(d) Draw the trellis diagram for this code with an information sequence of length $L = 4$.
(e) Draw the code tree for this code with $L=4$.

(f) For a BSC with $p < 0.5$, decode $\bar{r} = (10,01,00,00,11,11,00)$ to find both the estimated code sequence $\hat{c}$ and the message sequence $\hat{m}$ by a use of the Viterbi algorithm.

8.4 A rate 2/3 binary convolutional encoder is specified by the generator matrix,
$$G(D) = \begin{pmatrix} 1+D & 1 & D \\ 1+D+D^2 & 1+D & D^2 \end{pmatrix}.$$
Is the code, generated by G(D), catastrophic?

8.5 If a convolutional code can be implemented by the use of a systematic encoder, show that such a code is non-catastrophic.

8.6 For non-systematic encoders a pertinent question is whether one can recover the input sequence $\bar{m}(D)$ from the code stream in the absence of errors. The non-systematic convolutional- code system, defined by the generator matrix, $G(D)$ is called an inverse system if
$$\bar{c}(D)G'(D) = \bar{m}(D)G(D)G'(D) = m(D)D^l.$$
That is, this code recovers the original message sequence with a delay of l shift times. $G'(D)$ is the system function for this inverse. Suppose that a rate-1/2 non-systematic code is specified by $G(D) = (1+D+D^3, 1+D+D^2+D^3)$.

(a) Determine an inverse function $G'(D)$.
(b) Draw the block diagram for such an inverse circuit.
(c) Prove that if an inversion circuit doesn't exist for a convolutional code, that it is catastrophic.

8.7 Consider the rate-1/2 convolutional code given in 8.1.

(a) Using the puncturing matrix $(P_1, P_2) = (1111010, 1000101)$ generate a rate-7/8 punctured code from the mother code of rate 1/2.
(b) Determine the trellis diagram.

8.8 Construct an encoder for the convolutional code which is specified by the following generator matrix,
$$G(D)\begin{pmatrix} 1+D+D^2 & 1+D+D^3 & 1+D^2+D^3 \\ 1+D^3 & 1+D^2 & 1+D+D^3 \end{pmatrix},$$

such that this encoder has a minimum memory.

8.9 Show the log likelihood functions of the normalized transition probabilities given in Eq.(8.23) are equivalent to the log likelihood functions $\sum_{u=0}^{l+m-1} \log \left(P\left(r_u^q = b_l | s_u = a_i \right) \right)$ of the transition probabilities. Why is this likelihood function preferred to Eq.(8.23) ?

Bibliography

[1] P. Elias, "Coding for Noisy Channels", IRE Conv. Rec., Part 4, pp.37-47, 1955.
[2] E. A. Bucher and J. A. Heller, "Error probability bounds for systematic convolutional codes", IEEE Trans. Inform. Theory, Vol. IT-16, pp.219-224, March 1970.
[3] J. L. Massey and M. K. Sain, "Codes, automata, and continuous systems : explicit interconnections", IEEE Trans. Automatic Control, Vol. AC-12, pp.644-650, Dec. 1967.
[4] D. G. Forney, Jr., "Convolutional codes I : algebraic structure", IEEE Trans. Inform. Theory, Vol. IT-16, pp. 720-738, Nov. 1970.
[5] L.R. Bahl, C. D. Cullum, W. D. Frazer, and F. Jelinek, "An efficient algorithm for computing free distance", IEEE Trans. Inform. Theory, Vol. IT-18, pp. 437-439, May 1972.
[6] K. J. Larsen, "Comments on 'An efficient algorithm for computing free distance' ", IEEE Trans. Inform. Theory, Vol. IT-19, pp.577-579, July 1973.
[7] J. M. Wozencraft and B. Reiffen, Sequential Decoding, Cambridge, Mass.: MIT Press, 1961.
[8] R. M. Fano, "A heuristic discussion of probabilistic decoding", IEEE Trans. Inform. Theory, Vol. IT-9, pp. 64-74, Apr. 1963.
[9] F. Jelinek, "A fast sequential decoding algorithm using a stack", IBM J. Research and Development, Vol. 13, pp.675-685, Nov. 1969.
[10] J. L. Massey, Threshold Decoding, Cambridge, Mass. : MIT Press, 1963.
[11] I. S. Reed, "A class of multiple-error-correcting codes and the decoding scheme", IRE Trans., Vol. IT-4, pp.38-49, Sept. 1954.
[12] A. J. Viterbi, " Error bounds for convolutional codes and an asymptotically optimum decoding algorithm", IEEE Trans. Inform. Theory, Vol. IT-13, pp. 260-269, April 1967.

Fundamentals of Convolutional Codes

[13] J. K. Omura, "On the Viterbi decoding algorithm", IEEE Trans. Inform. Theory, Vol. IT-15, pp.177-179, Jan. 1969.
[14] G. D. Forney, Jr., " The Viterbi algorithm", Proc. IEEE, 61, pp. 268-278, March 1973.
[15] G. D. Forney, Jr., " Convolutional Code II : Maximum likelihood decoding", Information and Control, 25, pp.222-266, July 1974.
[16] Shu Lin and D. J. Costello, Jr. Error Control Coding : Fundamentals and Applications, Englewood Cliffs, NJ : Prentice-Hall, 1983.
[17] E. Paaske, "Short binary convolutional codes with maximal free distance for rate 2/3 and 3/4", IEEE Trans. Inform. Theory, vol. IT-20, pp.683-688, Sept. 1974.
[18] J. B. Cain, G. C. Clark, and J. M. Geist, "Punctured convolutional codes of rate (n-1)/n and simplified maximum likelihood decoding", IEEE Trans. on Inform. Theory, Vol. IT-25, No. 1, pp.97-101, Jan. 1979.
[19] A. J. Viterbi, "Convolutional codes and their performance in communication systems", IEEE Trans. on Communication, Vol . COM-19, pp.751-772, Oct. 1971.
[20] J. M. Wozencraft and I. M. Jacobs, Principles of Communication Engineering, New York : John Wiley & Sons, 1965.
[21] J. A. Heller and I. M. Jacobs, "Viterbi decoding for satellite and space communication," IEEE Trans. on Communications, vol. COM-19, pp.835-848, Oct. 1971.
[22] http://hideki.iis.u-tokyo.ac.jp/~robert/codes.html.
[23] G. C. Clark,Jr. and J. B. Cain, Error-Correction Coding for Digital Communications, Plenum Press, New York, 1981.

9 ARQ and Interleaving Techniques

Modern communication systems usually need some form of error control to achieve a low bit-error rate and to guarantee a stipulated constant quality of service. All of the error-control systems discussed in the preceding eight chapters can be used on channels in which the message flow is in only one direction. Such error-control systems are called forward error-correcting (FEC) systems. However, in many applications data flow is bi-directional. In such cases messages are sent over the "forward" channel and acknowledgment signals are sent back from the "return" channel. The simplest error control for channels with feedback is the technique, called automatic repeat-request (ARQ) protocol[2]. In these protocols the transmitted data are encoded for error detection; detected errors at the receiver result in the generation of a retransmission request. ARQ provides the best performance on pure-burst channels with memory while most FEC schemes work efficiently only on memoryless channels.

9.1 Automatic Repeat Request

In many applications the communications channel is error-free except for occasional bursts of noise of short duration. Such noise bursts may be caused, for example, by nearby electric power machinery, electrical storms, and single-event upsets in digital hardware. In such situations a simple error-detecting ARQ protocol can provide a great deal of protection. The goal is to design a ARQ system that detects the error burst, discards the affected data packet and requests a retransmission.

In a typical ARQ protocol the message sequence is broken up into packets of length k. Each of these packets is encoded using a high-rate binary error-detecting code C of length n. If the code C is a linear block code of dimension k and is capable of detecting any error pattern that is not in itself a codeword, it is known from Chapter 3 that there are a total of $2^n - 2^k$ detectable error patterns. The most frequently used error-detecting codes are the CRC codes discussed in Chapter 4. CRC encoders and decoders are implemented easily by the use of shift-registers. These codes are also very flexible in the sense that a single encoder/decoder pair can be used with packets of widely varying lengths. On the other hand, the error-detecting capabilities of most CRC codes for most other applications tend to be suboptimal.

To design an ARQ protocol the trade-off between reliability and throughput needs to be evaluated. In a FEC system, reliability is measured by the bit- or symbol-error rate. In an ARQ system reliability is often measured by the packet-error rate. The packet-error rate is defined as the percentage of error packets among the total number of received packets.

The computation of the packet error rate for an ARQ protocol is obtained next. An error packet is received if it arrives at the receiver, with an undetectable error pattern. Let P_{ud} be the probability of an undetected error of the code C and let P_d be the probability of a detected error, which also is the probability that a retransmission request is generated for a given transmission. Assume that the feedback (return) channel is error-free. Then the packet-error rate P_e is computed by a summation of the probabilities of the various events that result in the acceptance of an erroneous packet, i.e.

$$P_e = \sum_{j=0}^{\infty} P_d^j P_{ud} = \frac{P_{ud}}{1-P_d}, \qquad (9.1)$$

where $P_{ud} + P_d \leq 1$ and $P_d < 1$. Eq. (9.1) implies that for a given (fixed) undetected error probability the packet-error rate increases with the probability of a retransmission request. This is true since each additional retransmission increases the possibility of further errors.

In a FEC system the message throughput equals the code rate $R = k/n$. However, the throughput of an ARQ system is defined as the ratio k/T_p where T_p is the average time in bit-times (or bits) needed to transmit a single k-bit message packet.

ARQ and Interleaving Techniques

Let n_t be the average time in terms of bits for a successful single transmission of a k-bit message packet and also let N_p be the average number of times a packet has to be transmitted before it is successfully accepted by the receiver. Then, the throughput of an ARQ system satisfies $\lambda = k/T_p = k/(n_t N_p)$. Given that the probability P_d of a detected error is known, the average number of repeated packet transmissions N_p can be computed as follows:

$$N_p = \sum_{j=1}^{\infty} j P_d^{j-1}(1-P_d) = \frac{1}{1-P_d}. \tag{9.2}$$

From the above the throughput of an ARQ system is inversely proportional to the time parameter n_t which varies from protocol to protocol. In selecting a retransmission protocol, the designer must strike a balance between the complexity and the throughput of the ARQ system. For most of the data-network systems there are four basic retransmission protocols that can be selected : the simplex stop-and-wait protocol, the 1-bit sliding window protocol, the go-back-n protocol, and the selective-repeat protocol.

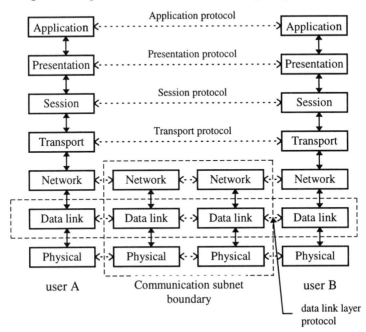

Figure 9.1 The OSI reference model.

A modern communications network has a layered architecture. One of commonly-used architecture is called the OSI (Open Systems Interconnection) reference model, shown in Figure 9.1. In layered networks, the retransmission protocols discussed above reside in the data-link layer while most of FECs reside in the physical layer. Four different ARQ protocols in the data-link layer are introduced next in the order of their increasing complexity. Before looking at these protocols, it is useful to make some explicit assumptions about the model for a communication system. First, assume that the physical layer, the data-link layer, and the network-layer are independent processes that communicate by transmitting messages back and forth. In some cases, the physical and data-link layer processes run on a processor which is inside a special network interface chip and the network layer on the main processor. But other implementations are also possible, e.g. the three processes are inside a single interface chip; the physical and data-link layers as procedures are called upon by the network-layer process.

The packets from the network layer are passed across the interface to the data-link layer. These data are to be delivered to the destination's network layer. The fact that the network layer of the destination might interpret part of the packet as a header is of no concern to the data-link layer.

When the data-link layer accepts a packet, it encapsulates the packet in a frame (a block of data) by adding a data-link header and trailer to it. Thus a frame consists of an embedded packet and some header (control) information. The frame is then transmitted to the other data-link layer. Assume that there exist some software procedures in a library such as : *send_to_physical_layer* to send a frame, and *get_from_physical_layer* to receive a frame. The transmitting device computes and appends the parity-check bits (or the checksum) to the data-link layer packets.

Initially, the receiver waits for the transmitted data to arrive. In this section it is assumed that such a process is indicated by the procedure call, *wait_for_event(&event)*. This procedure returns only when the event has happened (e.g. a frame has arrived). Upon a return the variable, *event*, tells what happened. The set of possible events differs for the various protocols to be described in what follows.

When a frame arrives at the receiver, the receiving device computes the parity-check bits (or the checksum). If the parity-check is incorrect, which means there is a transmission error, the data-link layer is then informed as follows : *event=parity_check_error*. If the parity-check is correct, the frame that has arrived is assumed to be undamaged, also the data-link layer is informed by the information : *event=frame_arrival*, so it can acquire the frame for inspection by the function call, *get_from_physical_layer*. As soon as the data-link layer in the receiving side has acquired an undamaged frame, the packet portion of the frame is passed to the network layer.

Some declarations of procedures in pseudo-code (C-language type[7][8]) are shown below. These declarations are common to four of the protocols to be discussed later. Assume that the maximum packet size in bytes is defined by MAX_SIZE, e.g. MAX_SIZE=512.

typedef enum {false, true} boolean;   /* boolean type definition */

```
struct packet {
   unsigned char data_array[MAX_SIZE];
};                                            /* packet definition */

typedef enum {data, ack, nak} frame_type;     /* frame type definition */

struct frame {                                /* frame are transported in this layer */
   frame_type type;                           /* frame type information */
   unsigned int seq;                          /* sequence number */
   unsigned int ack;                          /* acknowledgment   */
   packet net_packet;                         /* the network layer packet */
};

void wait_for_event(event_type *event);       /* Wait for an event to happen */

void get_from_network_layer(packet *p);   /* Get a packet from the network layer */

void send_to_physical_layer(frame *s);
          /* Send the frame in s to the physical layer for transmission */

void get_from_physical_layer(frame *r);
          /* Get a received frame from the physical layer and store it in r. */

void send_to_network_layer(packet *p);
                    /*Send data from a received frame to the network layer */

void start_timer(unsigned int k);   /* Start the timer and enable the time-out event */

void stop_timer(unsigned int k);    /* Stop the timer and disable the time-out event */
```

#define inc(k) k = k<MAX_SEQ ? k+1 : 0
/* Macro inc is expanded in-line : Increment k circularly */

Five data structures are described next : *boolean, seq_nr, packet, frame_kind,* and *frame*. A *boolean* is an enumerated type which can take on the values, *true* and *false*. A *seq_nr* is a small integer used to number the frames. These sequence integer numbers run from 0 up to *MAX_SEQ*, which is defined in each protocol that needs it. A *packet* is the unit of information exchanged between the network layer and the data-link layer on the same machine, or between network-layer peers. A *frame* is composed of four fields : *kind, seq, ack,* and *info*. The first three of which contain control information, and the last of which may contain actual data to be transferred. These control fields are called collectively the frame header. The *kind* field tells whether or not there are any data in the frame, because some of the protocols distinguish frames that contain exclusively control information from those that contain data. The *seq* and *ack* fields are used for the sequence numbers and acknowledgments, respectively; their use is described in more detail later. The *info* field of a data frame contains a single packet.

The procedure, **wait_for_event**, sits in a tight loop waiting for something to happen, as mentioned earlier. The procedures, **send_to_network_layer** and **get_from_network_layer**, are used by the data-link layer to pass packets to the network layer and to accept packets from the network layer, respectively. Note that the function calls, **get_from_physical_layer** and **send_to_physical_layer**, are used for passing frames between the data-link and physical layers, whereas the procedures, **send_to_network_layer** and **get_from_network_layer**, are used to pass packets between the data link and network layers.

9.1.1 A Stop-and-Wait Protocol

This protocol consists of two distinct procedures, one for the sender and the other for the receiver. The sender runs in the data-link layer of the source computer, and the receiver runs in the data-link layer of the destination computer. Protocols in which the sender delivers one frame and then waits for an acknowledgement before proceeding are called stop-and-wait.

In this mode the sender is in, what is called, a while loop which simply pumps data out onto the line as fast as it can. The body of the loop consists of three actions: fetch a packet from the (always obliging) network layer,

ARQ and Interleaving Techniques

construct an outgoing frame using the variable *s* (for send), and send the frame on its way. The sender must wait until an acknowledgement frame arrives before looping back and fetching the next packet from the network layer.

In the stop-and-wait protocol the receiver is equally simple. Initially, it waits for something to happen. The three possible types of events are *frame_arrival*, *parity_check_error*, and *timeout*. Eventually, the frame arrives and the procedure, *wait_for_event*, returns with *event=frame_arrival*. The function call, *get_from_physical_layer*, removes the newly arrived frame from the buffer and labels it with the variable *r*. Finally, the data portion is passed on to the network layer and the data-link layer settles back to wait for the next frame, thereby effectively suspending itself until the frame arrives.

Although data traffic is simplex, in the sense of going only from a sender to a receiver, frames do travel in both directions. Consequently, the communications channel between the two data-link layers needs to be capable of bi-directional information transfer. However, this protocol entails a strict alternation of the flow: first the sender sends a frame, then the receiver sends a frame, and finally the next sender sends other frames in sequence, etc. What is called, a half-duplex physical channel usually suffices for this type of protocol.

It is assumed that if a frame is damaged by errors while in transit, the receiving device detects this fact when it computes the parity-check (or checksum). If the frame is damaged in such a manner that the parity-check is correct, but, in fact, wrong, which is an exceedingly unlikely occurrence, this protocol (and all other such protocols) fail and an incorrect packet is delivered to the network layer).

Clearly, what is needed is some method for the receiver to be able to distinguish a frame, that it is observing for the first time from a re-transmitted frame. The obvious way to achieve this is to have the sender put a sequencing number in the header of each frame it sends. Then the receiver can check the sequence number of each arriving frame to see if it is a new frame or a duplicate of some previously received, but corrupted, frame.

An example of this kind of protocol is given in the procedure, given in the next example.

Example 9.1
/* Protocol 1 (stop-and-wait) provides for a one-directional flow of data from sender to receiver. */

```
#define MAX_SEQ 1                              /* set to 1 for this protocol */
typedef enum {frame_arrival, parity_check_error, timeout} even_type;

void sender( )
{
  unsigned int next_frame_to_send=0;
                              /*initialize seq number of next sending frame */
  frame send;
  packet buffer:                              /* buffer for a sending packet */
  event_type event;

  get_from_network_layer(&buffer);                     /* get first packet */
  while (buffer != NULL) {
        send.net_packet = buffer;    /* prepare a frame of data for transmission */
        send.seq = next_frame_to_send;   /* assign sequence number to the frame */
        send_to_physical_layer(& send);       /* send the constructed frame */
        start_timer(send.seq);            /* start the counting for time out */
        wait_for_event(& event);    /* frame_arrival parity_check_error, timeout */

        if (event == frame_arrival) {
          get_from_physical_layer(& send);       /* get the acknowledgement */
          if (send.ack == next_frame_to_send) {
            get_from_network_layer(& buffer);    /* get the next one to send */
            inc(next_frame_to_send);      /* advance count next_frame_to_send */
          }
        }
    }
}

void receiver( )
{
  unsigned int frame_expected=0;
  frame receive, send;
  event_type event:

  for(;;) {
        wait_for_event(& event);
                           /* possibilities : frame_arrival, parity_check_error */
```

```
        if (event == frame_arrival) {              /* a valid frame has arrived */
            get_from_physical_layer(& receive);    /* get the newly arrived frame */
            if (receive.seq == frame_expected) {   /* expect frame */
                send_to_network_layer(& receive.net_packet);
                                /* pass the data of the frame to the network layer */
                inc(frame_expected);   /* next time expect the other sequence number */
            }
            send.ack = 1 - frame_expected;    /* generate negative ack for the frame */
            send_to_physical_layer(& send);       /* none of the fields are used */
        }
    }
}
```

Protocols in which the sender waits for a positive acknowledgement before advancing to the next item of data are also, sometimes, called Positive Acknowledgement with Retransmission (PAR). Although it can handle lost frames (by a "timing out"), it requires the timeout interval be long enough to prevent premature time-outs. If the sender times out too early, while the acknowledgement is still on the way, it sends a duplicate.

When the previous acknowledgement finally arrives, the sender mistakenly could think that the just-sent frame is the one being acknowledged and would not realize that there is potentially another acknowledgement frame somewhere "in the pipe". If the next frame that is sent is lost completely, but the extra acknowledgement arrives correctly, the sender would not know to retransmit the lost frame, and the protocol would fail. To prevent this kind of failure one needs to put in some mechanism for the sender to know which frame is being acknowledged. The simplest approach is to put a sequence number in the acknowledgments, indicating which frame is being acknowledged.

In this protocol both the sender and the receiver have a variable whose value is remembered while the data-link layer is in the wait state. The sender remembers the sequence number of the next frame in order to send the instruction, *next_frame_to_send;* and the receiver remembers the sequence number of the next frame expected in the variable, *frame_expected.* Each protocol has a short initialization phase before it enters a loop.

After transmitting a frame, the sender starts the timing-clock (timer) running. If it was already running, it is reset to allow another full timing interval. This time interval must be chosen to allow enough time for the frame to get to the receiver in order for the receiver to process it in the worst case, and finally to allow the acknowledgement frame to propagate back to the sender.

After a frame is transmitted and the timer started, the sender waits for something to happen. There are three possibilities: an acknowledgement frame arrives undamaged. a damaged acknowledgement frame" staggers in", or the timer goes off. If a valid acknowledgement comes in, the sender fetches the next packet from its network layer and puts it in the buffer, overwriting the previous packet. It also advances the sequence number. If a damaged frame arrives or no frame arrives at all, neither the buffer nor the sequence number are changed, so that a duplicate is in a position to be sent.

When a valid frame arrives at the receiver, its sequence number is checked to see if it is a duplicate. If not, it is accepted, passed to the network layer, and an acknowledgement generated. Duplicates and damaged frames are not passed to the network layer.

9.1.2 A Sliding-Window Protocol

In the previous protocol data frames were transmitted in one direction only. In most practical situations there is a need to transmit data in both directions. One way to achieve full-duplex data transmission is to have two separate communication channels and to use each for simplex data traffic (in different directions). If this is done, one has two separate physical circuits, each with a "forward" channel (for data) and a "reverse" channel (for acknowledgments). In both cases the bandwidth of the reverse channel is almost entirely wasted. In effect, the user is paying for two circuits but using primarily only the capacity of one.

A better idea is to use the same circuit for transmitting/receiving data in both directions. In *Protocol* 1, given above, already frames are transmitted in both directions; the reverse channel is assumed to have the same capacity as the forward channel. In the present model the data frames from sender A to receiver B are intermixed with their acknowledgement frames transmitted from A to B. In the *"type"* field of the header of an incoming frame it is assumed that the receiver can tell whether the frame is data or an acknowledgement.

Although the interleaving of data and the control frames on the same circuit is an improvement over having two separate physical circuits, yet another improvement is possible. When a data frame arrives, instead of immediately sending a separate control frame, the receiver restrains itself and waits until the network layer can pass it the next packet. Then an acknowledgement is attached to the outgoing data frame (using the *"ack"* field in the frame header). In effect, the acknowledgement gets a free ride on the next outgoing data frame. This technique of temporarily delaying outgoing acknowledgments so that they can be attached to the next outgoing data frame is known as piggybacking.

The principal advantage of piggybacking over having other distinct acknowledgement frames is that it makes a better use of the available channel bandwidth. The *ack* field in the frame header costs only a few bits, whereas a separate frame would need an extra header, an acknowledgement and a checksum. In addition, the fewer frames sent means the fewer, "frame-arrived", interrupts and perhaps fewer buffers in the receiver, depending on how the receiver's software is organized. In the next protocol to be examined, the piggyback field costs only 1 bit in the frame header. In practice it rarely costs more than a few bits.

However, piggybacking introduces a complication not present in protocols which use separate acknowledgments. How long should the data-link layer wait for a packet on which the acknowledgement is to be piggybacked? If the data-link layer needs to wait longer than the sender's timeout period, the frame is retransmitted, defeating the purpose of acknowledgments. If the data-link layer were an oracle and could foretell the future, it would know when the next network layer packet was going to arrive, and could decide either to wait for it or to send immediately a separate acknowledgement, depending on how long the projected wait lasted. Of course, the data-link layer cannot foretell the future, so it must resort to some ad hoc scheme, such as waiting for a fixed number of milliseconds. If a new packet arrives quickly, the acknowledgement is piggybacked onto it; otherwise, if no new packet arrives by the end of this time period, the data-link layer simply sends a separate acknowledgement frame.

In addition to it being only simplex, **Protocol** 1, an given in Example 9.1 can fail under certain peculiar conditions which involve an early timeout. It would be nice to have a protocol that remained synchronized in the face of any combination of garbled frames, lost frames, and premature time-outs. The next three protocols are more robust and continue to function even under pathological conditions. All three belong to the class of protocols, called sliding-window protocols. The three differ among themselves in terms of their efficiency, complexity, and buffer requirements, as discussed next.

In all sliding-window protocols each outbound frame contains a sequence number, which ranges from 0 up to some maximum. The maximum is usually $2^n - 1$ so that the sequence number is forced to fit into an n-bit field. The most-used, stop-and-wait, sliding window protocol uses $n = 1$, thereby restricting the sequence numbers to be 0 and 1. However, more sophisticated versions do use an arbitrary n.

The essence of all sliding-window protocols is that at any instant of time, the sender maintains a set of sequence numbers which correspond to the frames it is permitted to send. These frames are said to be within the sending window. Similarly, the receiver also maintains a receiving window which corresponds to the frames that it is permitted to accept. The sender's window and the receiver's window need not have the same lower and upper limits, or even have the same size. In some protocols they are fixed in size, but in others they can grow or shrink as the frames are sent and received.

Although these protocols give the data-link layer more freedom about the order in which frames can be sent and received, it must be emphasized, that the requirement is not dropped, that the protocol must deliver packets to a destination network layer in the same order that they were passed to the data link layer on the sending machine. Nor does the requirement change, that the physical communication channel be "wire-like," that is, it must deliver all frames in the order sent.

The sequence numbers within the sender's window represent the frames sent, but as yet, not acknowledged. Whenever a new packet arrives from the network layer, it is given the next highest sequence number, and the upper edge of the window is advanced by one. When an acknowledgement arrives, the lower edge of the window is advanced by one. In this way the window continuously maintains a compact list of unacknowledged frames.

Since frames currently within the sender's window ultimately may be lost or damaged in transit, the sender must keep all of these frames within its memory for possible retransmission. Thus, if the maximum window size is n, the sender needs n buffer registers to hold the unacknowledged frames. If the window ever grows to its maximum size, the sending data-link layer must forcibly shut off the network layer until another buffer register becomes free.

The receiving data-link layer's window corresponds to the frames it may accept. Any frame that falls outside the window is discarded without comment. When a frame whose sequence number is equal to the lower edge of the window is received, it is passed to the network layer, an

ARQ and Interleaving Techniques 381

acknowledgement is generated, and the window is rotated by one. Unlike the sender's window, the receiver's window always remains at its initial size. Note that a window size of 1 means that the data-link layer only accepts frames in order, but for larger windows this is not so. The network layer, in contrast, is always fed data in the proper order, regardless of the window size of the data-link layer.

Example 9.2
/* Protocol 2 (sliding window) is bidirectional and is more robust than Protocol 1. */
#define MAX_SEO 1   /* set to 1 for protocol 2 */
typedef enum {frame_arrival, parity_check_error, timeout} event_type;

void protocol2 ( )
{
  unsigned int next_frame_to_send=0;   /* initialize next sending frame */
  unsigned int frame_expected = 0;   /* initialize the number of frame arriving */
  frame receive, send;
  packet buffer;   /* current packet being sent */
  event_type event;

  get_from_network_layer(& buffer);   /* get a packet from the network layer */

  send.net_packet=buffer;   /*prepare to send the initial frame */
  send.seq=next_frame_to_send;   /* insert sequence number into frame */
  send.ack=1-frame_expected;   /* piggybacked ack */
  send_to_physical_layer(& send);   /* transmit the frame */
  start_timer(send.seq);   /* start the timer */

  for(;;) {
        wait_for_event(& event);
                                           /*frame_arrival, parity_check_error timeout */
        if (event == frame_arrival) {   /* an undamaged frame. */
          frame_physical_layer(& receive);   /* get the frame */
          if (receive.seq == frame_expected) { /* handle expected-received frames */
                send_to_network_layer(& receive.net_packet);
                                           /* pass packet to network layer */
              inc(frame_expected);   /* sequence number expected next */
          }
          if (receive.ack == next_frame_to_send) {
                                           /* handle next sending frame */
                get_from_network_layer(&buffer);
                                           /* get new packet from network layer */
                inc(next_frame_to_send);   /* sender's next sequence number */

```
            }
        }
        send.net_packet = buffer;              /* data for a sending frame */
        send.seq = next_frame_to_send;
                                               /* assign sequence number to the frame */
        send.ack = 1 - frame_expected:         /* seq number of last received frame */
        send_to_physical_layer(& send);        /* transmit a frame */
        start_timer(send.seq);                 /* start the timer */
    }
}
```

9.1.3 A Protocol Using Go-Back-n

Until now a tacit assumption was made, that the transmission time needed for a frame to arrive at the receiver plus the transmission time for the acknowledgement to return, is negligible. Sometimes this assumption is not true. In such situations the long round-trip time can severely impact the efficiency of the bandwidth utilization. As an example, consider a 50-kbps satellite channel with a 500-ms round-trip propagation delay. Imagine that *Protocol* 2 is used to send a sequence of 1000-bit frames via a satellite. At t = 0, the sender starts sending the first frame. At t = 20 ms the frame has been completely sent. However, not until t = 270 ms has this frame fully arrived at the receiver, and not until t = 520 ms is the acknowledgement received by the sender. This happens only under the best of circumstances with no waiting in the receiver and only a short acknowledgement frame. This means, in this example, that the sender is blocked during 500/520 or 96 percent of the time (i.e., only 4 percent of the available bandwidth is used). Clearly, the combination of a long transit time, high bandwidth, and a short frame length is disastrous in terms of efficiency.

The problem, described above, can be viewed to be a consequence of the rule, the sender needs to wait for an acknowledgement before sending another frame. If this restriction is relaxed a much better efficiency can be achieved. Basically the solution lies in allowing the sender to transmit up to w frames before blocking, instead of sending just 1 frame. With an appropriate choice of w the sender would be able to continuously transmit frames for a time, equal to the round-trip transit time, without filling up the window. In the example, given above, w should be at least 26. Again for this case the sender begins sending frame-0 as before. By the time this sender has finished sending 26 frames, at t = 520, the acknowledgement for frame 0 has just

ARQ and Interleaving Techniques

arrived. Thereafter, acknowledgments arrive every 20 msec, so that the sender always gets permission to continue just when he needs to send more frames. At all times, 25 or 26 unacknowledged frames are still outstanding. In other words, the sender's maximum window size is 26.

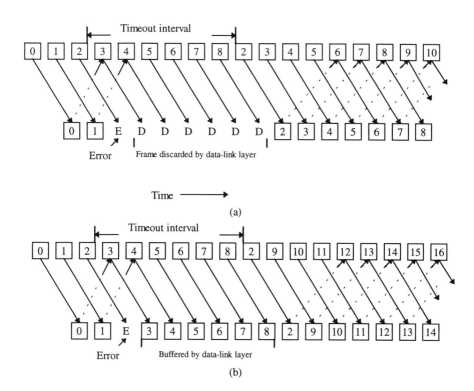

Figure 9.2 (a) Effect of an error when the receiver window size is 1.
(b) Effect of an error when the receiver window size is large.

The above-described technique is known as pipelining. To understand pipelining first let the channel capacity be b bits/second. Also let the frame size be l bits, and the round-trip propagation time be R seconds. Then the time required to transmit a single frame is l/b seconds. Also, after the last bit of a data frame is sent, there is a delay of $R/2$ seconds before this last bit arrives at the receiver, and another delay of at least $R/2$ seconds. for the acknowledgement to come back, a total delay of R seconds. In the stop-and-wait protocol the line is busy for l/b seconds and idle for R seconds. This yields a line utilization of $l/(l + bR)$ seconds so that if $l < bR$ the efficiency is less than 50 percent. Since there is always a nonzero delay for the acknowledgement to propagate back to the sender, in principle, pipelining can be used to keep the line busy during this interval, but if the interval is small, the additional complexity is not worth the trouble.

Pipelining frames over an unreliable communication channel raises some serious issues. First, if a frame in the middle of a long stream is damaged or lost, large numbers of succeeding frames will arrive at the receiver before the sender even finds out that anything is wrong. In this case, a damaged frame arrives at the receiver, it obviously should be discarded. However, how the receiver handles all the correct frames that follow a damaged frame becomes a problem. Remember that the receiving data-link layer is obligated to hand packets to the network layer in sequence.

There are two basic approaches to deal with errors in the presence of pipelining. One, called go-back-n, is for the receiver simply to discard all subsequent frames, sending no acknowledgments for the discarded frames. In other words, the data-link layer refuses to accept any frame except the next frame it must give to the network layer. If the sender's window fills up before the timer runs out, the pipeline begins to empty. Eventually, the sender will time-out and retransmit all unacknowledged frames in order, starting with the damaged or lost frame. This approach, shown in Fig. 9.3(a) where the receiver window size is 1, can waste considerable bandwidth if the error rate is high.

The other general strategy for handling errors when frames are pipelined, called selective repeat, is to have the data-link layer of the receiver store all of the correct frames that follow the bad frame. When the sender finally notices that something is wrong, it just retransmits only the one bad frame, and not all of its successors, as shown in Fig. 9.3(b). If the second try succeeds, the data-link layer of the receiver now has many correct frames in sequence, so that all of these frames can be handed-off to the network layer quickly and to have the highest number acknowledged.

Example 9.3
/*   Protocol 3 (pipelining) allows multiple sending frames. The sender can transmit up to MAX_SEQ frames without waiting for an ack. In addition, unlike the previous protocols, the network layer is not assumed to have a new packet all of the time. Instead, the network layer generates a network_layer_ready event whenever there is a packet to be sent. */

#define MAX_SEQ 7   /* it usually satisfies $2^n - 1$ */

ARQ and Interleaving Techniques

typedef enum {frame_arrival, parity_check_error, timeout, network_layer_ready}
event_type;

static boolean circular(unsigned int a, unsigned int b, unsigned int c)
{
  reture (((a <= b) && (b < c)) || ((c <a) && (a <= b)) || ((b < c) && (c <a)));
                                   /* Return true if a <=b <c circularly; false otherwise */
}

static void send_data(unsigned int frame_num, unsigned int frame_expected, packet
buffer[])   /* Construct and send a data frame. */
{
frame send;

send.net_packet = bufter[frame_num];   /*include packet into frame */
send.seq = frame_num;   /* assign sequence number into frame */
send.ack = (frame_expected + MAX_SEQ) %(MAX_SEQ+1);   /* piggyback ack */
send_to_physical_layer(& send);   /* transmit the frame */
start_timer(frame_num);   /* start the timer */
}

void protocol3( )
{
  unsigned int next_frame_to_send=0;
                                /* initialize for counting of the sending frame */
  unsigned int ack_expected=0;   /* initialize the earliest unacknowledged frame */
  unsigned int frame_expected=0;   /* initialize the next expected frame */
  frame receive;
  packet buffer[MAX_SEQ + 1];   /* buffers for the sending frame */
  unsigned int nbuffered=0;
     /* number of output buffers currently in use and initally no packets are buffered*/
  unsigned int i;   /* temp. index for the buffer array */
  event_type event;

  for(;;) {
    wait_for_event(& event);   /* all four possible cases */

    switch(event) {
       case network_layer_ready: /* the network layer has a packet to send */
                                        /* Accept, save, and transmit a new frame */
       get_from_network_layer(&buffer[next_frame_to_send]);
                                              /* get new packet */
       nbuffered++ ;   /* increase the sender's window */

```
        send_data(next_frame_to_send, frame_expected, buffer) ;
                                              /* transmit the frame */
        inc(next_frame_to_send);        /* advance sender's upper window edge */
        break;

    case frame_arrival:                 /* a data or control frame has arrived */
        get_from_physical_layer(& receive);
                                        /* get incoming frame from physical layer */
        if (receive.seq == frame_expected) {    /* check if it is the expected frame */
                                        /* frames are accepted only in order. */
            send_to_network_layer(& receive.net_packet);
                                        /* pass packet to network layer */
            inc(frame_expected);        /* advance lower edge of receiver's window */
        }

        while (circular(ack_expected, receive.ack, next_frame_to_send)) {
                                        /* Handle piggybacked ack. */
            nbuffered- -;               /* one frame fewer buffered */
            stop_timer(ack_expected);   /* frame arrived correctly; stop timer */
            inc(ack_expected);          /* advance sender's window */
        }
        break;

    case parity_check_error:
        break;                          /* ignore damaged frames */

    case timeout:                       /* retransmit all sending frames */
        next_frame_to_send = ack_expected;      /* start re-transmitting */
        for (i= 1; i<= nbuffered; i++) {
            send_data(next_frame_to_send, frame_expected, buffer);
                                        /* resend 1 frame */
            inc(next_frame_to_send);    /* prepare to send the next one */
        }
    }

    if (nbuffered < MAX_SEO)
        enable_network_layer();
    else
        disable_network_layer();
    }
}
```

The above strategy in Protocol 3 corresponds to a receiver window which is larger than 1. Any frame within the window can be accepted and buffered until all of the preceding frames are passed to the network layer. However, this approach can require a large memory for the data-link layer if the window is long.

The go-back-n approach (Figure 9.2a) and the selective-repeat approach (Figure 9.2b) is trade-off between bandwidth and the buffer space needed for the data-link layer. The go-back-n approach is often used in the case that bandwidth is cheaper while the selective-repeat approach is often applied in the case that memories are cheaper depending on which resource is more valuable, one approach or the other can be used. *Protocol* 3 is a pipelining protocol in which the data-link layer of the receiver only accepts frames in order; the frames following an error are discarded. In this protocol, the assumption, that the network layer always has an infinite supply of packets to send, is dropped. When the network layer has a packet it wants to send, it simply causes the *network_layer_ready* event to happen. However, in order to enforce the flow-control rule, that no more than $v=MAX\_SEQ$ unacknowledged frames are outstanding at any time, the data-link layer must be able to prohibit the network layer from performing this transmission with more work. The library procedures, *enable_network_layer* and *disable_network_layer*, perform this function.

9.1.4 A Protocol Using Selective Repeat

Protocol 3 works well if errors are rare, but if the line is poor, a lot of bandwidth is wasted on retransmitted frames. An alternative strategy for handling errors is to allow the receiver to accept and buffer the frames following a damaged or lost frame. Such a protocol does not discard frames merely because an earlier frame is damaged or lost.

In this protocol, both the sender and receiver maintain a window of acceptable sequence numbers. The sender's window size starts out at 0 and grows to some predefined maximum, MAX_SEQ. The receiver's window, by contrast, is always fixed in size and equal to MAX_SEQ. The receiver has a buffer reserved for each sequence number within its window. Associated with each buffer is an "arrived" bit telling whether or not the buffer is full or empty. Whenever a frame arrives, its sequence number is checked to see if it falls within the window. If so, and it has not already been received, it is accepted and stored. This action is taken without regard to whether or not it is certain that a next packet is expected by the network layer. Of course, it must be kept within the data-link layer and not passed to the network layer until all of the lower numbered frames already are delivered to the network layer in the correct order. A protocol that uses this algorithm is given as Protocol 4.

Example 9.4
/* Protocol 4 (non-sequential receive) accepts frames out of order, but passes packets to the network layer in order. Associated with each outstanding frame is a timer. When the timer goes off, only that frame is retransmitted, not all of the outstanding frames, as in protocol 4. */

```
#define MAX_SEQ 7                              /* it usually satisfies 2^n -1 */
#define HALF ((MAX_SEQ + 1)/2)
typedef enum {frame_arrival parity_check_error, timeout, network_layer_ready,
ack_timeout} event_type;
boolean no_nak = true:                         /* no nak has been sent yet */
unsigned int  oldest_frame = MAX_SEQ + 1;
                                               /* initial value is only for the simulator */

static boolean circular(unsigned int a, unsigned int b, unsigned int a)
{
return ((a <= b) && (b <c)) | | ((c <a) && (a <= b)) | | ((b <c) && (c< a));
                     /* Return true if a <=b <c circularly; false otherwise */
}


static void send_frame(frame_type ftype, unsigned int frame_num, unsigned int
frame_expected, packet buffer[])
{
                     /* Construct and send a data, ack, or nak frame. */
frame    send;
send.type= ftype;                              /* type == data, ack, or nak */
if (ftype == data)
       send.net_pactet = buffer[frame_num % HALF];
send.seq = frame_num            /* only meaningful for data frames */
send.ack = (frame_expected + MAX_SEQ) % (MAX_SEQ + 1);
if (ftype == nak) no_nak = false;              /* one nak per frame, please */
send_to_physical_layer(& send);                /* transmit the frame */
if (ftype == data)
       start_timer(frame_num% HALF);
stop_ack_timer();                              /* no need for separate ack frame */
```

ARQ and Interleaving Techniques

}

```
void protocol4( )
{
unsigned int ack_expected;              /* lower edge of sender's window */
unsigned int next_frame_to_send;        * upper edge of sender's window + 1 */
unsigned int frame_expected;            /* lower edge of receiver's window */
unsigned int too_far;                   /* upper edge of receiver's window + 1 */
int i;                                  /* temp. index for into buffer loop */
frame receive;
packet out_buf[HALF];                   /* buffers for the sending frames */
packet in_buf[HALF];                    /* buffers for the receiving frames */
boolean arrived[HALF];                  /* received frame bit map */
unsigned int nbuffered;                 /* number of output buffers currently used */
event_type event;
enable_network_layer();                 /* initialization */
ack_expected = 0;                       /* next ack expected on the received frames */
next_frame_to_send = 0;                 /* number of next sending frame */
frame_expected = 0;
too_far = HALF;
nbuffered = 0;                          /* initially no packets are buffered */

memset(arrived, 0, HALF*sizeof(boolean));
for( ; ; ) {
   wait_for_event(& event);             /* five possible cases */
   switch(event) {
      case network_layer_ready          /* accept, save, and transmit a new frame */
         nbuffered ++;                  /* expand the window */
         get_frame_network_layer(&out_buf[next_frame_to_send % HALF]);
                                        /* get new packet */
         send_frame(data, next_frame_to_send frame_expected, out_buf);
                                        /* transmit the frame */
         inc(next_frame_to_send);       /* advance upper window edge */
         break;

      case frame_arrival:               /* a data or control frame has arrived */
         get_from_physical_layer(&receive);
                                        /* get incoming frame from physical layer */
         if (receive.type == data) {
                                        /* An undamaged frame has arrived. */
            if ((receive.seq != frame_expected) && no_nak)
               send_frame(nak, 0, frame_expected, out_buf);
            else
```

```
        start_ack_timer();
    if (circular(frame_expected, receive.seq, too_far) && (arrived[r.seq%HALF] ==
false)) {                              /* Frames may be accepted in any order. */
        arrived[receive.seq % NA_BUFS] = true;         /* mark buffer as full */
        in_buf[receive.seq % HALF] = receive.net_packet:   /* insert data into buffer */
        while (arrived[frame_expected % HALF]) {
                                       /* Pass frames and advance window. */
            send_to_network_layer(& in_buf[frame_expected % HALF]);
            no_nak = true;
            arrived[frame_expected % HALF] = false;
            inc(frame_expected);        /* advance lower edge of receiver's window */
            inc(too_far);               /* advance upper edge of receivers window */
            start_ack_timer();          /* check if a separate ack is needed */
        }
    }
}

if(receive.type==nak) &&
        circular(ack_expected,(receive.ack+1)%(MAX_SEQ+1),next_frame_to_send))
    send_frame(data, (receive.ack+1} % (MAX_SEQ + 1), frame_expected,
out_buf);

while (circular(ack_expected, receive.ack, next_frame_to_send)) {
    nbuffered - - ;                    /* handle piggybacked ack */
    stop_timer(ack_expected % HALF);   /* frame arrived correctly */
    inc(ack_expected):                 /* advance lower edge of sender's window */
}
break;

case parity_check_error :
    if (no_nak)
            send_frame(nak, 0, frame_expected, out_buf);    /* damaged frame */
    break;

case timeout:
    send_frame(data, oldest_frame, frame_expected, out_buf);  /* it is timed out */
    break;

case ack_timeout:
    send_frame(ack,0,frame_expected, out_buf);   /* ack timer expired; send ack */
}
if (nbuffered < HALF)
        enable_network_layer();
```

ARQ and Interleaving Techniques

else
       disable_networl_layer();
}
}

In protocol 3, there is an implicit assumption that the channel is heavily loaded. When a frame arrives, no acknowledgement is sent immediately. Instead, the acknowledgement is piggybacked onto the next outgoing data frame. If the reverse traffic is light, an acknowledgement can be postponed for a long period of time. If there is a lot of traffic in one direction and no traffic in the other direction, only MAX_SEQ packets are sent, and then the protocol blocks.

In protocol 4 this problem is fixed. After an in-sequence data frame arrives, an auxiliary timer is started by *start_ack_timer*. If no reverse traffic presents itself before this timer goes off, a separate acknowledgement frame is sent. An interrupt, due to an auxiliary timer, is called an *ack_timeout* event. With this arrangement one-directional traffic flow is now possible, because of the lack of reverse data frames onto which acknowledgments can be piggybacked is no longer an obstacle. Only one auxiliary timer exists, and if *start_ack_timer* is called while the timer is running, it is reset to a full acknowledgement-timeout interval.

It is essential that the timeout associated with the auxiliary timer be appreciably shorter than the timer used for the timing out of the data frames. This condition is required to make sure that the acknowledgement for a correctly received frame arrives before the sender times-out and retransmits the frame.

Protocol 4 uses a more efficient strategy than protocol 3 to deal with errors. Whenever the receiver detects an error, it sends a negative acknowledgement (NAK) frame back to the sender. Such a frame is a request for the retransmission of the frame, specified in the NAK. There are two cases in which the receiver will send a NAK: a corrupted frame has arrived or a frame other than the expected frame arrives (a potential lost frame). To avoid making multiple requests for the retransmission of the same lost frame, the receiver keeps track of whether a NAK already has been sent for a given frame. The variable *no_nak* in protocol 4 is set, true, if no NAK has been sent yet for *frame_expected*. If the NAK gets damaged or lost, the sender eventually times out and retransmits the missing frame. If the wrong frame arrives after a NAK has been sent and lost, an ACK is sent to resynchronize the sender to the receiver's current status.

9.2 Interleavers

Physical channels are always a combination of both memory- and memoryless-channel types. Examples of memory-channel types are the "fast" and "slowly" fading channels. The terms, "fast" and "slow" fading, can be understood best only when considered in relation to their channel-symbol transmission rates. It is assumed here that a slowly fading channel exhibits fades with durations which exceed the times required to transmit several channel symbols. Thus a fade which affects one channel symbol is very likely to affect temporally a number of adjacent channel symbols. As a consequence for slow fading transmission errors often occur in bursts. Therefore, a sophisticated error- control scheme must be capable of correcting both random and burst errors. Many communication systems contain ARQ along with FEC. Such a hybrid ARQ is one potential solution, but it requires a feedback channel and has the disadvantage of introducing a variable information rate to the system. However, for some applications, such as the compact-disc player, digital radio or television broadcasting, where ARQ is not applicable, alternative techniques of ARQ need to be introduced to combat burst errors that do not require a feedback channel. These techniques are the concerns of this section.

Most of the block codes discussed previously in this text were random-error correcting. A random error-correcting code corrects up to t symbol errors per codeword, regardless of the placement of those errors. A problem arises with these codes whenever the channel, encountered in the application, is bursty. An error burst focuses several symbol errors within a small number of received codewords. The error-correction capacity, required by the words corrupted by bursts, is thus not used by the words that are not affected.

The performance of convolutional codes also is sensitive to bursty channels, though the reasons are more subtle than in the block-code case. In the decoding of convolutional codes error events occur whenever the received codeword is closer in distance to an incorrect codeword than it is to the codeword that was transmitted. It is shown in Chapter 8 that convolutional codewords can be viewed as branches in its trellis diagram. Thus in such a case any pair of codewords differs by one or more branches in the trellis diagram, starting at the all-zero state. Since these differing branches often consist of several consecutive branches in the trellis diagram, the error patterns that cause one codeword to look like another are bursty. Thus a convolutional code may be able to correct an arbitrarily large number of well-spaced errors while, at the same time, is unable to handle a short burst.

Much effort has gone into the development of codes or coding techniques that can specifically deal with channel-burst errors. Such codes include the RS codes discussed in Chapter 6. Another technique is the ARQ technique discussed in the last section. If no return channel is available, the combination of FEC with interleaving, as shown in Fig. 9.2, is a common solution to combat burst errors[1,2,3,4,5]. Interleaving does not change the constant throughput of a FEC system. However, it does produce an additional constant delay. Two interleaving techniques are considered : interleaving at the channel-symbol level and interleaving at the codeword symbol level.

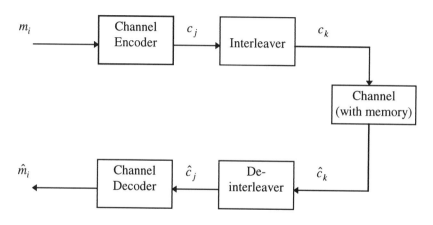

Figure 9.3: A FEC scheme with interleaving.

An interleaver reorders the channel-encoded symbols c_j in time before transmitting them over the channel. This operation is neutralized or reversed by the deinterleaver at the receiver before channel decoding. If the parameters of the interleaver are chosen properly with regard to the channel characteristics, this reordering results in the spreading of the bursts of errors over a long time interval in the channel-encoded symbol sequence. Bursts then appear to be single errors to the channel decoder. However, the resulting additional transmission delay for such a system can become quite large if a large channel memory exists. However, this delay is usually not of importance when a broadcast-communication system is envisaged, which is typical of a communication system without a return channel.

Two principle classifications of interleavers are the *periodic* and the *pseudo-random* [2]. Periodic interleavers have a lower complexity and smaller memory requirements than the pseudo-random types and therefore institute a better match to the channel, when used in a highly-integrated mobile-communication system with a restricted processing power.

This section presents two commonly used types of periodic interleavers, the periodic block interleaver and the periodic convolutional interleaver. After giving some basic definitions in Subsection 9.2.1 both types of interleavers are described in Subsection 9.2.2 with regard to their overall delay and memory requirements when applied to different FEC schemes in Subsection 9.2.3. Subsection 9.2.4 describes the integration of the interleaving tool with a communication system and certain aspects of interleavers that are related the problems of synchronization. Also, Subsection 9.2.5 provides a C-language implementation of an optimum interleaver/deinterleaver structure.

9.2.1 Definitions of Interleaver Parameters

The usual type of interleaver belongs to the class of (I, N) interleavers, where I is called the interleaver depth, and N is called the interleaver constraint length or, more simply, the length. The (I, N) interleaver, when concatenated with its inverse, the (I, N) deinterleaver, has the following properties [1,2,3]:
- Each symbol in the input sequence appears only once in the re-ordered output sequence of the interleaver. The interleaver and its deinterleaver in sequence are equivalent to a delay of D symbols.
- Two symbols in the input sequence, which are separated by less than N symbols, are re-ordered in such a manner that their minimum separation at the interleaver output is I symbols. This implies that a single burst of b<I errors, inserted by the channel errors, affects only the output symbols which are separated by at least N symbols after the inverse re-ordering by the deinterleaver.

The combination of the (I, N) interleaver with its corresponding (I, N) deinterleaver introduces an additional overall symbol delay D into the communication system and requires a total memory of S symbols. It is evident that an optimum (I, N) interleaver/deinterleaver system should achieve the minimum values of the parameters D and S for a required performance.

The parameters I and N are often selected with regard to the channel-burst characteristics and the particular FEC scheme as follows:
- I should be selected to be larger than the length of the longest expected error burst.
- N should be larger than the codeword length of the block code used for correction or larger than the constraint length of the convolutional code.

ARQ and Interleaving Techniques

If both rules are observed, a burst error results in only, at most, one single error in any code-block length (for a block code) or constraint length (for a convolutional code) at the channel decoder.

9.2.2 Periodic interleavers

In the following two subsections a brief description is given of two basic periodic-interleaver structures and their characteristics.

Periodic-block interleaver

A simple realization of an (I, N) interleaver is a periodic-block interleaver that consists of a I×N storage matrix [3,4,5]. The symbols are written column-wise into the matrix and then read out by rows. This column and row interleaving process is performed periodically. The deinterleaver performs the same sequence of operations on the corresponding (I, N) deinterleaver matrix which is the transpose of the (I, N) interleaver matrix. The start of each interleaver block must be known in order to accomplish the deinterleaving process. Hence, a block-wise synchronization is required to achieve a correct deinterleaving.

In a (I,N) block interleaver, any two adjacent symbols at the input are separated by $I-1$ other symbols at the output. Since N is larger than the codeword length, a burst of b symbol errors causes a maximum of $\lfloor b/I \rfloor + 1$ symbol errors in each codeword. Therefore, for a t-error correction block code, the maximum correctable burst length b is $b_{max} = I \cdot t - I + 1$.

The memory requirements in symbols are $I \cdot N$ for interleaver and $I \cdot N$ for deinterleaver. Thus, the overall memory is

$$S_{BI} = 2 \cdot I \cdot N. \qquad (9.3)$$

The resulting overall delay for both the interleaving and deinterleaving processes is

$$D_{BI} = 2 \cdot I \cdot N \qquad (9.4)$$

symbols.

The following subsection shows that a periodic-block interleaver/deinterleaver system is suboptimum in both the overall delay and memory requirements.

Periodic convolutional (or synchronous) interleaver

The reordering of a sequence of symbols in time can be interpreted also as a tapped delay-line operation with various delays. This implies that it has the form of the general interleaver structure, shown in Figure 9.3 [2,3,4]. Accordingly, the inverse operation is accomplished by the deinterleaver structure that is depicted also in Figure 9.3. To ensure a correct deinterleaving process, the two commutators must operate synchronously. This is carried out usually by their incorporation with the frame synchronization procedures.

To obtain the (I, N) interleaver and its corresponding (I, N) deinterleaver with the properties, defined in Subsection 9.2.1, the parameters I, N, and the number of symbols M for each delay-unit need to be selected with regard to their relation [2,3,4],
$$N = M \cdot I. \tag{9.5}$$

For a given I and N, M is called the minimum dispersion distance and is determined by $M = \lfloor N/I \rfloor + 1$. When transmission takes place via the system in Figure 9.3, all symbols receive the total delay of
$$D_{CI} = N \cdot (I-1) \tag{9.6}$$
symbol times, excluding the channel delay. Also the total memory requirements in symbols is evidently the same, i.e.
$$S_{CI} = N \cdot (I-1). \tag{9.7}$$

Suppose that N is larger than or equal to the codeword length of a block code. Each symbol in a codeword is thus sent to a different delay line. At the output of the interleaver, any two adjacent symbols of a codeword are separated by $N \cdot M$ symbols. A burst of b errors may cause a maximum of $\lfloor b/(N \cdot M + 1) \rfloor + 1$ errors in each codeword. Therefore, for a t-error correction block code, the maximum correctable burst length b is $b_{max} = (N \cdot M + 1)(t-1) + 1$.

Since the delay and the required memory of the (I,N) interleaver/deinterleaver are equal, the system is optimum in the sense of the minimum achievable memory requirement. It is shown in [1] that the resulting delay D is also the minimum possible delay for any (I,N) interleaver/deinterleaver. Hence, the periodic convolutional (I,N) interleaver/deinterleaver, presented in Figure 9.4, is optimum. Compared with the (I, N) periodic-block interleaver, described in the previous subsection, it achieves less than half of the overall delay and exactly one-half of the required memory.

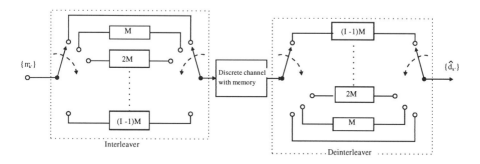

Figure 9.4 : Periodic convolutional interleaver and its corresponding deinterleaver.

Consider next the realization of the convolutional interleaver with $I=128$ and $M=8$.

Example 9.5
Convolutional interleaving is illustrated in the left portion of Figure 9.3. The interleaver commutator increments its position at a pre-determined symbol frequency with a single output from each position. With a convolutional interleaver the input symbols are sequentially shifted into the bank of $I=128$ registers. Let each successive register have $M=8$ symbols of more storage than the preceding register. Then the first interleaver path has zero delay, the second has a $M=8$ symbol period of delay, the third a $2*M$ symbol period of delay, and so on, up to a $I=128$ symbol path that has $127*M$ symbol periods of delay. M is selected by a 3-bit field to have the integer value 8. A sync trailer is inserted at the end of a block. This is sometimes called a framing signal. Such an insertion occurs after the interleaving.

The convolutional interleaver can be implemented by the repeated reuse of a memory of size $M*I*(I-1)/2$. Therefore, for $I=128$ and $M=8$, the memory required is 8x128x(128-1)/2=65024 symbols. The storage element for each commutator-arm position can be memory-mapped in the manner, given in Table 9.1.

Associated with each row of the above table are the first and the last memory locations and positions that obey the following relationships :
Initial First Address = 0;
Initial Last Address = 7;
Next First Address = Previous Last Address + 1;
Next Last Address = Previous Last Address + (Row+2)*8.

Row	Address	
	first address	last address
0	0000	0007
1	0008	0017
2	0018	002F
⋮	⋮	⋮
⋮	⋮	⋮
127	FA08	FDFF

Table 9.1 Memory mapping for commutator-arm

In the design of the convolutional interleaver, the above equations are used to calculate the first and last memory locations of each commutator-arm position, and a 128x16 SDRAM is used to store the next-address locations. At every clock cycle the last address is compared with the current memory address. If they are equal, the next memory location for that commutator-arm position is reset to the first address. If the last address has not been reached, the current address is incremented by 1 and stored in the SRAM. It is by this means that the memory address SDRAM is constantly updated and the proper memory location is accessed each clock cycle.

Figure 9.5 shows the detailed block diagram of the convolutional interleaver. The row counter (I counter) is incremented on the rising edge of the clock pulse. The output of this counter then accesses the memory address RAM, where the external SRAM addresses are stored. During the first half of the clock cycle, memory-read is performed and the contents of the external SRAM is stored in a register on the falling edge of the clock. During the second half of the clock cycle, memory-write is performed, and the input data is written to the same memory location. At the same time, the next address location for the current commutator position is calculated which is based on the first and the last address. Then this value is stored in the memory address RAM to wait for the commutator arm to return to this position I clock cycles later. Notice that when the sync signal is inserted that the I counter is reset to zero and the commutator arm goes to position I=1 on the next clock edge. Also, the first address location is used to access the memory.

The de-interleaver structure is exactly the opposite or reversal of the interleaver. The detailed implementation of the de-interleaver is similar to that of interleaver. The only major difference is the assignment of the memory address map. The first and the last address is calculated by the following equations :
  Initial First Address = 0;
  Initial Last Address = (I-1)*M-1;
  Next First Address = Previous Last Address + 1;
  Next Last Address = Previous last Address + (Row-1)*8;
Figure 9.5 shows the detailed implementation of the de-interleaver.

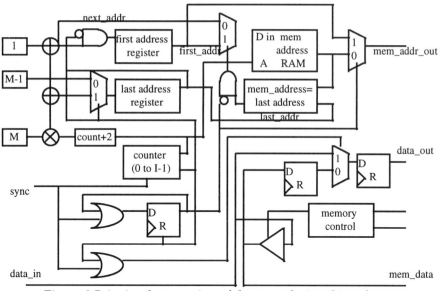

Figure 9.5 An implementation of the convolutional interleaver

9.2.3 Comparsion of the Two Interleavers

The two basic periodic interleaver/deinterleaver structures, the periodic (I,N) block interleaver/deinterleaver and the periodic (I,N) convolutional (or synchronous) interleaver/deinterleaver, are now compared. A consideration of the characteristics of both structures shows that the convolutional

interleaver/deinterleaver combination has the following general advantages over the periodic interleaver/deinterleaver:
- Minimum overall symbol delay.
- Minimum memory requirements.

Note that the required memory per symbol at the receiver is about four times greater than that of the sender when the soft decision channel-decoding strategies, e.g. the Viterbi decoding, are utilized. This fact must be considered when using Eqs.(9.3) and (9.7).

In both structures the synchronization of the interleaver and its deinterleaver is necessary. When block interleaving is applied to FEC with block codes, the synchronization problems of the deinterleaver and the channel decoder are similar and can be treated in common. The same is true for the use of the convolutional channel-codes together with convolutional interleaving.

If synchronization is viewed from a more general point of view, one can identify another advantage of convolutional interleaving over that of block interleaving: The uncertainty of the convolutional interleaver/deinterleaver being synchronized is on the order I, while on the order $N \cdot I$ for the corresponding block interleaver/deinterleaver. But the usefulness of this advantage depends on the synchronization strategy and the overall system structure to which it is applied.

9.2.4 Integration of an Interleaver in a Communication System

In a communication system the interleavers often are applied between the digital modulator and the FEC system or between two FECs (in the later case it is called a cross-Interleaver). Such interleaving sometimes can be performed on a communication network. In the case of FEC it is useful to have the error control in the adaptation layer of the communication system so that different sources can be treated differently.

One of the "optimum" places for the interleaver is between the output of the FEC encoder and the input of the modulator. In this situation the interleaver gets the FEC symbols and sends the interleaved output symbols to the modulator. In general the total number of bits is not changed by the interleaver.

In order to reverse the process, the deinterleaver has to be synchronized to the interleaver period of I bits. To accomplished this, the length of each FEC needs to be restricted to a multiple of I. For a typical example I=32bits and the codeword length N= n * 4 bytes, where n is a positive integer. In this

ARQ and Interleaving Techniques

case the interleaver and the deinterleaver can be synchronized by the convention that their switches start at position 0 at the beginning of each packet.

Since the interleaving/deinterleaving process introduces the delay D, the first packets, given to the adaptation layer of the demultiplexer, contain garbage (or zeros if the interleaver memory is initialized with zeros). This latter fact needs to be taken into account in the application.

9.2.5 A C-Language Program

This subsection provides a C-language implementation of an interleaver and deinterleaver as proposed originally by Ramsey in [1]. These two functions are called once per symbol by the following:

int interl_out = interleaver(int I, int M, int input_symbol);
int deinterl_out= deinterleaver(int I, int M, int received_symbol);

Symbols *I* and *M* are the above described interleaver/deinterleaver parameters. Note that this C-language program is intended to be more readable than to be run-time optimized.

```
#include <stdio.h>
#include <stdlib.h>

int interleaver(int I, int M, int input_symbol)
{
  int i, out_symbol;
  static int *rlength;
  static int *column_pointer;
  static int row_pointer;
  static int **memo_i;
  static int init=0;

  if (init==0)                                    /* Memory allocation: */
  {
    rlength = (int *) calloc(I, sizeof(int));     /* Memory length */

    column_pointer = (int *) calloc(I, sizeof(int));  /* Column pointers */
    memo_i = (int **) calloc(I, sizeof(int *));       /* Allocate memory */

    for (i=1; i<I; i++)
```

```
    {
      rlength[i]=M*i;
      memo_i[i] = (int *) calloc(rlength[i], sizeof(int));
    }

    row_pointer=0;                              /* row index for the first row */
    init=1;
  }

  if (row_pointer==0)                           /* Symbol processing */
    out_symbol=input_symbol;                    /* No delay for the first row */
  else
  {
    out_symbol=memo_i[row_pointer][column_pointer[row_pointer]];
                                                /* Reading in */

    memo_i[row_pointer][column_pointer[row_pointer]]=input_symbol;
                                                /* Writing to memory */

    column_pointer[row_pointer] =
          (column_pointer[row_pointer]
          +rlength[row_pointer]-1)%rlength[row_pointer];
                                                /* Shift memory contents */
  }

  row_pointer=(row_pointer+1)%I;                /* point to the next row */

  return(out_symbol);
}


int deinterleaver(int I, int M, int received_symbol)
{
  int i, out_symbol;
  static int *rlength;
  static int *column_pointer;
  static int row_pointer;
  static int **memo_d;
  static int init=0;

  if (init==0)                                  /* Memory allocation: */
  {
    rlength = (int *) calloc(I, sizeof(int));   /* Memory length */
```

```
  column_pointer = (int *) calloc(I, sizeof(int));      /* Column pointers */
  memo_d = (int **) calloc(I, sizeof(int *));           /* Allocate memory */
  for (i=1; i<I; i++)
  {
    rlength[i]=M*i;
    memo_d[i] = (int *) calloc(rlength[i], sizeof(int));
  }

  row_pointer=I-1;                                      /* row index for the first row */
  init=1;
}

if (row_pointer==0)                                     /* Symbol processing */
  out_symbol=received_symbol;                           /* No delay for the first row */
else
{
  out_symbol=memo_d[row_pointer][column_pointer[row_pointer]];
                                                        /* Reading in */
  memo_d[row_pointer][column_pointer[row_pointer]]=received_symbol;
                                                        /* Writing to memory */
  column_pointer[row_pointer]
              =(column_pointer[row_pointer]
              +rlength[row_pointer]-1)%rlength[row_pointer];
                                                        /* Shift memory contents */
}

row_pointer=row_pointer-1;                              /* point to the next row */
if (row_pointer==-1)
  row_pointer=I-1;

return(out_symbol);
}
```

Problems

9.1 In data-link protocols the CRC is always put in a data trailer, rather than in a header. Why ?

9.2 A block of message-bits with n rows and k columns uses horizontal and vertical parity bits for error detection. Suppose that 4-bits of error occur in the transmission. Derive an expression for the probability that the error is undetected.

9.3 Consider the operation of the selective- repeat protocol, given in Section 9.1.4, over a 1 Mbps error-free line. The maximum frame size is 1000 bits. New packets are generated about 1-second apart. The time-out interval is 10 msec. If the special acknowledgement timer were eliminated, an unnecessary time-out would occur. How many times on the average would a message be transmitted ?

9.4 In the selective-repeat protocol given in Section 9.1.4, assume that MAX_SEQ= $2^n - 1$. This condition is evidently desirable in order to make an efficient use of the header bits. Does this protocol work correctly, for example, for MAX_SEQ=4 ?

9.5 Frames of 2000 bits are sent over a 1 mbps satellite channel. Acknowledgments are always piggybacked onto data frames. The headers are very short. 5-bit sequences of numbers are used. What is the maximum achievable channel utilization for the following:
(a) Stop-and-wait.
(b) Go-Back-n.
(c) Selective Repeat.

9.6 Consider an FEC+interleaver, error-protection, system for a binary channel. Assume that error bursts up to length 8 bits can occur over the channel and a (24,12,8) block code is used. Design an interleaving scheme such that it can handle up to three such bursts. What are the total memory requirements and the end-to-end delay in units of information bit-times ?

9.7 Consider a coding system that consists of a (7,4) Hamming code and a block interleaver with I=3 and M=1. Show that, by sending the all-zeros sequence, that the insertion of any 3-bit burst-error pattern is scrambled into three separate codewords at the receiver, and thus that the burst is correctable.

9.8 Consider a binary block code of length $n=63$ that transmits over a Rayleigh-fading channel whose decorrelation time is 0.01 second.
(a) If the channel symbol rate is 16 kbps, design a block interleaver to effectively produce a memoryless channel as seen by the decoder. What is the total end-to-end delay of the system due to interleaving ?
(b) Repeat (a) for a convolutional interleaver.

9.9 For a given M and I show that the convolutional interleaver will save more than half of memory as compared to a block interleaver with the same parameters. Usually, block interleavers are implemented in a RAM table. The saving of memory in a convolutional interleaver comes at the expense of a more complex address generator than the RAM-based interleaver.

9.10 Consider a block interleaver with $N=3$ and $I=3$ and a convolutional interleaver with $M=1$ and $I=3$. For a input sequence
$$(\ldots, x_0, x_1, x_2, x_3, x_4, x_5, x_6, x_7, x_8, x_9, x_{10}, x_{11}, x_{12}, x_{13}, \ldots),$$
determine the interleaved sequences for both interleavers.

Bibliography

[1] John L. Ramsey, Realization of Optimum Interleavers, IEEE Transactions on information theory, VOL. IT-16, NO. 3, May 1970.
[2] G. David Forney, Jr. Burst-Correction Codes for the Classic Bursty Channel, IEEE Transactions on communications technology, VOL. COM-19, NO. 5, October 1971.
[3] George C. Clark, Jr.and J. Bibb Cain, Error-Correction Coding for Digital Communications, Plenum Press, New York, 1981.
[4] Marvin K. Simon, Sami M. Hinedi, William C. Lindsey , Digital Communication Techniques, Signal Design and Detection, Prentice-Hall, Englewood Cliffs, New Jersey, 1995.
[5] John G. Proakis, Digital Communications, Third Edition, McGraw-Hill, Inc., New York, 1995.
[6] Shu Lin and Daniel J. Costello, Jr. Error Control Coding: Fundamentals and Applications, Prentice-Hall, Englewood Cliffs, New Jersey, 1983.
[7] B. W. Kernighan and D. M. Ritchie, The C Programming Language (2nd ed.), Prentice Hall, Englewood Cliffs, New Jersey, 1988.
[8] A. S. Tanenbaum, "Computer Networks", Prentice Hall, Upper Saddle, New Jersey, 1996.

10 Applications of Convolutional Codes in Mobile Communications

Convolutional codes have many applications in various communication systems. Two important applications in mobile communications are introduced in this chapter.

10.1 Convolutional Codes used in the GSM systems

The Global System for Mobile communications(GSM) is the digital cellular-radio system, specified by the European Telecommunications Standards Institute(ETSI). The GSM system provides speech, telefax, and data services such as telephony, paging, etc..

To send messages through the GSM system, several types of signals need to be transmitted over the physical channels. Among them are the speech signals, the user data, the control signals (for signaling), etc.. The arrangement needed to transmit data from a single signal type over a physical channel is called a single logic (or logical) channel. In the concept of a logic channel, the information carried on the channel is usually more significant than the just physical nature of the signals in a GSM system. Different types of information can exist in the system on different types of logic channels. The contents of a logic channel can appear in many different forms, such as a pulse-code signal in one frequency and time slot. Such a signaling system usually is called time-division multiplex. However, once a physical channel is assigned the task of transporting the contents of a particular logic channel, this assignment does not change.

A logic channel can carry either signaling data or a user's data. The data of either kind are mapped onto physical channels. The manner in which the data are mapped to the physical resources depends on the type of data.

The GSM system distinguishes between traffic channels that are reserved for user data, and the control channels that contain the network-management messages and some of the channel-maintenance tasks. The traffic channels allow a user to transmit either speech or data, e.g., computer data. Depending on the type, either speech or data, different channels are utilized. Control channels move the network-management related messages to make sure that all of the traffic moves about the system reliably and efficiently.

The protection of the data in a GSM system is based on an elaborate set of channel-coding mechanisms that utilize logic channels. Such a channel-coding mechanism is illustrated in the following subsections by the use of logic channels.

10.1.1 Channel Coding of the Speech Data in a GSM System

Usually, the most important service offered to the user of a cellular mobile network is speech transmission. The general technical requirement is simple: transmit voice signals at an acceptable level of quality. This is made difficult by the dramatic destructiveness that a mobile channel can have on the speech data. Therefore, it is important to have the proper protection of the transmitted speech data.

In conventional analog-radio systems the continuous low-frequency voice signal, often called the baseband signal, modulates a radio-frequency carrier. The information (the analog-voice signal) is carried by the variations in (1) the amplitude, (2) the phase, or (3) the frequency of the transmitted radio signal. These modulation techniques are called, respectively, amplitude modulation(AM), phase modulation(PM), and frequency modulation(FM). In the receiver the modulated signal is demodulated in order to retrieve the original continuous baseband signal. Additive disturbances in the radio channel result in the modification of one or more of the characteristics of these modulated radio signals. The modulation errors caused by channel noise often yield audible distracting noise to the user.

In the GSM digital transmission system, what is called, Time-Division Multiple Access (TDMA) is utilized. This system does not transmit information continuously on one selected channel. The information is transmitted as pulse modulation, so that its content, a sampled-data representation of the original continuous audio signal, is compressed in digital form in the time domain and transmitted over the radio path. In the receiver the information content is decoded and decompressed, i.e. re-generated, in order to recreate a replica of the original continuous-audio signal. The distinct logic states in such a digital system, the configurations of the ones and zeros (in bits), are transmitted by a use of some digital modulation, instead of the classical analog modulation of the carrier.

To accomplish digital modulation a binary representation of the voice signal is needed. Due to the somewhat restricted transmission capabilities of a mobile radio channel, it is usually desirable to minimize the number of bits used in such a binary representation of voice. The device, that transforms the voice signal into a digital stream of data, suitable for the transmission over the radio interface and capable of being used to regenerate the original voice signal, is called a speech codec (speech coder/decoder).

The block diagram in Figure 10.1 shows that the required components needed by speech coding or the codec.

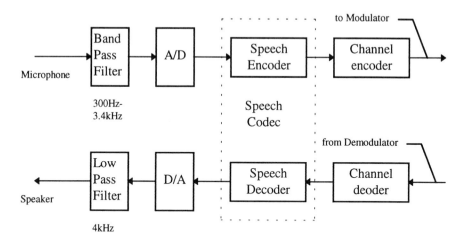

Figure 10.1. The process of speech coding in GSM

In Figure 10.1 the sound of the human voice is converted in GSM into an electrical signal by a microphone. Such a signal is sampled by a 13-bit analog-to-digital(A/D) converter at the sampling frequency (or rate) of 8kHz or 8,000 samples per second. Thus, in the GSM system the A/D converter delivers a data rate of 8,000x13bits/s=104kbits/s. The 104kbits/s data rate is far too high to be economically transmitted directly over the radio interface.

The speech codec in the GSM system uses the two techniques described next to reduce the redundancy in the data, that represents voice traffic.

Linear predictive coding(LPC) and the analysis of regular pulse-excitation (RPE). In every 20ms, 160 sampled values from the A/D converter are taken and stored in an intermediate memory. An analysis of this set of data samples produces eight filter coefficients and an excitation signal for a time-invariant digital filter. This filter can be regarded as a digital imitation of the human vocal tract, where the filter coefficients represent vocal modifiers, that is, the teeth, tongue, larynx, and the excitation signal represents sound, e.g. the pitch, loudness or the absence of sound that passes through the vocal tract (filter). The correct setting of the filter coefficients and the appropriate excitation signal yield a sound of the human voice.

Certain attributes of the human ear and the vocal tract yield sound levels that are perceived on a logarithmic scale. These are used in the data reduction process. In order to transform the 160 samples into filter coefficients, these samples are first divided into four blocks of 40 samples each. Each block represents a 5-ms period of voice. These blocks of digital data are sorted into four sequences, where each sequence contains every fourth sample from the original 160 samples. Sequence 1 contains the samples numbered 1,5,9,13,...,37, sequence 2 contains the samples numbered 2,6,10, 14, ..., 38, sequence 3 contains the samples numbered 3,7,11,15,...,39, and Sequence 4 contains the samples numbered 4, 8,12,16,...,40. The first reduction in the data comes when the speech encoder selects the sequence with the most energy in the above four sequences.

The LPC and RPE have a short memory of approximately 1 ms. A more long-term consideration of the neighboring (or adjacent) blocks in time is not performed at this point in the process. There are numerous correlations in the human voice, especially in the long vowels, such as the *a* in *car* or the *oo* in *tool*, where the same sound re-occurs in succeeding 5-ms samples. If one takes the similarity of sounds between adjacent samples (the adjacent 5-ms blocks) into account, the amount of data required to describe the human voice can be reduced significantly. This second reduction is performed by a long-term prediction(LTP) function.

Long-term prediction analysis. The LTP function accepts a sequence that is selected by the LPC/RPE analysis. Upon the acceptance of a sequence it then looks among all of the previous sequences, which still reside for 15 ms in another intermediate memory, for the earlier sequence that has the highest correlation with the current sequence. In other words, the LTP function looks for that one sequence from among those already received in memory that is most similar to the sequence just received from the LPC/RPE. Then it is only necessary to transmit a value, which represents the difference between these two sequences, along with a pointer, to tell the receiver which sequence it should select for comparison among the recently received sequences. Thus the receiver knows the differential values it needs to apply to which sequences. The transmission of the entire sequence of samples is not necessary, only the differences between sequences are needed. This further reduces the amount of data on the channel.

The speech encoder generates a block of 260 bits, namely a speech frame, once every 20ms. This corresponds to a data rate of 13 Kbits/s. Then, these speech frames are channel-coded before they are forwarded to the modulator of the transmitter. The channel encoder also inserts some new redundancy into the data stream in an orderly manner so that the receiver at the other end of the noisy channel is capable of correcting the bit errors caused by the channel. The type of channel coding used almost doubles the data rate to 22.8 kb/s.

The bits of data in a speech frame contain enough information to enable the speech decoder in the receiver to reconstruct the speech signal. Before channel coding the 260 bits in each speech frame are sorted into different types of data in accordance with their functions and importance. These different classes receive differing amounts of attention from the channel coder in such a way that the excess redundancy is efficiently inserted into the data stream. There are three classes of speech data bits. The most important bits are the class-I bits which constitute 50 bits out of the 260 in each speech frame. The class-I bits describe the filter coefficients, the block amplitudes and some of the other parameters that are used in the speech codec. Since these parameters are regarded as the most important, they require the greatest amount of protection needed in the channel-coding process. Next in importance are the class-II bits (132 bits from the 260 speech frame bits), which consist of the RPT pointers, RPE pulses and the LTP parameters. The least important, but still necessary to have, are the class-III bits (78 out of the 260 bits) which contain the RPE pulse and the filter parameters. The class-III bits do not receive any protection from the channel-coding process.

Experiments demonstrated that the speech quality depends primarily on an error-free transmission of the class-I bits. Their importance meant that these bits need to be protected in such a manner that errors in their transmission can be detected, and that certain specified errors be corrected. The classes II and III bits are not as important and as a consequence were not protected as carefully as the class-I bits.

Speech data are channel-coded in two steps. First the class I bits are block-coded with a cyclic code(a CRC code) which is used only for error detection. This is accomplished by the addition of three parity-check bits to the speech data which detects errors and gives an error indication to the decoder. This is the first stage of the coding function. The second stage of the decoder takes care of error detection; it audits the preceding error-correction function of the decoder. If the block-code decoder detects an error in the class-I bits, then the entire 260 bits in the block are abandoned rather than passed on to the speech codec. When a block is discarded in this way, the speech codec is informed that a speech block was discarded and that it should try to interpolate the subsequent speech data. Such a criterion is shown to give better speech quality than if the speech codec had to reproduce a sound from the corrupted data.

The second function of the channel-coding process is to utilize convolutional coding. This code is applied to both of the class-I and -II bits, including the parity bits generated from the class-I bits. For a code to be able to correct errors, a certain number of additional bits need to be added to the payload bits. The convolutional code, employed for this application, used the rate $r = 1/2$ and a delay of $m=4$. This meant that five consecutive bits were used for the calculation of the redundancy bits. Before the information bits are encoded, four bits are added and set to zero in order to reset the convolutional code. Since five bits are used to calculate the appropriate redundancy bits, four trailing zeros are appended after the last data bit.

The encoder of this convolutional code is shown in Figure 10.2. When the encoder starts, all of the memory cells are reset to zero. In the absence of the data block some additional zeros are entered, which , in normal circumstances, would come from the subsequent block. Figure 10.3 illustrates the complete channel-coding scheme for all of the speech bits. In this procedure, 189 bits enter the convolutional encoder, 2x189 = 378 bits exit the encoder, and 78 class-III bits are added to the 378 bits to yield 456 bits. This is exactly 4 times 114 with 114 being the number of coded bits (or 57 bytes) within one burst.

Transmission errors can be compensated to a certain degree in the speech decoder by interpolation. This interpolation is accomplished with the filter coefficients that are kept in memory and reused when new coefficients are not available. Uncorrected bit errors found among the class I bits cause the entire speech frame to be discarded. The discarded and missing frames are extrapolated with the data from the last valid speech frame. Short-term disturbances are bridged by the speech codec. The above-described channel decoding process (the error-correction and concealment) enhances the system's resistance to reception problems in such a manner that any degradation in quality is not noticeable by the user.

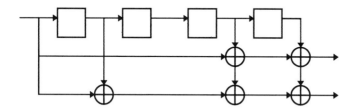

Figure 10.2 A convolutional encoder.

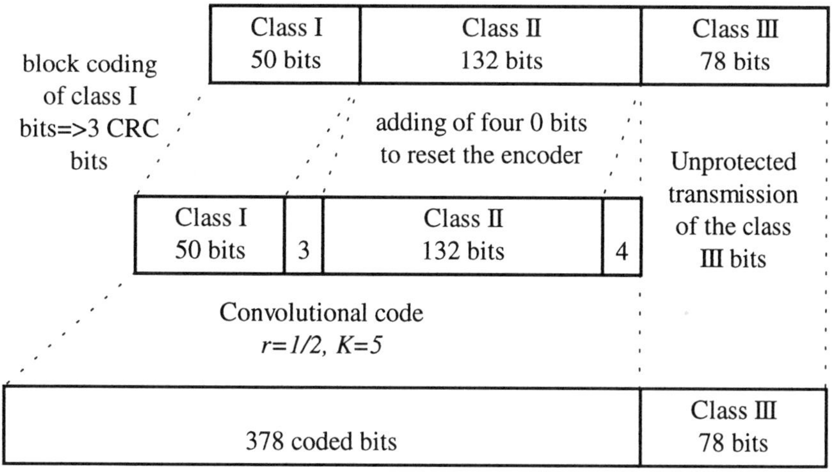

block coding of class I bits=>3 CRC bits	Class I 50 bits		Class II 132 bits		Class III 78 bits
		adding of four 0 bits to reset the encoder			Unprotected transmission of the class III bits
	Class I 50 bits	3	Class II 132 bits	4	
	Convolutional code r=1/2, K=5				
378 coded bits					Class III 78 bits

Figure 10.3 Block and convolutional coding of full-rate speech data.

10.1.2 Channel Coding of the Data Channel in the GSM System

The coding scheme for the user-data channel of the GSM system is quite complex. The reason for this is that the user bits must be protected even better than the traffic channel speech data. There are five different data channels in the GSM system. Each channel has a different data rate and a different interleaving scheme. For protection it uses convolutional codes with different parameters.

Figure 10.4 shows the coding scheme for the 9.6-kbits/s information data transmission. Even though the rate, 9.6-kbits/s, refers to the user's transmission rate, the actual rate is increased to 12-kbits/s by means of the channel coding of the terminal equipment (at the base stations); 12-kbits/s is the rate delivered to the mobile station. The user's bit stream is divided into four blocks of 60 bits each, for a total of 240 bits, which are coded together in a convolutional code. In contrast to the coding of the speech data, no block code is applied prior to the convolutional coding because error detection is performed already by the terminal equipment. Again if convolutional codes are used, four zero bits need to be added to the 240 data bits in order to reset the decoder. The parameters for the convolutional code are the same as those used for speech coding ($K=5$, $r=1/2$). The convolutional code accepts 244 bits and generates 488 coded bits. Unfortunately, these 488 bits do not fit easily into the 456 bits-per-block scheme discussed previously for the coded speech data. Thus, the excess 488-456=32 bits are punctured in accordance with a certain rule in order to obtain the correct number of bits.

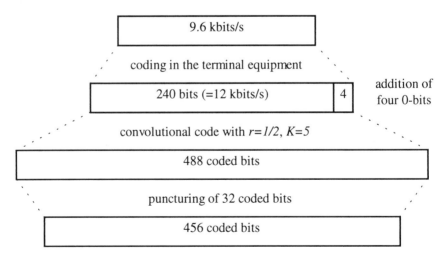

Figure 10.4 Channel-coding for 9.6 kbits/s of data transmission.

The rate, 2.4 kbits/s, for the data channel is shown in Figure 10.5. For the data channel a rate-1/6 convolutional code with K=5 is used. The data that comes from the terminal equipment has a rate of 3.6 kbits/s, including the error-correction coding. The customer's coded data are divided into 72-bit blocks to which, since K=5, four zero bits are added. The convolutional code transforms the 72+4=76 input bits into 76x6=456 coded output bits, which map onto eight subblocks in the same way that they were mapped in the speech-frame case.

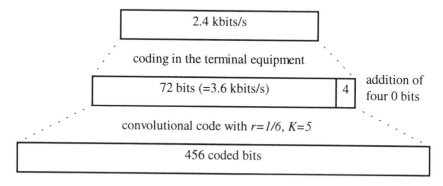

Figure 10.5 Convolutional coding for the 2.4-kbits/s of data channel.

10.1.3 Channel-Coding of the Signaling Channels

The signaling data are the most important data in the network. The signaling-information frame in a GSM system contains a maximum of 184 bits that need to be encoded. It makes no difference whether or not the type of signaling information to be transmitted is mapped onto a specific logic control channel. The format always stays the same.

The significance of each of the 184 bits is assumed to be the same, and no distinction is made between these bits as it was for the case with the three classes of speech bits. Again the coding scheme is divided into two steps as it is shown in Figure 10.6. The first step utilizes a block code, which is dedicated to the detection and correction of burst errors. There are errors that occur when parts of a burst, or a complete burst, corrupts the mobile radio channel. For this case the block code belongs to the family of Fire codes. It adds 40 parity bits to the 184 data bits which results in 224 Fire-coded bits.

The second step involves the use of a convolutional code which has the same parameters that were used for the trellis channel with r=1/2, K=5 and 4 additional zero bits to yield a block of 228 bits. The convolutional code doubles the number of bits to be transmitted to 456 bits.

Figure 10.7 summarizes the different channel-coding processes used in the different logic channels for both signal and data on the physical layers of the GSM system. Although not explained in detail, it illustrates the various combinations of convolutional codes and other channel-coding schemes, used for the error control of a GSM system.

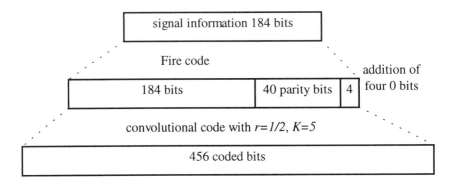

Figure 10.6 Channel-coding scheme for signal channel

10.2 Convolutional Codes Specified in CDMA Cellular Systems

CDMA stands for code-division multiple access, which implies that users in the CDMA-based digital cellular system are distinguished from each other by a code rather than by frequency-band-allotted time slots. There are many CDMA systems in the world, but the CDMA system discussed next is the one described in IS-95, which is the standard specified for cellular mobile-telecommunications systems by the Electronics Industries Association and Telecommunication Industries Association.

The techniques used in CDMA have been known since the late 1940's. They were used in secure military communications since the early 1950s. Cellular systems allow access to their resources by various techniques. One of these techniques is to deliberately spread the spectrum of a radio which shares access in a system by the use of a particular high-speed code word that is unique to that radio. This code, which is usually some type of pseudo-random noise(PN) code, further modulates the normal payload data, thus spreading the transmitter's output in frequency with the identifying code.

Applications of Convolutional Codes in Mobile Communications

This class of techniques is therefore called spread-spectrum radio. If each radio has its own radio-specific PN code, then all radios, except the one it is desired to receive, look like noise to all other receivers in the system. A receiver's correlator in such a system distinguishes the emissions of a particular radio from the many others in the same radio band by regarding all others in the occupied spectrum, which don't have the same PN code, to be noise.

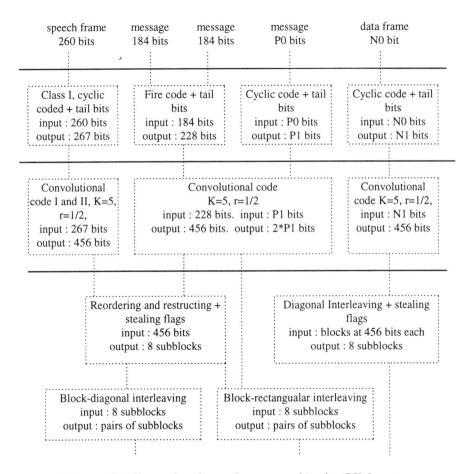

Figure 10.7 Channel-coding schemes used in the GSM system

In a manner that is used in the previously discussed GSM system the logic channels in a CDMA system are mapped into both the forward and reverse physical channels. The logical channels are separated from each other by different spreading codes rather than in frequency and time as it was in the case for GSM. In the following subsections, applications of convolutional codes in both the reverse and forward CDMA channels are presented.

10.2.1 Error Protection for Reverse CDMA Channel

The reverse channel is transmitted from all the mobiles within a particular cell site's coverage area to the base station. In the IS-95, the Reverse CDMA Channel is composed of logic-access channels and reverse-logic traffic channels. These channels share the same CDMA frequency assignments used in direct-sequence CDMA. Each channel is spread by the use of a Walsh function which is specific to a particular channel. This spreading process, sometimes, is called Walsh chipping. Finally, all of the channels that make up the reverse-channel output are spread by a common PN code. Figure 10.8 shows an example of the signals received by a base station on the reverse CDMA Channel. Each Traffic Channel is identified by a long-code-user sequence; each Access Channel is identified by a long-code Access-Channel sequence. Multiple Reverse CDMA Channels also can be used by a base station in a frequency-division multiplexed manner.

The Reverse CDMA Channel has the detailed overall structure shown in Figure 10.9. Data that is transmitted on the Reverse CDMA Channel are grouped into 20 ms frames. All data transmitted on the Reverse CDMA Channel are convolutionally encoded, block interleaved, modulated by a 64-ary orthogonal modulation, and direct-sequence spread prior to transmission.

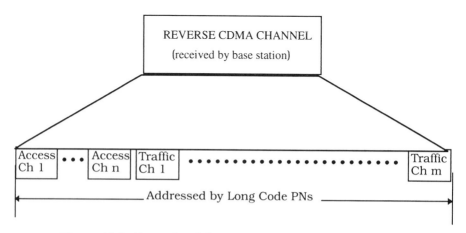

Figure 10.8. Example of the Reverse Logic CDMA Channels Received at a Base Station

Applications of Convolutional Codes in Mobile Communications 419

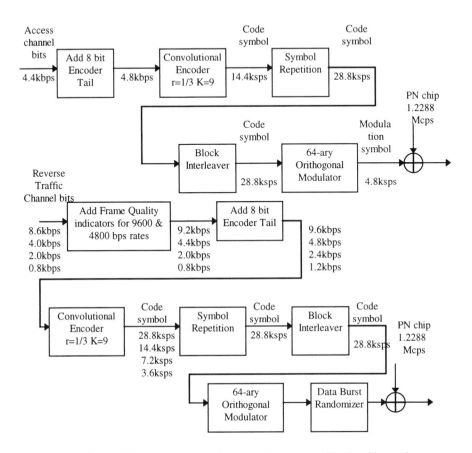

Figure 10.9. Error-control coding for reverse CDMA Channel

After the addition of the frame-quality indicators for both the 9600 bps and 4800 bps rates and the addition of eight Encoder tail-bits, data frames can be transmitted on the Reverse Traffic Channel at data rates of 9600, 4800, 2400, and 1200 bps. The Reverse-Traffic Channels can use any of these data rates for transmission. The transmission duty cycle on a Reverse-Traffic Channel varies with the transmission data-rate. Specifically, the transmission duty cycle of 9600 bps frames is 100 percent, the transmission duty cycle of 4800 bps frames is 50 percent, the transmission duty cycle of 2400 bps frames is 25 percent, and the transmission duty cycle of 1200 bps frames is 12.5 percent as shown in Table 10.1. Since the duty cycle for transmission varies proportionately with the data rate, the actual burst-transmission rate is fixed at 28,800 code symbols per second. Also since the six code symbols are modulated as one of 64 modulation symbols for transmission, the modulation symbol-transmission rate is fixed at 4800 modulation symbols per second. This results in a fixed Walsh-chip rate of 307.2 kcps. The rate of the spreading PN sequence is fixed at 1.2288 Mcps, so that each Walsh chip is spread by four PN chips. Table 10.1 defines the signal rates and their relationships with the various transmission rates of the Reverse Traffic Channel.

Data rate	9600	4800	2400	1200	bits/second
Information rate	8600	4000	2000	800	bits/second
Information bit per frame	172	80	40	16	bits/frame
Parity bits per frame	12	8	0	0	bits/frame
Data bits per frame	192	96	48	24	bits/frame
Code rate	1/3	1/3	1/3	1/3	bits/code sym
Data symbols per frame	576	288	144	72	code sym/frame
Repetitions	0	1	3	7	times
Total bits per frame	576	576	576	576	bits/frame
Modulation	6	6	6	6	code sym/mod symbol
Modulation Symbol Rate	4800	4800	4800	4800	mod sym/sec.
Code symbol rate	28800	28800	28800	28800	code sym/sec.

Table 10.1. Contents of a 20ms frame for reverse traffic-channels.

The above rates are the same for the Access Channel except that the transmission rate is fixed at 4800 bps after adding eight Encoder Tail Bits. Each code symbol is repeated once, and the transmission duty cycle is 100 percent. Table 10.2 defines the signal rates and their relationships with the Access Channel.

Applications of Convolutional Codes in Mobile Communications

The modulation parameters for the Reverse Traffic Channel and the Access Channel are shown in Tables 10.1 and 10.2, respectively.

Data rate	4800	bits/seconds
PN chip rate	1.2288	Mcps
Code Rate	1/3	bits/code sym
Repetition	1	times
Transmit duty cycle	100.0	%
Code symbol rate	28,800	Code sym /second
Modulation	6	code sym/mod symbol
Modulation symbol rate	4800	Mod sym/secomd
Walsh chip rate	307.20	kcps
Mod symbol duration	208.33	μs
PN chips/Code symbol	42.67	PN chip/code sym
PN chips/Mod symbol	256	PN chip/mod symbol
PN chips/Walsh chip	4	PN chips/Walsh chip

Table 10.2. Access Channel Modulation Parameters

The access channel supports a fixed data-rate operation at 4800 bps. The reverse traffic channel supports a variable data-rate operation at 9600, 4800, 2400, and 1200 bps.

The mobile station convolutionally encodes the data transmitted on the reverse-traffic channel and the access channel prior to interleaving. The convolutional code used for this application has a rate of 1/3 and has a constraint length of 9. The generator functions for this code are $g_0 = 557$ (octal), $g_1 = 663$ (octal), and $g_2 = 711$(octal). This is a rate-1/3 convolutional code that generates three code symbols for each data-bit input of the encoder. These code symbols constitute the output so that the code symbol (c_0)

encoded with the generating function g_0 is outputted first, the code symbol (c_1) encoded with the generating function g_1 is outputted second, and the code symbol (c_2) encoded with the generating function g_2 is outputted last. The state of the convolutional encoder upon initialization is the all-zero state. The first code-symbol output after initialization is a code symbol encoded by the generating function g_0.

Convolutional encoding involves the modulo-2 addition of selected taps of a serially, time-delayed, data sequence. The length of the data-sequence delay is equal to $v-1$, where v is the constraint length of the code. Figure 10.10 illustrates the encoder for the code specified in this section.

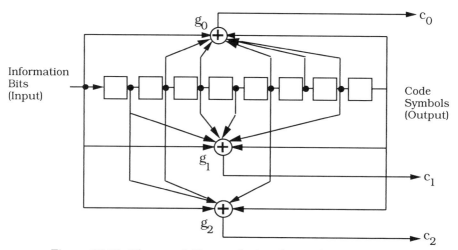

Figure 10.10. The rate-1/3 convolutional encoder with $v = 9$.

Output code symbols from the convolutional encoder are repeated before being sequentially interleaved when the data rate is lower than 9600 bps.

The code-symbol repetition rate on the Reverse-Traffic Channel varies with the data rate. Code symbols are not repeated for the 9600 bps data rate. Each code symbol at the 4800 bps data rate is repeated one time (each symbol occurs over 2 consecutive time periods). Each code symbol at the 2400 bps data rate is repeated 3 times (each symbol occurs over 4 consecutive time periods). Each code symbol at the 1200 bps data rate is repeated 7 times (each symbol occurs over 8 consecutive time periods). All of the data rates (9600, 4800, 2400, and 1200 bps) result in a constant code-symbol rate of 28800 code symbols per second. On the Reverse Traffic Channel these repeated code symbols are not transmitted multiple times. Rather, the repeated code symbols are input to the block-interleaver function, and all but one of the code-symbol repetitions is deleted prior to actual transmission due to the variable transmission duty-cycle.

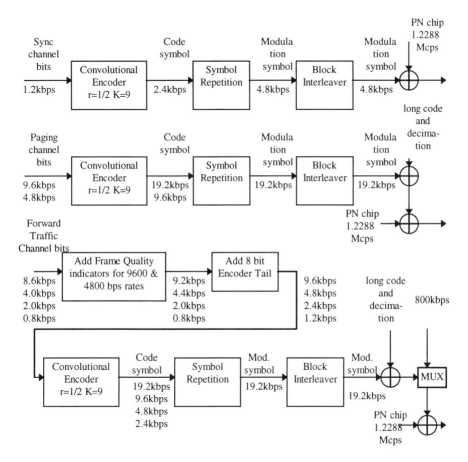

Figure 10.11. Error-control coding for forward CDMA channels

For the Access Channel, which has a fixed data rate of 4800 bps, each code symbol is repeated one time (each symbol occurs over 2 consecutive time periods). On the Access Channel both of the repeated code symbols are transmitted.

10.2.2 Error Protection for Forward CDMA Channel

The Forward CDMA Channel is transmitted from the cell site base station to mobile users. Such a channel has the detailed overall structure, shown in Figure 10.11. The Forward CDMA Channel consists of the following logical channels: the Pilot Channel from one Sync Channel to seven Paging Channels and a number of Forward-Traffic Channels. These logical channels are distinguished from each other by their spreading codes. Each of these code channels is orthogonally spread by the appropriate Walsh functions and is then spread by a quadrature pair of PN sequences at a fixed chip rate of 1.2288 Mcps (million chips/sec). Multiple Forward CDMA Channels also can be used within a base station in a frequency-division multiplexed manner.

An example assignment of the code channels that are transmitted by a base station is shown in Figure 10.12. Of the 64 code channels available for use this example depicts the Pilot Channel (always required), one Sync Channel, seven Paging Channels (the maximum number allowed), and 55 Traffic Channels. In another possible configuration one could replace all of the Paging Channels and the Sync Channel, one-for-one with the Traffic Channels, for a maximum of one Pilot Channel, zero Paging Channels, zero Sync Channels, and 63 Traffic Channels.

Data Rate	1200	bits/second
PN Chip Rate	1.2288	Mcps
Code Rate	1/2	bits/code symbol
Repetition	1	mod sym/code sym*
Modulation Symbol Rate	4,800	Mod Symbols/sec.
PN Chips/Modulation Symbol	256	PN chips/mod sym
PN Chips/Bit	1024	PN chips/bit

*Each repetition of a code symbol is a modulation symbol.
Table 10.3 Sync Channel Modulation Parameters

Applications of Convolutional Codes in Mobile Communications   425

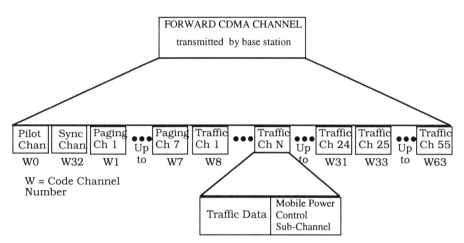

Figure 10.12. Example of a Forward CDMA Channel,

Transmitted by a Base Station

The modulation parameters for the Forward CDMA Channel are as shown in Tables 10.3, 10.4, and 10.5.

Data Rate	9600	4800	bits/second
PN Chip Rate	1.2288	1.2288	Mcps
Code Rate	1/2	1/2	bits/code symbol
Repetition	0	1	mod sym/code sym*
Modulation Symbol Rate	19,200	19,200	Mod Symbols/sec.
PN Chips/Modulation Symbol	64	64	PN chips/mod sym
PN Chips/Bit	128	256	PN chips/bit

*Each repetition of a code symbol is a modulation symbol.
Table 10.4. Paging Channel Modulation Parameters

The Sync Channel operates at a fixed rate of 1200 bps. The Paging Channel supports a fixed data-rate operation at 9600 or 4800 bps. The Forward Traffic Channel supports a variable data-rate operation at 9600, 4800, 2400, and 1200 bps.

The message data over the Sync Channel, Paging Channels, and Forward Traffic Channels are convolutionally encoded prior to transmission. The convolutional code is rate- 1/2 with a constraint length of 9. The generator functions of the code are $g_0 = 753$ (octal) and $g_1 = 561$ (octal). This is a rate-1/2 code that generates two code symbols for each data-bit input to the encoder. These code symbols are outputted in such a manner that the code symbol (c_0) which is encoded with the generating function g_0 is outputted first, and the code symbol (c_1) encoded with the generating function g_1 is outputted last. The state of the convolutional encoder, upon initialization, is the all-zero state. The first code-symbol output after initialization is a code symbol that is encoded with the generating function g_0.

Data Rate	9600	4800	2400	1200	bits/second
Information rate	8600	4000	2000	800	bits/second
Info. bits per frame	172	80	40	16	bits/frame
Parity bits per frame	12	8	0	0	bits/frame
Data bits per frame	192	96	48	24	bits/frame
Code rate	1/2	1/2	1/2	1/2	bits/coded symbol
Coded bits per frame	384	192	96	48	coded sym /frame
Repetition	0	1	3	7	times*
Total bits per frame	384	384	384	384	bits/frame
Modulation Symbol Rate	19,200	19,200	19,200	19,200	mod sym/sec.

*Each repetition of a code symbol is a modulation symbol.

Table 10.5. Forward Traffic Channel Modulation Parameters

Convolutional encoding involves the modulo-2 addition of selected taps of a serially time-delayed data sequence. The length of the data sequence delay is equal to v -1, where v is the constraint length of the code. Figure 10.13 illustrates the particular v =9, rate 1/2, convolutional encoder that is used for these channels.

For the Sync Channel each convolutionally encoded symbol is repeated one time (each symbol occurs over 2 consecutive time periods) prior to block interleaving.

For the Paging and Forward Traffic-Channels, each convolutionally encoded symbol is repeated prior to block interleaving whenever the information rate is lower than 9600 bps. Each code symbol at the 4800 bps rate is repeated 1 time (each symbol occurs over 2 consecutive times). Each code symbol at the

2400 bps data rate is repeated 3 times (each symbol occurs over 4 consecutive times). Each code symbol at the 1200 bps data rate is repeated over 7 times (each symbol occurs over 8 consecutive times). For all of the data rates (9600, 4800, 2400, and 1200 bps) this results in a constant modulation symbol rate of 19200 modulation symbols per second.

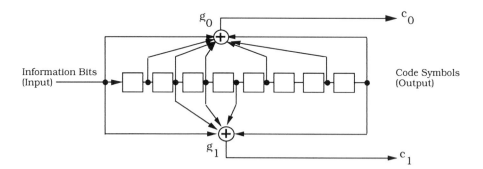

Figure 10.13. The rate-1/2, convolutional encoder with $v = 9$.

All symbols after repetition on the Sync Channel, Paging Channels, and Forward-Traffic Channels are block interleaved. The Sync Channel uses a block interleaver that spans 26.666 ms which is equivalent to 128 modulation symbols at the symbol rate of 4800 sps.

Bibliography

[1] G. L Stuber, Principle of Mobile Communication, Kluwer Academic Publishers, Boston, 1996.
[2] W.C.Y. Lee, "Overview of cellular CDMA", IEEE Trans. Veh. Technology, Vol. 40, pp.291-302, May 1991.
[3] GSM Technical Specification, ETSI, Sophia Antipolis, Vols. 04.05 and 04.06.
[4] EIA/TIA IS-95, "Mobile station – base station compatability standard for dual-mode wideband spread spectrum cellular system," 1995.
[5] K. Raith and J. Uddenfelt, "Capacity of digital cellular TDMA systems," IEEE Trans. Veh. Technology, Vol. 40, pp. 323-332, May 1991.
[6] David J. Goodman, Wireless Personal Communications Systems, Addison-Wesley, 1997.

11 Trellis-Coded Modulation

As it is discussed in Chapter 1, power and bandwidth are limited resources in modern communication systems, and the efficient exploitation of these resources invariably involves an increase in complexity. It has become apparent over the past few decades that when there are strict limits on the power and bandwidth resources, the complexity of communication systems still can be increased to obtain efficiencies that are even closer to the theoretical limits.

From a classical perspective coding for error control usually necessitates an increase in bandwidth which, in turn, decreases the required signal-to-noise ratio. The reason for this bandwidth expansion is that the rate of the encoder output in transmitted symbols-per-second is greater than the rate at the encoder input by a factor equal to the reciprocal of the code rate, R, assuming identical alphabets. Thus, a half-rate ($\frac{1}{2}-rate$) encoding increases the transmitted symbol rate by a factor of 2 relative to the uncoded symbol rate. For power-limited channels, such as the deep-space communication channels, one must trade the bandwidth expansion for transmitted power in order to achieve the desired system performance.

These facts have led to the misunderstanding, that error-control can provide only significant gains in energy efficiency, when the possibility of bandwidth expansion is present. In fact, for bandwidth-limited channels, such as the telephone channels, error-control codes were not popular in the past because of this apparent need for bandwidth expansion.

Actually, in bandwidth-limited channels several bandwidth-efficient signaling techniques, such as Pulse- Amplitude Modulation(PAM),

Quadrature-Amplitude Modulation(QAM), and Multi-Phase-Shift-Keying (MPSK) modulation, were employed to support a high bandwidth efficiency (in bits/s/Hz). Traditionally, error-control coding and modulation were considered to be two separate parts of the digital communication system. The modulator and demodulator were designed usually to convert an analog waveform channel into a discrete channel, and the channel encoder and decoder were designed to correct the errors that occurred in this discrete channel. Higher improvements in performance were achievable only by lowering the code rate and by increasing the redundancy in the code, of course, at the cost of an increased decoding complexity and bandwidth expansion.

The consideration of coding and modulation as an integrated system was first suggested by Massey [1] in 1974. Two years later Ungerboeck presented a feasible means for integrating coding and modulation through what is now called trellis-coded modulation(TCM)[2]. In a paper, published in 1982, Ungerboeck [2] demonstrated that by the integration of the convolutional code with a bandwidth-efficient modulation scheme that significant coding gains could be achieved without expanding the bandwidth. At the time Ungerboeck's viewpoint on coding was considered to be quite revolutionary thinking about coding and its effect on bandwidth. Ungerboeck reasoned that the symbol set for the modulator could be enlarged, when coding is adopted, relative to that needed by uncoded signaling, and he selected those redundant modulation sequences which had memory and were chosen from a larger coding constellation.

If the signal set's "dimensions-per-information bit" remains unchanged, then at least to first order, the power spectrum of TCM modulation remains identical to that of the uncoded signal. The proper design of this code with an enlarged constellation can achieve coding gains on an AWGN channel, which are quite impressive, from 3 to 6 dB. The key to the realization of this gain is that TCM coding and modulation must be performed with full attention to the Euclidean distances between the coded modulator-output sequences, rather than to the more commonly used Hamming distances at the encoder output.

In this chapter we begin the discussion by considering one- and two-dimensional M-ary modulation and the trade-off between spectral efficiency and power efficiency. Then, an example of a two-dimensional trellis-coded modulation (TCM) system is introduced to illustrate the concept and fundamentals of a TCM scheme. The emphasis is placed on three key concepts: expanded signal constellations, signal constellation partitioning,

Trellis-Coded Modulation

and the selection of partitions by convolutional encoders. This is followed by a discussion of decoding by the Viterbi decoder and a presentation of various performance-analysis techniques. The chapter concludes with a discussion of implementation considerations and a look at the some unequal error-protection techniques.

11.1 M-ary Modulation, Spectral Efficiency and Power Efficiency

Usually, a digital communication system is called "digital" in the sense that it transmits information that is in the form of discrete symbols. In a binary modulation system, such as the binary phase-shift keying (BPSK) system, two transmitted symbols are assigned. The Nyquist bandwidth for a communication signal is equal to the rate at which the symbols are transmitted. If the symbols are binary, an r bits-per-second data rate requires, at least, an r-Hz Nyquist bandwidth. The spectral efficiency of a system is defined to be the number of bits-per-second transmitted per-one-Hz of bandwidth. The binary system, mentioned above, thus has a spectral efficiency of 1 bit/sec/Hz.

The binary signal constellation is generally represented by a pair of points on the real line. The position of a signal on the line is proportional to its magnitude (square root of signal energy) and its relative phase. Figure 11.1 shows the signal diagram for the BPSK constellation. Normalized integer labels are frequently used in such diagrams; they preserve the relative differences in signal magnitude but also are less cumbersome than the absolute values of the received or transmitted-signal magnitudes.

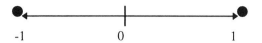

Figure 11.1. BPSK Signal Constellation

When one needs to evaluate the performance of such a system for some received noise energy, the labels in the signal constellation diagram are weighted by $\sqrt{E_s}$, where E_s is the average received signal energy. The symbol-error rate is often computed by determining the probability that the channel noise causes the received signal to be closer to one of the incorrect signals than to the signal that was actually transmitted.

As an example, consider an additive white-Gaussian-noise(AWGN) channel with the two-sided spectral density $N_0/2$ at the input to the receiver. On such a channel the noise can be represented by a complex-Gaussian random variable n that is added to the transmitted signal; if the signal s has been transmitted, the receiver will see a complex signal of the form, $r = s + n$. If the different signals are transmitted equally likely (with equal probability), the maximum-likelihood receiver selects that signal which is closest in Euclidean distance to the received signal, i.e. the signal $\hat{s}$ is determined such that

$$\hat{s} = \begin{cases} \sqrt{E_s}, & \text{if } d_E(\sqrt{E_s}, r) < d_E(-\sqrt{E_s}, r), \\ -\sqrt{E_s}, & \text{if } d_E(-\sqrt{E_s}, r) < d_E(\sqrt{E_s}, r), \end{cases} \quad (11.1)$$

where $d_E(s,r) = \sqrt{\|s-r\|^2}$ denotes the Euclidean distance in the complex plane.

Consider the BPSK constellation given in Figure 11.1 and assume that +1 has been transmitted, i.e. $s = \sqrt{E_s}$. The bit-error rate(BER), P_b, is then computed as follows:

$$P_b = P\{d_E(s,r) \geq d_E(-s,r)\} = P\{n \geq \sqrt{E_s}\}$$
$$= Q\left(\sqrt{\frac{2E_s}{N_0}}\right), \quad (11.2)$$

where $Q(x) = \dfrac{1}{\sqrt{2\pi}} \int_x^\infty e^{-y^2/2} dx$. $\quad (11.3)$

In many applications the desired data rate (in bits/second) far exceeds the available bandwidth (in Hz). In such cases it is necessary to increase the spectral efficiency of the communication system. This is often accomplished by increasing the size of the signaling constellation in such a manner that the symbol transmission rate (and thus the Nyquist bandwidth) remains the same. However, the number of possible values taken on by these symbols is increased.

The most obvious expansion of the constellation in Figure 11.1 is amplitude modulation (AM), as exemplified by the 8-AM constellation in Figure 11.2. The 8-AM constellation provides 3 bits/sec (double-sideband AM), three times the spectral efficiency of the BPSK signal.

Trellis-Coded Modulation

Figure 11.2. 8-AM Signal Constellation

Define the minimum Euclidean distance d_{min} for a constellation to be the shortest Euclidean distance between any pair of distinct signals. For the example shown in Figure 11.2 $d_{min} = 2$, which equals the Euclidean distance between the symbols -1 and +1, and also the Euclidean distance between symbols +1 and +3, etc.

At high signal-to-noise ratios (SNRs), the most likely error events are those in which the transmitted symbol is confused with only one of its nearest neighbors. Then, if one assumes that there are always two nearest neighbors, e.g. for the signal point +1, the symbol-error rate for AM on an AWGN channel is approximated easily by

$$P_e \approx 2Q\left(d_{min}\sqrt{\frac{2E_s}{N_0}}\right). \tag{11.4}$$

The amount of energy required to transmit one of the 8-AM signals in Figure 11.2 is proportional to the square of the distance of the signal from the origin. If the distance between adjacent signals in Figure 11.2 is the same at the receiver input as that in Figure 11.1, then the average energy-per-transmitted symbol is significantly more for 8-AM than it is for BPSK. Let the normalized average-received energy for the BPSK case be $S_m = 1$. Here the notation "S_m" denotes normalized average-received energy, while the subscript m is the logarithm (to the base 2) of the cardinality of the signal constellation. For a 2^m-ary AM constellation with the same minimum free distance as BPSK, the normalized average energy is

$$S_m = \frac{(4^m - 1)}{3}. \tag{11.5}$$

8-AM thus requires $S_3 = 21$ times (13.22dB) more energy than BPSK to maintain the same minimum distance. Equation (11.5) also can be used to derive the additional energy required for each additional bit/sec/Hz. Thus,

$$S_{m+1} = 4S_m + 1. \tag{11.6}$$

From Eq. (11.6) it is clear that, asymptotically, each additional bit/sec/Hz requires approximately 4 times (6 dB) more average energy per transmitted symbol. This provides a rough estimate of the cost to obtain this additional spectral efficiency.

The AM constellations discussed above are one-dimensional. By extending these constellations to the two-dimensional or "quadrature-amplitude" modulation (QAM),i.e. to the complex plane, one can obtain an even better performance at the expense of only a small additional complexity. Figure 11.3 shows examples of complex-valued rectangular QAM constellations as well as some M-ary phase-shift keying (MPSK) constellations. MPSK signals can be parameterized in terms of a single variable (the phase), so it can be argued that they are also one-dimensional constellations. However, most of the practical MPSK receiver designs project the received signal onto a pair of orthogonal signals before making a phase estimate. Thus MPSK can be treated as a form of QAM. Both types have their respective advantages and disadvantages. The rectangular constellations provide a better minimum-distance performance versus an average-energy performance but are subject to distortion when they pass through nonlinear devices (e.g., traveling-wave tubes and other amplifiers that are operated in or near saturation). For the MPSK constellations the minimum distances are relatively small, but the modulated signals have a constant envelope and are thus not distorted by channel nonlinearities.

Two-dimensional constellations can be implemented by modulating a pair of orthogonal carriers as it is shown in Figure 11.4. *m* bits are taken from the source and mapped onto an ordered pair (*x,y*). These two coordinates are then used to modulate a pair of orthogonal carriers which, when added, form a single complex signal *z*. The modulated sine and cosine carriers are orthogonal and do not interfere with each other on a linear channel. Two-dimensional rectangular constellations thus can be viewed as a pair of orthogonal one-dimensional constellations. In the receiver, the received complex signal is decomposed into the pair ($\hat{x}, \hat{y}$) and then demodulated and remapped back to the binary sequence.

The block and convolutional codes discussed in this book provide coding gain by the addition of the appropriate redundant bits. Redundancy makes it possible for the receiver to detect and, in some cases, to correct the errors caused by channel noise. Redundancy is usually measured by the code rate *r*. If the coded system is to retain the same information-transmission rate as the uncoded system, then the symbol transmission rate R_s for the coded

Trellis-Coded Modulation

system must be $1/r$ times that of the uncoded system. In a band-limited system this would be impossible if one encodes in the classical manner since such coding for error control necessitates an increase in bandwidth. Trellis-coded modulation provides a powerful, additional, alternative.

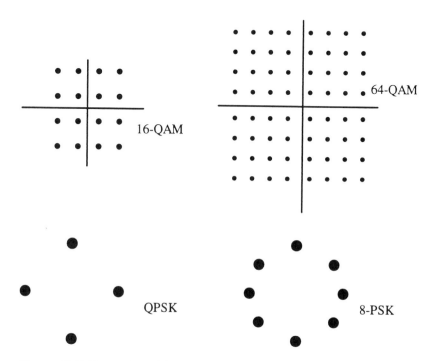

Figure 11.3. Examples of Signal Constellations for QAM and MPSK

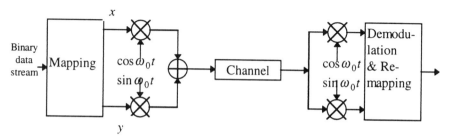

Figure 11.4. QAM Modulator and Demodulator

11.2 TCM Schemes

In this section, QPSK and 8-PSK are used to introduce trellis-coding schemes.

11.2.1 Encoding of Trellis Codes

First, let us begin the discussion with an illustrative example of Ungerbock's method of combining coding and modulation.

Example 11.1
Assume that a standard QPSK modulation scheme is considered. This allows one to transmit two information bits per symbol. That is, QPSK has a spectral efficiency of 2 bits/sec/Hz. It is known from elementary communication theory[2] that the most likely error, which a decoder makes, is to confuse the two neighboring signals, say signals $\bar{s}_0$ and $\bar{s}_1$, as shown in Figure 11.5. For the transmission over an AWGN channel, this error happens with probability,

$$P_{\bar{s}_0 \to \bar{s}_1} = Q\left(\sqrt{\frac{d_E^2(\bar{s}_0, \bar{s}_1)}{2N_0}}\right), \tag{11.7}$$

where for the present example $d_E^2(\bar{s}_0, \bar{s}_1) = 2E_s$ and E_s is the average energy per symbol.

Instead of having a system which transmits only one symbol at a time for 8-PSK, let the "trellis" encoder, shown in Figure 11.6, be used. This encoder consists of two parts, the first of which is a rate 2/3 convolutional encoder with the four binary states $\bar{S}_j = (S_j^{(1)}, S_j^{(2)})$. The second part of the encoder is called a signal mapper, and its function is the memoryless mapping of the 3 output bits $\bar{c}_j = (c_j^{(1)}, c_j^{(2)}, c_j^{(3)})$ of the convolutional encoder onto one of the eight symbols of the 8-PSK signal set, shown in Figure 11.5(b). Some commonly-used mapping functions are listed in Table 11.1. Although the expansion of one signal set from QPSK to 8-PSK provides the redundancy required for coding, it also reduces the distance between the signal points if the average signal energy is kept constant. For the present example, this is shown in Figure 11.5(b). The reduction in the distances between the signal points increases the error rate which needs to be compensated by the use of coding if the encoding scheme is to provide a benefit.

Trellis-Coded Modulation

Name of the mapping function	QPSK	8-PSK
Natural Binary Mapping	$\bar{s}_0 \to (00)$	$\bar{s}_0 \to (000)$
	$\bar{s}_1 \to (01)$	$\bar{s}_1 \to (001)$
	$\bar{s}_2 \to (10)$	$\bar{s}_2 \to (010)$
	$\bar{s}_3 \to (11)$	$\bar{s}_3 \to (011)$
		$\bar{s}_4 \to (100)$
		$\bar{s}_5 \to (101)$
		$\bar{s}_6 \to (110)$
		$\bar{s}_7 \to (111)$
Gray-Code Mapping	$\bar{s}_0 \to (00)$	$\bar{s}_0 \to (000)$
	$\bar{s}_1 \to (01)$	$\bar{s}_1 \to (001)$
	$\bar{s}_2 \to (11)$	$\bar{s}_2 \to (011)$
	$\bar{s}_3 \to (10)$	$\bar{s}_3 \to (010)$
		$\bar{s}_4 \to (110)$
		$\bar{s}_5 \to (111)$
		$\bar{s}_6 \to (101)$
		$\bar{s}_7 \to (100)$

Table 11.1 Some signal mapping functions

The convolutional encoder accepts 2 information bits $\bar{m}_j = (m_j^{(1)}, m_j^{(2)})$ at each symbol time j and transits from a state $\bar{S}_j$ to one of four possible successor states $\bar{S}_{j+1}$, and output a 3-bits symbol. In this fashion the trellis encoder maps each 3-bits symbol into a complex sequence of symbols $\bar{x} = (\ldots, x_{-1}, x_0, x_1, x_2, \ldots)$. On the assumption that the encoder is operated in a continuous fashion, there are four choices at each time j, which allows one to transmit two information bits/symbol, the same as it is for QPSK.

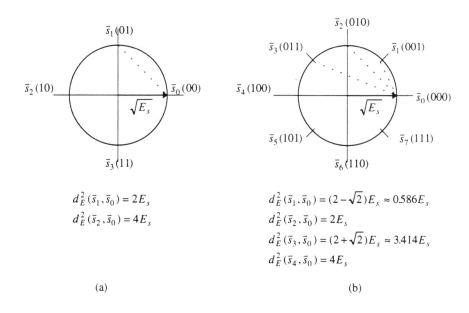

Figure 11.5 Signal sets of (a) QPSK and (b) 8PSK in signal amplitude-phase planes with the natural mapping

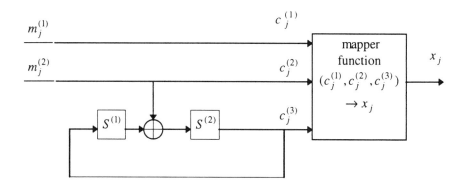

Figure 11.6 Encoder of a rate 2/3 convolutional coded 8PSK

At this point a graphical interpretation of the operations of the convolutional encoder is useful. Since the encoder in Figure 11.6 is time invariant, its operations can be represented by the state-transition diagram, shown in

Trellis-Coded Modulation

Figure 11.7. The nodes in this transition diagram are the states of the convolutional encoder, and the branches represent the possible transitions between the state pairs $(S_j^{(1)}, S_j^{(2)})$. Each branch is labeled as an *input/output* pair. The pair of input bits $\bar{m} = (m^{(1)}, m^{(2)})$ which cause the transitions of the states that result in the output triple $\bar{c} = (c^{(1)}, c^{(2)}, c^{(3)})$, i.e. the output signal $x(\bar{c})$. In Figure 11.3 the output $x(\bar{c})$ for the natural mapping in Table 11.1 is represented in decimal form as $x(\bar{c}) = c^{(1)} 2^2 + c^{(2)} 2^1 + c^{(3)} 2^0$.

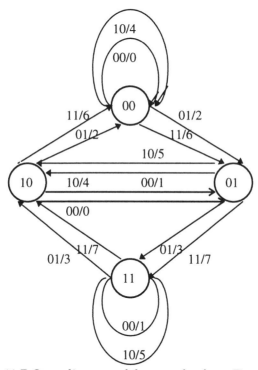

Figure 11.7. State diagram of the encoder from Figure 11.6.

If the states $(S_j^{(1)}, S_j^{(2)})$ are indexed both by their values and the time index j, then Figure 11.7 expands into the trellis diagram given in Figure 11.8. In this Figure, the branches are also labeled as input/output pairs. The input bits $\bar{m} = (m^{(1)} m^{(2)})$ of this input pair cause the transition of states and output symbols that map onto the output signals $\bar{s}_i$ consisting of the 8-PSK signals shown in Figure 11.5(b). The trellis diagram of a trellis encoder is therefore a two-dimensional representation of the operations of the encoder.

The trellis captures all possible state transitions, starting from the originating state (usually state (0,0)) and terminating at the final state (usually also state (0,0)). Assume that the trellis in Figure 11.8 is terminated at length L. This requires that the convolutional encoder be driven back into state (0,0) at time $j=L$. As a consequence, the branches at times $j = L-2$ and $L-1$ are predetermined, and at these times no information is transmitted. In practice, the length of the trellis can be several hundred, thousands, or possibly even an infinite number of time units for a continuous operation. When and where to terminate the trellis are matters determined by practical considerations.

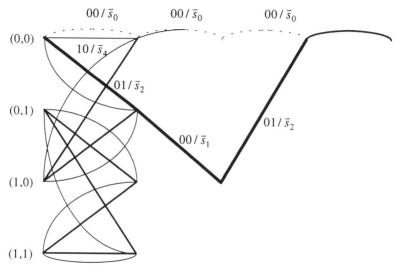

(a) Trellis diagram with two code-symbol paths : bold-lines and dashed-line

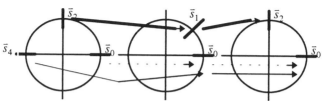

(b) Signal point changes of the two code-symbol paths

Figure 11.8 Trellis diagram for the encoder in Figure 11.7.

Trellis-Coded Modulation

Each path through the trellis in Fig. 11.8 corresponds to a unique message sequence which, in turn, corresponds to a unique 8-PSK signal sequence (see Figure 11.5b). The term trellis-coded modulation originates from the fact that these encoded sequences consist of modulated symbols(in this case phase-modulated symbols) rather than binary digits.

11.2.2 Decoding Trellis Codes

At first sight it is not clear, what has been gained by the rather complex encoding process in Example 11.1, since the signals in the 8-PSK signal sets are closer in phase angle than those in the QPSK signal set and have a higher symbol-error rate. The convolutional encoder in Figure 11.6, however, puts restrictions on the symbols that are in a sequence, and these restrictions can be exploited by a smart decoder. In fact, what counts in decoding is the Euclidean distance between the encoded or modulated signal sequences $\bar{z} = \{\cdots, \bar{s}_i, \cdots, \bar{s}_j, \cdots\}$, not the Hamming distance between the binary encoded sequences. Notice that the modulated signal $\bar{s}_i$ is expressed by two quadrature components, i.e. $\bar{s}_i = (s_{x_i}, s_{y_i})$.

In order to decode optimally note first that any use of hard-decision demodulation, prior to the ML decoding process of a received coded sequence, causes an irreversible loss of information which translates into a loss of SNR. Recall from Chapter 8 for a convolutional code with BPSK that this loss is about 2-3dB. For QAM, the hard-decision demodulation maps each received symbol $\hat{s} = (\hat{s}_x, \hat{s}_y)$ into a signal point $\bar{s}_i = (s_{x_i}, s_{y_i})$ by the use of the criterion $\min_i d_E^2(\hat{s}, \bar{s}_i)$. In a bandwidth-efficient encoding scheme with an expanded signal set, such as the 8PSK shown in Fig. 11.6b, a compensation of the power penalty is required. In order to avoid the hard-decision loss, a QAM soft-decision criterion is used in the trellis-decoding process. To achieve this both the demodulation and decoding stages are combined into a single step so that the decoder operates directly on the received "soft-output" samples of the channel.

To realize a soft-decision decoder which operates directly on the channel outputs advantage is taken of the fact that a trellis encoder generates a coded sequence $\bar{z}^{(j)} = (\ldots\ldots, z_0^{(j)}, z_1^{(j)}, \ldots\ldots)$ of signal constellations $z_i^{(j)}$ for message j. This coded sequence is a complex sequence of points in the signal constellation on the two-dimensional plane. Then this sequence is modulated by a pulse waveform $p(t)$ into the following form :

$$s^{(j)}(t) = \sum_i z_i^{(j)} p(t - iT), \qquad (11.8)$$

where T is the sampling time period. This encoded output signal is transmitted over some noisy channel. The purpose of the receiver is to recover this message from this possibly corrupted received signal sequence.

It is known from Chapter 2 that the optimal receiver to detect one of a set (constellation) of messages is a bank of matched filters. A matched filter selects that message j which corresponds to the signal waveform $s^{(j)}(t)$ that produces the largest sampled output value. The time-response of the filter that matches the encoded sequence $s^{(j)}(t)$ in Eq.(11.8) is given by

$$s^{(j)}(-t) = \sum_i z_i^{(j)} p(-t - iT), \qquad (11.9)$$

and if $r(t)$ is the received signal, the sampled response function of the matched filter to $r(t)$, at the time that the filter is matched, is given by

$$\int_{-\infty}^{\infty} r(t) s^{(j)}(t) dt = \sum_i z_i^{(j)} y_i, \qquad (11.10)$$

where $y_i = \int_{-\infty}^{\infty} r(x) p(x - iT) dx$.

For time-orthogonal pulses of unit energy, i.e. for pulses with the property,

$$\int_{-\infty}^{\infty} p(t - uT) p(t - vT) dt = \begin{cases} 1, & u = v, \\ 0, & u \neq v, \end{cases} \qquad (11.11)$$

one has that the energy of the signal $s^{(j)}(t)$ is given by

$$E_s^{(j)} = \int_{-\infty}^{\infty}\int_{-\infty}^{\infty} s^{(j)}(x) s^{(j)}(y) dx dy = \sum_i \left|z_i^{(j)}\right|^2. \qquad (11.12)$$

For AWGN the maximum-likelihood receiver, which is equivalent to the above matched-filter receiver, selects that sequence $\bar{z}^{(j)}$ which minimizes

$$D^{(j)} = \left\|\bar{z}^{(j)} - \bar{y}\right\|^2. \qquad (11.13)$$

By the use of Eq. (11.10), such a minimization process is equivalent to the maximization of

$$C^{(j)} = 2\sum_i \text{Re}\left\{z_i^{(j)} y_i^*\right\} - \sum_i \left|z_i^{(j)}\right|^2. \qquad (11.14)$$

Trellis-Coded Modulation

Equation (11.14) is called the metric of the sequence $\bar{z}^{(j)}$, and as such is the metric to be maximized over all possible choices of signal sequences. To accomplish this maximization define the partial metric at time i by

$$C_i^{(j)} = 2\sum_{l=-r}^{i} \text{Re}\{z_l^{(j)} y_l^*\} - \sum_{l=-r}^{i} |z_l^{(j)}|^2 \qquad (11.15)$$

Evidently, the partial metric $C_i^{(j)}$ in Eq.(11.15) can be expressed recursively as follows:

$$C_i^{(j)} = C_{i-1}^{(j)} + 2\text{Re}\{z_i^{(j)} y_i^*\} - |z_i^{(j)}|^2. \qquad (11.16)$$

With the partial metric in Eq.(11.16) a modification of the Viterbi algorithm, discussed in Chapter 9 for decoding convolutional codes, can be used for the soft decoding of trellis codes as follows:

Step 1 : Initialize the S states of the ML decoder with the metric $C_{-r}^{(j)} = -\infty$ and let the starting state of the encoder, usually state $j=0$, have the metric $C_{-r}^{(0)} = 0$. Also let $i=-1$.

Step 2 : Compute the branch metric,

$$d_i^{(j)} = 2\text{Re}\{z_i^{(j)} y_i^*\} - |z_i^{(j)}|^2, \qquad (11.17)$$

and the partial metric,

$$C_i^{(j)} = C_{i-1}^{(j)} + d_i^{(j)},$$

for each state $S_i^{(j)}$ and each constellation extension $z_i^{(j)}$.

Step 3 : Shift the states $S_i^{(j)} \to S_{i+1}^{(j)}$ in accordance with the trellis transitions, determined by the encoder and from the 2^k merging paths. Store the survivor path $\hat{z}^{(j)}$ for which $C_i^{(j)}$ is maximized.

Step 4 : If $i<l$, let $i=i+1$ and go to Step 2.

Step 5 : Output the survivor path $\hat{z}^{(j)}$ that maximizes $C_l^{(j)}$ as the ML estimate of the transmitted sequence.

In practice, it can be shown that one does not need to wait until the entire trellis-coded received sequence is decoded before starting to output the Viterbi-decoded symbols $z_i^{(j)}$ or the corresponding message. Since the underlying codes are convolutional codes the decoding constraint length of the resulting trellis code is very close to $i_t \approx 5v$. This fact is verified for rate-1/2 convolutional codes[3], and the argument extends to the more general trellis codes. Therefore, the Viterbi algorithm can be approximated by a fixed-delay decoder by modifying Steps 4 and 5 as follows :

Step 4 : If $i \geq i_t$, output $z_{i-i_t}^{(j)}$ from the survivor $\hat{z}^{(j)}$ with the largest metric $C_i^{(j)}$ as the estimated symbol at time $i - i_t$. If $i < l-1$, let $i = i+1$ and go to Step 2.

Step 5 : Output the remaining estimated symbols $z_i^{(j)}, l - i_t < i \leq l$, from the survivor $\hat{z}^{(j)}$ that maximizes $C_l^{(j)}$.

Notice now that in the above fixed-delay decoder one can let $l \to \infty$ so that the complexity of this decoder is no longer determined by the length of the sequence. Another issue for soft-decision decoding requires that finite precision (e.g. 256-level (8-bits) quantization) arithmetic can be used in any practical soft-decision decoder.

Example 11.2
Consider the trellis code specified in Example 11.1. The recursive steps of the Viterbi algorithm are illustrated in Figure 11.9. At each step the path metrics, $C_i^{(j)}$, are the correlation between the best paths that lead to State j and the received sequence at time i. At time j assume that the best paths to states (0,0) and (0,1) are the bold paths.

11.2.3 Coding Gain

If the maximum-likelihood criterion is applied in Examples 11.1 and 11.2 to the soft-decision decoding process on an AWGN channel, the optimum-decision rule must depend on the Euclidean distances between the sequences of the demodulated phase samples. That is, the optimum decoder chooses that coded sequence of phase samples which is nearest in Euclidean distance to the received sequence. For such a trellis decoder the performance of the TCM coded scheme is determined primarily by the minimum Euclidean distance between the phase-coded sequences. In other words, the most efficient TCM schemes are achieved by a maximization of the minimum Euclidean distance between coded sequences, rather than of the minimum Hamming distance.

Soft-decision decoding of a TCM code is performed by the use of the modified soft-decision Viterbi algorithm developed in Sec. 11.2.2. The Viterbi algorithm searches for the most likely path through the trellis diagram and chooses that path which is closest in Euclidean distance to the

Trellis-Coded Modulation

received sequence. Recall from Chapter 9 that the performance of this decoding algorithm depends on the minimum Euclidean distance, called the free-Euclidean distance d_{free}, between any pair of paths which can form an error event. The free-Euclidean distance can be computed either from its trellis diagram or the state diagram, given in Figure 11.7.

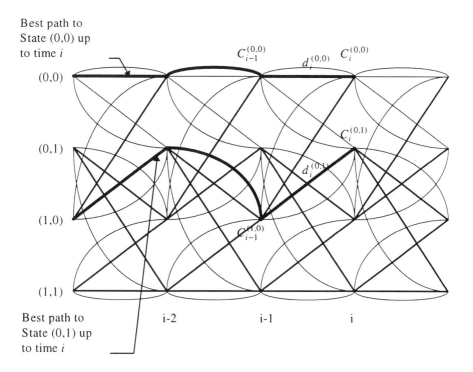

Figure 11.9 An example of the recursive steps of the Viterbi algorithm.

Example 11.3

Assume that the transmitted signal sequence is $\bar{z} = (\bar{s}_0, \bar{s}_0, \bar{s}_0, \bar{s}_0 ...)$. It is seen from Figure 11.8 that the possible error sequences could be $\bar{r}_1 = (\bar{s}_4, \bar{s}_0, \bar{s}_0, \bar{s}_0 ...)$, $\bar{r}_2 = (\bar{s}_2, \bar{s}_1, \bar{s}_2, \bar{s}_0, ...)$, and so on. Consider now the error events associated with the particular received sequence $\bar{r}_1 = (\bar{s}_4, \bar{s}_0, \bar{s}_0, \bar{s}_0 ...)$. It is clear that the squared-Euclidean distance, associated with this error event, is given by $d_E^2(\bar{z}, \bar{r}_1) = d_E^2(\bar{s}_4, \bar{s}_0) + d_E^2(\bar{s}_0, \bar{s}_0) + d_E^2(\bar{s}_0, \bar{s}_0) + \cdots = 4E_s$. Next, consider the error events, shown by the bold-line-segments in Figure 11.8, namely those events that were associated with the sequence $\bar{r}_2 = (\bar{s}_2, \bar{s}_1, \bar{s}_2, \bar{s}_0, ...)$ that diverges from state 00 and re-merges with the same state after more than a one-state transition. The squared-Euclidean distance, associated with the error event of $\bar{r}_2$, is expressed generally by

$$d_E^2(\text{diverging pairs of paths}) + \cdots + d_E^2(\text{re-merging pairs of paths})$$
$$= d_E^2(\bar{s}_2, \bar{s}_0) + d_E^2(\bar{s}_1, \bar{s}_0) + d_E^2(\bar{s}_0, \bar{s}_2) = 4.586 E_s,$$

which is larger than the squared-Euclidean distance between sequences $\bar{z}$ and $\bar{r}_1$ (see Figure 11.8(b)).

It can be verified by an enumeration of all cases that the error events associated with the sequence $\bar{r}_1$ has the minimum squared-Euclidean distance among all of the error events, and hence the free-Euclidean distance is determined to be

$$d_{free}^2 = d_E^2(\bar{s}_4, \bar{s}_0) = 4E_s.$$

The error sequence $\bar{r}_1$ differs in one symbol branch from the all-zero-symbol code sequence $\bar{z}$. Thus, an optimal decoder makes an error between these two sequences with probability,

$$P_{\bar{z} \to \bar{r}} = Q\left(\sqrt{\frac{d_{free}^2}{2N_0}}\right), \tag{11.18}$$

where $d_{free}^2 = 4E_s$, which incidentally is two times larger than the QPSK distance of $d_E^2(\bar{s}_0, \bar{s}_1) = 2E_s$. Going through all possible sequence pairs, one finds that the distance between $\bar{r}_1$ and $\bar{z}$ has the smallest squared-Euclidean distance, and hence the probability of an error between those two sequences is the most probable error.

It is now seen that sequence decoding, rather than symbol decoding, can increase the distances between alternatives, even though the signal constellation used for the sequence coding have a smaller minimum distance between the signal points.

To compare the coded and uncoded schemes it is common to use the coding-gain parameter G, which is defined as the difference between the values of the coded and uncoded SNR (in dB), required to achieve the same error-event probability, namely

$$G = SNR|_{uncoded} - SNR|_{coded}. \tag{11.19}$$

Trellis-Coded Modulation

At a high SNR this gain is expressed by

$$G_\infty = G|_{SNR \to \infty} = 10 \log \frac{(d_{free}^2 / E_s)_{coded}}{(d_{free}^2 / E_s)_{uncoded}}, \tag{11.20}$$

where G_∞ represents the asymptotic coding gain and $(E_s)_{coded}$ and $(E_s)_{uncoded}$ represent the average signal energy for the coded and uncoded systems, respectively. Also, $(d_{free}^2)_{coded}$ and $(d_{free}^2)_{uncoded}$ are the free distances for the coded and uncoded systems, respectively. Note that for the uncoded scheme, d_{free} is simply the minimum Euclidean distance between signal points.

Example 11.4
For a bandwidth-efficient coding scheme the asymptotic coding gain is evaluated by comparing it with the uncoded scheme with the same throughput. For the QPSK and the 8-PSK coded scheme, presented in Example 11.1, $(E_s)_{coded} = (E_s)_{uncoded}$. Also for the 8-PSK scheme $(d_{free}^2)_{coded} = 4E_s$ and for the QPSK scheme $(d_{free}^2)_{uncoded} = 2E_s$. Thus, the asymptotic coding gain is computed to be
$$G_\infty = 10 \log(4/2) = 3dB.$$
Hence, for this code one may decrease the symbol power by about 3 dB; that is, it needs somewhat less than half of the power needed for uncoded QPSK.

Examples 11.1 through 11.4 show that if the integrated coding and modulation scheme is designed by increasing the minimum Euclidean distance between coded sequences, then the loss from the expansion of the signal set often can be overcome to achieve a significant coding gain.

For a bandwidth-efficient coding scheme the asymptotic coding gain is evaluated by comparing it with the uncoded scheme with the same throughput. Not all coded systems can provide coding gains. In fact, some coded systems can have a degraded performance when compared with the uncoded system. Only a very carefully designed trellis-coding system can provide an actual coding gain.

In the following section a signal-set partitioning technique and a complete procedure for designing trellis-code schemes are discussed.

11.3 Set Partitioning and Construction of Codes

Simple trellis codes which are designed by hand already offer impressive coding gains. Such a design procedure is not suggested for exhaustive coding studies but is an important illustrative tool. It is seen from Example 11.1 that a good trellis code is one in which different symbol sequences are separated by large squared-Euclidean distances. Of particular importance is the minimum squared-Euclidean distance d_{free}^2. A code with a large d_{free}^2 is generally expected to perform well, and d_{free}^2 has become the major design criterion for trellis codes.

One heuristic design rule[2], which was used successfully to design codes with a large d_{free}^2, is based on the following observation : If the branches of a trellis which leave a state are assigned to signals in a subset, where the distances between points are large, and also if such signals are assigned to those branches which merge into a state, one is assured that the total distance is at least the sum of the minimum distances between the signals in these subsets.

The design begins with the selection of a trellis size which is measured by the number of encoder states. A typical trellis size has $S = 2^v$ states so that v equals the number of binary-memory elements in the trellis encoder. Also assume that a throughput k in units of bits-per-trellis level (or, say, signal points) is specified. Then, 2^k is the number of trellis branches(arcs) that leave and merge with each trellis state. In general this number of arcs can be allocated in many different ways. For example, if $k = 3$, the eight required graph transitions could be divided as eight single branches, four groups of two, or two groups of four. Knowing which is best is not immediately obvious, but given a choice, one merely assigns signal constellation subsets of the proper size to each of these various state transitions. The design rules are then as follows:
1. Employ all signal-constellation subsets equally often in labeling the trellis.
2. For these subset assignments, that share a common splitting state or merging state, choose those signal- constellation subsets for which the minimum interset distance is largest.

For small trellises, where subsets are assigned only once, it can be shown that rule 2 generally cannot be satisfied. However, these rules still allow

Trellis-Coded Modulation   449

considerable latitude in the labeling of a large trellis, thereby pointing to the need to examine many different codes.

11.3.1 Set Partitioning

An essential step in the design of a trellis code is to expand the complete signal-constellation set into several signal-constellation subsets. Such a step often is called set partitioning. Consider a modulation constellation having M points in an N-dimensional space. The constellations may be M-ary amplitude modulation(AM) in one dimensional space, *M-ary* quadrature amplitude modulation(QAM) or PSK in two dimensions, or more generally any regular arrangement of points in N dimensions, such as the subsets of an N-dimensional lattice. This latter process is referred to as lattice partitioning.

To begin with, the original set is partitioned into p_1 equal-sized disjoint subsets, called *cosets*, each of size M/p_1 points. These subsets are denoted by $A_1, A_2, ..., A_{p_1}$. This partitioning is done in such a manner that within each subset the minimum Euclidean distance between signal points is maximal for the adopted subset size and is uniform across the subsets. This is always possible with those initial constellations which have a standard symmetry, or if the constellation is a lattice, the lattice coset decomposition provides the recipe. Next, each of these subsets is split further into p_2 subsets, denoted by $B_1, B_2, ..., B_{p_2}$. Proceeding recursively in this manner, the original constellation is decomposed eventually into single-point subsets. Typically, the splitting factors p_i are equal, and often $p_i = 2$.

The initial descriptions of Ungerboeck were restricted to one- and two-dimensional spaces. To clarify set partitioning, consider now the following two-dimensional space example :

Example 11.5
A partitioning of an 8-PSK constellation and its natural partitioning sequence are shown in Figure 11.10. At each step, a splitting by two is employed so that the subset sizes also shrink by a factor of two. Note that the first level of partitioning produces two rotated QPSK sets, the next level produces four binary PSK sets, and the final level has single-point subsets. The sequence of the right/left branches in the partitioning tree produces 3-bit binary labels on each of the original 8-PSK points. Clearly, there are many equivalent labelings, and the label selected amounts to a natural binary progression as one moves around the constellation circle. In effect, such a labeling produces

a mapping from binary 3-tuples to signal points. Ungerboeck dubbed this technique, "mapping by set partitioning". However, labeling of the signal points is not crucial, until one decides to design an actual encoder and decoder.

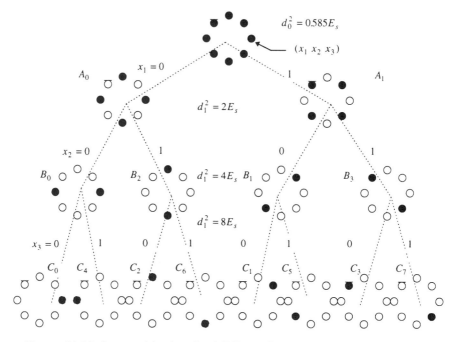

Figure 11.10 Set partitioning for 8-PSK with its associated bit labeling.

More importantly, as the partitioning proceeds, it should be noted that the intra-subset minimum distance increases from the 8PSK distance to the QPSK distance (subsets A_i in Figure 11.10), further to the BPSK distance (subsets B_i in Figure 11.10). The sequence of squared-minimum distances is thus $0.565 E_s$, $2 E_s$ and $4 E_s$, where E_s is the energy attached to each symbol. One can attach infinite intraset distances to singleton sets.

In the above example one is assured that the minimum squared-free distance between any two paths is at least twice that of the original QPSK signal set. A general structure of a trellis encoder with 2 information bits-per-modulator symbol is shown in Figure 11.11.

Trellis-Coded Modulation

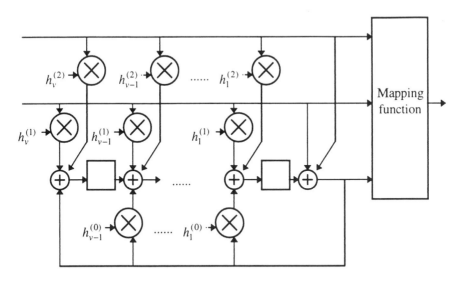

Figure 11.11 A general structure for a trellis encoder

The sets of the connector coefficients $h_i^{(j)}$ for $i=1,2,...,v-1$ and $j=0,1,2$, which result in good trellis codes for a given mapping, are quite difficult to find. These sets are usually obtained by a use of computer searches or heuristic construction algorithms. Table 11.2 shows some of the best 8-PSK trellis codes which use 8-PSK with the natural mapping. This table provides the connector coefficients $\bar{h}^{(j)} = \{h_1^{(j)},...,h_{v-1}^{(j)}\}$, the squared-free distance d_{free}^2 and the asymptotic coding gain.

Number of states	$\bar{h}^{(0)}$	$\bar{h}^{(1)}$	$\bar{h}^{(2)}$	d_{free}^2	Asymptotic Coding Gain(dB)
4	5	2	-	4.00	3.0
8	11	2	4	4.59	3.6
16	23	4	16	5.17	4.1
32	45	16	34	5.76	4.6
64	103	30	66	6.34	5.0
128	277	54	122	6.59	5.2
256	435	72	130	7.52	5.8
512	1525	462	360	7.52	5.8
1024	2701	1216	574	8.10	6.1

Table 11.2 The best 8-PSK trellis codes with number of states ≤ 1024.

The general trellis coding for band-limited channels is depicted schematically in Figure 11.12. One begins with the desire to communicate k information bits per-modulator symbol and adopts a modulator constellation that is slightly larger than 2^k. Often the size is exactly 2^{k+1}, and one speaks of expanding the constellation by a factor of 2 over that needed for an uncoded transmission. Next, one adopts some $k_0 \leq k$ to be that number k_0 of input bits that enter the encoder and influences the state. This leaves $k - k_0$ bits as uncoded bits. The encoder is typically a binary convolutional-type encoder, which adds $r=1$ bits of redundancy, thereby producing $k_0 + 1$ output bits.

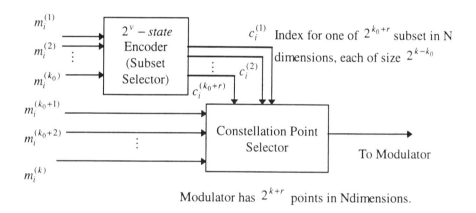

Figure 11.12 Generic trellis-coded modulation structure.

To combine trellis encoding with the modulator, one partitions the original modulator constellation in order to obtain 2^{k_0+1} subsets, each with 2^{k-k_0} points. Note that the total number of points in the constellation is thus 2^{k+1}. Furthermore, the subsets are labeled by a vector of $k_0 + 1$ bits which are produced by the encoder. That is, to say, the encoder chooses a subset for its modulation points in a constellation. Its labels are exactly the $k_0 + 1$ highest bits of the set-partitioning tree; that is, the encoder specifies the modulator input bits $b_1, b_2, ..., b_{k_0+1}$ at time j. The remaining $k - k_0$ uncoded bits then merely select a member of this chosen subset for transmission. To put this another way, the coset sequence selection has both memory and redundancy. However, the selection of a point within a coset doesn't have memory. These types of signals are relatively easy to distinguish in noise.

Trellis-Coded Modulation

11.3.2 The Error-Event Probability

A performance evaluation of trellis codes is discussed in this section. The encoder for a trellis code is a convolutional code which is combined with a signal-mapping function. The optimal trellis decoder is one that retraces the encoded sequence (path). A decoding error occurs if the path it follows through its trellis does not coincide with the one taken by the encoder. Such a scenario is illustrated in Figure 11.13. Here the decoder starts with an error at time interval i, followed by an erroneous path which re-merges with the correct sequence at time interval $i+L$.

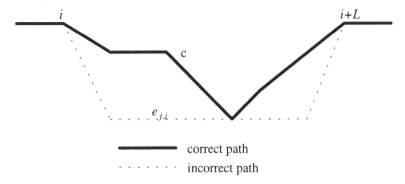

Figure 11.13 The correct path and an error path of length L in a trellis diagram

An important performance measure for a trellis code is the bit-error rate(BER) associated with the decoder. Unfortunately, the calculation of BER is very complex, and as a consequence no efficient methods to compute it exist. One alternate performance measure is the probability that a code-sequence error occurs. Such an error happens when the decoder follows a path that diverges from the correct path somewhere in the trellis.

Figure 11.14 shows an example trellis where the correct path is indicated by a solid line and all possible error paths by dotted lines. Assume that the total length of the trellis and the code is l and that the code is started in the zero state and also is terminated in the zero state.

The probability of error, P_e, is then the probability that any of the dotted paths in Figure 11.14 is chosen, i.e. P_e is the probability of the union of the individual errors $e_{j,i}$ given by

$$P_e = \Pr\left(\bigcup_i \bigcup_j e_{j,i} | c\right), \qquad (11.21)$$

where $e_{j,i}$ is the j-th error path that departs from the correct path c at time unit i. To obtain the average probability of error, the probability P_e needs to be averaged over all correct paths as follows:

$$P_{avg} = \sum_c P(c) \Pr\left(\bigcup_i \bigcup_j e_{j,i} | c\right), \qquad (11.22)$$

where $P(c)$ is the apriori probability that the encoder chooses path c.

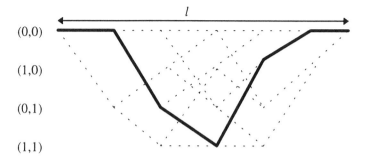

Figure 11.14 A trellis diagram that shows the correct path and its possible error paths

The probability in Eq. (11.22) is still difficult to evaluate and the union bound, which is given by $\Pr\left(\bigcup_j E_j\right) \leq \sum_j \Pr(E_j)$, can be used to bound and estimate P_{avg} as follows:

$$P_{avg} \leq \sum_c P(c) \sum_i \Pr\left(\bigcup_j e_{j,i} | c\right) \leq \sum_c P(c) \sum_i \sum_j \Pr(e_{j,i} | c). \qquad (11.23)$$

If the total length l of the encoded sequence is very large, P_{avg} can be so large that, $\lim_{l \to \infty} P_{avg} = 1$. Thus, this probability in Eq.(11.23) is not a good measure of the performance of a trellis code. One way out of this dilemma is to normalize P_{avg} per unit time. This yields the following bound for this new measure of performance, namely,

Trellis-Coded Modulation

$$P_{avg}(e) = \lim_{l \to \infty} \frac{P_{avg}}{l} \le \sum_c P(c) \sum_{e_j} P(e_j|c), \qquad (11.24)$$

where e_j is the event that an error starts at an arbitrary but fixed time unit, say i. Also, the correct sequence c up to node i and after node $i+L$ is irrelevant for the error path e_j of length L. The right side of Eq.(11.24) can also be interpreted as, what is called, the first event error probability --- that is, the probability that the decoder starts its first error event at node i. $P_{avg}(e)$ is bounded above by the first-event error probability.

Denote $\Pr(e_j|c)$ by $P_{c \to e_j}$ which, since there are only two hypotheses involved, can be computed for an AWGN channel by

$$P_{c \to e_j} = Q\left(\sqrt{d^2(c,e_j)\frac{RE_b}{2N_0}}\right),$$

where $R=k/n$ is the code rate in bits/symbol, N_0 is the one-sided noise-power spectral density, E_b is the energy-per-information bit and $d^2(c,e_j)$ is the normalized squared-Euclidean distance between the signals on the error path e_j and the signals on the correct path c. In the normalization of Eq.(11.24) it is assumed that the signals of the constellation have a unit average energy. Thus a substitution of $P_{c \to e_j}$ in the upper bound for $P_{avg}(e)$ yields

$$P_{avg}(e) \le \sum_c P(c) \sum_{e_j|c} Q\left(\sqrt{d_E^2(c,e_j)\frac{RE_b}{2N_0}}\right), \qquad (11.25a)$$

which can be rearranged in the form,

$$P_{avg}(e) \le \sum_{d_j^2} W_{d_j^2} Q\left(\sqrt{d_j^2 \frac{RE_b}{2N_0}}\right), \qquad (11.25b)$$

(by counting how often each of the distances d_j^2 occurs in Eq.(11.25a)), where $W_{d_j^2}$ is the average number of times that the distance d_j^2 occurs. The smallest d_j^2 that can be found in the trellis is called d_{free}^2, the free squared-Euclidean distance.

A bound on the average BER is obtained next from the average error event probability in Eq.(11.25b). Note first that error event $c \to e_j$ causes a certain number of bit errors. Thus one can let $W^{(b)}_{d_j^2}$ be the average number of bit errors on the error paths with the distance d_j^2. A bound on the bit-error rate per unit time $P_b(t)$ is obtained from Eq.(11.25b) by substituting $W^{(b)}_{d_j^2}$ for $W_{d_j^2}$. This yields :

$$P_b(t) \leq \sum_{d_j^2} W^{(b)}_{d_j^2} Q\left(\sqrt{d_j^2 \frac{RE_b}{2N_0}}\right).$$

Since the trellis code processes k bits per time unit, a bound on the average bit-error probability is given by :

$$P_b \leq \sum_{d_j^2} \frac{1}{k} W^{(b)}_{d_j^2} Q\left(\sqrt{d_j^2 \frac{RE_b}{2N_0}}\right). \tag{11.26}$$

Figure 11.15 shows the bound Eq.(11.26) for some popular 8PSK trellis codes, given by Ungerbock [2]. This bound is compared with results obtained in [2] by simulation. In Figure 11.15 bounds labeled, Series 1, Series 2, and Series 3 correspond to the three 8-PSK Ungerbock trellis codes with constraint length $v=8,6$, and 4, respectively.

11.4 Rotational Invariance

In this section an important implementation issue for trellis codes is considered. It arises when there is a possibility of, what are called, significant phase offsets in the receiver.

Ideally in the coherent detection of a DSB-SC signal (see Appendix A), the absolute carrier phase of the transmitted signal is assumed to be known at the receiver in order to determine the correct phase orientation of the signal constellation in the complex plane. If the M-PSK signal $s_m(t) = A\cos(2\pi f_0 t + \phi_m(t))$ is received with a phase-offset of angle ϕ, then the received signal is given by

$$r_m(t) = A\cos(2\pi f_0 t + \phi_m(t) + \phi),$$

where A is the signal amplitude and $\phi_m(t)$ is the time-varying phase of the TCM data signals. The TCM receiver generates an estimate of the carrier phase $\phi_m(t)$, then it uses that estimate to demodulate the incoming signal.

Trellis-Coded Modulation

For example, for QPSK the phase $\phi_m(t)$ of the data can assume the angles $\pi/2, \pi, 3\pi/2$ or 2π. The carrier phase-tracking loop, which is usually a phase-locked loop (PLL) circuit, first needs to eliminate the data-dependent part of the phase. In the case of a QPSK signal, such a process can be accomplished by raising $r_m(t)$ to the fourth power, i.e. let

$$r_m(t) = \frac{A^4}{8}\cos(2\pi 4 f_0 t + 4\phi_m(t) + 4\phi) + \frac{A^4}{2}\cos(2\pi 2 f_0 t + 2\phi_m(t) + 2\phi) + \frac{3A^4}{8}.$$

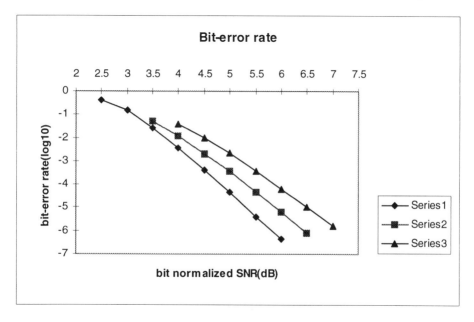

Figure 11.15 Bounds on the average bit-error rate of some 8PSK trellis codes

This produces a spectral line at $4f_0$, given by $\cos(2\pi 4 f_0 t + 4\phi_m(t) + 4\phi)$. However $4\phi_m(t)$ is always a multiple of 2π for QPSK modulation so that the fourth-power spectral line is always free of the data-dependent phase changes and is given simply by $\cos(2\pi 4 f_0 t + 4\phi)$. Thus, following this nonlinear operation a PLL is used to track the phase of this signal, and a local oscillator signal of frequency f_0 is generated by dividing the frequency of this tracking signal by 4. Notice that the phase offsets of the multiples of $\pi/2$ cannot be identified by this procedure so that this carrier-phase is ambiguous with respect to multiples of $\pi/2$.

These phase ambiguities often have an undesirable effect on the decoding of TCM codes. In the direct (uncoded) carrier-phase acquisition with its associated tracking circuits, such phase ambiguities often cause phase jumps that result in a loss of synchronization. Re-synchronization can be expected to take some time. For the normal TCM system the received-signal constellation can be a phase rotated version of the transmitted constellation, which the trellis decoder usually cannot decode properly. This requires that all possible phase rotations of a constellation need to be tried until the correct phase orientation is found.

It is therefore desirable to design TCM systems in such a manner that they have as much phase invariance as possible in their rotation angles in order not to affect the decoder operation. Such phase invariance helps also to ensure a more rapid resynchronization after a temporary loss of signal because of a phase jump. A trellis code is called rotationally invariant with respect to the rotation of the constituent constellation by the carrier phase angle ϕ if the decoder can correctly decode the transmitted information sequence when the local oscillator phase differs by the carrier phase ϕ. If $\bar{c}$ is a sequence of coded symbols, then denote by $\bar{c}^\phi$ the symbol sequence obtained by rotating the phase of each symbol c_i by the angle ϕ, i.e. $\bar{c}^\phi = e^\phi \cdot \bar{c}$. Then, a TCM code is rotationally invariant with respect to the rotation of the phase angle ϕ if $\bar{c}^\phi$ is a code sequence in the set of all valid code sequences $\bar{c}$.

There are two requirements that need to be considered to make a trellis code transparent to a phase rotation of the signal constellation. First, the code must be rotationally invariant; that is, the decoder is able to find a valid code sequence after a rotation. Secondly, the rotated sequence needs to map back into the same information bits as the unrotated sequence. The latter requirement is achieved by the differential encoding of the information bits.

A differential encoding method for digital modulation maps the data sequence into a continuous-time waveform in such a way that the data sequence can be recovered despite an ambiguity in the received waveform, usually a phase ambiguity. The simplest example is differential PSK. This modulation technique protects against a 180⁰ phase ambiguity in the received waveform.

The differential trellis codes discussed next are a generalization of differential QPSK. These codes are used in passband channels in which the

Trellis-Coded Modulation

carrier phase is known only to within a multiple of 90^0. These codes can simplify the carrier-acquisition problem because after a temporary interruption in carrier-phase synchronization, the phase lock loop (PLL) re-locks the carrier phase to any one of four phase positions. The phase of the received waveform may be offset by 0^0, 90^0, 180^0, or 270^0 from the original phase of the transmitted waveform. Nevertheless, for each of these four possibilities, the demodulated data sequence remains the same. What is needed is a technique to design a code such that every 90^0 rotation of a codeword is another codeword. Then, by differential encoding, the end-to-end transmission is made transparent to these different offsets of carrier phase.

The idea of differential trellis codes is easier to understand if one first examines differential modulation without trellis coding. Consider the 16-ary square-signal constellation, shown in Figure 11.16a, that can be visualized as four subconstellations, each with four signal points. Each subconstellation is invariant under a rotation by any multiple of 90^0. Each of its points is given an identical two-bit label characterizing that subconstellation.

Two bits of data are encoded into a point of the signal constellation using the indicated labels. Then two more bits are used to select the quadrant. The quadrant is selected differentially by the rule,

$$00 \to \Delta\phi = 0^0, \quad 10 \to \Delta\phi = 90^0, \quad 11 \to \Delta\phi = 180^0, \quad 01 \to \Delta\phi = 270^0.$$

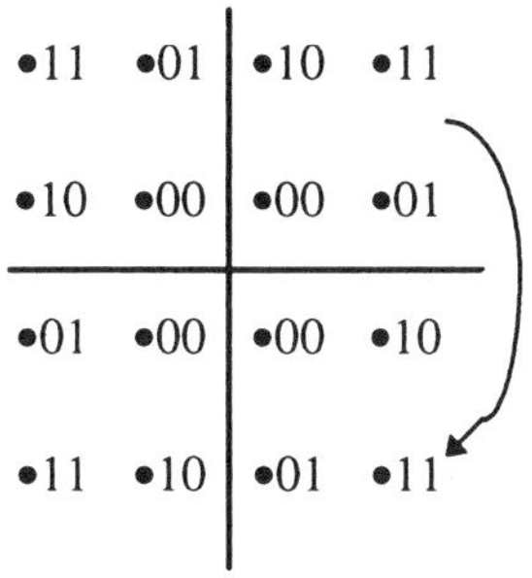

```
 •11  •01 | •10  •11
 •10  •00 | •00  •01
----------+----------
 •01  •00 | •00  •10
 •11  •10 | •01  •11
```

Figure 11.16a A symmetrically labeled signal constellation

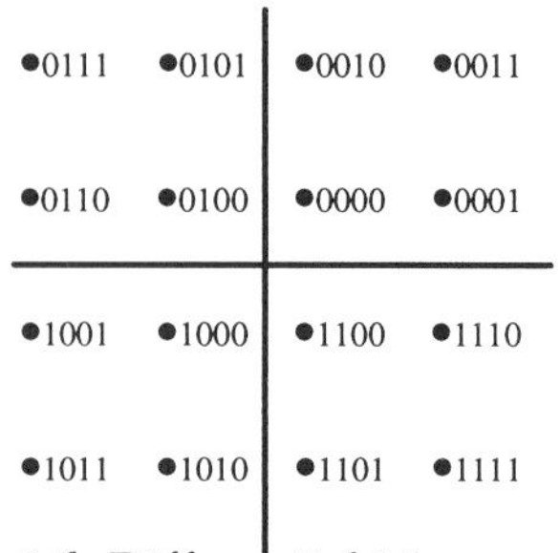

```
•0111   •0101  |  •0010   •0011

•0110   •0100  |  •0000   •0001
────────────────┼────────────────
•1001   •1000  |  •1100   •1110

•1011   •1010  |  •1101   •1111
```

Figure 11.16b Differential 16-ary modulation

The differentially-coded 16-ary signal constellation is shown in Fig. 11.16b. The two data bits on the right define a point in the subconstellation, and two data bits on the left define a phase rotation of one of the four multiples of 90^0. The demodulator is designed to undo what the modulator does, by first detecting the proper points in the signal constellation and then mapping the sequence of labels into the sequence of data. A primary requirement of differential encoding is that the signal constellation be labeled in a special manner. A secondary requirement is to express the data in a special differential form which is easily taken care of by a proper pre-coding and postcoding. If these requirements are met, the method of differential coding can be combined with the method of trellis coding. Some trellis codes are invariant under a 90^0 phase rotation; others are not. If a code utilizes differential coding, it can be exploited in the design of the encoder and decoder so as to make the system transparent to 90^0 rotations in the channel. Therefore, the encoder and decoder should employ some form of differential encoding.

Some trellis codes have a structure that is already compatible with differential encoding. In particular, trellis codes with two or more uncoded bits may have this property. Differential encoding can be used with such a partition by changing the way that the two uncoded data bits are mapped into signal points of the coset. These two bits can be viewed as an integer modulo-4 and, instead of encoding this integer directly, encode the modulo-4 sum of it and the previous encoded value (see Example 11.6 below). This amounts to a precoding of the uncoded data bits which is undone by a postcoder after demodulation. Figure 11.17 illustrates this type of differential trellis coding.

Considerable effort is required to achieve rotational invariance in a trellis code. Sometimes, it is possible to achieve rotational invariance by the use of a higher-dimensional TCM system. To illustrate this fact there are several four- and eight-dimensional TCM systems, which are based on the two-

Trellis-Coded Modulation

dimensional MPSK. The constellation of these TCM codes demonstrate the same rotational invariance as the MPSK constellations themselves. In these systems several consecutive two-dimensional MPSK signals are used to construct a single, higher-dimensional, signal. Regrettably, space limitations preclude a treatment of multidimensional trellis codes in this text. The interested reader is referred to [3] and [4].

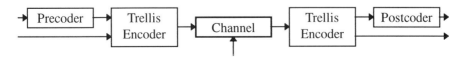

Figure 11.17 Differential trellis coding

Though rotational invariance is difficult to achieve, it is sometimes possible to obtain it in a TCM system by the use of two-dimensional QAM with a nonlinear coding.

Example 11.6
Consider the rotationally invariant TCM system specified in Figure 11.18. Note that in this figure the differential coding is performed by means of the modulo-4 adder, applied to the recursive equations : $c_j^{(i)} = m_j^{(i)} + c_{j-1}^{(i)}$ for $i=1,2$, and $j=1,2,...$

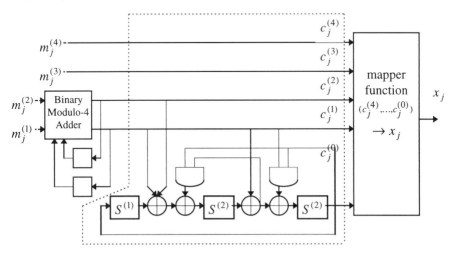

Figure 11.18(a) Encoder for the 90^0 phase-invariant TCM system

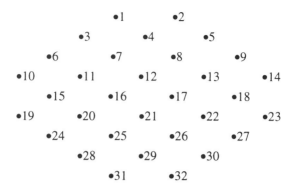

Figure 11.18(b) 32-CROSS Labeling j for v.32.

The QAM constellation for this system is called 32-CROSS, and the encoder consists of an eight-state, nonlinear convolutional encoder, combined with certain nonlinear logic operations(AND gates). The mapping function for the points of the constellation is given in Table 11.3.

The constellation in Figure 11.18 is partitioned into the eight subsets in Figure 11.19 : A_0 ={all points which are marked a} , A_1 ={all points which are marked b}, ... , A_7 ={all points which are marked h}. Note that successive 90^0 phase rotations yield $A_0 \to A_1 \to A_2 \to A_3 \to A_0$ and $A_4 \to A_5 \to A_6 \to A_7 \to A_4$. If a code sequence is chosen from the sequence of subsets { A_4, A_0, A_3, A_6, A_4 }, the 90^0-rotation sequence for this code sequence falls into the sequence of subsets { A_5, A_1, A_0, A_7, A_5 }. It is easy to verify that this rotated sequence is another valid code sequence.

(11111) → 1	(10100) → 9	(11010) → 17	(10111) → 25
(11000) → 2	(00000) → 10	(11101) → 18	(10001) → 26
(01000) → 3	(01111) → 11	(00111) → 19	(10110) → 27
(00101) → 4	(00010) → 12	(01001) → 20	(01110) → 28
(01010) → 5	(01101) → 13	(00110) → 21	(00001) → 29
(10010) → 6	(00011) → 14	(01011) → 22	(01100) → 30
(10101) → 7	(11001) → 15	(00100) → 23	(11100) → 31
(10011) → 8	(11110) → 16	(10000) → 24	(11011) → 32

Table 11.3 Signal mapping of 32-CROSS constellation.

Trellis-Coded Modulation 463

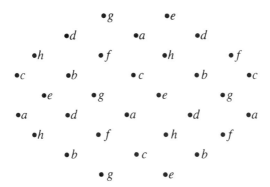

Figure 11.19. The set partitioning of the 32-CROSS constellation into the eight subsets.

In general, a code is rotationally invariant if one concatenates successive trellis paths, i.e. for each path $i \to j \to k ,...,$ there exists a valid path $f_\phi(i) \to f_\phi(j) \to f_\phi(k),...,$ of the angle ϕ-rotated symbols through the trellis, where f_ϕ is a function. For all transitions from some state i to another state j, denote the associated subset by A. Denote by B the subset obtained when A is rotated by the angle ϕ. Then B is the subset associated with all transitions from states $f_\phi(i)$ to states $f_\phi(j)$ (see Figure 11.20).

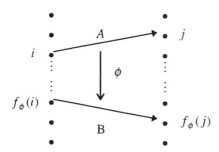

Figure 11.20 A trellis branch and its rotated branch.

The three leading bits $c_j^{(4)}, c_j^{(3)}, c_j^{(2)}$ in Figure 11.18(a) are not affected by any 90^0 rotation of the constellation. Only the last two bits $c_j^{(1)}, c_j^{(0)}$ are modified and need to be encoded differentially. In Figure 11.18(a) this function is realized by the XOR gate on the input line $m_j^{(1)}$ and the second delay cell; that is, the differential encoder is integrated into the trellis encoder. This integrated differential encoder is indicated by the dotted block in Figure 11.18(a).

The trellis code in this example was found to have the same 4-dB coding gain as the best non-invariant code of the same complexity. This rotationally invariant system was later adopted in the CCITT V.32 Recommendation [5]. V.32 was intended for 9.6 Kbps traffic over two-wire telephone lines and 14.4 Kbps traffic over four-wire circuits. The same nonlinear encoder was later adopted for use with 64-QAM and 128-CROSS in V.17 (14.4 Kbps FAX traffic over standard phone lines) and V.33 [6J, [7].

Wei [3] demonstrates the following conclusions about the design of such rotational invariant TCM systems.
1. A nonlinear convolutional encoder needs to be used in a TCM encoder. This follows from the fact that a 90^0 rotation induces a nonlinear transformation on the signal labels.
2. Rotationally invariant TCM systems may not provide the same coding gain or have the same complexity as comparable noninvariant systems.

There are, however, a few examples of rotationally invariant, two-dimensional, systems whose performance and complexity match those of the best-known non-invariant codes.

11.5 Unequal Error Protection(UEP) Codes and the Pragmatic Approach to TCM Systems

UEP codes are families of error-control codes that can change their coding strengths flexibly and adaptively. These codes are useful for protecting information sources such as audio, image, and video that are typically characterized by unequal importance of bits (due to the differing importance of the data to be encoded and properties of practical source-coding algorithms). These codes are also useful for dealing with time-varying channel conditions, where the requirement of needed protection varies. The vast majority of radio channels experience variations in noise and/or attenuation over time. For example, there are many possible different types of noise on a mobile wireless communication channel. In some cases these are very rapid fluctuations in the channel conditions (e.g., fading and

Trellis-Coded Modulation

multipath reflections), while in other cases the changes occur very slowly (e.g.. geosynchronous satellite channels). It is frequently useful to design a communication system that adapts to such changing conditions, thereby maximizing the information throughput, given some fixed reliability constraint.

UEP has been studied for both block codes and convolutional codes. One such convolutional code is known as the rate-compatible punctured convolutional (RCPC) code. RCPC codes have become very popular because of the flexibility and efficiency of these codes. The basic idea of RCPC codes is that one can specify an array of codes of differing strengths by appropriately "puncturing" certain bits from these codewords in order to control the rate of the code. Such a puncturing process has already been discussed in Section 8.7. However, the main goal in Section 8.7 for punctured convolutional codes was to reduce the computational complexity of the high-rate codes. The RCPC coding technique is attractive to implementation because the code comes from a single family that is specified by the underlying convolutional encoder with the various decoding resolutions being based on the same Viterbi-decoding trellis structure. As an example, consider a RCPC code family with rates ranging from 8/9 to 8/24. All decoding algorithms for this family of codes are derived from the Viterbi algorithm for a rate- 8/24 convolutional code. RCPC codes have become popular lately for protecting audio, image and video sources, where different components of the bitstream have different sensitivities and therefore deserving of different degrees of protection.

Other UEP techniques for time-varying channel conditions utilize, what is called, the pragmatic approach for TCM[8][9]. TCM Systems allow for reliable communication at high spectral efficiencies, but as the channel conditions vary, making it necessary to vary the number of bits/sec/Hz being transmitted in order to maintain reliability. Viterbi et al. [8] have demonstrated that it is possible to construct TCM systems with an adjustable spectral efficiency. The technique proposed in [8] has been termed pragmatic trellis codes wherein only slight modifications are made in the usual convolutional encoder and decoder. In particular, an adaptive TCM system is designed in [8] with the de facto industry and government standard for a, rate-1/2, 64-state, convolutional code and its modified Viterbi decoder.

This adaptive TCM system uses 4-, 8-, and 16-PSK and provides 1, 2, and 3 bits/sec/Hz with the asymptotic coding gains of 7.0 dB, 3.0 dB, and 5.3 dB, respectively. These pragmatic trellis codes perform almost as well as the best trellis codes of comparable complexity for an AWGN channel [8]. Figure 11.21 illustrates the "pragmatic" encoder proposed in [8]. Note that Gray coding is used to label the signal sets instead of the "natural" ordering used earlier in this chapter. The Gray code makes it possible for the convolutional encoder to select a phase within a sector. The Gray code mapping for this application is given as follows for 2^m-ary PSK: $00 \to 0\ radians$, $01 \to \pi/2^{m-1}\ radians$, $11 \to 2\pi/2^{m-1}\ radians$, $10 \to 3\pi/2^{m-1}\ radians$. Based on Figure 11.21, a pragmatic trellis code has been developed on a chip for 8-PSK and 16-PSK modulation [9]. Pragmatic codes for PAM and QAM signal constellations are described and discussed in detail in [8], [9] and [10].

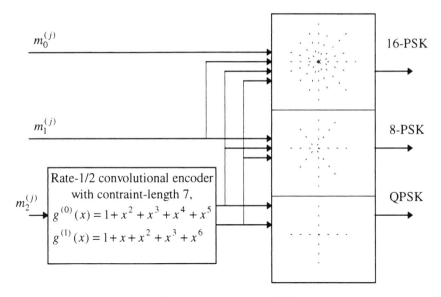

Figure 11.21 Viterbi's Pragmatic TCM systems

While RCPC codes and pragmatic trellis codes are powerful, they are not optimally suited to those applications, where the transmission is unable to alter its configuration dynamically or where one needs to have a code that is simultaneously decodeable at different rates. Examples are broadcast and multicast communications. Multicasting multimedia data over communication networks often require multi-resolution coding to adaptively deliver selectable contents to each receiver which usually have a different available bandwidths. One specific technique for these applications is called

Trellis-Coded Modulation

the embedded error-control code. For a detailed discussion of embedded codes, the reader is referred to [11].

Problems

11.1 Consider the 4-state trellis code generated by the circuit in Figure 11.22. This trellis code is optimal in free-distance for transmitting 2-bits per 8-PSK symbol.

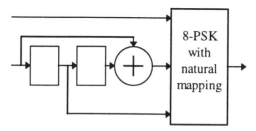

Figure 11.22 A 4-state trellis code

(a) Draw the trellis diagram.
(b) Construct a partition tree for the following signal constellations and compute the intersignal distance at each level of the partition tree.
(c) Determine d_{free} and the asymptotic coding gain for the system relative to an uncoded QPSK system.
(d) Suppose that the encoder begins in state 0 and ends in state 0 after the transmission of 10 bits. Let the received sequence be the sequence $\bar{r} = (1.1e^{j80°}, 0.9e^{-j30°}, 1.2e^{j160°}, 0.7e^{j60°}, 1.1e^{j200°})$ of complex numbers. Decode this sequence.

11.2 Consider the 32-point cross constellation given in Fig. 11.18. Perform a set partitioning in a divide-by-2 manner until subsets of size 4 are produced.

11.3 A commonly-used modulation scheme for dealing with a fixed unknown phase offset is differential 8PSK. Partition an 8PSK constellation to give an 8-ary PSK code that is invariant under 180° offsets of carrier phase.

Bibliography

[1] J. L. Massey, "Coding and modulation in digital communications", Proceedings of the 1974 International Zurich Seminar on Digital Communications, Zurich, Switzerland, pp.E2(1)-E2(4), March 1974.

[2] G. Ungerboeck, "Channel coding with multilevel/phase signals", IEEE Trans. Inform. Theory, vol. IT-28, pp.55-67, Jan. 1982.

[3] L. F. Wei, "Rotationally invariant trellis coded modulation with multi-dimensional MPSK," IEEE Journal on Selected Areas in Communications, vol. SAC-7, No. 9, pp.1281-1295, Dec. 1989.

[4] S. S. Pietrobon, R. H. Deng, A. Lafanechere, G. Ungerboeck, and D. J. Costello, Jr., " Trellis coded multidimensional phase modulation," IEEE Trans. on Inform. Theory, vol. IT-36, No. 1, Jan. 1990.

[5] IBM Europe, "Trellis-coded modulation schemes with 8 state symmetric encoder and 90^0 symmetry for use in data modems transmitting 3-7 bits per modulation interval", CCITT SG XVII Contribution COM XVII, No. D180, Oct. 1983.

[6] U. Black, The V series recommendations, protocols for data communications over the telephone network, McGraw-Hill, New York, 1991.

[7] G. Ungerboeck, "Trellis-coded modulation with redundant signal sets, Part II : State of the art," IEEE Communications magazine, Vol. 25, No. 2, pp.12-21, Feb. 1987.

[8] A. J. Viterbi, J. K. Wolf, E. Zehavi and R. Padovani, " Pragmatic approach to trellis-coded modulation", IEEE Communications Magazine, pp. 11-19, July 1989.

[9] J. K. Wolf and E. Zehavi, "P codes : pragmatic trellis codes utilizing punctured convolutional codes", IEEE communications Magazinw, pp. 11-19, July 1995.

[10] W. H. Paik, S. A. Lery, and C. Heegard, "Method and apparatus for communicating digital data using trellis coded QAM", US. Patent No. 5,233,629, Aug. 93.

[11] M. C. Lin, C. -C Lin, and S. Lin, "Computer Search for binary cyclic VEP codes of odd length upto 65", IEEE Trans. Inform. Theory, vol. 36, No.4, pp.924-935, July 1990.

12 Concatenated Coding Systems and Turbo Codes

12.1 Concept of Concatenated Coding System

As discussed in Chapter 1, one of the goals of coding research is to find a class of codes and associated decoders such that the probability of error could be made to decrease exponentially at all rates less than channel capacity while the decoding complexity increased only algebraically. Thus the discovery of such codes would make it possible to achieve an exponential tradeoff of performance vs. complexity.

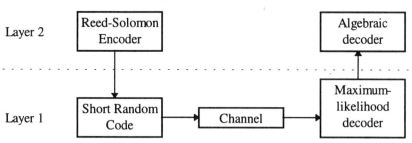

Figure 12.1 A Concatenated Coding

One solution to this quest is called concatenated coding. Concatenated coding has the multilevel coding structure[1], illustrated in Figure 12.1. In the lowest physical layer of a data network, a relatively short random "inner code" can be used with maximum-likelihood decoding to achieve a modest error probability, say, a bit-error rate of $P_b \approx 10^{-2}$, at a code-rate which is near channel capacity. Then in a second layer, a long high-rate algebraic nonbinary Reed-Solomon(RS) "outer code" can be used along with a powerful algebraic error-correction algorithm to drive down the error probability to a level as low as desired with only a small code-rate loss.

RS codes have a number of characteristics that make them quite popular. First of all they have a very efficient bounded-distance decoding algorithms such as the Berlekamp-Massey algorithm or the Euclidean algorithm. Being nonbinary, RS codes also provide a significant burst-error-correcting capability. Perhaps the only disadvantage in using RS codes lies in their lack of an efficient maximum-likelihood soft-decision decoding algorithm. The difficulty in finding such an algorithm is in part due to the mismatch between the algebraic structure of a finite field and the real-number values at the output of receiver demodulator.

In order to support reliable transmission over a Gaussian channel with a binary input, it is well-known that the required minimum $\frac{E_b}{N_0}$ is - 1.6 dB for soft-decision decoders, which increases to 0.4 dB for hard-decision decoders. Here $\frac{E_b}{N_0}$ is the ratio of the received energy per information bit to the one-sided noise power spectral density. For binary block codes the above result assumes that the code-rate approaches zero asymptotically with code length. For a rate-1/2 code the minimum $\frac{E_b}{N_0}$ necessary for reliable transmission is 0.2 dB for soft-decision decoders and 1.8 dB for hard-decision decoders. These basic results suggest the significant loss of performance when soft-decision decoding is not available for a given code.

The situation is quite different for convolutional codes that use Viterbi decoding. Soft decisions are incorporated easily into the Viterbi decoding algorithm in a very natural way, providing an increase in coding gain of over 2.0 dB with respect to the comparable hard-decision decoder over an additive white Gaussian noise channel. Unfortunately convolutional codes present their own set of problems. For example, they cannot be implemented easily at high coding rates. They also have an unfortunate tendency to generate burst errors at the decoder output as the noise level at the input is increased.

A "best-of-both-worlds" situation can be obtained by combining RS codes with convolutional codes in a concatenated system. The convolutional code

Concatenated Coding Systems and Turbo Codes

(with soft-decision Viterbi decoding) is used to "clean up" the channel for the Reed-Solomon code, which in turn corrects the burst errors emerging from the Viterbi decoder. Therefore, by the proper choice of codes the probability of error can be made to decrease exponentially with overall code length at all rates less than capacity. Meanwhile, the decoding complexity is dominated by the complexity of the algebraic RS decoder, which increases only algebraically with the code length.

Generally, the "outer" code is more specialized in preventing errors generated by the "inner" code when it makes a mistake. The "inner" code can also be a binary block code other than a binary convolutional code. For a band-limited channel, trellis codes often are selected as "inner" codes. To further improve error-correction performance, interleavers are usually applied between the "inner" and "outer" codes to provide resistance to burst errors.

In the next section, an overview of the state-of-the-art in concatenated coding systems is provided in a variety of applications. Among these applications are the CCSDS standard for telemetry-channel coding, the ATSC standard for digital television and the ITU-T recommendation J.83 for television and sound transmission.

12.2 Concatenated Coding Systems with Convolutional (or Trellis) Codes and RS Codes

12.2.1 Concatenated Codes in the NASA Deep-Space Standard

Downlinks of deep-space communications are generally characterized as power-limited channels. The size and weight of on-board batteries and solar cells contribute significantly to launch costs. This is a particularly troublesome issue with deep-space probes. For deep-space communications, a bit-error rate of 10^{-5} is desired. However, the poor channel conditions of deep-space communications often cause a larger bit-error rate and a low signal-to-noise ratio. Thus, it is a requirement to adopt powerful error-control codes that operate efficiently at extremely low signal-to-noise ratios. Convolutional codes have been particularly successful in these applications, beginning with the convolutional encoder/sequential decoder used in the *Pioneer* missions [2]. A NASA Planetary Standard was later developed for other missions, including a portion of the *Voyager* mission. The *Voyager* mission allowed for the use of the two convolutional codes listed in Table 12-

1. Both offered the maximal d_{free} for their respective rates and constraint lengths.

Convolutional Codes	Rate	Constraint Length	d_{free}	Generator Polynomials
C_1	1/2	7	10	$g^{(0)}(D) = 1 + x + x^3 + x^4 + x^6$
				$g^{(1)}(D) = 1 + x^3 + x^4 + x^5 + x^6$
C_2	1/3	7	15	$g^{(0)}(D) = 1 + x + x^3 + x^4 + x^6$
				$g^{(1)}(D) = 1 + x^3 + x^4 + x^5 + x^6$
				$g^{(2)}(D) = 1 + x^2 + x^4 + x^5 + x^6$

Table 12.1 Convolutional codes used in deep-space communications

With 3-bit soft-decision decoding, codes C_1 and C_2 provide 5.1 dB and 5.6 dB of coding gain, respectively, at a decoded bit-error rate of 10^{-5} [3]. C_1 has since become an "industry standard" and has been used in applications from military satellite communications to commercial cellular telephony. Figure 12.2 shows the bit-error-rate performance of C_1 as compared to an uncoded system and a (54, 32) Reed-Solomon error-control system. The modulation format for all three curves is coherently demodulated BPSK. The C_1 performance curve is an upper bound that remains tight until the SNR becomes very small (<- 2.5 dB). The Reed-Solomon curve treats decoder failures as errors. Note that at low channel SNRs (<4.5 dB), the convolutional code outperforms the Reed-Solomon code. It is only at higher SNRs that the Reed-Solomon code provides better performance. This is an example of a general tendency shown by convolutional and Reed-Solomon codes. The convolutional codes are good choices for applications that require a moderate amount of reliability over poor channels. This is particularly true when soft-decision decoding is used in conjunction with channel side information. Reed-Solomon codes are a good choice for applications requiring extremely high levels of reliability on channels that are only moderately noisy.

When the channel condition is extremely poor, it is often necessary to use more than one error-control code. In Section 7.2.2, a pair of shortened RS codes is used in order to improve the fidelity of a digital-audio system. Such a system is generally one type of concatenated coding system. Given the bursty nature of the digital-audio channel, it was found that a pair of RS codes was the best choice. In deep-space satellite applications the primary problem is random noise. Figure 12.3 shows the NASA/European Space Agency standard adopted in 1987 for deep-space missions. The rate-1/2 Planetary-Standard convolutional code is joined by a (255,223) RS code in a concatenated system. The overall code-rate for this case is 0.437.

Concatenated Coding Systems and Turbo Codes

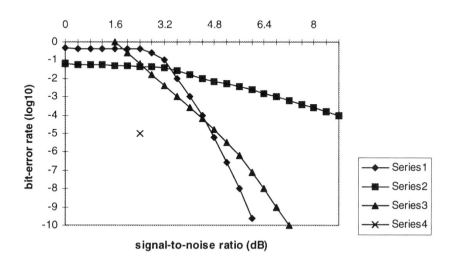

Figure 12.2. Bit-Error-Rate Performance for
- a rate-1/2, $K = 7$ convolutional code (Series3),
- a (64,32) Reed-Solomon code(Series1),
- the NASA/ESA concatenated system(Series4),
- and an uncoded BPSK channel(Series2)

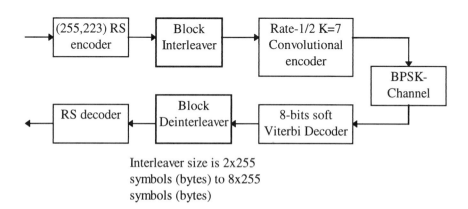

Interleaver size is 2x255 symbols (bytes) to 8x255 symbols (bytes)

Figure 12.3. NASA/ESA deep-space error-control coding standard

The Viterbi decoder in this system is designed to improve the effective channel quality to the point where the RS code could be used efficiently. The RS code also provides some additional advantages in this scheme. When the Viterbi decoder suffers a decoder error, a code word is outputted that usually differs from the transmitted word over a few consecutive trellis branches. It follows that, though the input to the Viterbi decoder is corrupted by random noise, the output of the decoder tends to have burst errors. The RS code, with its inherent burst-error correcting capability, is thus a natural choice to "clean up" the channel after the Viterbi decoder.

An RS code corrects certain burst-errors that occur in the length (in bits) of the RS code symbols. In the NASA/ESA system, the RS code can correct up to 16 bytes per codeword. Bytes (Eight-bit symbols) were chosen, because the smallest nonzero path through the $K = 7$ convolutional code traverses 7 branches of its trellis is the most frequently occurring error bursts at the output of the Viterbi decoder. The error bursts thus have lengths of 7 or 8 bits and can be trapped by a single RS symbol (in the best case).

If the Viterbi decoder output has bursts whose lengths exceed 128 bits, the RS decoder probably fails or, what is less likely, a decoder error occurs. In order to correct longer bursts, a symbol interleaver is placed between the two encoders. Currently, NASA/ESA missions use block interleavers that hold from $n = 2$ to $n = 8$ RS codewords. For example, the NASA *Galileo* mission to Jupiter used $n = 2$, while ESA's *Giotto* mission to Halley's Comet used $n = 8$. The impact of the interleaver size on the performance of the NASA/ESA system is discussed in detail in [4].

Computer-simulation results show that the NASA/ESA system with interleaver sizes $n = 2, 4$, and 8 provide a decoder BER of 10^{-5} at an E_b/N_o of 2.6dB, 2.45dB, and 2.35dB, respectively. The uncoded BPSK system requires an E_b/N_o of 9.6 dB to provide the same BER performance. The coding gain with respect to an uncoded system is thus 7.0 to 7.25dB, depending on the size of the interleaver. The concatenated system thus provides an increase of 2 dB in coding gain over the rate-1/2 Planetary Standard convolutional code by itself (see Figure 12.2).

12.2.2 Digital Television-Transmission System

Digital broadcasting systems are developed for the delivery of digital-compressed video, audio and data services. The general requirements for such systems are high reliability and efficiency. Since digital television and

Concatenated Coding Systems and Turbo Codes

broadcasting service channels are band-limited, TCM and RS codes are considered usually for most of the applications.

Digital television-transmission systems usually use concatenated forward-error correction(FEC) codes that consist of an RS code, an interleaver, and a trellis (convolutional) code. At the decoder, the Viterbi decoding, deinterleaving, and RS decoding are used. As an example of a digital television transmission system, the Grand-Alliance High-Definition TeleVision(HDTV) system, is shown in Table 12.2.

Grand Alliance HDTV system Parameters	Terrestrial Transmission Mode	High Data Rate (Cable) Mode
Channel Bandwidth	6MHz	6MHz
Symbol Rate	10.76Msymbols/second	10.76Msymbols/second
Bits per symbol	3	4
Trellis FEC	2/3 rate	None
Interleavers	I=52, M=4	
RS FEC	(208,188,21)	(208,188,21)

Table 12.2 Digital television transmission systems

The Grand Alliance vestigial-sideband(VSB) digital-transmission system [5] provides the basis for a family of transmission systems suitable for data transmission over a variety of media. This family of transmission systems shares the same pilot, symbol rate, data-frame structure, interleaver, RS code, and synchronization pulses. The VSB system offers two modes: a terrestrial-broadcast mode (8 VSB), and a high data-rate mode (16 VSB).

Terrestrial-broadcast mode

The terrestrial-broadcast mode (known as 8-VSB) supports a payload-data rate of 19.28 Mbps in a 6 MHz channel. A functional block diagram of a representative 8-VSB terrestrial broadcast transmitter is shown in Figure 12.4. For completeness, a few of the relevant parts of the transmitter have been included as well. The input to the transmission subsystem is a 19.39Mbps serial data stream, comprised of 188-byte MPEG-compatible data packets (including a sync byte and 187 bytes of data which represent a payload-data rate of 19.28 Mbps).

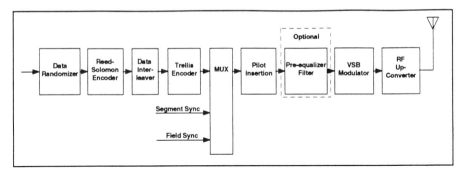

Figure 12.4 VSB transmitter.

First the source data are randomized (here, a randomizer is used to ensure a high-data transition density for purposes of bit-timing recovery at the decoder). Then these data processed for error correction in the form of an RS coding (20 RS parity bytes are added to each packet), a 1/6 data-field interleaving (explanation later) and finally a 2/3-rate trellis encoding. Figure 12.5 shows the format in which the data are organized for transmission. Each Data Frame consists of two Data Fields, each containing 313 Data Segments. The first Data Segment of each Data Field is a unique synchronizing signal (Data-Field Sync) which includes the training sequence used by the equalizer in the receiver. The remaining 312 Data Segments each carry the equivalent of the data from one 188-byte transport packet plus its associated FEC overhead. The actual data in each Data Segment comes from several transport packets because of data interleaving.

Each Data Segment consists of 832 symbols. The first 4 symbols are transmitted in binary form and provide segment synchronization. The Data- Segment Sync signal also represents the sync-byte of the 188-byte MPEG-compatible transport packet. The remaining 828 symbols of each Data Segment carry the data that constitute the remaining 187 bytes of the transport packet and its associated FEC overhead. These 828 symbols are transmitted as 8-level signals and therefore carry three bits per symbol. Thus, 828 x 3 = 2484 bits of data are carried in each Data Segment. This exactly matches the requirement needed to send a protected transport packet (see Table 12.3).

The RS code used in the VSB transmission subsystem is the (207,187) code. The RS data-block size is 187 bytes, with 20 RS parity bytes added for error correction. A total RS block size of 207 bytes is transmitted per Data Segment. In creating bytes from the serial bit stream, the Most-Significant-Bit (MSB) is the first serial bit. The 20 RS parity bytes are sent at the end of the Data

Concatenated Coding Systems and Turbo Codes

Segment. The parity-generator polynomial and the primitive field-generator polynomial are shown in Figure 12.6.

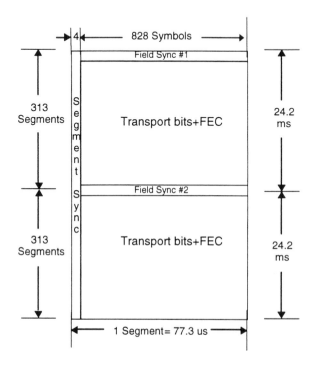

Figure 12.5 VSB data frame.

Although the RS code naturally protects against burst errors, the data is interleaved for further protection. Burst errors result from both impairment and trellis-decoding mistakes where a single error often propagates through the trellis decoder, multiplies, and becomes a burst error. Trellis decoders with a large constraint length often cause correspondingly long burst errors to be created.

187 data bytes + 20 RS parity bytes = 207 bytes
207 bytes x 8 bits/byte = 1656 bits
2/3 rate trellis coding requires 3/2 x 1656 bits = 2484 bits.

Table 12.3 Coding Parameters

The goal of the interleaver is to separate adjacent data bytes from the same RS block over time so that a long burst of noise or interference would be necessary to encompass more than 10 data bytes and overrun the protection of the RS code. Since this RS code can correct up to 10 byte errors per block, regardless of how many individual bit errors have occurred within each byte, it is advantageous to group consecutive symbols from the trellis decoder into bytes to get the best burst-error performance.

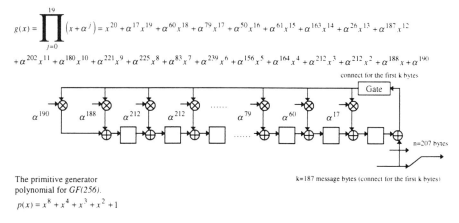

Figure 12.6. The (207,187) RS Encoder.

In the VSB transmission system, a convolutional interleaver is used to spread the output symbols of the RS encoder. Such an interleaver has $I=52$ taps and an increment-per-tap of $M=4$ as shown in Figure 12.7. The first byte of the RS encoded-output block is the input and output of the zero-delay interleaver commutator arm. The k-th commutator arm consists of $k \cdot M$ byte delays for $k=0,1,...,51$ and $M=4$. An output byte is read from the k-th FIFO or circular buffer, an input byte is written or shifted into the k-th buffer, and the commutator advances to the $(k+1)$-th interleaver arm. After reading and writing from the last commutator arm, the commutator advances to the zero-delay arm for its next output. The depth $I=52$ and $M=4$ deinterleaver consists of an $I(I-1)M/2=5304$ RS symbol memory. Such interleaving is provided to a depth of about $312/52 \approx 1/6$ of a data field (4 ms deep). Only data bytes are interleaved. The interleaver is synchronized to the first data byte of the data field. Intrasegment interleaving is performed for the benefit of the trellis-coding process.

The 8-VSB transmission system employs a 2/3 rate (R=2/3) trellis code (with one unencoded bit which is precoded). That is, one input bit is encoded into

Concatenated Coding Systems and Turbo Codes

two output bits using a 1/2 rate convolutional code while the other input bit is precoded. The signaling waveform used with the trellis code is a 8-level (3 bit) one-dimensional constellation. The transmitted signal is referred to as 8-VSB. A 4-state trellis encoder is used. Also trellis-code intrasegment interleaving is used. This uses twelve identical trellis encoders and precoders that operate on the interleaved data symbols. The code interleaving is accomplished by the encoding symbols (0, 12, 24, 36 ...) as one group, symbols (1, 13, 25, 37, ...) as a second group, symbols (2, 14, 26, 38, ...) as a third group, and so on for a total of 12 groups.

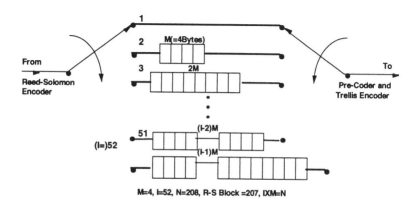

Figure 12.7. Convolutional interleaver (byte shift register illustration).

Each byte in a RS codeword is sent serially in the order (7, 6, 5, 4, 3, 2, 1, 0), i.e. the MSB is transmitted first. These eight serial bits are then split into two subsequences : $X_2 = (7, 5, 3, 1)$ and $X_1 =(6, 4, 2, 0)$. The subsequence X_1 is input to a feedback convolutional encoder while the subsequence X_2 is pre-coded by an interference filter. A standard 4-state optimal Ungerboeck code is used for the encoding (Y_1, Y_2). The 4-state feedback trellis encoder is shown in Figure 12.8. Also shown is the precoder and the symbol mapper. The trellis coder and precoder intrasegment interleaver is shown in Figure 12.9 which feeds the mapper shown in Figure 12.8. Referring to Figure 12.9, data bytes are fed from the byte interleaver to the trellis coder and precoder, and they are processed as whole bytes by each of the twelve encoders. Each byte produces four symbols from a single encoder.

The output multiplexer shown in Figure 12.9 advances by four symbols on each segment boundary. However, the state of the trellis encoder is not advanced. The data coming out of the multiplexer follow normal ordering from the encoder, 0 through 11, for the first segment of the frame, but on the second segment the order changes and symbols are read from the encoders, 4 through 11, and then, 0 through 3. The third segment reads from the encoders, 8 through 11, and then, 0 through 7. This three-segment pattern repeats through the 312 Data Segments of the frame. Table 12.4 shows the interleaving sequence for the first three Data Segments of the frame.

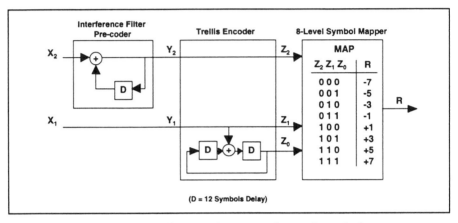

Figure 12.8. 8-VSB trellis encoder, precoder, and symbol mapper.

A complete conversion of bytes to serial bits needs 828 bytes to produce 8x828=6624 bits. Data symbols are created from 2 bits sent in the order of MSB first, so a complete conversion operation yields 3312 data symbols, which corresponds to 4 segments of 828 data symbols. The 3312 data symbols, when divided by the 12 trellis encoders, produce 276 symbols per trellis encoder. Finally the 276 symbols, when divided by 4 symbols per byte, gives 69 bytes per trellis encoder.

The conversion starts with the first segment of the field and proceeds in groups of 4 segments until the end of the field. 312 segments per field, when divided by 4, gives 78 conversion operations per field.

During Segment-Sync the input to 4 encoders is skipped, and the encoders cycle with no input. The input is held until the next multiplex cycle, and then it is fed to the correct encoder.

For more details of the byte-to-symbol conversion and the associated multiplexing of the trellis encoders, the reader is referred to [5].

Concatenated Coding Systems and Turbo Codes

The encoded trellis data are passed through a multiplexer that inserts the various synchronization signals (the Data-Segment Sync and Data Field Sync). Also a two-level (binary) 4-symbol Data-Segment Sync is sandwiched into the 8-level digital data stream at the beginning of each Data Segment. (The MPEG sync byte is replaced by the Data-Segment Sync.) The Data-Segment Sync embedded in random data is illustrated in Figure 12.10.

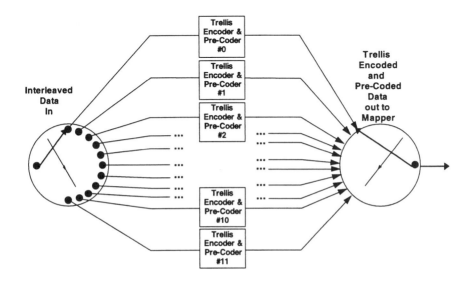

Figure 12.9. Trellis code interleaver.

Segment	Block 0	Block 1	...	Block 68
0	D0 D1 D2 ... D11	D0 D1 D2 ... D11	...	D0 D1 D2... D11
1	D4 D5 D6 ... D3	D4 D5 D6 ... D3	...	D4 D5 D6... D3
2	D8 D9 D10...D7	D8 D9 D10...D7	...	D8 D9 D10...D7

Table 12.4 Interleaving Sequence

A complete segment consists of 832 symbols: 4 symbols for the Data-Segment Sync and 828 data-plus-parity symbols. The Data-Segment Sync is binary (2-level). The same sync pattern occurs regularly at 77.3 μs intervals and is the only signal that repeats at this rate. Unlike the data, the four symbols for the Data-Segment Sync are not Reed-Solomon or trellis encoded, nor are they interleaved. The Data-Segment Sync pattern is a 1001 pattern, as shown in Figure 12.9.

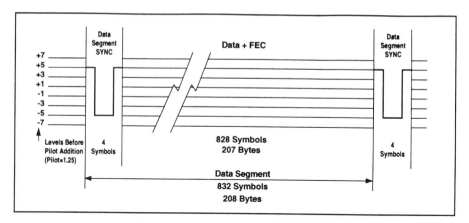

Figure 12.10. 8-VSB data segment.

The data are not only divided into Data Segments, but also into Data Fields, each consisting of 313 segments. Each Data Field (of 24.2 ms) starts with one complete the Data Segment of Data Field Sync as shown in Figure 12.11. Each symbol represents one bit of data (2-level). The 832 symbols in this segment are defined in Figure 12.11.

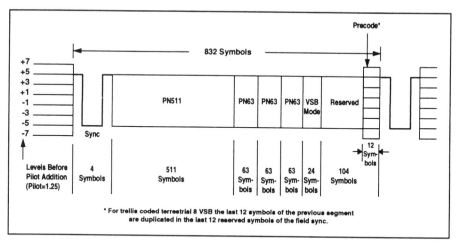

Figure 12.11. VSB data field sync.

High data-rate mode

The high data-rate mode of digital television trades off transmission robustness (of a 28.3 dB signal-to-noise ratio) for a payload data rate (of 38.57 Mbps). Most parts of the high data-rate-mode VSB system are identical or

Concatenated Coding Systems and Turbo Codes

similar to the terrestrial system. The primary difference is the number of transmitted levels (8 versus 16) and the use of trellis coding and NTSC interference rejection filtering in the terrestrial system.

Figure 12.12 illustrates a typical data segment, where the number of data levels is now 16, due to the doubled data rate. Each portion of the 828 data symbols represents 187 data bytes and 20 Reed-Solomon bytes followed by a second group of 187 data bytes and 20 Reed-Solomon bytes (before convolutional interleaving).

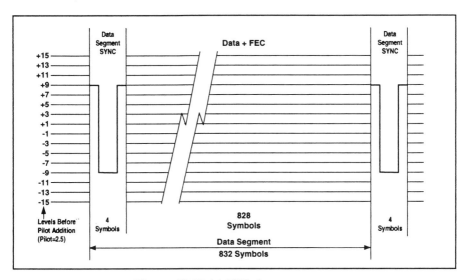

Figure 12.12 16-VSB data segment.

Figure 12.13 shows the block diagram of the transmitter. It is identical to the terrestrial VSB system, except the trellis coding is replaced with a mapper which converts data to multi-level symbols. See Figure 12.14.

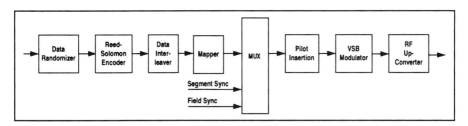

Figure 12.13. 16-VSB transmitter.

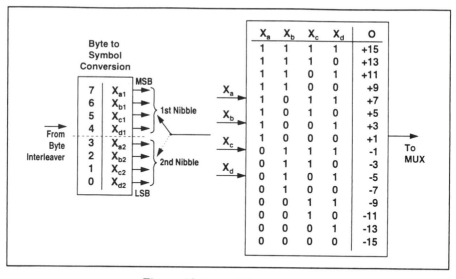

Figure 12.14. 16 VSB mapper.

The interleaver is a 26 data segment inter-segment convolutional byte-interleaver. Interleaving is provided to a depth of about $26/312 \approx 1/12$ of a data field (2 ms deep). Again, only data bytes are interleaved. Figure 12.13 shows the mapping of the outputs of the interleaver to the nominal signal levels (-15, -13, -11, ..., 11, 13, 15).

12.2.3 ITU-T Recommendation J.83

The ITU-T recommendation J.83 [6][40] is an international standard of digital multi-programmed systems for television, sound and data services for cable distribution. This standard defines the framing structure, channel coding, and channel modulation for a digital multi-service television-distribution system that is specific to a cable channel.

The design of the modulation, interleaving, and channel-coding is based on the testing and characterization of cable systems in North America. The modulation is Quadrature-Amplitude Modulation(QAM) with a 64-point signal constellation (64-QAM) and with a 256-point signal constellation (256-QAM), which is user selectable. The forward error correction (FEC) is based on a concatenated coding approach that produces high coding gains at a moderate complexity and overhead. Concatenated coding offers improved performance over a block code with a similar overall complexity. The system FEC is optimized for an almost error-free operation at a threshold output error- event rate of one error-event per 15 minutes[7].

Concatenated Coding Systems and Turbo Codes

The data format input to the modulation and coding is assumed to be an MPEG-2 transport, as defined in reference [8]. Here the MPEG-2 transport is a 188-byte data-packet assembled from compressed video and audio bit-streams. Channel coding and transmission are specific to a particular medium or communication channel. The expected channel-error statistics and distortion characteristics are critical in determining the appropriate error correction and demodulation. The cable channel, including fiber trucking, is primarily regarded as a bandwidth-limited linear channel with a balanced combination of white noise, interference, and multi-path distortion. The QAM technique used, together with adaptive equalization and concatenated coding, is well suited to this application and channel.

The basic block diagram of cable transmission processing is shown in Figure 12.15.

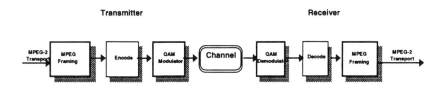

Figure 12.15. Cable Transmission Block Diagram

In this section only FEC blocks are discussed. The FEC is composed of four processing layers. As illustrated in Figure 12.16.

The FEC section uses various types of error correcting algorithms and deinterleaving techniques to transport data reliably over the cable channel.
- **RS Coding** – Provides block encoding and decoding to correct up to three symbols within an RS block.
- **Interleaving** – Evenly disperses the symbols, that protect against a burst of symbol errors from being sent to the RS decoder.
- **Randomization** – Randomizes the data on the channel to allow effective QAM demodulator synchronization.
- **Convolutional Coding** – Provides convolutional encoding and soft decision trellis decoding of random channel errors.

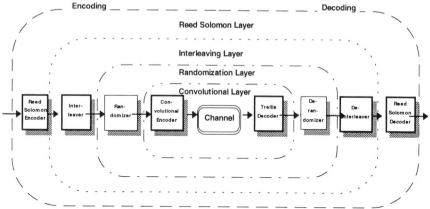

Figure 12.16. Layers of Processing in the FEC

RS Coding

The data stream (MPEG-2 transport, etc.) is Reed-Solomon encoded using a (128,122) code over GF(2^7). This code has the capability of correcting up to t=3 errors per RS block. The same RS code is used for both 64-QAM and 256-QAM. However, the FEC frame format is different for each modulation type.

For 64-QAM, an FEC frame consists of a six-symbol sync trailer which is appended to the end of 60 RS blocks, with each block containing 128 symbols. Each RS and sync symbol consists of 7-bits. Thus, there is a total of 53,760 data bits and 42 frame-sync trailer bits in an FEC frame.

For 256-QAM, an FEC frame consists of a 40-bit sync trailer which is appended to the end of an 88 RS block FEC frame. Each symbol in a RS block is a 7 bit symbol. The total number of data bits in a 256-QAM FEC frame is 78,848 bits.

The Reed-Solomon encoder is implemented as follows: A systematic encoder is utilized to implement a t=3, (128,122) extended Reed Solomon code over GF(128). The primitive polynomial used to form the field over GF(128) is: $p(x) = x^7 + x^3 + 1$.

The generator polynomial used by the encoder is:

$$g(x) = (x+\alpha)(x+\alpha^2)(x+\alpha^3)(x+\alpha^4)(x+\alpha^5)$$
$$= x^5 + \alpha^{52}x^4 + \alpha^{116}x^3 + \alpha^{119}x^2 + \alpha^{61}x + \alpha^{15}.$$

The message polynomial input to the encoder consists of 122, 7-bit symbols, and is described as follows:
$$m(x) = m_{121}x^{121} + m_{120}x^{120} + \ldots m_1 x + m_0.$$
This message polynomial is first multiplied by x^5, then divided by the generator polynomial $g(x)$ to form a remainder, described by the following:
$$r(x) = r_4 x^4 + r_3 x^3 + r_2 x^2 + r_1 x + r_0.$$
This remainder constitutes five parity symbols which are then added to the message polynomial to form a 127-symbol code word that is an even multiple of the generator polynomial..

The generated code word is now described by the following polynomial:
$$c(x) = m_{121}x^{126} + m_{120}x^{125} + m_{119}x^{124} + \ldots$$
$$+ r_4 x^4 + r_3 x^3 + r_2 x^2 + r_1 x + r_0$$
By construction a valid code word has roots at the first through fifth powers of the primitive field element a.

An extended parity symbol c_p is generated by evaluating the code word at the sixth power of alpha as
$$c_p = c(\alpha^6)$$
This extended symbol is used to form the last symbol of a transmitted RS codeword. The extended code word then appears as follows:
$$\bar{c}(x) = m_{121}x^{127} + m_{120}x^{126} + \ldots + r_1 x^2 + r_0 x + c_p$$
The structure of the RS codeword that illustrates the order in which the symbols are transmitted from the output of the RS encoder is shown as follows:
$$m_{121} \ m_{120} \ m_{119} \ \ldots \ r_1 \ r_0 \ c_p$$
Note that the order that symbols are sent is from left to right.

Interleaving
Interleaving is included in the modem between the RS block coding and the randomizer to enable the correction of burst-noise-induced errors. In both 64-QAM and 256-QAM a convolutional interleaver is employed. The depth of the interleaver employed is I=128 field symbols.

Convolutional interleaving is illustrated in Figure 12.16. The interleaving commutator position is incremented at the RS symbol frequency, with a single symbol output from each position. In the convolutional interleaver the

R-S code symbols are sequentially shifted into a bank of 128 registers. Each successive register has M-symbols more storage than the preceding register. The first interleaver path has zero delay, the second has a M symbol period of delay, the third 2*M-symbol period of delay, and so on, up to the 128th path which has 127*M-symbol period of delay. This is reversed for the deinterleaver in the Cable Decoder in such a manner that the net delay of each RS symbol is the same through the interleaver and deinterleaver. Burst noise in the channel causes a series of incorrect symbols. These are spread over many RS codewords by the deinterleaver in such a manner that the resultant in a symbol errors per codeword are within the range of the RS decoder-correction capability.

For 64-QAM, in the absence of all other impairments and ignoring any error propagation from the trellis decoder, the burst tolerance at the input of the RS decoder is 95 µsec. To accomplish this, M is set equal to 1 in the interleaver.

For 256-QAM, the interleaver continues to employ a depth of I=128, but here M may be selected by a 3 bit field in the frame-sync trailer to have an integer value between 1 and 8. The burst tolerance can be from 67 µsec. to 539 µsec. for M= 1 to 8, respectively.

Randomization

The randomizer is the third layer of processing in the FEC block diagram. The randomizer provides for even distribution of the symbols in the constellation, which enables the demodulator to maintain proper lock. The randomizer adds a pseudo-random noise (PN) sequence to the transmitted signal to assure a random transmitted sequence.

For both 64 and 256-QAM, the randomizer is initialized during the FEC-frame trailer, and is enabled at the first symbol after the trailer. Thus the trailer itself is not randomized. Initialization is defined as a preloading to the 'all-ones' state, for the randomizer structure shown in Figure 12.18. The randomizer uses a linear feedback shift register specified over the finite field $GF(2^7)$ defined as follows:

$$f(x) = x^3 + x + \alpha^3,$$

where α satisfies $\alpha^7 + \alpha^3 + 1 = 0$.

Concatenated Coding Systems and Turbo Codes

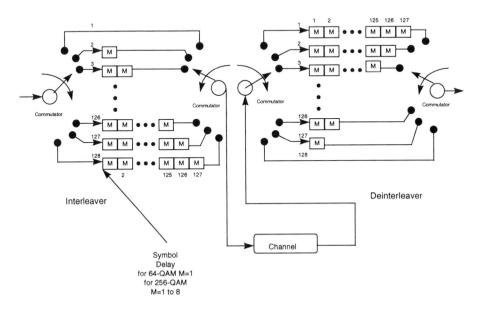

Figure 12.17. Interleaving Functional Block Diagram

The structure of this randomizer is shown in Figure 12.17.

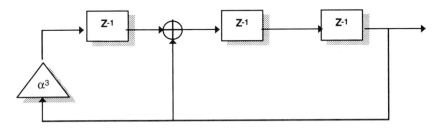

The randomizer polynomial
$$f(x) = x^3 + x + \alpha^3$$

Figure 12.18 Randomizer (7-bit Byte Scrambler)

Trellis Coded Modulation

As part of the concatenated coding scheme, trellis coding is employed as the inner code. It allows for the utilization of redundancy to improve the SNR by increasing the size of the symbol constellation without increasing the symbol rate. As such, it is more properly termed, "trellis-coded modulation", as discussed in Chapter 11. A block diagram for the trellis-coded modulator used for 64-QAM is shown in Figure 12.19.

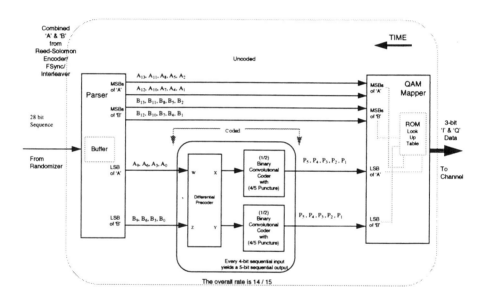

Figure 12.19 A 64-QAM trellis coded modulator block diagram

The input to the trellis coded modulator is a 28-bit sequence of four, 7-bit R-S symbols, which are labeled in pairs of 'A' symbols and 'B' symbols. All 28 bits are assigned to a trellis group, where each trellis group forms 5 QAM symbols, as shown in Figure 12.20. Each trellis group is further divided into two groups: two upper or MSB uncoded-bit streams, and one lower or LSB coded-bit stream.

Of the 28 bits in the trellis group, the 4 LSB's of the trellis group are fed to the BCC for coding, thereby, producing 5 coded bits. The remaining 10 bits are

Concatenated Coding Systems and Turbo Codes

sent uncoded to the mapper. This produces an output of 15 bits.. Thus, the overall trellis-coded modulation, for 64-QAM, yields a 14/15 rate.

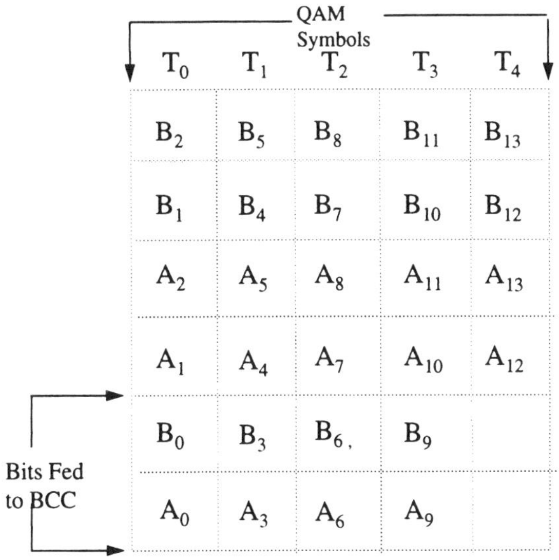

	T_0	T_1	QAM Symbols T_2	T_3	T_4
	B_2	B_5	B_8	B_{11}	B_{13}
	B_1	B_4	B_7	B_{10}	B_{12}
	A_2	A_5	A_8	A_{11}	A_{13}
	A_1	A_4	A_7	A_{10}	A_{12}
Bits Fed to BCC	B_0	B_3	B_6	B_9	
	A_0	A_3	A_6	A_9	

Figure 12.20 64-QAM Trellis Group

The trellis group is formed from RS symbols as follows: For the "A" symbols, the RS symbols are read, from MSB to LSB, A_{10}, A_8, A_7, A_5, A_4, A_2, A_1 and A_9, A_6, A_3, A_0, A_{13}, A_{12}, A_{11}. The four MSB's of the second symbol are fed into the BCC, one bit at a time, with the LSB first. The remaining bits of the second symbol and all the bits of the first symbol are fed into the mapper, uncoded, with the LSB first one bit at a time. The four bits sent to the BCC produces 5 coded bits, labeled P_1, P_2, P_3, P_4, P_5. The same process is used as well for the "B" bits. The complete process is shown in Figure 12.20. With 64-QAM, 4 of the R-S symbols conveniently fit into one trellis group.

The trellis-coded modulator includes a punctured rate-1/2 binary convolutional encoder that is used to introduce redundancy in the LSB's of the trellis group. The coding provides the appropriate SNR gain. The convolutional encoder is a 16-state non-systematic rate-1/2 encoder with the generator: G1 = 010 101, G2 =011 111 (25,37$_{octal}$), or equivalently the generator matrix [$1 \oplus D^2 \oplus D^4$, $1 \oplus D \oplus D^2 \oplus D^3 \oplus D^4$]. The outputs of the encoder are fed into the puncturing matrix: 0001 1111 (where "0" denotes NO transmission, "1" denotes transmission), which produces a single serial bit stream. The puncturing matrix essentially converts the rate-1/2 encoder to rate-4/5 encoder. The internal structure of the punctured encoder is illustrated in Figure12.21.

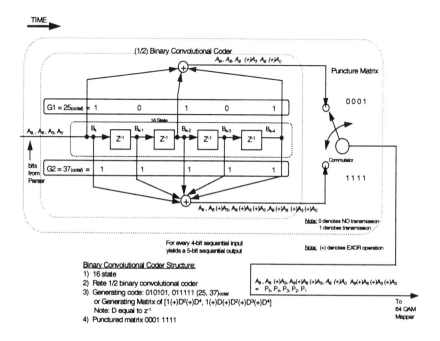

Figure 12.21 Punctured binary-convolutional coder

For the 256-QAM, a trellis-coded modulator is also employed; however, there are two different types of trellis groups: a non-sync group and a sync group. Each trellis group produces 5 QAM symbols, but the non-sync group contains 38 data bits and the sync group contains 30 data bits and 8 sync bits, Figure 12.22 shows both a non-sync trellis group and a sync trellis group. Since there are 88 RS blocks per FEC frame, there are a total of 2,076 trellis groups per frame. Of these trellis groups, 2,071 are non-sync trellis groups and 5 are sync-trellis groups. The 5 sync-trellis groups come at the end of the frame. The frame-sync trailer is aligned to the trellis groups. The trellis group is further divided into two groups: one upper or MSB uncoded bit stream and one lower or LSB-coded bit stream. The first bit of the non-sync trellis group is assigned to the MSB of the first RS symbol in the FEC frame. The output of the BCC is the five parity bits, labeled P_1 through P_5.

Concatenated Coding Systems and Turbo Codes

To form trellis groups from the RS code words, the RS code words are sent serially beginning with the MSB of the first codeword. The RS code words have the order m_{121} m_{120} m_{119} ... r_1 r_0 c_-. The MSB of symbol m_{121} is placed in position A_0 of the non-sync trellis group, shown in Figure 12.22. Bits are then placed in trellis group locations, from R-S symbols, in the following order: A_0 B_0 A_1 ... B_3 A_4 B_4 ... B_{13} B_{14} B_{15}. For the sync trellis groups, the RS bits begin at location A_1 instead of A_0.

The same BCC as 64-QAM is used, and it is also punctured to the rate 4/5. From each trellis group 4 bits are coded by the BCC producing 5 coded bits. Fifteen bits are left uncoded to obtain a rate 19/20 trellis-coded modulator. The 256-QAM trellis-coded modulator is shown in Figure 12.23.

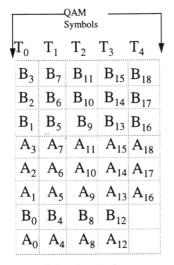

Figure 12.22 256-QAM Sync and Non-Sync Trellis Groups

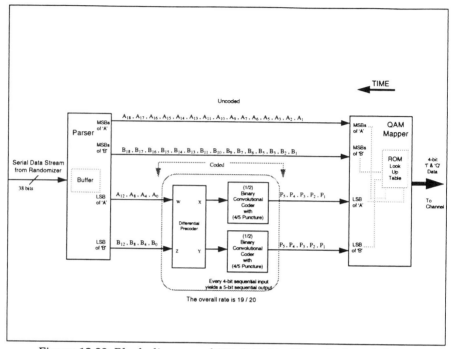

Figure 12.23. Block diagram of 256-QAM trellis-coded modulator.

The differential precoder, shown in Figure 12.24, performs a 90° rotationally-invariant trellis coding. Rotationally-invariant coding is employed in both the 64 and 256-QAM modulators. The key for robust modem designs is to have a very fast recovery from a carrier phase slip. In a non-rotationally invariant design, a carrier phase slip can cause the need for a major resynchronization of the FEC that leads to a burst of errors at the FEC output.

The differential precoder allows the information to be carried by changes in phase, rather than by the absolute phase. For 64-QAM the 3rd and the 6th bits of the 6-bit symbols are differentially encoded. If one masks out the 3rd and the 6th bits in Figure 12.25, the 90° rotational invariance of the remaining 4 bits is inherent in the symbol constellation.

For 256-QAM the 4th and the 8th bits of the 8-bit symbols are encoded differentially. If one masks out the 4th and the 8th bits in Figure 12.26, the 90° rotational invariance of the remaining 6 bits again is inherent in the symbol constellation.

Concatenated Coding Systems and Turbo Codes

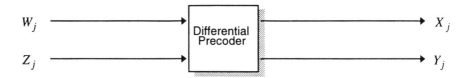

Differential Precoder Equations

$$X_j = W_j + X_{j-1} + Z_j(X_{j-1} + Y_{j-1})$$
$$Y_j = Z_j + W_j + Y_{j-1} + Z_j(X_{j-1} + Y_{j-1})$$

Figure 12.24. Differential Precoder

For 64-QAM, the QAM mapper receives the coded and uncoded 3-bit 'A' and 'B' data from the trellis-coded modulator. These bits are used to address a look-up table which produces the 6-bit constellation symbol. The 6-bit constellation symbol is then sent to the 64 QAM modulator where the signal constellation illustrated in Figure 12.25 is generated.

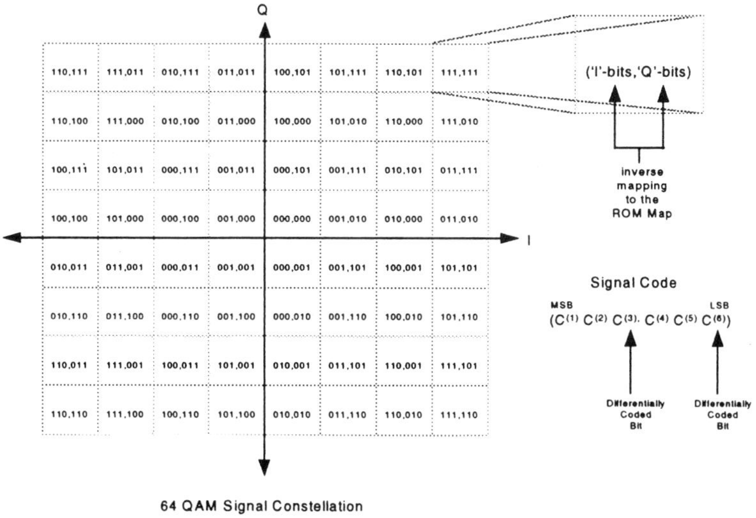

Figure 12.25 64-QAM Constellation

For 256-QAM, the QAM mapper receives the coded and uncoded 4-bit 'A' and 'B' data from the trellis- coded modulator. It uses these bits to address a look-up table which produces the 8-bit constellation symbol. The 8-bit constellation symbol is then sent to the 256-QAM modulator, where the signal constellation illustrated in Figure 12.26 is generated.

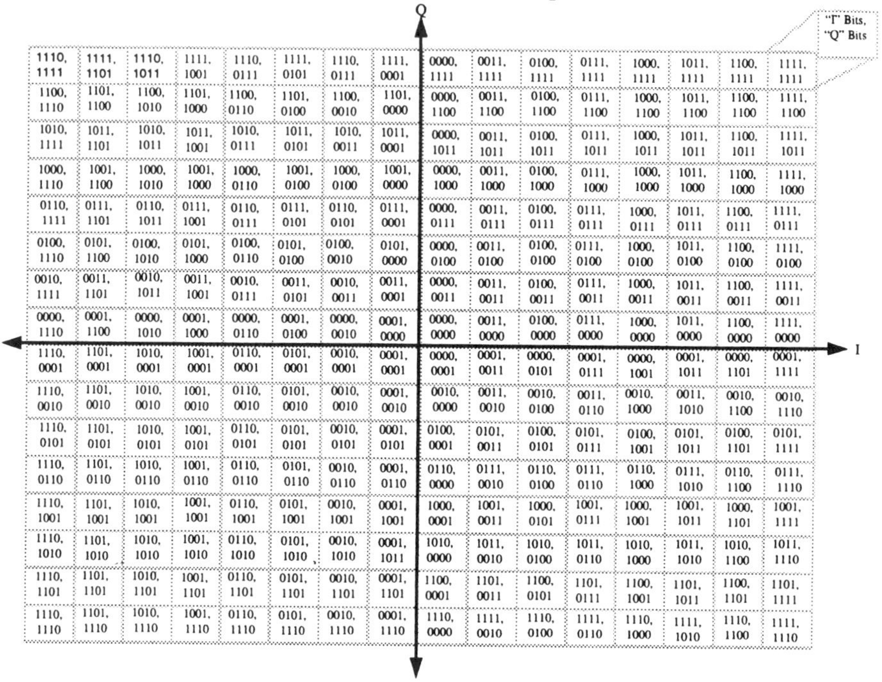

Figure 12.26 256-QAM Constellation

12.2.4 Performance Improvements for Concatenated Coding Systems

As a further improvement of the concatenated coding systems, errors-and-erasures decoding of the RS codes can be used to provide some performance improvement if the erased symbols are available from the inner decoder. One method of providing erased symbols is based on two facts : (1) a frame of several RS code blocks is interleaved prior to the encoding of the inner code, and (2) decoding errors from the inner decoder are typically bursty, resulting in strings of error symbols. Although long strings of error symbols usually cause problems for an RS decoder, after de-interleaving they are more spread out, making them easier to decode. In addition, once a symbol error has been corrected by the RS decoder, symbols in the corresponding positions of the other codewords in the same frame can be flagged as erasures, thus making them easier to decode. For details of this technique, the reader is referred to the paper[9].

Another method for improving a concatenated-coding system makes use of iterative decoding. In one approach, the RS codes in a given frame are assigned different rates such that the average rate is unchanged. After an initial inner decoding of one frame, the most powerful (lowest rate) outer code is decoded, and then its decoded information bits (correct with a very high probability) are fed back and treated as known information bits (side information) by the inner decoder in a second iteration of the decoding process. This procedure can be repeated until the entire frame is decoded with each iteration using the lowest-rate outer code that has not yet been decoded. The use of known information bits by the inner decoder has been termed "state pinning", and the technique is discussed further in [10]

A more general approach to iterative decoding of concatenated codes is the Soft-Output Viterbi Algorithm(SOVA) proposed in [11]. In the SOVA reliability information about each decoded bit is appended to the output of the Viterbi decoder. An outer decoder which accepts soft inputs can then use this reliability information to improve its performance. If the outer decoder also provides reliability information at its output, iterative decoding can proceed between the inner and outer decoders. In general, such iterative decoding techniques for concatenated systems can result in additional coding gains of about 1.0 dB. In the CCSDS system, however, the outer RS decoder cannot make full use of such reliability information. Nevertheless, several combinations of error forecasting, state pinning, and iterative decoding have been applied to the CCSDS system and have resulted in an additional coding gain of about 0.5 dB[12][13].

A significant new discovery called "turbo codes" is discussed in the next section. Such codes, also known as parallel concatenated convolutional codes, were first introduced in [14]. Turbo codes combine a convolutional code along with a pseudorandom interleaver and a maximum a posteriori-probability(MAP) iterative decoding to achieve performances that are very close to the Shannon limit.

12.3 Turbo Codes

It is well known from Shannon's theory, discussed in Chapter 1, that a randomly chosen code of sufficiently long block length n is capable of approaching channel capacity[15]. However, the maximum-likelihood (ML) decoding of such a code increases exponentially in complexity with n to the point, where decoding becomes practically nonrealizable. In previous chapters and sections a number of encoding and decoding techniques are discussed that achieve significant gains in terms of power efficiency and bandwidth efficiency. Although these techniques have resulted in widespread applications of FEC codes, the quest for practical techniques that achieve near channel-capacity performance has met with only limited success, particularly in power-limited applications.

Historically, the coding gains of most practical systems were limited by the available techniques for implementation. Even at the time of the earliest applications of FEC coding better codes were available, but usually it often was not feasible to implement them. As technology advanced, so was the ability to implement complex coding schemes. The use of higher performance coding schemes reduced the gap between real system performance and the theoretical limits. However, the gains of these complex codes often were not as dramatic as expected, and it appeared that the law of diminishing returns had manifested itself.

One of the goals of coding research in recent decades has been to develop codes that had large equivalent block lengths, yet contain enough structure that practical decoding was possible. In 1993 a new encoding and decoding scheme, called *turbo codes*, was introduced [14]. Turbo codes achieved near-capacity performance on an additive white Gaussian-noise channel. Also, due to the use of a pseudorandom interleaver, turbo codes can be made to appear random to the channel, and yet to possess enough structure so that practical decoding is still possible.

In their original paper [14], C. Berrou, A. Glavieux, and P. Thitimasjshima show by simulation that turbo codes are capable of achieving a bit-error rate (BER) of 10^{-5} at an E_b/N_o ratio of just 0.7 db. It appears that the 40-year effort to reach channel capacity was achieved by this latest discovery. However, in order to achieve this level of performance, long block sizes of 65,532 data bits are required. The simulation results of this turbo code are presented in Figure 12.25. Because of the prohibitively long latency involved with such large block sizes, the original turbo codes are not applicable to most real-time two-way communication systems.

In this section, the fundamentals of this turbo coding and decoding scheme are discussed briefly. The interested reader is referred to the literature [14-

Concatenated Coding Systems and Turbo Codes

29] for more details and current results on the decoding and performance analyses of Turbo codes.

12.3.1 Parallel Concatenated Coding

Turbo coding is accomplished by the parallel concatenation of two or more systematic codes. A generalized turbo encoder is shown in Figure 12.27.

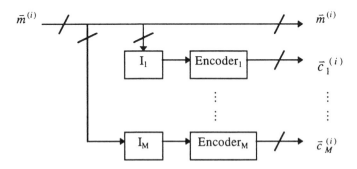

Figure 12.27. Generalized turbo encoder

At the i-th time unit, a k-bit message vector $\bar{m}^{(i)}$ is inputted in parallel into M sets of interleavers I_j and encoders, Encoder$_j$. Each interleaver scrambles the $\bar{m}^{(i)}$ in a pseudo-random fashion and feeds its output into a constituent rate-k/n systematic encoder. Each of the M constituent encoders generate a $(n-k)$-bit output vectors $\bar{c}_j^{(i)}$ at their output. The input vector $\bar{m}^{(i)}$ together with the M other output vectors $\bar{c}_j^{(i)}$ are concatenated to form the code word. The overall code rate of such an encoder is:

$$r = \frac{k}{(n-k)M + k}. \tag{12.1}$$

This type of encoder structure is called a parallel concatenation because all of the constituent encoders operate on the same *set* of input bits, rather than have any one encoder encode the output of the previous encoder.

For most applications, only two such encoders are used and the input to the first encoder is not interleaved. These constituent encoders themselves are identical, and are usually systematic recursive- convolutional encoders as it is discussed in the next section. Extra tail bits are added in order to ensure that the constituent convolutional encoders return to the all-zero state at the end of each block. This is discussed further in the next subsection. A block diagram of a Turbo encoder with two constituent convolutional encoders is shown in Figure 12.28. For the remainder of the section, only Turbo encoders with two constituent convolutional encoders are considered. For a discussion of the implications of using more than two constituent encoders, see reference [18] for details.

12.3.2 The Constituent Encoder

Although any systematic encoder can be used for the constituent encoders of the turbo encoder in Figure 12.27, the common practice is to use systematic recursive convolutional encoders [19]. The use of a convolutional encoder makes it possible for the decoder to utilize a modified version of the Viterbi algorithm.

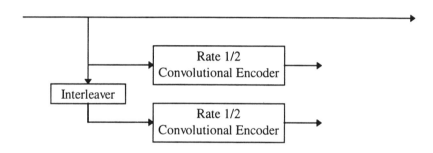

Figure 12.28. Rate 1/3 turbo encoder

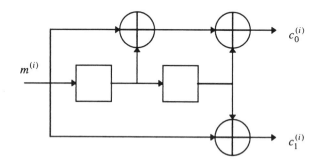

Figure 12.29 Example of a non-recursive nonsystematic convolutional encoder

A commonly discussed form of the convolutional encoder is the non-recursive nonsystematic convolutional encoder, shown in Figure 12.29. This type of encoder cannot be used as the constituent encoder for a turbo code

Concatenated Coding Systems and Turbo Codes

simply because it is not systematic. Use of a non-recursive systematic convolutional encoder is also unacceptable because of the poor distance properties of the resulting code [14]. As shown in Chapter 8, the encoder in Figure 12.29 can be modified in such a way that it becomes systematic and retains its distance properties. This modification yields a feedback systematic encoder of the form as shown in Figure 12.30. This encoder is recursive since the state of the internal shift register depends on past outputs. Such an encoder is called a Recursive Systematic Convolutional (RSC) encoder [14].

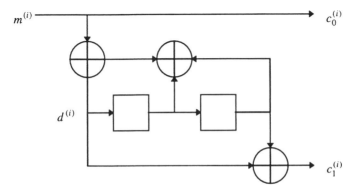

Figure 12.30. Example recursive systematic convolutional encoder.

A convolutional encoder can be used to generate a block code if the internal state of the encoder is known at the beginning and end of the codeword. The usual convention is to initialize the encoder to the all-zero state and then to encode the data. After the k data bits are encoded, a number of tail bits are added to the encoded word in order to force the encoder back to the all-zero state. The number of tail bits required to bring the encoder back to the all-zero state is on the order of the memory m of the convolutional encoder.

A tail of m zeros brings a nonrecursive convolutional encoder back to the all-zeros state. Due to the presence of feedback, a tail of zeros does not necessarily bring a recursive convolutional encoder back to the all-zeros state. The tail required to bring a recursive convolutional encoder to the all-zeros state is found by solving a state-variable equation at the feedback element. For the example in Figure 12.31, the state equation at the feedback element is given by:

$$d^{(i)} = m^{(i)} \oplus d^{(i-1)} \oplus d^{(i-2)} \quad (12.2)$$

Solving this equation for $m^{(i)}$ yields:

$$m^{(i)} = d^{(i)} \oplus d^{(i-1)} \oplus d^{(i-2)} \tag{12.3}$$

Since one wants to bring the encoder back to the all-zero state, $d^{(i)}$ is set to zero so that one obtains

$$m^{(i)} = d^{(i-1)} \oplus d^{(i-2)} \tag{12.4}$$

This is the required input to the encoder that is needed to force the encoder into the all-zero state.

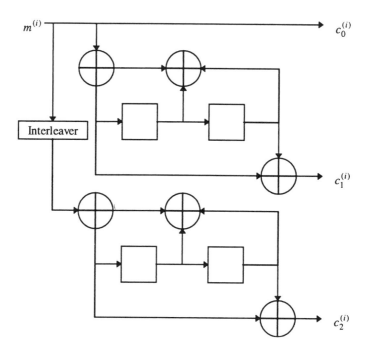

Figure 12.31: An example of rate 1/3 turbo encoder

12.3.3 Interleaver Design

A rate-1/3 turbo encoder, which uses the example RSC encoder in Figure 12.30, is shown in Figure 12.31. The pseudorandom interleaver in this circuit is used to permute the input bits in such a manner that the two encoders operate on the same set of input bits, but with different input sequences to the encoders. In order to generate a block code using a parallel concatenation of convolutional encoders it is desirable for both of the

Concatenated Coding Systems and Turbo Codes   503

constituent encoders to start and end in the all-zeros state. By properly terminating both trellises, joint maximum-likelihood estimation is made possible, at least in theory. Due to the presence of the pseudorandom interleaver it is seemingly difficult to simultaneously terminate the trellises of both of the convolutional encoders by the use of a single m-bit tail.

In [20] and [21] it is shown that by the proper design of the pseudorandom interleaver, it is possible to force both of the constituent encoders back to the all-zero state with a single m-bit tail. The condition on the interleaver is based on the fact that the impulse response of each RSC constituent encoder is periodic with some period given by

$$p \leq 2^m - 1. \tag{12.5}$$

l	$a(l)$	l	$a(l)$
1	16	18	30
2	29	19	13
8	9	20	2
4	10	21	21
5	14	22	25
6	6	28	26
7	31	24	3
8	8	25	28
9	12	26	20
10	22	27	27
11	17	28	7
12	33	29	5
13	34	80	15
14	23	31	4
15	24	32	11
16	19	88	18
17	32	84	1

Table 12.5: An example of the pseudorandom interleaver design

If the feedback polynomial is primitive of degree m, then the impulse response is a maximal length sequence, and equality holds in the above equation. The feedback polynomial for the example encoder is given by

$$g(x) = 1+x+x^2, \tag{12.6}$$

which is primitive. Thus, one concludes that the impulse response of this example constituent-RSC encoder is periodic with period $p = 2^2 - 1 = 3$.

In Chapter 9, It is described how an the interleaver is a function that maps an integer $0 < l < n+1$ onto another integer l_0 for $0 < l_0 < n+1$. A bit in position l at

the input to the interleaver is placed in position l_0 at the output of the interleaver. If this mapping function is called a, then the mapping can be expressed by
$$l_0 = a(l) \ . \tag{12.7}$$

A condition for the interleaver[21], that allows both of the RSC encoders to be terminated with the same tail, is the following:
$$l \bmod p = a(l) \bmod p \ . \tag{12.8}$$
for all integers l.

The condition in Eq.(12.8) states that taking l modulo p yields the same result as taking $a(l)$ modulo p. Thus any pseudo-random interleaver that meets this design condition can be used in a turbo encoder. In [22] it is shown that some interleavers perform better than others, and a procedure for eliminating poor interleaver designs is proposed. In essence, the good interleaver has the property of matching low-weight parity sequences with high-weight parity sequences in almost all cases, thus generating a code with very few low-weight ("bad") codewords.

The design condition is best illustrated with an example. If one uses the example turbo encoder of Figure 12.31 and encodes data words of length $k = 32$, then the interleaver must be capable of encoding blocks of length $n = k + m = 34$. An interleaver that meets the conditions of Eq. (12.8) is shown in Table 12.5.

For sufficiently large message-sequence block lengths a performance which is very close to the Shannon limit can be achieved at a moderate bit-error rate(BER), even though the free distance of the turbo codes is not large. In fact, for most interleavers, the free distance of the above code is only $d_{free} = 4$, even for very long block lengths. The excellent performance at moderate BER's is due rather to the quite drastic reduction in the number of nearest neighbor codewords compared to a conventional convolutional code. In a paper by Benedetto and Montorsi [19], it is shown that the number of nearest neighbors is reduced by a factor of N, where N is the blocklength. Sometimes this factor is called the "interleaver gain."

It is shown above in Sections 12.3.2 and 12.3.3 that turbo codes are block codes, and that the encoder is a parallel concatenation of systematic-recursive convolutional encoders. The solution to the problem of jointly terminating the trellises of two constituent encoders is presented in terms of their interleaver design.

12.3.4 Puncturing the Output

In Figure 12.31, the turbo encoder output consists of the information sequence and two parity sequences, thus it has a code rate of 1/3. For many applications, it is common practice to puncture the output of the encoder in order to increase the code rate to 1/2. Alternately puncturing (deleting) bits from the two parity sequences $\bar{c}_1$ and $\bar{c}_2$ produces a code rate of 1/2, and other code rates can be achieved by using additional parity generators and/or different puncturing patterns.

For a rate-1/2 punctured turbo code, the first output stream is the input stream itself (plus the necessary padding), while the second output stream is generated by multiplexing the M nonsystematic outputs of the RSC encoders. An example by a rate-1/2 punctured turbo encoder is shown in Figure 12.32. In this example, the multiplexer chooses its odd indexed outputs from the output of the upper RSC encoder and its even indexed outputs from the output of the lower RSC encoder.

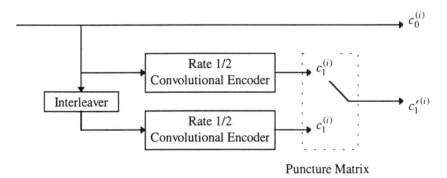

Puncture Matrix

Figure 12.32. An example of rate 1/2 punctured turbo encoder

Sometimes it is not necessary to take half of the parity stream's bits from the first encoder and half from the second. If n_1 bits per block are taken from the output of the first encoder and n_2 bits per block are taken from the output of the second encoder, then the punctured code rate of the first encoder is [14]:

$$r_1 = \frac{k}{2n_1 + n_2}, \tag{12.9}$$

and the punctured code rate of the second encoder is:

$$r_2 = \frac{k}{n_1 + 2n_2}. \tag{12.10}$$

If $n_1 \neq n_2$ then the overall turbo code can be viewed as the parallel concatenation of two codes of unequal code rate. It is suggested in [14] that using unequal code rates can improve the performance of the decoder. Unequal code rates are used in the turbo-encoding circuit proposed in reference [23]. For this proposed circuit, the multiplexer chooses three bits from the output of the first encoder for every one bit chosen from the output of the second encoder.

When puncturing is used, the joint effect of the interleaver and multiplexer on the code's distance properties must be taken into account. It is suggested in [24] that when puncturing is used, the interleaver should be designed in such a manner that even indexed bits are mapped to the even indexed bits and the odd indexed bits are mapped to odd indexed bits.

Note that puncturing can be used to set the code rate to any arbitrary value, not just 1/2. This feature has interesting applications in unequal error protection, as discussed in [25]. when the channel is relatively clean or the information bits relatively unimportant, the puncturing process is used to achieve a rate of 1/2. If the channel becomes more noisy, or if the information bits require extra protection, then the encoder could stop puncturing the output, thereby decreasing the bit-error rate at the cost of lowering the code rate to 1/3.

Next consider the generator matrix of a turbo code. Because turbo codes are linear block codes, the encoding operation can be considered to be a modulo-2 vector-matrix multiplication. A code vector $\bar{c}$ can be generated by multiplying an information vector m times a generator matrix G as follows:
$$\bar{c} = \bar{m} \cdot G \bmod 2 \tag{12.11}$$
Each row in the generator matrix is the code word associated with a particular information sequence of weight one. More specifically, row i of G is the code word associated with an input m that is identically zero in all positions except position i, where it equals one.

Because the code is systematic, the columns of the generator matrix can be rearranged into the form:
$$G = [P \; I_k], \tag{12.12}$$
where I_k is the $k \times k$ identity matrix, P is a $k \times (N-k)$ matrix, and N is the length of each of the codewords. The parity-check matrix, associated with this generator matrix, is:
$$H = [I_{N-k} \; P^T], \tag{12.13}$$
where I_{N-k} is the $(N-k) \times (N-k)$ identity matrix and P^T is the transpose of P.

Concatenated Coding Systems and Turbo Codes

Once the parity check matrix is found, the minimum distance of the code can be computed. As in the case of any linear block code, the minimum distance is the smallest number of columns in H that sum to zero [3]. The minimum distance is useful in determining the performance of the turbo code.

12.3.5 Decoding of Turbo Codes

The other important feature of turbo codes is the iterative decoder, which uses a soft-in/soft-out maximum aposteriori probability (MAP) decoding algorithm first applied to convolutional codes by Bahl, Cocke, Jelinek, and Raviv [28]. This algorithm is more complex than the Viterbi algorithm by about a factor of three, and for conventional convolutional codes it offers little performance advantage over Viterbi decoding. However, in turbo decoding, the fact that it gives the maximum MAP estimate of each individual information bit, is crucial in making it possible for the iterative decoding procedure to converge at very low SNR's.

In the original paper on turbo codes [14], Berrou et. al. proposed an iterative decoding scheme based on an algorithm presented by Bahl et. al. in 1974 [28]. The Bahl decoding algorithm differs from the Viterbi algorithm in the sense that it produces soft outputs. While the Viterbi algorithm outputs either a 0 or 1 for each estimated bit, the Bahl algorithm outputs a continuous value that weights the confidence or *log*-likelihood of each bit estimate. While the goal of the Viterbi decoder is to minimize the codeword error by finding a maximum likelihood (ML) estimate of the transmitted code word, the Bahl algorithm attempts to minimize the bi-error rate by estimating the a posteriori probabilities (APP) of the individual bits of the codeword.

The turbo decoder consists of M elementary decoders one for each encoder in Figure 12.27. Each elementary decoder uses the Bahl algorithm to produce a soft decision for each received bit. After each iteration of the decoding process, every elementary decoder shares its soft-decision output with the other $M-1$ elementary decoders.

For the encoder given in Figure 12.32 there are two elementary decoders that are interconnected as shown in Figures 12.33 and 12.34. In these latter figures it is assumed that the information stream $c_0^{(i)}$ in Figure 12.32 is received as $r_0^{(i)}$, while the parity stream $c_1'^{(i)}$ is received as $r_1'^{(i)}$. The received information stream r_0 is passed to both decoders DEC_1 and DEC_2, and the parity stream $r_1'^{(i)}$ is demultiplexed into the received parity-check bits $r_1^{(i)}$ and $r_2^{(i)}$, corresponding to $c_1^{(i)}$ and $c_2^{(i)}$, respectively. The demultiplexer sends the parity bits $r_1^{(i)}$ generated by the first encoder to DEC_1, and the parity bits $r_2^{(i)}$ generated by the second encoder to DEC_2. The first decoder produces a soft decision which is based on its received parity bits $r_1^{(i)}$ along with the received information stream $r_0^{(i)}$. This soft decision is interleaved by a_2 and passed as the systematic input to DEC_2. The second decoder uses its received parity bits $r_2^{(i)}$ along with the interleaved soft decision from the first decoder to produce a soft decision of its own.

The soft decision at the output of the second decoder does not constitute a maximum a posteriori (MAP) estimate of the codeword since the two decoders works to this point more or less independently. In order to find an estimate that approaches the MAP estimate, the information produced by the two elementary encoders must be shared with one another In an iterative fashion. This is accomplished in Figure 12.33. Here the soft-decision of DEC_2 is deinterleaved by I_2^{-1}; and re-introduced as the input to DEC_1. DEC_1 once again produces a soft estimate that is interleaved by I_2 and fed back into DEC_2. As the number of these iterations approaches infinity, the estimates $\hat{r}_1^{(i)}$ and $\hat{r}_2^{(i)}$ at the output of DEC_1 and DEC_2, respectively, approach the MAP solution. In practice, the number of iterations needs to only be about 18, and in many cases as few as 6 iterations can provide satisfactory performance [26].

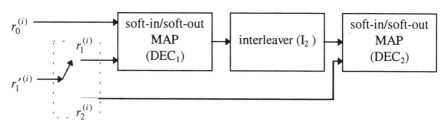

Figure 12.33. Initialization stage of the turbo decoder

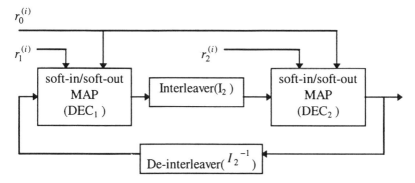

Figure 12.34. Iteration stages of turbo decoder

Up to this point it has been assumed that the elementary decoders in Figures 12.33 and 12.34 used the modified Bahl algorithm of [14]. However, the Bahl decoder suffers from a complexity that is significantly higher than that of the Viterbi algorithm. Currently, effort is under way to find reduced-complexity elementary decoders that can produce soft outputs that use a modification of the Viterbi algorithm [29], [30], [31], [32]. The algorithm that seems most promising is the Soft-Output Viterbi Algorithm, or SOVA algorithm[31].

Several bounds [26] on the probability of codeword error can be used in the analysis of short block-length turbo codes, but these probability bounds quickly become intractable as the block size increases. Although the probability of codeword error provides some insight into the performance of the code, this probability and its bounds do not consider the fact that the turbo decoder attempts to minimize the probability of a bit error rather than the probability of a codeword error. An analytic method that accurately predicts the performance of large block-length turbo codes in terms of *bit-error* rate is still a topic for future research.

In practice, at almost any bandwidth efficiency, the performance of turbo codes is less than 1.0 dB away from capacity. This is achievable with short constraint-length turbo codes, very long block lengths, and with about 10-20 iterations of decoding. It is shown in [14] that the performance of a Turbo code at a BER of 10^{-5} for rate $r = 1/2$ and information block length $N = 2^{16} = 65536$, with 10 iterations of decoding. This is a about 3.8 dB better than the CCDSS (2, 1, 7) code (see Sec. 12.2.1), with roughly the same decoding complexity. The major disadvantages of a turbo code are its long decoding delays, due to the large block lengths and iterative decoding, and its weaker performance at lower BER's, due to its low free distance. However, the long delays are not a major problem in many applications such as digital television and deep-space missions.

Several turbo-coding systems are already developed [39]. One such system is JPL's turbo decoder designed by David Watola. Basic technical specifications of this decoder are :
- Bit rate : < 64 Kbits,
- Number of States : < 64,
- Two component codes,
- Code rate : $\frac{1}{10} < R < 1$.
- MAP bitwise-decoding algorithm with 8-bit (soft) input symbols and 16-bit metrics.

The performance of this turbo codec is shown in Figure 12.35. In this algorithm an interleaver of length 16384 bits is used. Series 1 represents the bit-error rate of a rate 1/3 turbo code with 11 iterations and 32 states. Series 2 represents the bit-error rate of a rate 1/4 turbo code with 13 iterations and 32 states, and Series 3 represents the bit-error rate of a rate 1/2 turbo code with 18 iterations and 34 states.

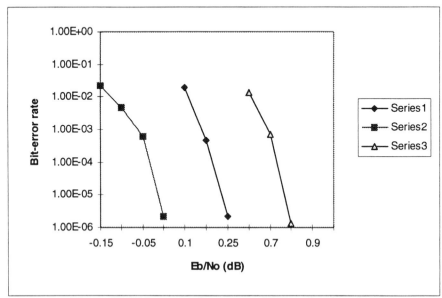

Figure 12.35 Some performance of JPL turbo decoder.

Concatenated Coding Systems and Turbo Codes 511

The reported turbo codes[39] outperform the concatenated codes used by the Deep Space Network(DSN) since Voyager at Neptune. All of the DSN concatenated codes use a powerful Reed-Solomon outer code in concert with a convolutional inner code, whereas the turbo codes are assumed to operate without concatenation. For example, turbo codes with rates 1/2, 1/4, and 1/6 can be constituted from two 10-state constituent codes. At bit error rate of 10^{-5}, these turbo codes are able to achieve an improvement of the bit signal-to-noise ratio(SNR). The SNR improvements achieved by turbo codes are approximately 1.5 dB, 1.3dB, and 1.0 dB, with respect to the Voyager, Galileo, and Cassini codes. In addition to providing improved performance, turbo decoders have a lower complexity than the Galileo and Cassini concatenated decoders.

12.3.6 Some Applications of Turbo Coding in Mobile Communications

Basic encoding and decoding structures of turbo codes are illustrated by the examples in the previous subsections. It is expected that the performance of a turbo code improves with the block length and/or the constraint length increases. However, the complexity of a decoder sets an upper limit on the constraint length, while the maximum allowable latency in the system sets a limit on the maximum block length. It is anticipated that a rate-1/2 turbo code with code length of 384 bits and constraint length 5 could offer better performance with lower complexity than a fixed length convolutional code of constraint length 9 with the same code length and rate. In order for this to be true, the problem of finding a practical reduced-complexity decoder must first be solved. If turbo codes are proven to perform well with code lengths of 384 bits, they could have a substantial impact on current digital cellular standards such as IS-95 (see Chapter 10). Even if turbo codes prove to perform poorly in digital cellular systems, it is quite likely that they should begin to find application in systems that allow larger block sizes, such as high definition television [33]. Another promising application of turbo codes is in high spectral efficiency modulation in the form of turbo trellis-coded modulation for telephone or cable MODEM and Digital TV[34].

In 1993, decoding of two or more product-like codes was proposed using iterative (Turbo) decoding. The basic concept of this new coding scheme is to use a parallel concatenation of at least two codes with an interleaver between the encoders. Decoding is based on alternately decoding the component codes and then passing the so-called extrinsic information to the next decoding stage. For good soft-in/soft-out decoders this extrinsic information is an additional part of the soft output. Even though very simple component codes are used, the "turbo" coding scheme is able to achieve performance close to Shannon's bound, at least for large interleavers and at BER's of approximately 10^{-5}.

Also, it turns out that the turbo method, originally applied to parallel concatenated codes, is more general [36] and can be successfully applied to many detection/decoding problems such as serial concatenation, equalization, coded modulation, multiuser detection, joint source and channel decoding, etc. We mention here two applications for mobile communications to show the potential of turbo decoding.

Parallel concatenated (Turbo) Codes for GSM Applications. The iterative turbo decoding was not available at the time when GSM was standardized. But it is interesting to consider what could be gained in the GSM System with turbo decoding. In [37], the FEC system for GSM is redesigned with parallel concatenated codes, while keeping the framing and the interleaver as in the full rate GSM System. This meant that a frame length of only 185 information bits could be used. This is quite different from the common usage of turbo codes, where several thousand bits are usually decoded in one frame. Nevertheless. it was shown that on GSM urban channels with frequency hopping that an improvement in the channel signal-to-noise ratio (i.e., in additional coding gain) of 0.8 to 1.0dB is possible. Although several iterations of the turbo decoder are necessary to achieve this gain, the total decoding complexity is comparable to the GSM system because the turbo component codes had only four states, compared to the 16-state code used in standard GSM.

Turbo Detection in the IS-95 Mobile System: In this section, we explain the principle of turbo detection for the IS-95 uplink when coherent demodulation is applied. An application to the noncoherent uplink communication can be found in [38]. For iterative decoding, it is necessary to replace the outer Viterbi decoder with a modified soft-output Viterbi algorithm (SOVA) or a MAP decoder which delivers soft-output information with each decoded bit The soft information from the outer decoding stage is interleaved and then fed back as apriori information in a re-decoding of the inner code. This leads to a serial turbo decoding scheme. A schematic overview of the receiver is given in Fig. 12.34.

To obtain the a priori information the outer code is decoded with a SOVA or MAP decoder. The extrinsic part of the soft output for the outer decoded bits is then used as the a *priori* information for the systematic bits of the inner code *in* the feedback loop, shown in Figure. 12.34. The reader is referred to [38] for the more detailed information about this decoding scheme.

Simulation results for the AWGN channel show that an additional coding gain of about 1.0-1.5 dB can be achieved with iterative decoding after five iterations.

Bibliography

[1] G. D. Forney, Jr., Concatenated Codes, Cambridge, MA : MIT Press, 1966.
[2] J. L. Massey, "Deep space communications and coding : a match made in heaven," in Advanced Methods for Satellite and Deep Space Communication, Lecture Notes in Control and Information Sciences, Vol. 182, Berlin : Springer-Verlag, 1992.
[3] S. Lin and D. Costello, Error Control Coding: *Fundamentals and Applications,* Englewood Cliffs, NJ: Prentice Hail, Inc., 1983.
[4] J. Hagenauer, E. Offer, and L. Papke, "Matching Viterbi decoders and Reed-Solomon decoders in a concatenated system," in Reed Solomon Codes and Their Applications, New York : IEEE Press, 1994.
[5] Grand Alliance HDTV System Specification, April 14, 1994.
[6] ITU-T Telecommunication Standardization Sector of ITU, "Digital multi-programmer systems for television sound and data services for cable distribution"---Television and sound transmission, ITU-T Recommendation J.83, Oct. 1995.
[7] Prodan, R. et al., "Analysis of Cable System Digital Transmission Characteristics," NCTA Technical Papers, 1994.
[8] ITU-T Recommendation H.222.0 (1995) | ISO/IEC 13818-1 : 1996, Information technology--generic coding of moving pictures and associated audio information systems.
[9] E. Paaske, "Improved decoding for a concatenated coding system recommended by CCSDS," IEEE Trans. Communication, Vol. 38, pp.1138-1144, Aug. 1990.
[10] O. M. Collins and M. Hizlan, "Determinate state convolutional codes", IEEE Trans. Communications, vol. 41, pp.1785-1794, Dec. 1993.
[11] J. Hagenauer and P. Hoeher, "A Viterbi algorithm with soft-decision outputs and its applications," in Proc. 1989 IEEE Global Communication Conference, pp.47.1.1-47.1.7, Dallas, TX, Nov. 1989.

[12] R. J. McEliece and L. Swanson, "Reed-Solomon codes and exploration of the solar system," in Reed-Solomon Codes and Their Applications, IEEE Press. pp.25-40, 1994.

[13] J. Hagenauer, E. Offer, and L. Papke, "Matching Viterbi decoders and Reed-Solomon decoders in concatenated systems," in Reed-Solomon Codes and Their Applications, IEEE Press. pp.242-271, 1994.

[14] C. Eerrou, A. Glaneux, and P. Thitimasjshima, "Near Shannon limit error-correcting coding and decoding: Turbo-codes(1)," in Proc., IEEE Int. Conf. on Communications (Geneva, Switzerland, May 1993), pp.1064-1070.

[15] T. Cover and J. Thomas, Elements of information Theory, New York:Wiley Interscience, 1991.

[16] P. Jung and M. Nahan, "Performance evaluation of turbo codes for short frame transmission Systems," Electronics Letters, vol.30, pp. 111-113, Jan.1994.

[17] P. Sung, "Comparison of turbo-code decoders applied to short frame transmission systems," IEEE J. Select. Areas Comm., vol-14, pp.530-537, Apr.1996.

[18] D. Divealar and F. Pollara, "Multiple turbo codes," in Proc., IEEE MILCOM (San Diego, CA, 1995), pp.279-285.

[19] S. Benedetto and U. Montorsi, "Unveiling turbo codes: Some results on parallel concatenated coding schemes," IEEE Trans. Inform. Theory, vol.42, pp. 409-428, Mar.1996.

[20] A. Barbulescu and S. Pietrobon, "Terminating the trellis of turbo-codes in the same state," Electronics Letters, vol.31, pp.22-23, Jan.1995.

[21] W. Blackert, E. Hall, and S. Wilson, "Turbo code termination and interleaver conditions," Electronics Letters, vol. 31, pp.2082-2083, Nov.1995.

[22] B. Robertson, "Improving decoder and code structure of parallel concatenated recursive systematic (turbo) codes," in Proc., IEEE Int. Conf on Universal Personal Communications (1994), pp.183-187.

[23] C. Berron, P. Combelles, P. Penard, B. Talibart. "An IC for turbo-codes encoding and decoding," in Proc., IEEE Int. Solid-State Circuits Conf. (1995), pp.90-91.

[24] A. Barbulescu and S. Pietrobon, "Interleaver design for turbo codes," Electronics Letters, vol.30, pp.2107-2108, Dec.1994.

[25] A. Barbulescu and S. Pietrobon, "Rate compatible turbo codes," Electronics Letters, vol.31, pp.535-536, Mar.1995.

[26] S. Proalds, Digital Communications, third edition. New York: McGraw-Hill, Inc., 1995.

[27] S. Wicker, Error Control Systems for Digital Communication and Storage. Englewood Cliffs, NJ: Prentice Hall, Inc., 1995.

[28] L. Bahl, S. Cocke, F. Jeinek, and J. Raviv, "Optimal decoding of linear codes for minimizing symbol error rate," IEEE Trans. Inform. Theory, vol.20, pp.248-287, Mar.1974.
[29] C. Berrou, P. Adde, E. Angui, and S. Faudeil, "A low complexity soft-output Viterbi decoder architecture," in Proc., IEEE Int. Conf. on Communications (Geneva, Switzerland, May 1993), pp. 737-740.
[30] P. Jung, "Novel low complexity decoder for turbo-codes," Electronics Letters, vol.31, pp.86-87, Jan.1995.
[31] J. Hagenauer, "Iterative decoding of binary block and convolutional codes," IEEE Trans. Inform. Theory, vol.42, pp.429-445,Mar.1996.
[32] J. Hagenauer and L. Papke, "Decoding turbo codes with the soft-output Vit₉rbi algorithm (SOYA)," in Proc., Int. Symp. on Inform. Theory (Trondheim, Norway, June 1994), p.164.
[33] C. Berron and P. Combelles, "Digital television: Hierarchical channel coding using turbo-codes," in Proc., IEEE Int. Conf. on Communications (New Orleans, LA, May 1994), pp.737-740.
[34] S. Goff, A. Glavietix, and C. Berrou, "Turbo-codes and high spectral efficiency modulation," in Proc., IEEE Int. Conf. on Communications (New Orleans, LA, May 1994), pp.645-649.
[35] O. M. Collins, "The subtleties and intricacies of building a constraint length 15 convolutional decoder," IEEE Trans. Communication, Vol. 41, pp.1785-1794, Dec. 1993.
[36] J. Hagenauer, "The turbo principle : Tutorial introduction and state of the art," in Proc. Int. Symp. Turbo Codes, pp.1-11, Brest, France, Sept. 1997.
[37] F. Burkert, G. Caire, J. Hagenauer, and G. Lechner, "Turbo decoding with unequal error protection applied to GSM speech coding," in Proc. 1996 IEEE Global Communications conf. , pp.2044-2048, London, U.K., Nov. 1996.
[38] R. Herzog, A. Schmidbauer, and J. Hagenauer, "Iterative decoding and despreading improves CDMA-systems using M-ary orthogonal modulation and FEC", in Proc. 1997 IEEE Int. Conf. Communication, Montreal, Canada, June 1997.
[39] D. Watola, "JPL Turbo decoder", http: // www331.jpl.nasa.gov /public /TurboDecoder.html
[40] IEEE Project 802.14/a, "Cable-TV access method and physical layer specification", 1997.

A Some Basics of Communication Theory

The process of sending (information) messages from a transmitter to a receiver is essentially a random experiment. The transmitter selects one message and sends it to the receiver. The receiver has no knowledge about which message is chosen by the transmitter, for if it did, there would be no need for the transmission. The transmitted message is chosen from a set of messages known to the transmitter and the receiver. If there were no noise, the receiver could identify the message by searching through the entire set of messages. However, the transmission medium, called the channel, usually adds noise to the message. This noise is characterized as a random process. In fact thermal noise is generated by the random motion of molecules or particles in the receiver's signal-sensing devices. Most noise has the property that it adds linearly to the received signal.

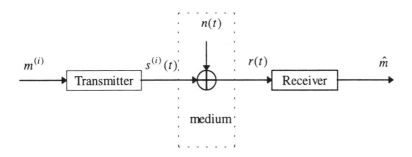

Figure A.1 A simplified block diagram of a communication process over a channel with additive noise.

Figure A.1 shows a simplified system block diagram of a transmitter/receiver communication system. The transmitter performs the random experiment of selecting one of the M messages in the message set $\{m^{(i)}\}$, say $m^{(i)}$, and then sends it corresponding waveform $s^{(i)}(t)$, chosen from a set of signals $\{s^{(i)}(t)\}$. A large body of literature is available on how to model channel impairments of the transmitted signal. This book concentrates primarily on the almost ubiquitous case of additive white-Gaussian noise (AWGN).

A.1 Vector Communication Channels

One commonly used method to generate signals at the transmitter is to synthesize them as a linear combination of N basis waveforms $\phi_j(t)$. That is, the transmitter selects

$$s^{(i)}(t) = \sum_{j=1}^{N} s_j^{(i)} \phi_j(t) \tag{A.1}$$

as the transmitted signal for the i-th message. Often the basis waveforms are chosen to be orthonormal; that is, they fulfill the condition,

$$\int_{-\infty}^{\infty} \phi_u(t)\phi_v(t)dt = \delta_{jl} = \begin{cases} 1, & u = v, \\ 0, & otherwise. \end{cases} \tag{A.2}$$

This leads to a vector interpretation of the transmitted signals, since, once the basis waveforms are specified, $s^{(i)}(t)$ is completely determined by the N-dimensional vector,

$$\bar{s}^{(i)} = \left(s_1^{(i)}, s_2^{(i)}, \ldots, s_N^{(i)}\right). \tag{A.3}$$

These signals can be visualized geometrically as the signal vectors $\bar{s}^{(i)}$ in the Euclidean N-space, spanned by the usual orthonormal basis vectors, where each basis vector is associated with a basis function. This geometric representation of a signal is called a signal constellation. The idea is illustrated for N = 2 in Figure A.2 for the signals $s^{(1)}(t) = \sin(2\pi f_0 t)u(t)$, $s^{(2)}(t) = -\cos(2\pi f_0 t)u(t)$, $s^{(3)}(t) = -\sin(2\pi f_0 t)u(t)$, and $s^{(2)}(t) = -\cos(2\pi f_0 t)u(t)$ where

Some Basics of Communication Theory

$$u(t) = \begin{cases} \sqrt{\dfrac{2E_s}{T}}, & 0 \leq t \leq T, \\ 0, & \text{otherwise}, \end{cases} \quad \text{(A.4)}$$

and $f_0 = \dfrac{k}{T}$ is an integer multiple of $\dfrac{1}{T}$. The first basis function is $\phi_1(t) = \sqrt{\dfrac{2}{T}}\sin(2\pi f_0 t)$, and the other basis function is $\phi_2(t) = \sqrt{\dfrac{2}{T}}\cos(2\pi f_0 t)$. The signal; constellation in Figure A.2 is called quadrature phase-shift keying (QPSK).

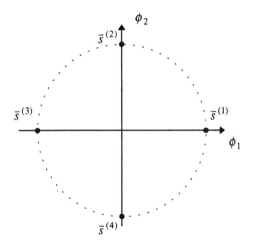

Figure A.2 Illustration of QPSK

There is a one-to-one mapping at the signal vector $\bar{s}^{(i)}$ onto the transmitted message $\bar{m}^{(i)}$. The problem of decoding a received waveform is therefore equivalent to the ability to recover the signal vector $\bar{s}^{(i)}$. This can be accomplished by passing the received signal waveform through a bank of correlators where each correlator correlates $s^{(i)}(t)$ with one of the basis functions to perform the operation,

$$\int_{-\infty}^{\infty} s^{(i)}(t)\phi_j(t)dt = \sum_{l=1}^{N} s_l^{(i)} \int_{-\infty}^{\infty} \phi_l(t)\phi_j(t)dt = s_j^{(i)}. \quad \text{(A.5)}$$

That is, the j-th correlator recovers the j-th component $s_j^{(i)}$ of the signal vector $\bar{s}^{(i)}$.

Next, define the squared Euclidean distance between two signals $\bar{s}^{(i)}$ and $\bar{s}^{(j)}$, given by $d_E^2(\bar{s}^{(i)}, \bar{s}^{(j)}) = |\bar{s}^{(i)} - \bar{s}^{(j)}|^2$, which is a measure of, what is called, the noise resistance of these two signals. Furthermore, the expression,

$$d_E^2(\bar{s}^{(i)}, \bar{s}^{(j)}) = \sum_{l=0}^{N}\left(s_l^{(i)} - s_l^{(j)}\right)^2 = \int_{-\infty}^{\infty}\left(s^{(i)}(t) - s^{(j)}(t)\right)^2 dt \qquad (A.6)$$

is, in fact, the energy of the difference signal $s^{(i)}(t) - s^{(j)}(t)$.

It can be shown that the correlation-receiver is optimal, in the sense that no relevant information is discarded and that the minimum error probability is attained, even when the received signal contains additive white Gaussian noise. In this latter case the received signal $r(t) = s^{(i)}(t) + n(t)$ produces the received vector $\bar{r} = \bar{s}^{(i)} + \bar{n}$ at the output of the bank of correlators. The statistics of the noise vector $\bar{n}$ are easily evaluated, using the orthogonality of the basis waveforms and the noise correlation function $E[n(t)n(t+\tau)] = \frac{N_0}{2}\delta(t)$ for white-Gaussian noise, where $\delta(t)$ is Dirac's delta function and N_0 is the one-sided noise power spectral density. Thus the correlation of any two noise components n_u and n_v is given by

$$E[n_u n_v] = \int_{-\infty}^{\infty}\int_{-\infty}^{\infty} E[n(\alpha)n(\beta)]\phi_u(\alpha)\phi_v(\beta)d\alpha d\beta = \begin{cases} \frac{N_0}{2}, & u = v, \\ 0, & otherwise. \end{cases} \qquad (A.7)$$

It is seen, that by the use of the orthonormal basis waveforms, the components of the random noise vector n are all uncorrelated. Since $n(t)$ is a Gaussian random process, the sample or component values n_u, n_v are necessarily also Gaussian. From the foregoing one concludes that the components of n are independent, Gaussian random variables with a common variance $N_0/2$ and zero mean value.

The advantages of the above vector point of view are manifold. First, one doesn't need to be concerned with the actual choices of the signal waveforms when discussing receiver algorithms. Secondly, the difficult problem of waveform communication, involving stochastic processes and continuous signal functions, has been transformed into the much more manageable vector communications system which involves only signal and random vectors.

A.2 Optimal Receivers

If the bank of correlators produces a received vector $\vec{r} = \vec{s}^{(i)} + \vec{n}$, then an optimal detector chooses that message hypothesis $\hat{m} = m^{(j)}$ which maximizes the conditional probability $P[\bar{m}^{(j)}|\vec{r}]$. This is known as a maximum aposteriori (MAP) receiver.

Evidently, a use of Bayes' rule yields

$$P[\bar{m}^{(j)}|r] = \frac{P[\bar{m}]p(\vec{r}|\bar{m}^{(j)})}{p(\vec{r})}. \quad (A.8)$$

Thus, if all of the signals are used equally often, the maximization of $P[\bar{m}^{(j)}|\vec{r}]$ is equivalent to the maximization of $p(\vec{r}|\bar{m}^{(j)})$. This is the maximum-likelihood (ML) receiver. It minimizes the signal-error probability only for equally-likely signals.

Since $\vec{r} = \vec{s}^{(i)} + \vec{n}$ and $\vec{n}$ is an additive Gaussian random vector which is independent of the signal $\vec{s}^{(i)}$, the optimal receiver is derived by the use of the conditional probability density $p(\vec{r}|\bar{m}^{(j)}) = p_n(\vec{r} - \vec{s}^{(j)})$. Specifically, this is the N-dimensional Gaussian density function given by

$$p_n(\vec{r} - \vec{s}^{(j)}) = \frac{1}{(\pi N_0)^{N/2}} \exp\left(-\frac{|\vec{r} - \vec{s}^{(j)}|^2}{N_0}\right). \quad (A.9)$$

The maximization of $p(\vec{r}|\bar{m}^{(j)})$ is seen to be equivalent to the minimization of the squared-Euclidean distance,

$$d_E^2(\vec{r}, \vec{s}^{(j)}) = |\vec{r} - \vec{s}^{(j)}|^2, \quad (A.10)$$

between the received vector and the hypothesized signal vector.

The decision rule in (A.10) implies that the decision regions $D^{(j)}$ for each signal point consists of all the points in Euclidean N-dimensional space that are closer to $\vec{s}^{(j)}$ than any other signal point. Such decision regions for QPSK are illustrated in Figure A.3.

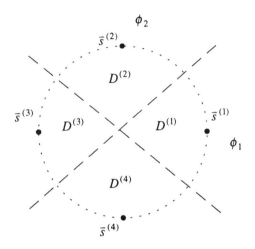

Figure A.3 Decision regions for ML detection of QPSK signals.

Hence the probability of error, given a particular transmitted signal $\bar{s}^{(i)}$, can be interpreted as the probability that the additive noise $\bar{n}$ carries the signal $\bar{s}^{(i)}$ outside its decision region $D^{(i)}$. This probability is calculated by

$$P\left(error|\bar{s}^{(i)}\right) = \int_{\bar{r} \in D^{(i)}} p_n(\bar{r} - \bar{s}^{(i)}) d\bar{r}. \tag{A.11}$$

Equation (A.11) is, in general, quite difficult to calculate in closed form, and simple expressions exist only for certain special cases. The most important such special case is the two-signal error probability. This is the probability that signal $\bar{s}^{(i)}$ is decoded as signal $\bar{s}^{(j)}$ on the assumption that there are only these two signals. To calculate the two-signal error probability all signals are disregarded except $\bar{s}^{(i)}$, $\bar{s}^{(j)}$, e.g. $\bar{s}^{(i)} = s^{(1)}$ and $\bar{s}^{(j)} = s^{(3)}$ in Figure A.3. The new decision regions are $D^{(i)} = D^{(1)} \cup D^{(4)}$ and $D^{(j)} = D^{(2)} \cup D^{(3)}$. The decision region $D^{(i)}$ is expanded to the half-plane, and the probability of deciding on message $\bar{m}^{(j)}$ when message $\bar{m}^{(i)}$ was actually transmitted is

$$P_{\bar{s}^{(i)} \to \bar{s}^{(j)}} = Q\left(\sqrt{\frac{d_E^2\left(\bar{s}^{(i)}, \bar{s}^{(j)}\right)}{2N_0}}\right) \tag{A.12}$$

Some Basics of Communication Theory

where $d_E^2(\bar{s}^{(i)}, \bar{s}^{(j)}) = |\bar{s}^{(i)} - \bar{s}^{(j)}|^2$ is the energy of the difference signal, and

$$Q(u) = \frac{1}{\sqrt{2\pi}} \int_u^\infty \exp\left(-\frac{v^2}{2}\right) dv \qquad (A.13)$$

is a nonelementary integral, called the (Gaussian) Q-function. The probability in (A.12) is known as the *pairwise error probability*.

The correlation operation in (A.5), used to recover the signal-vector components, can be implemented as a filtering operation. The signal $s^{(i)}(t)$ is passed through a filter with time impulse response $\phi_u(-t)$ to obtain

$$x(t) = s^{(i)}(t) * \phi_u(-t) = \int_{-\infty}^{\infty} s^{(i)}(y) \phi_u(y-t) dy. \qquad (A.14)$$

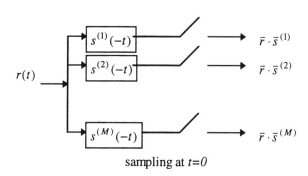

Figure A.4 Matched filter receiver, where M is the number of messages.

If the output of the filter $\phi_u(-t)$ is sampled at time $t = 0$, equations (A.14) and (A.5) are identical; i.e.,

$$x(t=0) = \int_{-\infty}^{\infty} s^{(i)}(y) \phi_u(y) dy \qquad (A.15)$$

Of course, some appropriate delay actually needs to be built into the system in order to guarantee that $\phi_u(-t)$ is a causal filter. Such delays are not considered further in this book.

The maximum-likelihood receiver minimizes $d_E^2(\bar{r}, \bar{s}^{(i)}) = |\bar{r} - \bar{s}^{(i)}|^2$ or, equivalently, it maximizes

$$2 \cdot \bar{r} \cdot \bar{s}^{(i)} - \left|\bar{s}^{(i)}\right|^2, \qquad (A.16)$$

where the term $|\bar{r}|^2$, which is common to all the hypotheses, is neglected. The correlation $\bar{r} \cdot \bar{s}^{(i)}$ is the central part of (A.16) and can be implemented as a basis-function matched-filter receiver, where the summation is performed after the correlation. That is,

$$\bar{r} \cdot \bar{s}^{(i)} = \sum_{j=1}^{N} r_j s_j^{(i)} = \sum_{j=1}^{N} s_j^{(i)} \int_{-\infty}^{\infty} r(t) \phi_j(t) dt \qquad (A.17)$$

Such a receiver is illustrated in Figure A.4. Usually the number of basis functions is much smaller than the number of signals so that the basis-function matched-filter implementation is the preferred realization.

A.3 Message Sequences

In practice, information signals $s^{(i)}(t)$ will most often consist of a sequence of identical, time-displaced waveforms, called pulses, described by

$$s^{(i)}(t) = \sum_{j=-u}^{u} a_j p(t - jT), \qquad (A.18)$$

where $p(t)$ is some pulse waveform, the a_j are the discrete symbol values from some finite signal alphabet (e.g., the binary signaling: $a_j \in \{-1,1\}$), and $2u + 1$ is the length of the sequence of symbols. The parameter T is the timing delay between successive pulses, also called the *symbol period*. The output of the filter matched to the signal $s^{(i)}(t)$ is given by

$$y(t) = \sum_{j=-u}^{u} a_j \int_{-\infty}^{\infty} r(v) p(v - jT - t) dv, \qquad (A.19)$$

and the sampled value of $y(t)$ at $t=0$ is given by

$$y(t=0) = \sum_{j=-u}^{u} a_j \int_{-\infty}^{\infty} r(v) p(v - jT) dv = \sum_{j=-u}^{u} a_j y_j \qquad (A.20)$$

where $y_j = \int_{-\infty}^{\infty} r(v) p(v - jT) dv$ is the output of the filter which is matched to the pulse $p(t)$ sampled at time $t = jT$. Thus the matched filter can be implemented by the pulse-matched filter, whose output $y(t)$ is sampled at multiples of the symbol time T.

Some Basics of Communication Theory

In many practical applications one needs to shift the certer frequency of a narrowband signal to some higher frequency band for purposes of transmission. The reason for this may lie in the transmission properties of the physical channel, which allows the passage of signals only in some unusually high-frequency bands. This occurs, for example, in radio transmission. The process of shifting a signal in frequency is called modulation by a carrier frequency. Modulation is also important for wire-bound transmissions, since it makes possible the coexistence of several signals on the same physical medium, all residing in different frequency bands; this is known as frequency division multiplexing (FDM). Probably the most popular modulation method for digital signals is quadrature double-sideband suppressed- carrier (DSB-SC) modulation.

DSB-SC modulation is a simple linear shift in frequency of a signal $x(t)$ with low-frequency content, called baseband, into a higher-frequency band by multiplying $x(t)$ by a cosine or sine waveform with carrier frequency f_0, as shown in Figure A.5, to obtain the signal,

$$s_0(t) = x(t)\cos(2\pi f_0 t), \qquad (A.21)$$

on carrier to, where the factor $\sqrt{2}$ is used to make the powers of $s_0(t)$ and $x(t)$ equal.

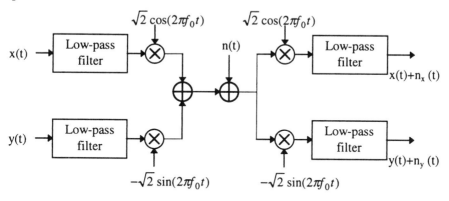

Figure A.5 Quadrature DSB-SC modulation and demodulation stages.

If the baseband signal x(t) occupies frequencies which range from 0 to W Hz, then $s_0(t)$ occupies frequencies from $f_0 - W$ to $f_0 + W$ Hz, an expansion of the bandwidth by a factor of 2. But we quickly note that another signal, $-y(t)\sqrt{2}\sin(2\pi f_0 t)$, can be put into the same frequency band and that both baseband signals $x(t)$ and $y(t)$ can be recovered by the demodulation operation shown in Figure A.5. This is the product demodulator, where the low-pass filters $W(f)$ serve to reject unwanted out-of-band noise and signals. It can be shown that this arrangement is optimal; that is, no information or optimality is lost by using the product demodulator for DSB-SC-modulated signals.

If the synchronization between the modulator and demodulator is perfect, the signals $x(t)$, the *in-phase* signal, and $y(t)$, the *quadrature* signal, are recovered independently without either affecting the other. The DSB-SC-modulated bandpass channel is then, in essence, a dual channel for two independent signals, each of which may carry an independent data stream.

In view of our earlier approach which used basis functions one may want to view each-pair of identical input pulses to the two channels as a two-dimensional signal. Since these two dimensions are intimately linked through the carrier modulation, and since bandpass signals are so ubiquitous in digital communications, a complex notation for bandpass signals has been widely adopted. In this notation, the in-phase signal $x(t)$ is real, and the quadrature signal $jy(t)$ is an imaginary signal, expressed by

$$s_0(t) = x(t)\sqrt{2}\cos(2\pi f_0 t) - y(t)\sqrt{2}\sin(2\pi f_0 t)$$
$$= \text{Re}\left[(x(t)+jy(t))\sqrt{2}e^{2\pi j f_0 t}\right], \quad (A.22)$$

where $s(t) = x(t) + jy(t)$ is called the complex envelope of $s_0(t)$.

B C-programs of Some Coding Algorithms

This appendix provides some sample C-program routines of some coding algorithms.

B.1 Encoding and decoding for Hamming codes

```
/************************************************
* Encoding and decoding of binary Hamming codes
*
* Parameters of Hamming codes : (See Chapter 3)
* Codeword length : n = 2**m - 1
* Dimension of the code : k= 2**m-1-m =n - m
* The minimum distance of the code : dmin=3
*
* This is a systematic encoder. The decoder can decode
* most of 1 bit errors.
*
* gen_poly[] = coefficients of the generator polynomial
* (see Chapter 4).
* coef_of_remd[] = coefficients of the remainder after
* division by p(x).
* num_of_err = number of errors.
* num_of_dec_err = number of decoding errors.
* err_pos[] = error positions.
* coef_of_recd[] = coefficients of the received
* polynomial.
```

```
 *
 *************************************************** /

#include <stdio.h>
#include <stdlib.h>
#include <memory.h>
#include <math.h>

int n,k,m,length,err_pos[10],seed;
int *gen_poly,coef_of_remd[21];
int coef_of_recd[1024], message[1024], syndrome_table[100000];
int num_of_err,num_of_dec_err = 0;
long position[21] = {
   0x00000001, 0x00000002, 0x00000004, 0x00000008,
   0x00000010, 0x00000020, 0x00000040, 0x00000080,
   0x00000100, 0x00000200, 0x00000400, 0x00000800,
   0x00001000, 0x00002000, 0x00004000, 0x00008000,
   0x00010000, 0x00020000, 0x00040000, 0x00080000,
   0x00100000};
long syndrome;

void generator(void);
void construct_table(void);
void encode(void);
void decode(void);

void main(void)
{
        int i;

        printf("An encoding and decoding program for binary cyclic Hamming codes\n");
/* Read m and determine n, k, and
        generator polynomial p(x) */
        generator();

        construct_table();

/* Randomly generate message */
        seed = 531073;
        srand(seed);
        for (i = 0; i < k; i++)
          message[i] = ( rand() & 65536 ) >> 16;
/* Encoding the message bits */
        encode();
```

C-Programs of Some Coding Algorithms

```
/*
 * coef_of_recd[] are the coefficients of
 * c(x) = x**(length-k)*message(x) + b(x)
 */
        for (i = 0; i < length - k; i++)
                coef_of_recd[i] = coef_of_remd[i];
        for (i = 0; i < k; i++)
                coef_of_recd[i + length - k] = message[i];
        printf("\nc(x) = ");
        for (i = 0; i < length; i++) {
                printf("%1d", coef_of_recd[i]);
                if (i && ((i % 20) == 0))
                        printf("\n");
        }
        printf("\n\n");

        printf("Enter the number of errors (0<v<10): ");
        scanf("%d", &num_of_err);

        printf("\nEnter error positions : ");
        for (i = 0; i < num_of_err; i++)
                scanf("%d", &err_pos[i]);

        printf("\n");
        if (num_of_err)
                for (i = 0; i < num_of_err; i++)
                        coef_of_recd[err_pos[i]] ^= 1;
        printf("\nr(x) = ");
        for (i = 0; i < length; i++) {
                printf("%d", coef_of_recd[i]);
                if (i && ((i % 20) == 0))
                        printf("\n");
        }
        printf("\n");

/* Decoding received codeword */
        decode();

        printf("Syndrome = %ld\n", syndrome);

/* Correcting errors */
        coef_of_recd[syndrome_table[syndrome]] ^= 1;

/* Show results */
        printf("\nResults:\n");
        printf("original message  = ");
```

```c
        for (i = 0; i < k; i++) {
                printf("%1d", message[i]);
                if (i && ((i % 20) == 0))
                        printf("\n");
        }
        printf("\nrecovered message = ");
        for (i = length - k; i < length; i++) {
                printf("%1d", coef_of_recd[i]);
                if ((i-length+k) && (((i-length+k) % 20) == 0))
                        printf("\n");
        }
        printf("\n");

        for (i = length - k; i < length; i++)
                if (message[i - length + k] != coef_of_recd[i])
                        num_of_dec_err++;
        if (num_of_dec_err)
           printf("Decoding failure ! There were %d decoding errors in message
positions\n", num_of_dec_err);
        else
           printf("Succesful decoding!\n");

}
void generator()
/*
 * (1) Construct the generator polynomial p(x) and
 * (2) Determining the codeword length and dimension.
 */
{
        int     i;

        printf("\nEnter a value of m such that the codeword length is ");
        printf("n <= 2**m - 1\n\n");
        do {
           printf("Enter m (between 3 and 10): ");
           scanf("%d", &m);
        } while (!((m <= 10)&&(m>2)));

        if(!(gen_poly=(int *)malloc(m+1)))
                printf("maloc failed\n");
        memset(gen_poly,0,(m+1)*sizeof(int));

        gen_poly[0] = gen_poly[m] = 1;
        if (m == 3)      gen_poly[1] = 1;
        else if (m == 4) gen_poly[1] = 1;
```

```
            else if (m == 5)    gen_poly[2] = 1;
            else if (m == 6)    gen_poly[1] = 1;
            else if (m == 7)    gen_poly[1] = 1;
            else if (m == 8)    gen_poly[4] = gen_poly[5] = gen_poly[6] = 1;
            else if (m == 9)    gen_poly[4] = 1;
            else if (m == 10)   gen_poly[3] = 1;

            printf("p(x) = ");
            for (i = 0; i <= m; i++)
              printf("%d", gen_poly[i]);

            printf("\n");
            n=(int)pow(2,m)-1;
  do {
            printf("Enter codeword length (n <= %d): ", n);
            scanf("%d", &length);
            k = length - m;
  } while ( !(length <= n) || !(k>0) );

    if ( length == n )
      printf("\nThis is a (%d,%d,3) Hamming code\n", length, k);
    else
      printf("\nThis is a (%d,%d,3) shortened Hamming code\n", length, k);
}

void construct_table(void)
{
   int i;

   memset(coef_of_recd,0,length*sizeof(int));

   syndrome_table[0] = 0;
   coef_of_recd[0] = 1;
   for (i=1; i< length; i++) {
     coef_of_recd[i-1] = 0; coef_of_recd[i] = 1;
     decode();

     syndrome_table[syndrome] = i;
   }
}

void encode()
/*
 * Compute redundacy coef_of_remd[], the coefficients of p(x). The redundancy
 * polynomial p(x) is the remainder after dividing x^(length-k)*message(x)
```

* by the generator polynomial g(x).
 */
{
        register int   i, j;
        register int   feedback;

        memset(coef_of_remd,0,(length-k)*sizeof(int));

        for (i = k - 1; i >= 0; i--) {
                feedback = message[i] ^ coef_of_remd[length - k - 1];
                if (feedback != 0) {
                        for (j = length - k - 1; j > 0; j--)
                                if (gen_poly[j] != 0)
                                        coef_of_remd[j] = coef_of_remd[j - 1] ^ feedback;
                                else
                                        coef_of_remd[j] = coef_of_remd[j - 1];
                        coef_of_remd[0] = gen_poly[0] && feedback;
                } else {
                        for (j = length - k - 1; j > 0; j--)
                                coef_of_remd[j] = coef_of_remd[j - 1];
                        coef_of_remd[0] = 0;
                }
        }
}

void decode()
{
   int   i, j;
   int   feedback;

   memset(coef_of_remd,0,(length-k+1)*sizeof(int));

   for (i = length - 1; i >= 0; i--) {
      feedback = coef_of_recd[i] ^ coef_of_remd[length - k - 1];
      if (feedback != 0) {
         for (j = length - k - 1; j > 0; j--)
            if (gen_poly[j] != 0)
               coef_of_remd[j] = coef_of_remd[j - 1] ^ feedback;
            else
               coef_of_remd[j] = coef_of_remd[j - 1];
         coef_of_remd[0] = gen_poly[0] && feedback;
      } else {
      for (j = length - k - 1; j > 0; j--)
         coef_of_remd[j] = coef_of_remd[j - 1];
      coef_of_remd[0] = 0;
```
```

```
        }
    }
  feedback = 1;
  syndrome = 0;
  for (i=0; i< length - k; i++) {
     syndrome += feedback*coef_of_remd[i];
     feedback = feedback << 1;
     }
}
```

B.2 Compute metric tables for a soft-decision Viterbi decoder

```
/***********************************************
* Compute metric tables for a soft-decision Viterbi decoder
* assuming AWGN on a BPSK channel (see Sec.8.5.2).
***********************************************/

#include <math.h>
#include <stdlib.h>

/***********************************************
* Compute log-likelihood metrics for Q-bit quantizer (Q=8)
***********************************************/
/* Integrated normal function : 1-Q(x) */
#define  normal(x) (0.5 + 0.5*erf((x)/1.414))

#define  MID_POINT 128   /* =2^Q/2 */

int compute_metric(
int metric_tab[2][256],      /* metric_tab[A][M] */
                             /* A -- number of sent symbols. */
                             /* M -- number of received symbols. M=2^Q */
int signal_amp,              /* Signal amplitude, units */
double noise_amp,            /* Relative noise amplitude */
int scale,         /* Scale factor */
int translation
)
{
```

```
            double n;
            int x;
            double metrics[2][256];
            double p1,p2;

            /*
             * samples of value x<0 are clipped to 0,
             * it includes the entire lower tail of the curve.
             */
            p1 = normal(((0-MID_POINT+0.5)/signal_amp + 1)/noise_amp);        /*
P(0|a1) */
            p2 = normal(((0-MID_POINT+0.5)/signal_amp - 1)/noise_amp);       /*
P(0|a2) */
            metrics[0][0] = log(p1/(p1+p2));
            metrics[1][0] = log(p2/(p1+p2));
            /* Scale and round to nearest integer */
            metric_tab[0][0] = floor(metrics[0][0] * scale + 0.5)+translation;
            metric_tab[1][0] = floor(metrics[1][0] * scale + 0.5)+translation;

            /*
             * samples of value x>255 are clipped to 255,
             * it includes the entire higher tail of the curve.
             */
            p1 = 1 - normal(((255-MID_POINT-0.5)/signal_amp + 1)/noise_amp);
            /*P(255|a1)*/
            p2 = 1 - normal(((255-MID_POINT-0.5)/signal_amp - 1)/noise_amp);
            /*P(255|a2)*/
            metrics[0][255] = log(p1/(p1+p2)) ;
            metrics[1][255] = log(p2/(p1+p2)) ;
            /* Scale and round to nearest integer */
            metric_tab[0][255] = floor(metrics[0][255] * scale + 0.5)+translation;
            metric_tab[1][255] = floor(metrics[1][255] * scale + 0.5)+translation;

            for(x=1;x<255;x++)
            {
                    /* P(x|a1), prob of receiving x given s0 transmitted */
                    p1 = normal(((x- MID_POINT+0.5)/signal_amp + 1)/noise_amp) -
                         normal(((x- MID_POINT-0.5)/signal_amp +
1)/noise_amp);

                    /* P(x|a2), prob of receiving x given s1 transmitted */
                    p2 = normal(((x- MID_POINT+0.5)/signal_amp - 1)/noise_amp) -
                         normal(((x- MID_POINT-0.5)/signal_amp -
1)/noise_amp);

                    metrics[0][x] = log(p1/(p1+p2)) ;
```

```
                    metrics[1][x] = log(p2/(p1+p2)) ;
                    /* Scale and round to nearest integer */
                    metric_tab[0][x] = floor(metrics[0][x] * scale + 0.5)+translation;
                    metric_tab[1][x] = floor(metrics[1][x] * scale + 0.5)+translation;
            }
}
```

Index

A

Abelian group,32
Add-compare-select(ACS) function,351
Additive noise channel,14,74
Additive white Gaussian noise (AWGN),16,334
Algebraic closure,31
Algebraic decoding algorithm,92
Amplitude modulation(AM),408,432
Analog-to-digital(A/D)converter,409
AND,107,161,219,238
Application-Specific Integrated Circuit(ASIC),238,285, 293
Architectural efficiency, 288
Area-Time(AT) product, 288
Associative law,41
Associativity,32
Asymptotic coding gain,447
Augmented generating function,332,354
Automatic-repeat-request (ARQ),75,167,369
  protocols,8
Automata theory,317
Asynchronous Transfer Mode (ATM),183,221-223

B

Bandwidth,21
  efficiency, 21
  Gabor,22
  z%,22
Bandwidth-limited channels, 15
Basis,60-70
  canonical,58, 60-61,63,65,
  dual,60
  normal,60
  triangular,61-62
BCH codes,189-227
  bound,196-199
  designed distance,193
  (true) minimum distance, 193, 198
  generator polynomial,190,192
  narrow-sense,190
  nonprimitive,191
  transform domain,189
Berlekamp's bit-serial multiplier,62-67
Blerlekamp-Massey (BM)
  Algorithm,234
  Decoding,235
Bhattacharyya bound,353
Binary phase shift keying (BPSK),25,90,431
Binary symmetric channel (BSC),74-75
Binary symmetric erasure channel (BSEC),347
Bit-error probability,352
Bit error rate, 13,20,25,101
Bit-error rate (BER), 303
Bit metric,349
Bit serial encoder,234
Block codes, 10,73-137
  linear,73-137
Boolean algebra,107
Boolean function, 107
Bose-Chaudhuri-Hocquenghem (BCH),25-26,
Bound:
  BCH,199
  Gallager's,18
  Hamming,105
  Singleton,245
  Union,352
Bounded distance decoder, 93,102
Branch gains,330
Branch metric computer,337,350
Burst error correction,119-121
Burst errors,119
  detection capability,171
  of length b,120

pattern,171
Burst-trapping, 307
Butterfly modules,351

C

canonical basis,58
canonical form,79
Catastrophic error propagation,323
CCSDS,471
Channel bandwidth,21,
Channel capacity,13,16,23
  theorem,13
Channel transition probability,336
Chien search, 207-208,214, 287
CMOS,219, 295
Coaxial-cable wires,6
Code-Division Multiple-Access
(CDMA),181,416
Code polynomial, 142
Codeword,73
Codes,
  cross-interleaved Reed-Solomon
(CIRC),236,300
  parent,170
  polynomial,155
Codeword,317
Coding gain,20-21,444
Combination logic circuit,11
Commutative law,54
Compact disk(CD), 286, 300
Compact disc player,1,233
Complete decoder, 93,101
Computational rate, 287
Concatenated codes, 26,236,469
Concurrency of operations, 287
Constituent encoder,500
Constraint length, 26,316,320
Consultative committee for space data
systems (CCSDS), 299
Convolutional codes,11,25, 315

non-systematic,26
Correct detection, probability of,102
Coset,94
  leader,94
CRC,370
CROSS constellations,463
Cyclic codes,139-186
  algebraic structure,140
  code polynomial,155
  dimension,142-143
  dual,149
  encoding, 151-154
  generator matrix,146-147
  generator polynomial,143
  linear,139
  parity-check matrix,146
  parity-polynomial of,143
  shift-register encoder and decoder,151
  syndrome-polynomial,154-155
  syndrome computation circuit,161
  syndrome decoder,159
  systematic encoding,145-149
Cyclic group,33-34
Cyclic redundancy check
(CRC),140,168
  codes,140
Cyclic shift,140

D

decibels(dB),20
Data-link layer,372
Decoder error,206,240,300
  probability of,101
Decoder failure,206,240, 305
Decoder latency, 294
Deep space telecommunication
system,235
Deinterleaver,119
Design distance, 198
Detected error event,102

Index

Detection-only decoder,100
Differential encoding,458
Differential trellis codes,458
Digital audio,1
Digital audio-tape(DAT), 309
Digital Compact Disk(CD),6,233,236
Digital Signal Processing (DSP),28,
Digital television,1,151,233
Digital transmission and storage system,2
Digital Video Disk(DVD),6, 307
Dimension, vector space,43
Direct-sequence spread spectrum (DSSS),241
Discrete Cosine Transform(DCT),215
Discrete memoryless channel (DMC),335,336
Distributive laws,41
Double-error-correcting,225
Double-sideband, suppressed-carrier(DSB-SC), 21,456,525
  amplitude modulation,21
Dual code,60 ,83,89
Division circuit,153
Dynamic progamming,333

E

Eight-ary Phase Shift Keying (8-PSK),
Eight-to-fourteen modulation(EFM), 307
Elementary symmetric functions, 203
Embedded error-control code, 467
Encoder latency, 294
Equivalent code, 79-80
Erasure channel, 86
  binary, 86
  nonbinary, 87
Error-and erasure decoding, 87, 306
Error-correction code (ECC), 308
Error-correction overhaed, 216
Error concealment, 6-7,9,75

Error-control performance, 88
Error-control rate,163
Error detection, 74,157
Error event probability, 453
Error function, 355
Error-locator polynomial, 202-214
Error location, 200
Error magnitudes, 212
Error patterns, 74
  coverage radius,170-171
  undetectable,74
Error probability, 14
Error-state diagrams, 331
Ethernet,183
Euclid's algorithm, 34,48-51,69-70,234
  division,34
  decoding,234-235
  polynomial,50-51
  recursive,50-51
Euclidean distance, 84, 441
Euclidean geometry, 84
Extended codes, 90

F

Fading channels, 392
Fano algorithm, 333
Fast-Fourier-transform (FFT),233
Fermat's theorem, 69-70
Fields,39-40,43-48
  arithmetic,51-70
  characteristic of,44
  complex,39
  extension,40
  finite,43-48
  ground,58
  a polynomial over, 52
  rational,39
  real,39
  ternary,44

Field-Programmable Gate Arrays (FPGAs),238,285, 296
Finite groups, 22
Finite-state,320
First-event error probability, 352
First-event probability, 354
Flip-flop(FF),53,151,238
Forward error correction (FEC), 7, 75
Frame, 374
Framing signal, 397
Free Euclidean distance, 445
Free Hamming distance, 328,329
Frequency division multiplexing (FDM),525
Frequency hopping,241
  spread spectrum (FHSS),241
Frequency modulation(FM),408
Full-duplex,378

G

Galileo mission,26,240
Gallager bound,18
Galois fields, (see finite-field)
Generating function,329,354
Generator matrix,77
Generator polynomials,319
Generator sequence,318
Gilbert-Varshamov bound,244
Global System for Mobile communication (GSM), 241,407
Go-back-n protocol,371,382
Golay codes,
  binary,164
Gray codes,139-140,
Greatest common divisor (GCD),33,48-50,287
Groebner basis,214
Groups, 31-37
  additive remainder,34
  commutative,47
  finite,32
  finite cyclic,33
  infinite,32
  infinite cyclic,33
  subgroup,32-33
  remainder,34-37
  additive remainder,34
Group of blocks(GOBs),215

H

Hadamard matrices, 117
Hadamard transform, 117
Half-duplex physical channel, 375
Halley's comet, 300
Hamming codes, 77,103-106
  binary, 77
  weight enumerator, 89
Hamming distance, 84
Hamming bound, 105
Hamming weight, 85,88
Hard decision decoding, 88,338
High data-rate cable mode, 286
High-Definition TeleVision, (HDTV), 21,237,286
High Frequency (HF), 28
Horner's rule, 220
H.261, 215
Hubble space telescope, 237, 295
Hybrid error-control, 8

I

Ideals,37-39
  principal,38-39
Identity element,32
I.432, 226

Index

Incomplete decoding,103
Information rate, 163
Inner code, 26,237
Inner-code parity (PI), 311
Inner product, 81
Interleavers, 119
  periodic,119
  pseudo-random,119
Integrated-Services Digital Networks(ISDNs), 216
intersymbol-interference,21
Intraframe prediction,215
Inverse element, 32
IS-95,181
iterative algorithms,233
ITU,215,226

J

Joint Tactical Data System (JTDS),241
Jupiter, 240

K

Key equation, 260
Kermit file-transfer protocol(FTPs),178

L

Layered communication networks, 286
likelihood function, 91
Linear block codes,73-137
Linear dependent,88
Linear feedback shift-register (LFSR),235
Linear independent,77
Linear predictive coding(LPC),410
Linear recursive shift-register,233
Linear sequential circuit,320
Log likelihood function, 92,349
Long-term prediction analysis,410

M

Mariner, 25,106
M-ary phase shift keying (MPSK),
Macroblocks(MBs), 215
Macwillams Identity, 89
Magnetic-disk storage, 4, 285
Magnetic-tape storage, 8
Majority logic decoding, 110,333
Matched-filter, 91
Matrix,
  multiplication,103
  nonsingular,212
  parity-check,82
  transpose,82
Maximum a posterior (MAP) decoder, 507
Maximum distance separable (MDS), 245
Maximum likelihood (ML) decoder, 15,76,90-92
Maximum likelihood decoding, 316,333
Maximum (random) error-correction capability, 86
Massey-Omura multiplier, 62,67-69
Matched filter, 523
Memoryless channels, 76
  nonbinary, 76
Memory order, 315
Message throughput,370
Microwave links, 6
Million operation per second (MOPS), 295

Minimal polynomials, 190-191
Minimum Hamming distance,85
  decoding,93
Minimum weight,94
Modem,27
Monic polynomial,141
Monomial function,108
Motion-compensation,215
Multi-phase-shift-keying(MPSK)
  modulation,430
Multiple-error-correction,189

N

NASA/ESA,26, 300
Negative acknowledgment (NAK),391
Network layer, 372
Newton's identities, 203,213
Neptune, 237
Noisy channel coding theorem, 13-17,
Noncatastrophic,325
Nonprimitive elements,191
Nonsystematic encoder,323
Nonsystematic feedforward-only
encoder,323
Normal basis,60
NOT,108
NP-complete,91
Null space,81
  basis,82
Nyquist bandwidth,432

0

Open System Interconnection
(OSI),177,372
Operation, binary, 43,47
Optical fibers,6,222-223
  communication,151

Optical memory units,6
Optimal code, 164
OR,107,238
Outer codes,237
Outer-code parity (PO), 311

P

Packet,374
Packet-error rate,370
Packets switched network,167
Packing radius,105
Pairwise error probability,352
Parity check,81
Parity check codes,81
Parity check equation,81
Parity check matrix,82
Parity polynomial,143
Partial path metric,348
Path, graph,
Path metric,337
Path metric updating and storage, 349
Perfect codes, 104,164
Performance - per - area metric, 290
Periodic-block interleaver,395
Periodic convolutional (or
synchronous)interleaver,396
Periodic interleaver,395
Periodic random,393
Personal Communication System
(PCS),28
Peterson's algorithm, 189,205,212,235
Peterson-Gorenstein-Zierler Decoding
Algorithm, 211-212
Phase-locked loop (PLL), 457
Phase modulation (PM), 408
Physical layer, 372
Piggybacking, 379
Pipelining, 383
Pipelining processing, 233

Index

Pioneer mission, 26
Pith metric, 348
Polynomial,
  degree of,38
  ideals,38-39
  irreducible,54
  rings,38-39
  primitive,55
Positive acknowledgment with retransmission(PAR),377
Power, 45-46
Power dissipation, 289
Power efficiency, 431
power-limited channels,15
Power-sum symmetric functions,201
Pragmatic approach, 465
Prime number, 44
Prime integer, 44
Primitive elements, Finite field, 46
Probability of, 175,227
  bit error,(see bit-error rate)
  discarding cells,227
  undetected error,175
  intercept,242
Product codes, 121-130
  systematic, 122
  maximum random-error-correction capability, 123
Programmable Logic Arrays (PLAs), 238
Programmable Logic Devices (PLDs), 238
Protocol, 382,387
Pseudo random, 393
Pseudo-random noise (PN),416
Pulse-amplitude-modulated (PAM),21,429
Punctured codes, 90
Punctured convolutional codes, 357
Puncturing matrix, 360
Phase-shift-keying (PSK),90

Q

q-ary set,10
q-ary uniform channel,93
Quadratic equation,220
Quadratic residue,162
Quadratic residue codes,162-164
  binary,162
  algebraic decoding, 213-214
  dimension,163
  generator polynomial,163
  minimum distance,163
Quadratic nonresidues, 162
Quadrature amplitude modulation (QAM),21,430
QPSK,27
Quotient,54

R

Radio links,6
Random access memory(RAM), 220, 294
Rate, block encoder,10
Rate, convolutional encoder,11
Rate, data-symbol transmission,20
Rate compatible punctured convolutional (RCPC) codes, 465
Read-Only Memory(ROM),69,81,294
Received word, 74
Reed decoding algorithm, 110
Reed-Muller (RM) codes, 9,103,105-119
  first order,105,115-119
  decoding by Hadamard Transform,118
  dimension,109
  encoding,
  encoding circuit,
  weight distribution,
  minimum distance,115
Reed-Solomon (RS) codes,25,233-281

decoding, 256
  generator polynomial,244
Reed-Solomon product code(RS-PC), 309
Regular pulse-excitation(RPE),410
Return channel,369
Rieger bound,120,245
Rings,37-39
  commutative,37-38
  polynomial,38-39
  quotient,38-39
Rotational invariance, 456
Row operation,80
Row space, 79-80

S

Sample-interpolation rate, 303
Satellite link, 6
Scrambler matrix,327
SDRAM,398
Selectable-rate decoder,363
Selective-repeat protocol, 371,389
Self-dual codes, 83
Sequential decoding, 26,333
Sequential logic circuit, 11
Set partitioning, 448
Shannon, C. E., 13
  Shannon Capacity, 23-24
  Shannon-Harley formula, 16
  Shannon bound, 22,24
  Shannon limit, 23
Shift register, 140,151
  cells,152
  circuit,140
Shortened codes,90
Signal-constellation,448
Signal-to-noise ratio(SNR),2,22,433
  energy,2
  power,22
Silicon area, 287

Simplex stop-and-wait protocol, 371
Single-error-correcting, 225
Singleton bound, 245
Sliding window protocol, 371,378
Soft-decision algorithm, 334
Soft decision decoding, 26,344
Soft-output Viterbi algorithm (SOVA), 240,509
Spatial redundancy, 215
Spectral density, 22,25
Spectral efficiency, 22,25,431
Sphere packing bound, 105
Spread-spectrum, 180
  cellular system,180
Stack algorithm, 333
Standard array decoder, 94-96
State diagrams, 320,439
Static-Random-Access-Memory (SRAM), 292
Stop-and wait (SW-ARO) protocols, 374
Subfields, 40,45
Subgroup, 32-33
Surviving paths, 342
Survivor, 337
Switch, 151
Systematic feedback encoder, 323
Synchronizer, 350
Syndrome, 97
  decoding, 97-99
  matrix,206
Systematic encoding,79
Systolic array,239

T

Table-look-up,103
Temporal redundancy,215
Terrestrial mode, 286
Throughput rate, 287, 291

Index

Time-division multiplex access (TDMA), 408
Time-domain, 316
Time-invariant, 320
Token-ring, 183
Trace function, 58
Transform-domain, 318
Transition probability, 348
Trellis coded modulation (TCM), 27, 429, 430
Trellis diagram, 322
Triangular basis, 61-62
Truncation length, 344
Truth table, 107
Turbo codes, 27, 469, 498
Twisted-pair telephone line, 6

U

Undetected error, probability of, 100
Unequal error protection (UEP), 464
Union bound, 352
Uranus, 236

V

Vandermonde determinant, 198
Vandermonde matrix, 211
Variable-length encoding, 215
Vectors,
 spaces, 40-43, 77
 subspaces, 42-43, 77
Very Large-Scale Integration (VLSI), 238, 285
Vestigial-sideband (VSB), 475
Video compression, 215
Video conferencing, 215
Videophone, 215
Viking mission, 25, 105

Viterbi algorithm, 316, 335
Viterbi decoder,
 algorithm, 19, 236
Viterbi decoding, 333
Voyager spacecraft, 26, 165, 236

W

Weight distribution, 88, 89
Weight enumerator polynomial, 89
Welch-Berlekamp decoding algorithm, 238

X

XOR, 12, 107, 219, 238, 316
XNOR, 51
X.25 protocol, 178
XYModem, 178

Z

Zmodem, 178, 180

About the Authors

Irving S. Reed received the B.S. and Ph.D. degrees from the California Institute of Technology (CalTech) in 1944 and 1949, respectively.

In 1949, Dr. Reed began his career as a research engineer at the Northrop Corporation, where he contributed to the development of the magnetic drum digital differential analyzer (MADDIDA) computer. He was part of the Northrop team which demonstrated the revolutionary capabilities of MADDIDA to John von Neumann at Princeton in 1950. In 1951, Dr. Reed joined MIT's Lincoln Laboratory. There he made primary contributions to the development of digital computers, culminating in the register transfer language (RTL) method of design. In 1960, Dr. Reed moved to the Rand Corporation as Senior Staff Member and in 1963 joined the University of Southern California, where he was appointed Charles Lee Powell Professor of Electrical Engineering and Computer Science.

Dr. Reed is the principal originator of the Reed-Muller (1954) and Reed-Solomon (1960) codes. His recent contributions to adaptive arrays, digital signal processing, detection theory and VLSI design are evidence of his continued intellectual leadership. Dr. Reed has authored and co-authored over 300 publications and is the holder of a dozen U.S. Patents.

Dr. Reed is the recipient of many honors, including: the IEEE Aerospace and Electronic Systems Society M. Barry Carlton Memorial Award for Best Paper (1985), Shannon Lecturer at the International Symposium on Information Theory (1982) as well as Shannon Award (1995). He also received the IEEE Hamming Award (1989) and the IEEE Masaru Ibuka Consumer Electronics Award (1995). Other awards include the distinguish alumni award from the CalTech (1992), and the honorary doctor of science degrees from universities of Alaska and Coronado (1998). He is a Fellow of the IEEE and a member of the National Academy of Engineering.

Xuemin Chen received the B. E. and M. E. degrees in electronics from National University of Defense and Technology of China in 1983 and 1986, respectively, and the Ph. D. degree in Electrical Engineering from University of Southern California (USC), Los Angeles, California, 1992.

From 1986 to 1988, he was a research staff in the State Key Laboratory of Automatic Target Recognition (ATR) in Hunan, China. He was employed by America On-Line (Johnson – Grace) as a research scientist from 1992 to 1994. He has been with General Instrument Corporation (GIC) since 1994. Dr. Chen is a Senior Staff Engineer of the Advanced Technology Department of GIC.

Dr. Chen is a Senior Member of IEEE and has authored and co-authored over 50 research articles and he holds more than a dozen patents and patent applications in digital communication and image/video processing. He has contributed three book chapters in channel coding and data compression. In 1990 and 1991, he received a National Science and Technology Award and a Ministry Science and Technology Award from China for his contributions in the development of ship target recognition systems. He received The Academic Achievement Award from USC in 1993. GIC also awarded Dr. Chen the inventor awards for 1997 and 1998.